S. Northshield

Grundlehren der mathematischen Wissenschaften 276

A Series of Comprehensive Studies in Mathematics

Grundlehren der mathematischen Wissenschaften

A Series of Comprehensive Studies in Mathematics

A Selection

Continued after Index

Thomas M. Liggett

Interacting Particle Systems

Springer-Verlag
New York Berlin Heidelberg Tokyo

Thomas M. Liggett
Department of Mathematics
University of California
Los Angeles, CA 90024
U.S.A.

AMS Subject Classifications: 60-02, 60K35, 82A05

Library of Congress Cataloging in Publication Data
Liggett, Thomas M. (Thomas Milton)
   Interacting particle systems.
   (Grundlehren der mathematischen Wissenschaften; 276)
   Bibliography: p.
   Includes index.
   1. Stochastic processes.   2. Mathematical physics.
3. Biomathematics.   I. Title.   II. Series.
QC20.7.S8L54   1985   530.1'5   84-14152

With 6 Illustrations.

Typeset by J. W. Arrowsmith Ltd., Bristol, England.
Printed and bound by R. R. Donnelley & Sons, Harrisonburg, Virginia.
Printed in the United States of America.

9 8 7 6 5 4 3 2 1

ISBN 0-387-96069-4 Springer-Verlag New York Berlin Heidelberg Tokyo
ISBN 3-540-96069-4 Springer-Verlag Berlin Heidelberg New York Tokyo

To my family:
Chris, Tim, Amy

Preface

At what point in the development of a new field should a book be written about it? This question is seldom easy to answer. In the case of interacting particle systems, important progress continues to be made at a substantial pace. A number of problems which are nearly as old as the subject itself remain open, and new problem areas continue to arise and develop. Thus one might argue that the time is not yet ripe for a book on this subject. On the other hand, this field is now about fifteen years old. Many important problems have been solved and the analysis of several basic models is almost complete. The papers written on this subject number in the hundreds. It has become increasingly difficult for newcomers to master the proliferating literature, and for workers in allied areas to make effective use of it. Thus I have concluded that this is an appropriate time to pause and take stock of the progress made to date. It is my hope that this book will not only provide a useful account of much of this progress, but that it will also help stimulate the future vigorous development of this field.

My intention is that this book serve as a reference work on interacting particle systems, and that it be used as the basis for an advanced graduate course on this subject. The book should be of interest not only to mathematicians, but also to workers in related areas such as mathematical physics and mathematical biology. The prerequisites for reading it are solid one-year graduate courses in analysis and probability theory, at the level of Royden (1968) and Chung (1974), respectively. Material which is usually covered in these courses will be used without comment. In addition, a familiarity with a number of other types of stochastic processes will be helpful. However, references will be given when results from specialized parts of probability theory are used. No particular knowledge of statistical mechanics or mathematical biology is assumed. While this is the first book-length treatment of the subject of interacting particle systems, a number of surveys of parts of the field have appeared in recent years. Among these are Spitzer (1974a), Holley (1974a), Sullivan (1975b), Liggett (1977b), Stroock (1978), Griffeath (1979a, 1981), and Durrett (1981). These can serve as useful complements to the present work.

This book contains several new theorems, as well as many improvements on existing results. However, most of the material has appeared in one form or another in research papers. References to the relevant papers are given in the "Notes and References" section for each chapter. The bibliography contains not only the papers which are referred to in those sections, but also a fairly complete list of papers on this general subject. In order to encourage further work, I have listed a total of over sixty open problems at the end of the appropriate chapters. It should be understood that these problems are not all of comparable difficulty or importance. Undoubtedly, some will have been solved by the time this book is published.

The following remarks should help the reader orient himself to the book. Some of the most important models in the subject are described in the Introduction. The main questions involving them and a few of the most interesting results about them are discussed there as well. The treatment here is free of the technical details which become necessary later, so this is certainly the place to start reading the book.

The first chapter deals primarily with the problem of existence and uniqueness for interacting particle systems. In addition, it contains (in Section 4) several substantive results which follow from the construction and are rather insensitive to the precise nature of the interaction. From a logical point of view, the construction of the process must precede its analysis. However, the construction is more technical, and probably less interesting, than the material in the rest of the book. Thus it is important not to get bogged down in this first chapter. My suggestion is that, on the first reading, one concentrate on the first four sections of Chapter I, and perhaps not spend much time on the proofs there. Little will be lost if in later chapters one is willing to assume that the global dynamics of the process are uniquely determined by the informal infinitesimal description which is given. The martingale formulation which is presented following Section 4 has played an important role in the development of the subject, but will be used only occasionally in the remainder of this book.

Many of the tools which are used in the study of interacting particle systems are different from those used in other branches of probability theory, or if the same, they are often used differently. The second chapter is intended to introduce the reader to some of these tools, the most important of which are coupling and duality. In this chapter, the use of these techniques is illustrated almost exclusively in the context of countable state Markov chains, in order to facilitate their mastery. In addition, the opportunity is taken there to prove several nonstandard Markov chain results which are needed later in the book.

In Chapter III, the ideas and results of the first two chapters are applied to general spin systems—those in which only one coordinate changes at a time. It is here, for example, that the general theory of attractive systems is developed, and that duality and the graphical representation are introduced. Chapters IV–IX treat specific types of models: the stochastic Ising model, the voter model, the contact process, nearest-particle systems, the exclusion process, and processes with unbounded values. These chapters have been written so that they are largely independent of one another and may be read separately. A good first exposure to this book can be obtained by lightly reading the first four sections of Chapter I, reading the first half of Chapter II, Chapter III, and then any or all of Chapters IV, V, and VI.

While I have tried to incorporate many of the important ideas, techniques, results, and models which have been developed during the past fifteen years, this book is not an exhaustive account of the entire subject of interacting particle systems. For example, all models considered here have continuous time, in spite of the fact that a lot of work has been done on analogous discrete time systems, particularly in the Soviet Union. Not treated at all or barely touched on are important advances in the following closely related subjects: infinite systems of stochastic differential equations (see, for example, Holley and Stroock (1981), Shiga (1980a, b) and Shiga and Shimizu (1980)), measure-valued diffusions (see, for example, Dawson (1977) and Dawson and Hochberg (1979, 1982)), shape theory for finite interacting systems (see, for example, Richardson (1973), Bramson and Griffeath (1980c, 1981), Durrett and Liggett (1981), and Durrett and Griffeath (1982)), renormalization theory for interacting particle systems (see, for example, Bramson and Griffeath (1979b) and Holley and Stroock (1978b, 1979a)), cluster processes (see, for example, Kallenberg (1977), Fleischmann, Liemant, and Matthes (1982), and Matthes, Kerstan, and Mecke (1978)), and percolation theory (see, for example, Kesten (1982) and Smythe and Wierman (1978)).

The development of the theory of interacting particle systems is the result of the efforts and contributions of a large number of mathematicians. There are many who could be listed here, but if I tried to list them, I would not know where to stop. In any case, their names appear in the "Notes and References" sections, as well as in the Bibliography. I would particularly like to single out Rick Durrett, David Griffeath, Dick Holley, Ted Harris, and Frank Spitzer, both for their contributions to the subject and for the influence they have had on me. Enrique Andjel, Rick Durrett, David Griffeath, Dick Holley, Claude Kipnis, and Tokuzo Shiga have read parts of this book, and have made valuable comments and found errors in the original manuscript.

Since this is my first book, this is a good place to acknowledge the influence which Sam Goldberg at Oberlin College, and Kai Lai Chung and Sam Karlin at Stanford University had on my first years as a probabilist. I would like to thank Chuck Stone for his encouragement during the early years of my work on interacting particle systems, and in particular for handing me a preprint of Spitzer's 1970 paper with the comment that I would probably find something of interest in it. This book is proof that he was right.

More than anyone else, it was my wife, Chris, who convinced me that I should write this book. In addition to her moral support, she contributed greatly to the project through her excellent typing of the manuscript. Finally,

I would like to acknowledge the financial support of the National Science Foundation, both during the many years I have spent working on this subject, and particularly during the past two years in which I have been heavily involved in this writing project.

Contents

Frequently Used Notation

S	A finite or countable set of sites.
Z^d	The d-dimensional integer lattice.
Y	The collection of finite subsets of S or $S \cup \{\infty\}$.
X	The state space of the process; usually $\{0, 1\}^S$.
$C(X)$	The continuous functions on X.
$D(X)$	The Lipschitz functions on X (see Section 3 of Chapter I).
$\mathcal{M}$	The increasing continuous functions on X.
$\mathcal{D}$	The functions on X which depend on finitely many coordinates.
$\mathcal{P}$	The probability measures on X.
$\mathcal{S}$	When $S = Z^d$, the translation invariant elements of $\mathcal{P}$.
$\mathcal{S}_e$	The extremal (or ergodic) elements of $\mathcal{S}$.
$\mathcal{I}$	The elements of $\mathcal{P}$ which are invariant for the process.
$\mathcal{I}_e$	The extremal elements of $\mathcal{I}$.
$\mathcal{R}$	The elements of $\mathcal{P}$ which are reversible for the process.
$\mathcal{G}$	The Gibbs measures corresponding to some potential.
$\mathcal{G}_e$	The extremal Gibbs measures.
δ_0, δ_1	The pointmasses on $\eta \equiv 0$ and $\eta \equiv 1$.
μ or ν	Typical elements of $\mathcal{P}$.
$\mu \leq \nu$	Stochastic monotonicity (see Definition 2.1 of Chapter II).
η_t or ζ_t	The Markov process which represents an interacting particle system.
$c(x, \eta)$	The flip rate at $x \in S$ when the configuration is $\eta \in X$.
$S(t)$	The semigroup corresponding to the process.
$\mu S(t)$	The distribution at time t when the initial distribution is $\mu \in \mathcal{P}$.
Ω	The generator or pregenerator of the process.
$\mathcal{D}(\Omega)$	The domain of Ω.
$\mathcal{R}(\Omega)$	The range of Ω.
Re	The real part of a complex number.
$p^{(n)}(x, y)$	The n-step transition probabilities for a discrete time Markov chain.
$p_t(x, y)$	The transition probabilities for a continuous time Markov chain.
$\mathcal{H}$	Harmonic functions; often with some additional constraints.

Introduction

The field of interacting particle systems began as a branch of probability theory in the late 1960's. Much of the original impetus came from the work of F. Spitzer in the United States and of R. L. Dobrushin in the Soviet Union. (For examples of their early work, see Spitzer (1969a, 1970) and Dobrushin (1971a, b).) During the decade and a half since then, this area has grown and developed rapidly, establishing unexpected connections with a number of other fields.

The original motivation for this field came from statistical mechanics. The objective was to describe and analyze stochastic models for the temporal evolution of systems whose equilibrium measures are the classical Gibbs states. In particular, it was hoped that this would lead to a better understanding of the phenomenon of phase transition. As time passed, it became clear that models with a very similar mathematical structure could be naturally formulated in other contexts—neural networks, tumor growth, spread of infection, and behavioral systems, for example.

From a more mathematical point of view, interacting particle systems represents a natural departure from the established theory of Markov processes. As such, with its different motivation, it has led to a large number of stimulating new types of problems. The solutions of many of these new problems has led in turn to the development of new tools, and to the exploitation on an entirely different level of tools which had earlier played only relatively minor roles in probability theory. A typical interacting particle system consists of finitely or infinitely many particles which, in the absence of the interaction, would evolve according to independent finite or countable state Markov chains. Superimposed on this underlying motion is some type of interaction. As a result of the interaction, the evolution of an individual particle is no longer Markovian. The system as a whole is of course Markovian. However, it is a large and complex Markov process which differs in many respects from the processes such as Brownian motion on Euclidean spaces which motivated much of the development of standard Markov process theory. Thus, while some connections with the Markovian universe are maintained, substantial departures from it occur as well.

Let us illustrate some of the differences between particle systems and the more standard Markov processes in a very simple context. Suppose that

$\{\eta_t(x), x \in S\}$ is a countable collection of independent irreducible continuous time Markov chains with state space $\{0, 1\}$. Of course, the analysis of this system is entirely elementary because of the independence assumption and the simple nature of the individual chains. Suppose however that the entire system η_t is viewed as a Markov process on the uncountable totally disconnected space $\{0, 1\}^S$. Then for any fixed initial configuration η_0, the distributions of η_t at different times are product measures which are mutually singular with respect to each other and with respect to the unique invariant measure for the process. This is of course vastly different from the much smoother behavior evidenced by Brownian motion or other more common Markov processes. If a simple interaction is superimposed on the underlying motion of this collection of two state Markov chains, these mutual singularity properties will in general remain, while the analysis based on independence is no longer available. Thus new techniques are required. The models to be treated in Chapters III–VII are obtained by superimposing various natural types of interactions on the simple systems described above. This is done by letting the flip rate of each coordinate depend on the values of other coordinates.

As might be expected, the behavior of an interacting particle system depends in a rather sensitive way on the precise nature of the interaction. Thus most of the research which has been done in this field has dealt with certain types of models in which the interaction is of a prescribed form. The unity of the subject comes not so much from the generality of the theorems which are proved, but rather from the nature of the processes which are studied, the types of problems which are posed about them, and the techniques which are used in their solution.

The main problems which have been treated involve the long-time behavior of the system. The first step in proving limit theorems is to describe the class of invariant measures for the process, since these are the possible limits as $t \to \infty$ of the distribution at time t. The next step is to determine to the extent possible the domain of attraction of each invariant measure. This means, to determine for each invariant measure, the class of all initial distributions for which the distribution at time t of the process converges to that measure as $t \to \infty$. In the case of the independent two state Markov chains, the answers to these questions are of course that there is a unique invariant measure for the process, which is the product of the stationary distributions for the individual two state chains. Its domain of attraction is the collection of all probability measures on $\{0, 1\}^S$.

In order to make the foregoing remarks more concrete, we will now describe informally some of the models which have received the most attention, and will specify the form which these problems take in each case. In the first three examples, only one coordinate of η_t changes at a time. In general, however, infinitely many coordinates will change in any interval of time. In the fourth example, two coordinates change at a time.

The Stochastic Ising Model. This is a model for magnetism which was introduced by Glauber (1963) and then first studied in some generality by Dobrushin (1971a, b). It is a Markov process with state space $\{-1, +1\}^{Z^d}$. The sites represent iron atoms, which are laid out on the d-dimensional integer lattice Z^d, while the value of ± 1 at a site represents the spin of the atom at that site. A configuration of spins η is then a point in $\{-1, +1\}^{Z^d}$. The dynamics of the evolution are specified by the requirement that a spin $\eta(x)$ at $x \in Z^d$ flips to $-\eta(x)$ at rate

$$\exp\left[-\beta \sum_{y:|y-x|=1} \eta(x)\eta(y)\right],$$

where β is a nonnegative parameter which represents the reciprocal of the temperature of the system. Note that the flip rate is higher when the spin at x is different from that at most of its neighbors than it is when it agrees with most of its neighbors. Thus the system "prefers" configurations in which the spins tend to be aligned with one another. In the language of statistical mechanics, this monotonicity is referred to as ferromagnetism. In the subject of interacting particle systems, such monotone systems are called "attractive." Of course when $\beta = 0$, the coordinates $\eta_t(x)$ are independent two-state Markov chains, so as observed earlier, the system has as its unique invariant measure the Bernoulli product measure ν on $\{-1, +1\}^{Z^d}$ with parameter $\frac{1}{2}$. Furthermore, for any initial distribution, the distribution at time t converges weakly as $t \to \infty$ to ν by the convergence theorem for finite-state irreducible Markov chains. Such a system, which has a unique invariant measure to which convergence occurs for any initial distribution, will be called ergodic. The first important problem to be resolved for the stochastic Ising model is to determine for which choices of β and d the process is ergodic. The first answer, as will be seen in Chapter IV, is that the process is ergodic for all β if $d = 1$. In fact, in one dimension the unique invariant measure is a stationary two-state Markov chain, which is regarded as a measure on $\{-1, 1\}^{Z^1}$. If $d \geq 2$, there is a critical $\beta_d > 0$ so that the process is ergodic if $\beta < \beta_d$ but not if $\beta > \beta_d$. If $d = 2$ and $\beta > \beta_2$, then there are exactly two extremal invariant measures. If $d \geq 3$ and β is sufficiently large, then there are infinitely many extremal invariant measures. Nonergodicity corresponds to the occurrence of phase transition, with distinct invariant measures corresponding to distinct phases.

The Voter Model. The voter model was introduced independently by Clifford and Sudbury (1973) and by Holley and Liggett (1975). Here the state space is $\{0, 1\}^{Z^d}$ and the evolution mechanism is described by saying that $\eta(x)$ changes to $1 - \eta(x)$ at rate

$$\frac{1}{2d} \sum_{y:|y-x|=1} 1_{\{\eta(y) \neq \eta(x)\}}.$$

In the voter interpretation of Holley and Liggett, sites in Z^d represent voters who can hold either of two political positions, which are denoted by zero and one. A voter waits an exponential time with parameter one, and then adopts the position of a neighbor chosen at random. In the invasion interpretation of Clifford and Sudbury, $\{x \in Z^d : \eta(x) = 0\}$ and $\{x \in Z^d : \eta(x) = 1\}$ represent territory held by each of two competing populations. A site is invaded at a rate proportional to the number of neighboring sites controlled by the opposing population. The voter model has two trivial invariant measures: the pointmasses at $\eta \equiv 0$ and $\eta \equiv 1$ respectively. Thus the voter model is not ergodic. The first main question in this case is whether there are any other extremal invariant measures. As will be seen in Chapter V, there are no others if $d \leq 2$. On the other hand, if $d \geq 3$, there is a one-parameter family $\{\mu_\rho, 0 \leq \rho \leq 1\}$ of extremal invariant measures, where μ_ρ is translation invariant and ergodic, and $\mu_\rho\{\eta : \eta(x) = 1\} = \rho$. This dichotomy is closely related to the fact that a simple random walk on Z^d is recurrent if $d \leq 2$ and transient if $d \geq 3$. In terms of the voter interpretation, one can describe the result by saying that a consensus is approached as $t \to \infty$ if $d \leq 2$, but that disagreements persist indefinitely if $d \geq 3$.

The Contact Process. This process was introduced and first studied by Harris (1974). It again has state space $\{0, 1\}^{Z^d}$. The dynamics are specified by the following transition rates: at site x,

$$1 \to 0 \quad \text{at rate 1,}$$

and

$$0 \to 1 \quad \text{at rate } \lambda \sum_{y:|y-x|=1} \eta(y),$$

where λ is a positive parameter which is interpreted as the infection rate. With this interpretation, sites at which $\eta(x) = 1$ are regarded as infected, while sites at which $\eta(x) = 0$ are regarded as healthy. Infected individuals become healthy after an exponential time with parameter one, independently of the configuration. Healthy individuals become infected at a rate which is proportional to the number of infected neighbors. The contact process has a trivial invariant measure: the pointmass at $\eta \equiv 0$. The first important question is whether or not there are others. As will be seen in Chapter VI, there is a critical λ_d for $d \geq 1$ so that the process is ergodic for $\lambda < \lambda_d$, but has at least one nontrivial invariant measure if $\lambda > \lambda_d$. The value of λ_d is not known exactly. Bounds on λ_d are available, however. For example,

$$\frac{1}{2d-1} \leq \lambda_d \leq \frac{2}{d}$$

for all $d \geq 1$. Good convergence theorems are known when $d = 1$. However, the limiting behavior of the system is much less well understood if $d \geq 2$. One of the key open problems, which remains a conjecture even when $d = 1$, is whether the critical contact process with $\lambda = \lambda_d$ is ergodic.

The Exclusion Process. The exclusion process was introduced by Spitzer (1970) as a model for a lattice gas at infinite temperature and by Clifford and Sudbury (1973) as a model in which two opposing species swap territory. The state space is $\{0, 1\}^S$ where S is a countable set. In the lattice gas interpretation, particles move on S in such a way that there is always at most one particle per site. If $\eta \in \{0, 1\}^S$, then $\{x: \eta(x) = 1\}$ is the set of occupied sites. The particles move on S according to the following rules:

(a)   a particle at $x \in S$ waits an exponential time with parameter one;
(b)   at the end of that time, it chooses a $y \in S$ with probability $p(x, y)$; and
(c)   if y is vacant, it goes to y, while if y is occupied, it stays at x.

Thus an exclusion interaction is superimposed on otherwise independent continuous time Markov chains on S with jump probabilities $p(x, y)$. This system is most completely understood when $p(x, y) = p(y, x)$. In this case, there is a one-parameter family of (trivial) invariant measures: the Bernoulli product measures with constant density $\rho \in [0, 1]$. The reason for the existence of a large number of extremal invariant measures is of course the fact that the evolution does not change the particle density. The first main question is whether or not these are the only extremal invariant measures. As will be seen in Chapter VIII, there are no others if and only if all bounded harmonic functions for $p(x, y)$ are constant. The domain of attraction of each invariant measure can be described completely in this symmetric case. Many open problems remain when $p(x, y)$ is not symmetric.

These four models and their generalizations will be treated in detail in Chapters IV, V, VI, and VIII respectively. Chapter VII is devoted to a class of one-dimensional spin systems called nearest-particle systems. One reason for their importance is the close connection between them and stationary renewal measures on $\{0, 1\}^{Z^1}$. This connection is analogous to the connection between the one-dimensional stochastic Ising model and stationary two-state Markov chains which was mentioned earlier. Chapter IX deals with a class of particle systems whose state space is $[0, \infty)^S$. The theory of these processes is substantially different from that treated in earlier chapters because of the fact that the state space is not compact, and hence the system may have no nontrivial invariant measures at all.

Chapter I

The Construction, and Other General Results

The interacting particle systems which are the subject of this book are continuous-time Markov processes on certain spaces of configurations of particles. These processes are normally specified by giving the infinitesimal rates at which transitions occur. In general there are infinitely many particles in the system, and infinitely many of them make transitions in any interval of time. The transition rates for individual particles can depend on the entire configuration. Consequently, it is not immediately clear whether the specification of the local dynamics determines the evolution of the system as a whole in a unique way. Before proceeding to analyze the infinite system, it is therefore necessary to examine this question of existence and uniqueness of the process.

In the first part of this chapter, interacting particle systems will be constructed under appropriate conditions in three steps:

(a) The given infinitesimal rates are used to write down an operator, whose closure will ultimately be the generator of the process.
(b) The Hille–Yosida theorem is used to construct the corresponding semigroup.
(c) General Markov process theory leads to the construction of the desired process from this semigroup.

The first three sections of this chapter will treat these three steps in reverse order. The major theorems in Sections 1 and 2 will not be proved, since they belong properly to the theory of Markov processes and functional analysis respectively, rather than to interacting particle systems. The conditions which are assumed in Section 3 are a quantitative version of the assertion that the transition mechanism in one part of the system does not depend too strongly on distant parts of the system.

The approach to the construction problem outlined above is particularly useful when the state space of the process is compact. When the state space of the process is noncompact, new problems arise involving the possibility of explosions at individual sites. Thus the construction of the process in the noncompact case will be handled differently in Chapter IX. The processes which are the subject of Chapter VII also fail in most cases of interest to satisfy the assumptions which will be made in Section 3. Thus they will be treated differently in that chapter.

The construction which is carried out in Section 3 and the estimates developed there have a number of important applications beyond the construction itself. The principal applications are:

(a) to give a sufficient condition for the ergodicity of the process;
(b) to prove that the process preserves asymptotic independence of distant coordinates over finite time periods; and
(c) to give a sufficient condition in the ergodic case for the unique invariant measure to have exponentially decaying correlations.

These are the subject of Section 4.

One unsatisfactory aspect of the treatment of the construction problem in Section 3 is that it fails to explain what, if anything, goes wrong when the assumed conditions on the rates do not hold. This situation is clarified by describing the process in terms of solutions to a martingale problem. From the point of view of the martingale approach, processes with prescribed local dynamics exist under minimal conditions. Attention is therefore focused on the question of uniqueness. The proof of uniqueness under the conditions of Section 3 involves essentially the same analytic techniques which are used in the semigroup approach, and in fact uniqueness follows immediately from the results of Section 3. Section 5 describes the martingale approach in some generality, and connects up the martingale problem with Feller processes and their generators. The existence of solutions to the martingale problem in the interacting particle system context is established under very weak assumptions in Section 6. Furthermore, it is shown there that when the solution is unique, it gives rise to a Feller process. Finally, Section 7 is devoted to a number of examples, including several which show how nonuniqueness for the martingale problem can occur.

The seven main sections depend on one another in the following way:

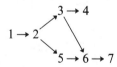

1. Markov Processes and Their Semigroups

Throughout this chapter, X will be a compact metric space with measurable structure given by the σ-algebra of Borel sets. Let $D[0, \infty)$ be the set of all functions $\eta.$ on $[0, \infty)$ with values in X which are right continuous and have left limits. This is the canonical path space for a Markov process with state space X. For $s \in [0, \infty)$, the evaluation mapping π_s from $D[0, \infty)$ to X is defined by $\pi_s(\eta.) = \eta_s$. Let $\mathscr{F}$ be the smallest σ-algebra on $D[0, \infty)$

relative to which all the mappings π_s are measurable. For $t \in [0, \infty)$, let $\mathcal{F}_t$ be the smallest σ-algebra on $D[0, \infty)$ relative to which all the mappings π_s for $s \le t$ are measurable.

Definition 1.1. A *Markov process* on X is a collection $\{P^\eta, \eta \in X\}$ of probability measures on $D[0, \infty)$ indexed by X with the following properties:

(a) $P^\eta[\zeta \in D[0, \infty): \zeta_0 = \eta] = 1$ for all $\eta \in X$.
(b)   The mapping $\eta \to P^\eta(A)$ from X to $[0, 1]$ is measurable for every $A \in \mathcal{F}$.
(c) $P^\eta[\eta_{s+\cdot} \in A | \mathcal{F}_s] = P^{\eta_s}(A)$ a.s. (P^η) for every $\eta \in X$ and $A \in \mathcal{F}$.

The expectation corresponding to P^η will be denoted by E^η. Thus

$$E^\eta Z = \int_{D[0, \infty)} Z \, dP^\eta$$

for any measurable function Z on $D[0, \infty)$ which is integrable relative to P^η. Let $C(X)$ denote the collection of continuous functions on X, regarded as a Banach space with

$$\|f\| = \sup_{\eta \in X} |f(\eta)|.$$

For $f \in C(X)$, write

$$S(t)f(\eta) = E^\eta f(\eta_t).$$

Definition 1.2. A Markov process $\{P^\eta, \eta \in X\}$ is said to be a *Feller process* if $S(t)f \in C(X)$ for every $t \ge 0$ and $f \in C(X)$.

Proposition 1.3. *Suppose $\{P^\eta, \eta \in X\}$ is a Feller process on X. Then the collection of linear operators $\{S(t), t \ge 0\}$ on $C(X)$ has the following properties*:

(a) $S(0) = I$, *the identity operator on $C(X)$.*
(b)   *The mapping $t \to S(t)f$ from $[0, \infty)$ to $C(X)$ is right continuous for every $f \in C(X)$.*
(c) $S(t+s)f = S(t)S(s)f$ *for all $f \in C(X)$ and all $s, t \ge 0$.*
(d) $S(t)1 = 1$ *for all $t \ge 0$.*
(e) $S(t)f \ge 0$ *for all nonnegative $f \in C(X)$.*

Proof. Part (a) is equivalent to statement (a) of Definition 1.1. The right continuity of $S(t)f(\eta)$ in t for fixed $\eta \in X$ is an immediate consequence of the right continuity of η_t and the continuity of f. The proof of the uniformity in η is somewhat more subtle. It can be found in Section 1 of Chapter IX of Yosida (1980). Part (c) is a consequence of statement (c) of Definition 1.1, as can be seen by the following computation:

$$\begin{aligned}
S(t+s)f(\eta) = E^{\eta}f(\eta_{t+s}) &= E^{\eta}[E[f(\eta_{t+s})|\mathscr{F}_t]] \\
&= E^{\eta}[E^{\eta_t}f(\eta_s)] \\
&= E^{\eta}[S(s)f](\eta_t) \\
&= S(t)S(s)f(\eta).
\end{aligned}$$

Properties (d) and (e) are immediate. $\square$

Definition 1.4. A family $\{S(t), t \geq 0\}$ of linear operators on $C(X)$ is called a *Markov semigroup* if it satisfies conditions (a)-(e) of Proposition 1.3.

Remark. Note that the mapping $t \to S(t)f$ is uniformly continuous for each $f \in C(X)$, since $\|S(t+\varepsilon)f - S(t)f\| = \|S(t)[S(\varepsilon) - I]f\| \leq \|S(\varepsilon)f - f\|$.

Why is $S(t)$ a contraction?

The importance of Markov semigroups lies in the fact that each one corresponds to a Markov process, as can be seen from the following converse to Proposition 1.3. Thus the problem of constructing a Feller process is reduced to that of constructing the corresponding semigroup. Proofs of the following theorem can be found in Chapter I of Blumenthal and Getoor (1968) and in Chapter I of Gihman and Skorohod (1975). These, together with Dynkin (1965), are excellent references on Markov processes.

Theorem 1.5. *Suppose* $\{S(t), t \geq 0\}$ *is a Markov semigroup on* $C(X)$. *Then there exists a unique Markov process* $\{P^{\eta}, \eta \in X\}$ *such that*

$$S(t)f(\eta) = E^{\eta}f(\eta_t)$$

for all $f \in C(X)$, $\eta \in X$, *and* $t \geq 0$.

Let $\mathscr{P}$ denote the set of all probability measures on X, with the topology of weak convergence: $\mu_n \to \mu$ in $\mathscr{P}$ if and only if $\int f d\mu_n \to \int f d\mu$ for all $f \in C(X)$. (An excellent reference on weak convergence is Billingsley (1968).) Note in particular that with respect to this topology, $\mathscr{P}$ is compact since X is compact. If $\mu \in \mathscr{P}$ and $\{P^{\eta}, \eta \in X\}$ is a Markov process, then the corresponding Markov process with initial distribution μ is a stochastic process η_t whose distribution is given by

$$P^{\mu} = \int_X P^{\eta} \mu(d\eta).$$

In view of this,

$$E^{\mu} f(\eta_t) = \int S(t) f \, d\mu$$

for $f \in C(X)$. This suggests the following definition.

Definition 1.6. Suppose $\{S(t), t \geq 0\}$ is a Markov semigroup on $C(X)$. Given $\mu \in \mathscr{P}$, $\mu S(t) \in \mathscr{P}$ is defined by the relation

$$\int f \, d[\mu S(t)] = \int S(t) f \, d\mu$$

for all $f \in C(X)$. (The probability measure $\mu S(t)$ is interpreted as the *distribution at time t of the process* when the initial distribution is μ.)

Much of this book is devoted to proving limit theorems for $\mu S(t)$ as $t \to \infty$ for processes of interacting particle system type. An important first step in proving such results is to identify all possible limits of $\mu S(t)$ as $t \to \infty$ for arbitrary initial distributions μ. This suggests the following definition.

Definition 1.7. A $\mu \in \mathscr{P}$ is said to be *invariant* for the process with Markov semigroup $\{S(t), t \geq 0\}$ if $\mu S(t) = \mu$ for all $t \geq 0$. The class of all invariant $\mu \in \mathscr{P}$ will be denoted by $\mathscr{I}$.

Proposition 1.8. (a) $\mu \in \mathscr{I}$ if and only if

$$\int S(t) f \, d\mu = \int f \, d\mu$$

for all $f \in C(X)$ and all $t \geq 0$.
  (b) $\mathscr{I}$ *is a compact convex subset of* $\mathscr{P}$.
  (c)   *Let $\mathscr{I}_e$ be the set of extreme points for $\mathscr{I}$. Then $\mathscr{I}$ is the closed convex hull of $\mathscr{I}_e$.*
  (d)   *If $\nu = \lim_{t \to \infty} \mu S(t)$ exists for some $\mu \in \mathscr{P}$, then $\nu \in \mathscr{I}$.*
  (e)   *If $\nu = \lim_{n \to \infty} T_n^{-1} \int_0^{T_n} \mu S(t) \, dt$ exists for some $\mu \in \mathscr{P}$ and some $T_n \uparrow \infty$, then $\nu \in \mathscr{I}$.*
  (f) $\mathscr{I}$ *is not empty.*

Proof. Part (a) is immediate from Definitions 1.6 and 1.7. The convexity of $\mathscr{I}$ is easy to see from (a). Since $S(t) f \in C(X)$ whenever f is, $\mu_n \to \mu$ implies that

$$\int S(t) f \, d\mu_n \to \int S(t) f \, d\mu$$

for all $f \in C(X)$. Therefore by (a), $\mathscr{I}$ is a closed subset of $\mathscr{P}$. Since $\mathscr{P}$ is compact, it follows that $\mathscr{I}$ is compact. Part (c) follows from (b) by an application of the Krein-Milman Theorem (see Royden (1968), for example). For part (d) use the following steps for $s \geq 0$ and $f \in C(X)$:

$$\int S(s)f \, d\nu = \lim_{t \to \infty} \int S(s)f \, d[\mu S(t)]$$

$$= \lim_{t \to \infty} \int S(t)S(s)f \, d\mu$$

$$= \lim_{t \to \infty} \int S(t+s)f \, d\mu$$

$$= \lim_{t \to \infty} \int S(t)f \, d\mu$$

$$= \lim_{t \to \infty} \int f \, d[\mu S(t)] = \int f \, d\nu.$$

The proof of (e) is similar: for $s \geq 0$ and $f \in C(X)$,

$$\int S(s)f \, d\nu = \lim_{n \to \infty} \int S(s)f \, d\left[T_n^{-1} \int_0^{T_n} \mu S(t) \, dt \right]$$

$$= \lim_{n \to \infty} T_n^{-1} \int_0^{T_n} \left[\int S(s+t)f \, d\mu \right] dt$$

$$= \lim_{n \to \infty} T_n^{-1} \int_s^{T_n+s} \left[\int S(t)f \, d\mu \right] dt$$

$$= \lim_{n \to \infty} \int f \, d\left[T_n^{-1} \int_0^{T_n} \mu S(t) \, dt \right]$$

$$= \int f \, d\nu.$$

The next to last equality follows from the fact that the difference between

$$\int_s^{T_n+s} \left[\int S(t)f \, d\mu \right] dt \quad \text{and} \quad \int_0^{T_n} \left[\int S(t)f \, d\mu \right] dt$$

is at most $2s \|f\|$. Finally, (f) follows from (e) and the compactness of $\mathscr{P}$, since there exist convergent subsequences of

$$\frac{1}{T} \int_0^T \mu S(t) \, dt$$

for any $\mu \in \mathscr{P}$ $\square$

Part (d) of Proposition 1.8 asserts that $\mathscr{I}$ is the set of all possible limits of $\mu S(t)$ as $t \uparrow \infty$. It is not necessarily the case that weak limits of subsequences $\mu S(t_n)$ with $t_n \uparrow \infty$ are in $\mathscr{I}$. For a counterexample, consider the process which moves at uniform speed around the unit circle. In this example, $\mathscr{I}$ consists only of normalized Lebesque measure on the circle, while any $\mu \in \mathscr{P}$ is a limit of subsequences $\mu S(t_n)$ with $t_n \uparrow \infty$.

Definition 1.7 was motivated by the problem of obtaining limit theorems for $\mu S(t)$ as $t \to \infty$. A closely related motivation is given by the following observation. Suppose $\mu \in \mathscr{P}$, and consider the corresponding Markov process η_t with initial distribution μ. As is easily checked, $\mu \in \mathscr{I}$ if and only if η_t is a stationary process in the sense that $\{\eta_t, t \geq 0\}$ and $\{\eta_{t+s}, t \geq 0\}$ have the same distribution for all $s \geq 0$. Once $\{\eta_t, t \geq 0\}$ is a stationary process, it is of course easy to extend the definition of η_t to $t < 0$ in such a way that $\{\eta_t, -\infty < t < \infty\}$ is a stationary process.

The following concept will play an important role throughout this book. It describes the nicest situation which one can have relative to convergence of $\mu S(t)$ as $t \to \infty$.

Definition 1.9. The Markov process with semigroup $\{S(t), t \geq 0\}$ is said to be *ergodic* if

(a) $\mathscr{I} = \{\nu\}$ is a singleton, and
(b) $\lim_{t \to \infty} \mu S(t) = \nu$ for all $\mu \in \mathscr{P}$.

2. Semigroups and Their Generators

The Hille–Yosida theory of semigroups of linear operators is naturally set in the context of a general Banach space. Our treatment is specialized to the Banach space $C(X)$, since only semigroups which correspond to Feller processes on X will arise in this chapter.

Definition 2.1. A (usually unbounded) linear operator Ω on $C(X)$ with domain $\mathscr{D}(\Omega)$ is said to be a *Markov pregenerator* if it satisfies the following conditions:

(a) $1 \in \mathscr{D}(\Omega)$ and $\Omega 1 = 0$.
(b) $\mathscr{D}(\Omega)$ is dense in $C(X)$.
(c)   If $f \in \mathscr{D}(\Omega)$, $\lambda \geq 0$, and $f - \lambda \Omega f = g$, then

$$\min_{\zeta \in X} f(\zeta) \geq \min_{\zeta \in X} g(\zeta).$$

Note by applying property (c) to both f and $-f$, that a Markov pregenerator has the following property: if $f \in \mathscr{D}(\Omega)$, $\lambda \geq 0$, and $f - \lambda \Omega f = g$, then $\|f\| \leq \|g\|$. In particular, g determines f uniquely. Normally, property (c) of the definition is verified by using the following result.

Proposition 2.2. *Suppose that the linear operator* Ω *on* $C(X)$ *satisfies the following property: if* $f \in \mathscr{D}(\Omega)$ *and* $f(\eta) = \min_{\zeta \in X} f(\zeta)$, *then* $\Omega f(\eta) \geq 0$. *Then* Ω *satisfies property* (c) *of Definition* 2.1.

Proof. Suppose $f \in \mathscr{D}(\Omega)$, $\lambda \geq 0$, and $f - \lambda \Omega f = g$. Let η be any point at which f attains its minimum. Such a point exists by the compactness of X and the continuity of f. Then

$$\min_{\zeta \in X} f(\zeta) = f(\eta) \geq f(\eta) - \lambda \Omega f(\eta) = g(\eta) \geq \min_{\zeta \in X} g(\zeta). \quad \square$$

Example 2.3. It can easily be checked by using Proposition 2.2 that the following are Markov pregenerators:

(a) $\Omega = T - I$, where T is a positive operator defined on all of $C(X)$ such that $T1 = 1$.

(b) $X = [0, 1]$, and $\Omega f(\eta) = \frac{1}{2} f''(\eta)$ with

$$\mathscr{D}(\Omega) = \{f \in C(X): f'' \in C(X), f'(0) = f'(1) = 0\}.$$

(c) $X = [0, 1]$, and $\Omega f(\eta) = \frac{1}{2} f''(\eta)$ with

$$\mathscr{D}(\Omega) = \{f \in C(X): f'' \in C(X), f''(0) = f''(1) = 0\}.$$

Definition 2.4. A linear operator Ω on $C(X)$ is said to be *closed* if its graph is a closed subset of $C(X) \times C(X)$. A linear operator $\bar{\Omega}$ is called the *closure* of Ω if $\bar{\Omega}$ is the smallest closed extension of Ω.

Not every linear operator has a closure. The difficulty which may arise in this context is that the closure of the graph of the operator may not be the graph of a linear operator. It may correspond instead to a "multivalued" operator. A simple example of a linear operator on $C[0, 1]$ which has no closure is given by $\mathscr{D}(\Omega) = \{f \in C[0, 1]: f'(0) \text{ exists}\}$ and $\Omega f(\eta) \equiv f'(0)$ for $f \in \mathscr{D}(\Omega)$. Fortunately, this type of problem does not arise in the case of Markov pregenerators.

Proposition 2.5. *Suppose* Ω *is a Markov pregenerator. Then* Ω *has a closure* $\bar{\Omega}$ *which is again a Markov pregenerator.*

Proof. Suppose $f_n \in \mathscr{D}(\Omega)$, $f_n \to 0$, and $\Omega f_n \to h$. Choose $g \in \mathscr{D}(\Omega)$. By the remark following Definition 2.1,

$$\|(I - \lambda \Omega)(f_n + \lambda g)\| \geq \|f_n + \lambda g\|$$

for $\lambda \geq 0$. Letting $n \to \infty$, this implies that

$$\|\lambda g - \lambda h - \lambda^2 \Omega g\| \geq \|\lambda g\|.$$

Dividing by λ and then letting $\lambda \downarrow 0$ gives

$$\|g - h\| \geq \|g\|.$$

Since $g \in \mathscr{D}(\Omega)$ is arbitrary and $\mathscr{D}(\Omega)$ is dense, it follows that $h = 0$. Therefore the closure of the graph of Ω is the graph of a (single-valued) linear operator $\bar{\Omega}$. To verify that $\bar{\Omega}$ is a Markov pregenerator, it suffices to check property (c) of Definition 2.1. Suppose $f \in \mathscr{D}(\bar{\Omega})$, $\lambda \geq 0$, and $f - \lambda \bar{\Omega} f = g$. By the definition of $\bar{\Omega}$, there exist $f_n \in \mathscr{D}(\Omega)$ so that $f_n \to f$ and $\Omega f_n \to \bar{\Omega} f$. Define g_n by

$$f_n - \lambda \Omega f_n = g_n.$$

Since Ω is a Markov pregenerator,

$$\min_{\zeta \in X} f_n(\zeta) \geq \min_{\zeta \in X} g_n(\zeta).$$

But $g_n \to g$ in $C(X)$, so we can pass to the limit to obtain

$$\min_{\zeta \in X} f(\zeta) \geq \min_{\zeta \in X} g(\zeta). \quad \square$$

Proposition 2.6. *Suppose Ω is a closed Markov pregenerator. Then the range of $I - \lambda \Omega$ is a closed subset of $C(X)$ for $\lambda > 0$.*

Proof. Suppose $g_n \in \mathscr{R}(I - \lambda \Omega)$ and $g_n \to g$. Here $\mathscr{R}$ denotes the range. Define f_n by

$$f_n - \lambda \Omega f_n = g_n.$$

Then

$$(f_n - f_m) - \lambda \Omega (f_n - f_m) = g_n - g_m,$$

so that $\|f_n - f_m\| \leq \|g_n - g_m\|$ by the remark following Definition 2.1. Since g_n is a Cauchy sequence, so is f_n, and f can be defined by

$$f = \lim_{n \to \infty} f_n.$$

Therefore

$$\lim_{n \to \infty} \Omega f_n = \lambda^{-1} \lim_{n \to \infty} (f_n - g_n) = \lambda^{-1}(f - g),$$

so since Ω is closed,

$$f - \lambda \Omega f = g.$$

Hence $g \in \mathscr{R}(I - \lambda \Omega)$ as required. $\quad \square$

Definition 2.7. A *Markov generator* is a closed Markov pregenerator Ω which satisfies

$$\mathscr{R}(I - \lambda\Omega) = C(X)$$

for all sufficiently small positive λ.

Proposition 2.8. (a) *A bounded (everywhere defined) Markov pregenerator is a Markov generator.*
(b)   *A Markov generator satisfies*

$$\mathscr{R}(I - \lambda\Omega) = C(X)$$

for all $\lambda \geq 0$.

Proof. A bounded operator is automatically closed. To check that a bounded operator Ω satisfies

$$\mathscr{R}(I - \lambda\Omega) = C(X)$$

for all sufficiently small positive λ it suffices to solve $f - \lambda\Omega f = g$ for $g \in C(X)$ and $0 < \lambda < \|\Omega\|^{-1}$ by setting

$$f = \sum_{n=0}^{\infty} \lambda^n \Omega^n g.$$

To prove part (b) it suffices to show that if $0 < \lambda < \gamma$, then $\mathscr{R}(I - \lambda\Omega) = C(X)$ implies that $\mathscr{R}(I - \gamma\Omega) = C(X)$. Suppose $g \in C(X)$, and we wish to solve

$$f - \gamma\Omega f = g$$

for $f \in \mathscr{D}(\Omega)$. Define $T: C(X) \to \mathscr{D}(\Omega)$ by

$$Th = \frac{\lambda}{\gamma}(I - \lambda\Omega)^{-1}g + \frac{\gamma - \lambda}{\gamma}(I - \lambda\Omega)^{-1}h,$$

which is well defined since $\mathscr{R}(I - \lambda\Omega) = C(X)$. By the remark following Definition 2.1,

$$\|Th_1 - Th_2\| = \frac{\gamma - \lambda}{\gamma}\|(I - \lambda\Omega)^{-1}(h_1 - h_2)\| \leq \frac{\gamma - \lambda}{\gamma}\|h_1 - h_2\|.$$

Therefore T has a unique fixed point, which will be called f. Then $f \in \mathscr{D}(\Omega)$ and

$$(I - \lambda\Omega)f = \frac{\lambda}{\gamma}g + \frac{\gamma - \lambda}{\gamma}f.$$

Rearranging these terms, it follows that

$$f - \gamma\Omega f = g. \quad \square$$

Theorem 2.9 (Hille–Yosida). *There is a one-to-one correspondence between Markov generators on $C(X)$ and Markov semigroups on $C(X)$. This correspondence is given by*:

(a)
$$\mathscr{D}(\Omega) = \left\{ f \in C(X) : \lim_{t \downarrow 0} \frac{S(t)f - f}{t} \ exists \right\}, \quad and$$

$$\Omega f = \lim_{t \downarrow 0} \frac{S(t)f - f}{t} \quad for f \in \mathscr{D}(\Omega).$$

(b)
$$S(t)f = \lim_{n \to \infty} \left(I - \frac{t}{n}\Omega \right)^{-n} f \quad for f \in C(X) \ and \ t \geq 0.$$

Furthermore,

(c) *if $f \in \mathscr{D}(\Omega)$, it follows that $S(t)f \in \mathscr{D}(\Omega)$ and $(d/dt)S(t)f = \Omega S(t)f = S(t)\Omega f$, and finally*

(d) *for $g \in C(X)$ and $\lambda \geq 0$, the solution to $f - \lambda\Omega f = g$ is given by*

$$f = \int_0^\infty e^{-t} S(\lambda t) g \, dt.$$

Ω is called generator of $S(t)$, and $S(t)$ is the semigroup generated by Ω. Proofs of the Hille–Yosida Theorem can be found in Chapter 1 of Dynkin (1965), in Chapter 2 of Gihman and Skorohod (1975), in Chapter IX of Yosida (1980), and in Chapter I of Ethier and Kurtz (1985).

The Hille–Yosida Theorem is usually used in the following way. The infinitesimal description of the process is used to define a Markov pregenerator Ω. Properties (a), (b), and (c) of Definition 2.1 are generally easy to verify with the aid of Proposition 2.2. The next step is to prove that $\mathscr{R}(I - \lambda\Omega)$ is dense in $C(X)$ for sufficiently small positive λ. This is usually the hard part of the program, and the part which requires the imposition of somewhat restrictive hypotheses on the transition rates. Then Propositions 2.5 and 2.6 imply that $\bar{\Omega}$ exists and is a Markov generator, which therefore generates a Markov semigroup by Theorem 2.9.

Example 2.10. In Example 2.3, each of the pregenerators is actually a Markov generator. In example (a) this follows from (a) of Proposition 2.8. Note that in examples (b) and (c), the verification of the condition $\mathscr{R}(I - \lambda\Omega) = C(X)$ for $\lambda \geq 0$ involves the solution of a simple ordinary differential equation. The Markov processes to which these examples correspond are:

(a)   The jump process which waits exponential times with parameter one between jumps. These are chosen with transition probabilities $p(\eta, d\zeta)$, where $p(\eta, d\zeta)$ are defined by $Tf(\eta) = \int f(\zeta)p(\eta, d\zeta)$ for $f \in C(X)$;
(b)   Brownian motion on $[0, 1]$ with reflecting barriers at 0 and 1; and
(c)   Brownian motion on $[0, 1]$ with absorbing barriers at 0 and 1.

A natural question is whether a Markov generator is determined by its values on a dense subset of $C(X)$. A negative answer is provided by examples (b) and (c) above, since in that case the two generators agree on the intersection of their domains, which is dense. This situation suggests that the following concept is important.

Definition 2.11. Suppose Ω is a Markov generator on $C(X)$. A linear subspace D of $\mathcal{D}(\Omega)$ is said to be a *core* for Ω if Ω is the closure of its restriction to D.

Of course, Ω is uniquely determined by its values on a core. In most cases, it is not possible to exhibit explicitly the full domain of a generator. On the other hand, if a generator is obtained via the procedure outlined following the statement of Theorem 2.9, then the domain of the original pregenerator is a core for the generator. Thus a core will be explicitly known. For most purposes, having a explicit description of a core for a generator is as useful as knowing the full domain. The following two results illustrate this remark. The first is a useful criterion for convergence of a sequence of semigroups in terms of the corresponding generators. For a proof of this theorem, see Kurtz (1969). The second result characterizes the invariant measures of the process in terms of the action of the generator on a core.

Theorem 2.12 (Trotter–Kurtz). *Suppose that Ω_n and Ω are the generators of the Markov semigroups $S_n(t)$ and $S(t)$ respectively. If there is a core D for Ω such that $D \subset \mathcal{D}(\Omega_n)$ for all n and $\Omega_n f \to \Omega f$ for all $f \in D$, then*

$$S_n(t)f \to S(t)f$$

for all $f \in C(X)$ uniformly for t in compact sets.

Proposition 2.13. *Suppose D is a core for the generator Ω of a Markov semigroup $S(t)$. Then*

$$\mathcal{I} = \left\{ \mu \in \mathcal{P}: \int \Omega f \, d\mu = 0 \text{ for all } f \in D \right\}.$$

Proof. Suppose that $\mu \in \mathcal{I}$ and $f \in \mathcal{D}(\Omega)$. Then

$$\int \Omega f \, d\mu = \int \lim_{t \downarrow 0} \left[\frac{S(t)f - f}{t} \right] d\mu$$

$$= \lim_{t \downarrow 0} \frac{\displaystyle\int S(t)f \, d\mu - \int f \, d\mu}{t} = 0.$$

Conversely, suppose that $\int \Omega f \, d\mu = 0$ for all $f \in D$. For $f \in \mathcal{D}(\Omega)$, there exist $f_n \in D$ such that $f_n \to f$ and $\Omega f_n \to \Omega f$. Therefore $\int \Omega f \, d\mu = 0$ for all $f \in \mathcal{D}(\Omega)$. If $f \in \mathcal{D}(\Omega)$ and $f - \lambda \Omega f = g$, then it follows that $\int f \, d\mu = \int g \, d\mu$. Rewriting this, one has

$$\int (I - \lambda \Omega)^{-1} g \, d\mu = \int g \, d\mu$$

for all $g \in C(X)$ and $\lambda \geq 0$. By (b) of Theorem 2.9,

$$\int S(t)g \, d\mu = \lim_{n \to \infty} \int \left(I - \frac{t}{n} \Omega \right)^{-n} g \, d\mu = \int g \, d\mu$$

for all $t \geq 0$ and $g \in C(X)$. Therefore $\mu \in \mathcal{I}$. $\square$

As examples of the application of Proposition 2.13, consider (b) and (c) of Examples 2.3 and 2.10. In case (b), Lebesgue measure on $[0, 1]$ is invariant, while in case (c), the pointmasses on $\{0\}$ and $\{1\}$ respectively are invariant.

It will often be the case that one wants to identify $\mathcal{I}$ for a process which is known to be the limit of simpler processes. For example, an infinite particle system may be the limit of finite particle systems. Even if one knows explicitly all of the invariant measures of the approximating processes, it is usually not the case that one can determine $\mathcal{I}$ completely from this information, as can be seen from the following result.

Proposition 2.14. *Suppose that Ω_n and Ω are the generators of Markov semigroups which converge in the sense of the hypothesis of Theorem 2.12. Let $\mathcal{I}_n$ and $\mathcal{I}$ be the invariant measures for these processes respectively. Then*

(a) $\mathcal{I} \supset \{ \mu \in \mathcal{P} : \text{there exist } \mu_n \in \mathcal{I}_n \text{ so that } \mu_n \to \mu \}$.
(b)   *The containment in (a) can be strict.*

Proof. Suppose $\mu_n \to \mu$ where $\mu_n \in \mathcal{I}_n$. Then for $f \in D$,

$$\int \Omega f \, d\mu = \lim_{n \to \infty} \int \Omega_n f \, d\mu_n = 0.$$

Therefore $\mu \in \mathscr{I}$. For (b), let Ω^* be any Markov generator for which the set of invariant measures $\mathscr{I}^*$ is not all of $\mathscr{P}$. Put $\Omega_n = n^{-1}\Omega^*$ and $\Omega = 0$. Then $\mathscr{I}_n = \mathscr{I}^*$ for all n, and $\mathscr{I} = \mathscr{P}$. $\square$

Part (c) of Theorem 2.9 asserts that $F(t) = S(t)f$ gives a solution to the following Cauchy problem:

$$F'(t) = \Omega F(t), \qquad F(0) = f \quad \text{for } f \in \mathscr{D}(\Omega).$$

In several instances, we will need to know that the semigroup provides a unique solution to problems of this type. The following result is appropriate for these applications.

Theorem 2.15. *Suppose that Ω is the generator of a Markov semigroup. Suppose $F(t)$ and $G(t)$ are functions on $[0, \infty)$ with values in $C(X)$ which satisfy:*

(a) $F(t) \in \mathscr{D}(\Omega)$ *for each $t \geq 0$,*
(b) $G(t)$ *is continuous on $[0, \infty)$, and*
(c) $F'(t) = \Omega F(t) + G(t)$ *for $t \geq 0$.*

Then

$$F(t) = S(t)F(0) + \int_0^t S(t-s)G(s)\,ds.$$

Proof.

$$\frac{S(t-s-h)F(s+h) - S(t-s)F(s)}{h}$$

$$= S(t-s)\left[\frac{F(s+h) - F(s)}{h}\right] + \left[\frac{S(t-s-h) - S(t-s)}{h}\right]F(s)$$

$$+ [S(t-s-h) - S(t-s)]F'(s)$$

$$+ [S(t-s-h) - S(t-s)]\left[\frac{F(s+h) - F(s)}{h} - F'(s)\right].$$

The first term on the right-hand side tends to $S(t-s)F'(s)$ as $h \to 0$ since $S(t-s)$ is a bounded operator; the second term tends to $-S(t-s)\Omega F(s)$ by (c) of Theorem 2.9 and assumption (a); the third tends to zero by the remark following Definition 1.4; and the fourth tends to zero because $S(t-s)$ and $S(t-s-h)$ are both contractions. Therefore for $0 < s < t$,

$$\frac{d}{ds}S(t-s)F(s) = S(t-s)F'(s) - S(t-s)\Omega F(s),$$

which by assumption (c), becomes

(2.16) $\dfrac{d}{ds} S(t-s)F(s) = S(t-s)G(s).$

By assumption (b) and the remark following Definition 1.4, $S(t-s)G(s)$ is a continuous function of s. Therefore (2.16) can be integrated from 0 to t to obtain

$$F(t) - S(t)F(0) = \int_0^t S(t-s)G(s)\, ds. \quad \square$$

3. The Construction of Generators for Particle Systems

The primary purpose of this section is to construct Markov generators for a large class of interacting particle systems. The class includes all of the systems which will be treated in this book, with the exception of those in Chapters VII and IX. In those two cases, new problems arise from the fact that the class of allowed configurations must be restricted to some extent in order for the process to be well defined and well behaved. The estimates developed in this section are not only useful for the construction, but have other important applications as well. These will be discussed in Section 4.

Throughout this section, S will denote a countable set of sites and W will be a compact metric space which will play the role of the phase space for each site. The state space for the process as a whole will then be $X = W^S$ with the product topology. The space X is of course compact and metrizable. Points in S will be denoted by letters such as x, y, u, and v, while points in X will be denoted by η or ζ. In many special cases of interest, $W = \{0, 1\}$ and $S = Z^d$, the d-dimensional integer lattice.

The local dynamics of the system are described by a collection of transition measures $c_T(\eta, d\zeta)$. For each $\eta \in X$ and finite $T \subset S$, $c_T(\eta, d\zeta)$ is assumed to be a finite positive measure on W^T. We will assume further that for each T the mapping

$$\eta \to c_T(\eta, d\zeta)$$

is continuous from X to the space of finite measures on W^T with the topology of weak convergence. The interpretation is that η is the current configuration of the system, $c_T(\eta, W^T)$ is the rate at which a transition will occur involving the coordinates in T, and $c_T(\eta, d\zeta)/c_T(\eta, W^T)$ is the distribution of the restriction to T of the new configuration after that transition has taken place.

Before proceeding to the general construction, we will pause to see the form which $c_T(\eta, d\zeta)$ takes for each of the examples described in the introduction. These are special cases of families of processes which will be studied in later chapters. The cardinality of T will be denoted by $|T|$.

Example 3.1. (a) *The Stochastic Ising Model.* $S = Z^d$, and $W = \{-1, +1\}$. If $|T| \geq 2$, then $c_T(\eta, d\zeta) = 0$. If $T = \{x\}$, then $c_T(\eta, d\zeta)$ puts mass

$$\exp\left[-\beta \sum_{y:|y-x|=1} \eta(x)\eta(y)\right]$$

on $\{-\eta(x)\}$. Recall that $\beta \geq 0$ in this model.

(b) *The Voter Model.* $S = Z^d$ and $W = \{0, 1\}$. If $|T| \geq 2$, then $c_T(\eta, d\zeta) = 0$. If $T = \{x\}$, then $c_T(\eta, d\zeta)$ puts mass

$$\frac{1}{2d} \sum_{y:|y-x|=1} 1_{\{\eta(y) \neq \eta(x)\}}$$

on $\{1 - \eta(x)\}$.

(c) *The Contact Process.* $S = Z^d$ and $W = \{0, 1\}$. If $|T| \geq 2$, then $c_T(\eta, d\zeta) = 0$. If $T = \{x\}$, then $c_T(\eta, d\zeta)$ puts mass 1 on $\{0\}$ if $\eta(x) = 1$ and mass

$$\lambda \sum_{y:|y-x|=1} \eta(y)$$

on $\{1\}$ if $\eta(x) = 0$.

(d) *The Exclusion Process.* S is a general countable set and $W = \{0, 1\}$. If $T = \{x, y\}$ with $\eta(x) = 1$ and $\eta(y) = 0$, then $c_T(\eta, d\zeta)$ puts mass $p(x, y)$ on $\{\zeta\}$ where $\zeta(x) = 0$ and $\zeta(y) = 1$. Here $p(x, y) \geq 0$ and $\sum_y p(x, y) = 1$. Otherwise, $c_T(\eta, d\zeta) = 0$.

For $f \in C(X)$ and $x \in S$, let

$$\Delta_f(x) = \sup\{|f(\eta) - f(\zeta)|: \eta, \zeta \in X \text{ and } \eta(y) = \zeta(y) \text{ for all } y \neq x\}.$$

This should be thought of as a measure of the degree to which f depends on the coordinate $\eta(x)$. Of course

$$\lim_{x \to \infty} \Delta_f(x) = 0$$

for all $f \in C(X)$. A function f can be thought of as being smooth if $\Delta_f(x) \to 0$ rapidly as $x \to \infty$. The following class of smooth functions will play the role of a core for the generators to be constructed:

$$D(X) = \left\{f \in C(X): \|f\| = \sum_x \Delta_f(x) < \infty\right\}.$$

Note that $D(X)$ is dense in $C(X)$. This can be obtained as an application of the Stone–Weierstrass Theorem (see Chapter 9 of Royden (1968)). It can also be seen directly by approximating $f \in C(X)$ by $f_T(\zeta) = f(\eta^\zeta)$ for some fixed $\eta \in X$, where

$$\eta^\zeta(x) = \begin{cases} \eta(x) & \text{if } x \notin T, \\ \zeta(x) & \text{if } x \in T, \end{cases}$$

for $\zeta \in W^T$.

The first result defines the basic pregenerator which will be used in this section. Two remarks should be made about its statement. First note that the form given below for Ω is a positive linear combination of generators of the type in Example 2.3(a) and 2.10(a). This motivates this form, since we want the process to be a superposition of the jump processes corresponding to the various $c_T(\eta, d\zeta)$. Secondly, the total transition rates have to be restricted in some way even to define Ω on $D(X)$. Assumption (3.3), which does this, formalizes the idea that the total transition rate for subsets of coordinates which involve a fixed site should be uniformly bounded. Let

$$c_T = \sup\{c_T(\eta, W^T) \colon \eta \in X\}.$$

Proposition 3.2. *Assume that*

(3.3) $$\sup_{x \in S} \sum_{T \ni x} c_T < \infty.$$

(a)   *For $f \in D(X)$, the series*

$$\Omega f(\eta) = \sum_T \int_{W^T} c_T(\eta, d\zeta)[f(\eta^\zeta) - f(\eta)]$$

converges uniformly and defines a function in $C(X)$. Furthermore,

$$\|\Omega f\| \le \left(\sup_{x \in S} \sum_{T \ni x} c_T \right) \|f\|.$$

(b) Ω *is a Markov pregenerator.*

Proof. The continuity assumption on the mapping $\eta \to c_T(\eta, d\zeta)$ implies that

$$\int_{W^T} c_T(\eta, d\zeta)[f(\eta^\zeta) - f(\eta)]$$

is in $C(X)$ for each T and each $f \in C(X)$. By regarding η^ζ as the result of changing the coordinates of η corresponding to sites in T one at a time, it is clear that

$$\left|f(\eta^{\zeta})-f(\eta)\right| \le \sum_{x \in T} \Delta_f(x).$$

Therefore

$$\left\|\int c_T(\eta,\, d\zeta)[f(\eta^{\zeta})-f(\eta)]\right\| \le c_T \sum_{x \in T} \Delta_f(x),$$

so

$$\sum_T \left\|\int c_T(\eta,\, d\zeta)[f(\eta^{\zeta})-f(\eta)]\right\| \le \left(\sup_x \sum_{T \ni x} c_T\right)\|\|f\|\|$$

for any $f \in D(X)$. Hence the series defining Ωf converges uniformly. Since the summands are continuous, it follows that $\Omega f \in C(X)$. For the proof of part (b) of the proposition, note that properties (a) and (b) of Definition 2.1 are immediate. Property (c) is verified by using Proposition 2.2 as follows. Suppose $f \in D(X)$ and $f(\eta) = \min\{f(\zeta): \zeta \in X\}$. Then $f(\zeta) \ge f(\eta)$ for all $\zeta \in X$, so $\Omega f(\eta) \ge 0$. $\square$

In order to use the theory described in Section 2 to construct a Markov generator from Ω, it is necessary to show that $\mathcal{R}(I - \lambda\Omega)$ is dense in $C(X)$ for all sufficiently small $\lambda > 0$. Additional assumptions on $c_T(\eta,\, d\zeta)$ will be needed in order to carry this out. The following brief outline of the proof should help motivate the next two results, which give an *a priori* bound on the smoothness of solutions of $f - \lambda\Omega f = g$. First Ω is approximated by a sequence of bounded pregenerators $\Omega^{(n)}$. Since bounded pregenerators are automatically generators by Proposition 2.8,

$$\mathcal{R}(I - \lambda\Omega^{(n)}) = C(X)$$

for each n and each $\lambda \ge 0$. Therefore, given a $g \in D(X)$, there are $f_n \in C(X)$ so that $f_n - \lambda\Omega^{(n)}f_n = g$. The *a priori* bound will imply that for small λ, f_n is smooth, uniformly in n. Thus if $g_n = f_n - \lambda\Omega f_n$ and sufficiently good estimates on the smoothness of f_n are available, it will follow that

$$\|g_n - g\| = \lambda\|(\Omega - \Omega^{(n)})f_n\| \to 0.$$

The fact that $\mathcal{R}(I - \lambda\Omega)$ is dense will then be a consequence of the fact that $g_n \in \mathcal{R}(I - \lambda\Omega)$ for each n, and that $D(X)$ is dense.

For $u \in S$ and finite $T \subset S$, let

$$c_T(u) = \sup\{\|c_T(\eta_1,\, d\zeta) - c_T(\eta_2,\, d\zeta)\|_T: \eta_1(y) = \eta_2(y) \text{ for all } y \ne u\},$$

where $\|\cdot\|_T$ refers to the total variation norm of a measure on W^T. This is a measure of the amount that $c_T(\eta, d\zeta)$ depends on the coordinate $\eta(u)$. Also let

$$\gamma(x, u) = \sum_{T \ni x} c_T(u) \quad \text{for } x \neq u, \ \gamma(x, x) = 0, \quad \text{and}$$

$$\varepsilon = \inf_{\substack{u \in S}} \inf_{\substack{\eta_1 = \eta_2 \\ \text{off } u \\ \eta_1(u) \neq \eta_2(u)}} \sum_{T \ni u} [c_T(\eta_1, \{\zeta : \zeta(u) = \eta_2(u)\})$$

$$+ c_T(\eta_2, \{\zeta : \zeta(u) = \eta_1(u)\})].$$

Of course, ε may very well be zero. There are many cases in which it is positive, however. In these cases it turns out, as we shall see in the next section, that it is important to include ε in the estimates below.

Lemma 3.4. *Assume* (3.3).

(a)   *Suppose $f \in D(X)$ and $f - \lambda \Omega f = g$ for some $\lambda \geq 0$. Then for $u \in S$,*

$$\Delta_f(u)[1 + \lambda\varepsilon] \leq \Delta_g(u) + \lambda \sum_{x \neq u} \gamma(x, u)\Delta_f(x).$$

(b)   *Suppose in addition that $\sum_T c_T < \infty$, so that Ω extends to a bounded generator on $C(X)$. Then the statement in (a) holds for all $f \in C(X)$.*

Proof. Part (b) follows immediately from part (a) since $D(X)$ is dense in $C(X)$. To prove part (a), fix $u \in S$ and choose $\eta_1, \eta_2 \in X$ so that $\eta_1(x) = \eta_2(x)$ for all $x \neq u$ and $f(\eta_1) - f(\eta_2) = \Delta_f(u)$. Such a choice is possible by the continuity of f and the compactness of X. Of course we may assume $\eta_1(u) \neq \eta_2(u)$, since otherwise the result is obvious. Take a finite T so that $T \not\ni u$. Then by the choice of η_1 and η_2, if $\zeta \in W^T$,

$$f(\eta_1^\zeta) - f(\eta_2^\zeta) \leq f(\eta_1) - f(\eta_2).$$

Rewriting this, we obtain

$$f(\eta_1^\zeta) - f(\eta_1) \leq f(\eta_2^\zeta) - f(\eta_2).$$

Therefore

$$\int c_T(\eta_1, d\zeta)[f(\eta_1^\zeta) - f(\eta_1)] \leq \int c_T(\eta_1, d\zeta)[f(\eta_2^\zeta) - f(\eta_2)],$$

so that

$$\int c_T(\eta_1, d\zeta)[f(\eta_1^\zeta) - f(\eta_1)] - \int c_T(\eta_2, d\zeta)[f(\eta_2^\zeta) - f(\eta_2)]$$

(3.5)
$$\leq \int [c_T(\eta_1, d\zeta) - c_T(\eta_2, d\zeta)][f(\eta_2^\zeta) - f(\eta_2)]$$

$$\leq c_T(u) \sum_{x \in T} \Delta_f(x).$$

Now take a finite T so that $T \ni u$. For $\zeta \in W^T$, regard $\zeta(u)$ as being an element of $W^{\{u\}}$, so that $\eta^{\zeta(u)}$ is η modified to be equal to $\zeta(u)$ at the u coordinate. Then define h on W^T by

$$h(\zeta) = f(\eta_i^\zeta) - f(\eta_i^{\zeta(u)}).$$

Note that this is independent of $i = 1, 2$ since $\eta_1 = \eta_2$ off u. Then

$$\sup_{\zeta \in W^T} |h(\zeta)| \leq \sum_{\substack{x \in T \\ x \neq u}} \Delta_f(x),$$

so that

$$\int c_T(\eta_1, d\zeta)[f(\eta_1^\zeta) - f(\eta_1)] - \int c_T(\eta_2, d\zeta)[f(\eta_2^\zeta) - f(\eta_2)]$$

$$= \int h(\zeta) c_T(\eta_1, d\zeta) - \int h(\zeta) c_T(\eta_2, d\zeta)$$

(3.6) $$+ \int c_T(\eta_1, d\zeta)[f(\eta_1^{\zeta(u)}) - f(\eta_1)] - \int c_T(\eta_2, d\zeta)[f(\eta_2^{\zeta(u)}) - f(\eta_2)]$$

$$\leq c_T(u) \sum_{\substack{x \in T \\ x \neq u}} \Delta_f(x)$$

$$- [f(\eta_1) - f(\eta_2)][c_T(\eta_1, \{\zeta: \zeta(u) = \eta_2(u)\}) + c_T(\eta_2, \{\zeta: \zeta(u) = \eta_1(u)\})]$$

since

$$f(\eta_2) \leq f(\eta_i^{\zeta(u)}) \leq f(\eta_1),$$

which is true by the choice of η_1 and η_2. Combining (3.5) and (3.6), it now follows that

$$\Delta_f(u) = f(\eta_1) - f(\eta_2) = g(\eta_1) - g(\eta_2) + \lambda \Omega f(\eta_1) - \lambda \Omega f(\eta_2)$$

$$\leq \Delta_g(u) + \lambda \sum_T c_T(u) \sum_{\substack{x \in T \\ x \neq u}} \Delta_f(x) - \lambda \varepsilon \Delta_f(u).$$

Therefore

$$(1+\lambda\varepsilon)\Delta_f(u) \leq \Delta_g(u) + \lambda \sum_x \gamma(x, u)\Delta_f(x). \quad \square$$

Let $l_1(S)$ be the Banach space of all functions β on S for which

$$\|\beta\|_{l_1(S)} = \sum_x |\beta(x)| < \infty.$$

Whenever $f \in D(X)$, Δ_f may be regarded as an element of $l_1(S)$, and then

$$\|\|f\|\| = \|\Delta_f\|_{l_1(S)}.$$

Lemma 3.7. *Assume* (3.3) *and*

(3.8) $$M = \sup_{x \in S} \sum_{T \ni x} \sum_{u \neq x} c_T(u) = \sup_{x \in S} \sum_u \gamma(x, u) < \infty.$$

(a)   *Then $\Gamma\beta(u) = \sum_x \beta(x)\gamma(x, u)$ defines a bounded operator on $l_1(S)$ with norm M.*

(b)   *If $f - \lambda\Omega f = g$ for some $f, g \in D(X)$ and $\lambda \geq 0$ such that*

$$\lambda M/(1+\lambda\varepsilon) < 1,$$

then

$$\Delta_f \leq [(1+\lambda\varepsilon)I - \lambda\Gamma]^{-1}\Delta_g$$

pointwise.

Proof. For part (a), take $\beta \in l_1(S)$ and compute

$$\sum_u |\Gamma\beta(u)| \leq \sum_u \sum_x |\beta(x)|\gamma(x, u) \leq M \sum_x |\beta(x)|,$$

so that $\|\Gamma\| \leq M$. To show that the norm of Γ is M, fix a $v \in S$ and let

$$\beta(x) = \begin{cases} 1 & \text{if } x = v, \\ 0 & \text{if } x \neq v. \end{cases}$$

Then $\|\beta\|_{l_1(S)} = 1$ and

$$\|\Gamma\beta\|_{l_1(S)} = \sum_u \Gamma\beta(u) = \sum_u \gamma(v, u),$$

which can be made arbitrarily close to M by an appropriate choice of v.

To prove part (b), rewrite part (a) of Lemma 3.4 as

$$\Delta_f \le \frac{1}{1+\lambda\varepsilon}\Delta_g + \frac{\lambda}{1+\lambda\varepsilon}\Gamma\Delta_f.$$

Since Γ is a positive operator and $\Delta_f, \Delta_g \in l_1(S)$, this relation can be iterated to obtain

$$\Delta_f \le \sum_{k=0}^{n-1} \frac{\lambda^k}{(1+\lambda\varepsilon)^{k+1}}\Gamma^k\Delta_g + \frac{\lambda^n}{(1+\lambda\varepsilon)^n}\Gamma^n\Delta_f.$$

Since $\lambda M/(1+\lambda\varepsilon) < 1$, the last term above tends to zero as $n \to \infty$, so that

$$\Delta_f \le \sum_{k=0}^{\infty} \frac{\lambda^k}{(1+\lambda\varepsilon)^{k+1}}\Gamma^k\Delta_g = [(1+\lambda\varepsilon)I - \lambda\Gamma]^{-1}\Delta_g. \quad \square$$

We now come to the main construction result.

Theorem 3.9. *Assume* (3.3) *and* (3.8).

(a) *The closure $\bar{\Omega}$ of Ω is a Markov generator of a Markov semigroup $S(t)$.*

(b) *$D(X)$ is a core for $\bar{\Omega}$.*

(c) *For $f \in D(X)$,*

$$\Delta_{S(t)f} \le e^{-\varepsilon t}\exp(t\Gamma)\Delta_f.$$

(d) *If $f \in D(X)$, then $S(t)f \in D(X)$ for all $t \ge 0$ and*

$$\||S(t)f\|| \le \exp[(M-\varepsilon)t]\||f\||.$$

Proof. The main step is to show that $\mathcal{R}(I - \lambda\Omega)$ is dense in $C(X)$ for all sufficiently small $\lambda \ge 0$. In order to do this, let S_n be finite sets which increase to S and define

$$c_T^{(n)}(\eta, d\zeta) = \begin{cases} c_T(\eta, d\zeta) & \text{if } T \subset S_n, \\ 0 & \text{if } T \not\subset S_n. \end{cases}$$

Then $c_T^{(n)} \le c_T$, $\gamma^{(n)}(x, u) \le \gamma(x, u)$, and

$$\Omega^{(n)}f \to \Omega f$$

as $n \to \infty$ for all $f \in D(X)$. In addition,

(3.10)
$$\sum_T c_T^{(n)} = \sum_{T \subset S_n} c_T < \infty, \quad \text{and}$$

(3.11) $$\sum_{x,u} \gamma^{(n)}(x, u) = \sum_{T \subset S_n} |T| \sum_u c_T(u) < \infty.$$

By (3.10) and Proposition 3.2, $\Omega^{(n)}$ extends to a bounded Markov pregenerator. Therefore by Proposition 2.8, $\Omega^{(n)}$ is a Markov generator and so

$$\mathscr{R}(I - \lambda\Omega^{(n)}) = C(X)$$

for all $\lambda \geq 0$. Fix $g \in D(X)$ and $\lambda > 0$, and define $f_n \in C(X)$ by

$$f_n - \lambda\Omega^{(n)}f_n = g.$$

By part (b) of Lemma 3.4,

$$\Delta_{f_n}(u) \leq \Delta_g(u) + \lambda \sum_x \gamma^{(n)}(x, u)\Delta_{f_n}(x).$$

Therefore, since $g \in D(X)$, $\sup_x \Delta_{f_n}(x) < \infty$, and (3.11) holds, it follows that $f_n \in D(X)$. Consequently, g_n can be defined by

(3.12) $$g_n = f_n - \lambda\Omega f_n.$$

Now take $\lambda < M^{-1}$. Since $\Gamma^{(n)} \leq \Gamma$, an application of part (b) of Lemma 3.7 for the generator $\Omega^{(n)}$ gives

(3.13) $$\Delta_{f_n} \leq [I - \lambda\Gamma^{(n)}]^{-1}\Delta_g \leq [I - \lambda\Gamma]^{-1}\Delta_g.$$

Therefore

$$\|g - g_n\| = \lambda\|(\Omega - \Omega^{(n)})f_n\|$$

$$\leq \lambda \sum_{T \not\subset S_n} c_T \sum_{x \in T} \Delta_{f_n}(x)$$

$$\leq \lambda \sum_{T \not\subset S_n} c_T \sum_{x \in T} [I - \lambda\Gamma]^{-1}\Delta_g(x)$$

$$= \lambda \sum_x [I - \lambda\Gamma]^{-1}\Delta_g(x) \sum_{\substack{T \not\subset S_n \\ T \ni x}} c_T,$$

which tends to zero as $n \to \infty$ by (3.3) and the Bounded Convergence Theorem. Since $g_n \in \mathscr{R}(I - \lambda\Omega)$, g is in the closure of $\mathscr{R}(I - \lambda\Omega)$. Thus we have shown that

$$D(X) \subset \overline{\mathscr{R}(I - \lambda\Omega)}$$

for $0 \leq \lambda < M^{-1}$. Since $D(X)$ is dense in $C(X)$, it follows that

$$\overline{\mathscr{R}(I - \lambda\Omega)} = C(X)$$

for λ in the same interval. Therefore $\bar{\Omega}$ is a Markov generator by Propositions 3.2, 2.5, and 2.6. It generates a Markov semigroup by Theorem 2.9. It is clear that $D(X)$ is a core for $\bar{\Omega}$ since Ω is the restriction of $\bar{\Omega}$ to $D(X)$. To prove part (c) of the theorem, we return to the construction in the first part of the proof and note that since $g_n \to g$ and $(I - \lambda\Omega)^{-1}$ is a contraction, $f_n = (I - \lambda\Omega)^{-1}g_n$ is a Cauchy sequence and hence has a limit which will be called f. By (3.12), Ωf_n has a limit as well, so $f \in \mathscr{D}(\bar{\Omega})$ and

$$\bar{\Omega}f = \lim_{n \to \infty} \Omega f_n$$

by the definition of the closure $\bar{\Omega}$. Passing to the limit in (3.12), we see that

$$g = f - \lambda\bar{\Omega}f.$$

Passing to the limit in (3.13) also, it follows that

$$\Delta_f \leq [I - \lambda\Gamma]^{-1}\Delta_g,$$

so in particular, $f \in D(X)$. Thus $(I - \lambda\bar{\Omega})^{-1}$ maps $D(X)$ into $D(X)$. Furthermore, by part (b) of Lemma 3.7 applied to Ω,

$$\Delta_{(I-\lambda\bar{\Omega})^{-1}g} \leq [(1+\lambda\varepsilon)I - \lambda\Gamma]^{-1}\Delta_g$$

whenever $g \in D(X)$ and λ satisfies $\lambda M/(1+\lambda\varepsilon) < 1$. Since

$$[(1+\lambda\varepsilon)I - \lambda\Gamma]^{-1}$$

is a positive operator, this relation can be iterated to yield

$$\Delta_{(I-\lambda\bar{\Omega})^{-n}g} \leq [(1+\lambda\varepsilon)I - \lambda\Gamma]^{-n}\Delta_g$$

for the same class of g and λ. Setting $\lambda = t/n$ in this inequality and passing to a limit on n using (b) of Theorem 2.9 gives part (c) of the theorem. Part (d) follows from part (c) by taking the $l_1(S)$ norm of both sides of that inequality. $\square$

Corollary 3.14. *Suppose* $c_T^{(n)}(\eta, d\zeta)$ *and* $c_T(\eta, d\zeta)$ *are transition rates, each of which satisfy* (3.3) *and* (3.8). *Let* $\Omega^{(n)}$ *and* Ω *be the corresponding pregenerators defined on* $D(X)$ *as in part* (a) *of Proposition 3.2, and let* $S_n(t)$ *and* $S(t)$ *be the semigroups generated by* $\bar{\Omega}^{(n)}$ *and* $\bar{\Omega}$ *respectively. Suppose that*

$$\Omega f = \lim_{n \to \infty} \Omega^{(n)}f$$

for all $f \in D(X)$. Then

$$S(t)f = \lim_{n \to \infty} S_n(t)f$$

for all $f \in C(X)$, uniformly for t in compact sets.

Proof. This is an immediate consequence of Theorems 2.12 and 3.9. □

Example 3.15 (Continuation of Example 3.1). In Examples (a), (b), and (c), $c_T = 0$ if $|T| \geq 2$, while $c_{\{x\}}$ is finite and independent of x. Also, $\gamma(x, u)$ is translation invariant (i.e. $\gamma(x, u) = \gamma(0, u - x)$) and is zero unless $|x - u| = 1$. Therefore conditions (3.3) and (3.8) hold for these examples, and hence Theorem 3.9 applies to them. In Example (d) on the other hand,

$$c_{\{x,y\}} = \max\{p(x, y), p(y, x)\}$$

for $x \neq y$, $c_T = 0$ if $|T| \neq 2$, and

$$\gamma(x, y) = \max\{p(x, y), p(y, x)\} \quad \text{if } x \neq y.$$

Therefore, in order for Theorem 3.9 to apply to this example, we need to assume that

$$\sup_{y} \sum_{x} p(x, y) < \infty.$$

(Recall that it was already assumed that $\sum_y p(x, y) = 1$.) Of course this is a rather mild assumption. It is satisfied, for example, whenever $p(x, y)$ is symmetric, or is translation invariant on Z^d.

Of the two assumptions in Theorem 3.9, (3.8) is by far the most important. The uniform bound which appears in (3.3) can be removed in many situations without much effort. For examples of this, see Sullivan (1974) and Gray and Griffeath (1976).

4. Applications of the Construction

In this section, several consequences of the construction will be obtained. The first of these is a sufficient condition for an interacting particle system to be ergodic. (Recall Definition 1.9.) This criterion is of interest primarily in case W is a finite set. It is for this application that the factor ε was introduced in Theorem 3.9. The sufficient condition for ergodicity is the assertion that the dependence of $c_T(\eta, d\zeta)$ on η as measured by $c_T(u)$ is weak relative to an underlying minimal transition rate which is measured by the quantity ε. One way of thinking about this is to let $W = \{0, 1\}$ and then consider the family of transition rates

$$c_T(\eta, d\zeta) = c_T^{(1)}(\eta, d\zeta) + Kc_T^{(2)}(\eta, d\zeta),$$

where $c_T^{(1)}$ satisfies the assumptions of Theorem 3.9, $c_T^{(2)}(\eta, d\zeta) = 0$ if $|T| \geq 2$, and for each $u \in S$, $c_{\{u\}}^{(2)}(\eta, d\zeta)$ is a probability measure on W which can depend on u, but not on η. A consequence of the next result is then that the process corresponding to $c_T(\eta, d\zeta)$ is ergodic for sufficiently large K. To see this, it is enough to note that $\gamma(x, u) = \gamma^{(1)}(x, u)$, while $\varepsilon \geq K$.

Theorem 4.1. *Assume conditions (3.3) and (3.8). If $M < \varepsilon$, then the process is ergodic. Furthermore, for $g \in D(X)$,*

$$\left\| S(s)g - \int g \, d\nu \right\| \leq \left(\sup_x \sum_{T \ni x} c_T \right) \frac{e^{-(\varepsilon - M)s}}{\varepsilon - M} \|g\|,$$

where ν is the unique element of $\mathcal{I}$.

Proof. Fix $0 < s < t$ and $g \in D(X)$. Then $S(r)g \in D(X)$ by (d) of Theorem 3.9, so by (c) of Theorem 2.9,

$$S(t)g - S(s)g = \int_s^t \Omega S(r)g \, dr.$$

By part (d) of Theorem 3.9 and part (a) of Proposition 3.2,

$$\|\Omega S(r)g\| \leq \left(\sup_x \sum_{T \ni x} c_T \right) e^{-(\varepsilon - M)r} \|g\|.$$

Therefore

(4.2) $$\|S(t)g - S(s)g\| \leq \left(\sup_x \sum_{T \ni x} c_T \right) \frac{e^{-(\varepsilon - M)s}}{\varepsilon - M} \|g\|$$

for all $t \geq s$, so

$$\lim_{t \to \infty} S(t)g$$

exists. This limit must be constant by part (d) of Theorem 3.9. Now suppose μ is any element of $\mathcal{I}$. (Recall that $\mathcal{I}$ is nonempty by part (f) of Proposition 1.8.) Then

$$\int S(t)g \, d\mu = \int g \, d\mu$$

by part (a) of Proposition 1.8, so that the value of the constant limit of $S(t)g$ must be $\int g \, d\mu$. Since

$$\lim_{t\to\infty} S(t)g = \int g \, d\mu$$

for all $g \in D(X)$, and $D(X)$ is dense in $C(X)$, it follows that μ is uniquely determined, and therefore $\mathscr{I}$ must be a singleton. The final statement of the theorem now follows by letting t tend to ∞ in (4.2). □

The above result gives not only convergence to equilibrium for weakly dependent interacting particle systems, but gives a type of exponential convergence as well. It is important to note that it is $\|\!|\!|g|\!|\!|$ and not $\|g\|$ which appears in the exponential estimate in Theorem 4.1. If $\|g\|$ had appeared there instead, then it would have followed that $\mu S(t)$ converges to ν in total variation. However, this essentially never occurs for interacting particle systems of the type discussed in this book. This observation becomes clearest in the case of countably many independent $\{0, 1\}$ valued Markov chains, where $\mu S(t)$ and ν are typically singular relative to each other. In this case, the total variation distance between $\mu S(t)$ and ν is equal to two for all t.

Example 4.3. ((a) through (d) are a continuation of Examples 3.1 and 3.15.)

(a) *The Stochastic Ising Model.* Here $\varepsilon = 2$ and $M = 2d \, e^{2\beta d}(1 - e^{-2\beta})$, so we obtain ergodicity from Theorem 4.1 for sufficiently small $\beta \geq 0$, where the cut-off depends on the dimension d. In case $d = 2$, for example, the theorem applies when $\beta < \frac{1}{2}\log((1 + \sqrt{3})/2) \approx .16$. As will be seen in Chapter IV, the stochastic Ising model is ergodic for all $\beta \geq 0$ if $d = 1$, and not ergodic for large β if $d \geq 2$. For $d = 2$, the critical β is known exactly: $\beta_2 = \frac{1}{2}\log(1 + \sqrt{2}) \approx .44$. Thus, while Theorem 4.1 does yield nontrivial information in this example, it does not capture the full dependence on dimension of the behavior of the process, nor does it work all the way up to the critical value when the critical value is finite. In fact, one should not expect a general result such as Theorem 4.1, whose assumptions say nothing about the geometry of the situation, to yield sharp answers in special cases. Other tools, which are more closely tied to the specifics of the model must be used in order to obtain the more refined results.

It is interesting to compare the above results with those which Theorem 4.1 yields when applied to an alternative version of the stochastic Ising model which is often used. In this version, the flip rate $\exp[-\beta\eta(x)\sum_{y:|y-x|=1}\eta(y)]$ is replaced by

$$\left\{1 + \exp\left[2\beta\eta(x)\sum_{y:|y-x|=1}\eta(y)\right]\right\}^{-1}.$$

The connection between the two versions, as will be seen in Chapter IV, is that they have the same value of β_d and they have the same invariant measure when $\beta < \beta_d$. For the alternative version, $\varepsilon = 1$ and

$$M = d\left(\frac{e^{2\beta} - e^{-2\beta}}{e^{2\beta} + e^{-2\beta}}\right).$$

Therefore Theorem 4.1 gives ergodicity for all $\beta \geq 0$ in one dimension and for $\beta < \frac{1}{4}\log 3 \approx .27$ in two dimensions in this case.

(b) *The Voter Model.* In this example, $\varepsilon = M = 1$, so Theorem 4.1 yields no information. Of course, this is as it should be, since the voter model is nonergodic in all dimensions. (It has at least two invariant measures: the pointmasses on $\eta \equiv 0$ and $\eta \equiv 1$ respectively.)

(c) *The Contact Process.* For the contact process, $\varepsilon = 1$ and $M = 2d\lambda$. Thus Theorem 4.1 implies that the contact process is ergodic for $\lambda < (2d)^{-1}$, and hence that $(2d)^{-1}$ is a lower bound for the critical value λ_d. In this example, the information provided by Theorem 4.1 is quite good, since λ_d is finite and positive for all $d \geq 1$ as will be seen in Chapter VI. In fact, in Chapter IX we will see that

$$\lim_{d \to \infty} 2d\lambda_d = 1,$$

so that the lower bound given by Theorem 4.1 is asymptotically correct in high dimensions.

(d) *The Exclusion Process.* Here

$$\varepsilon = \inf_{x} \sum_{y;\, y \neq x} \min\{p(x, y), p(y, x)\}, \quad \text{and}$$

$$M = \sup_{x} \sum_{y;\, y \neq x} \max[p(x, y), p(y, x)]$$

so that $\varepsilon \leq M$, with $\varepsilon = M$ in the symmetric case, provided that $p(x, x) = 0$ for all x. Thus again Theorem 4.1 gives no information, which is not surprising since the exclusion process is never ergodic.

(e) *The Majority Vote Process.* In this example, $S = Z^d$ and $W = \{0, 1\}$. If $|T| \geq 2$, then $c_T\{\eta, d\zeta\} = 0$. There are two parameters in the description of this process: a number $0 < \delta < 1$ and a finite set $N \subset Z^d$ which contains 0 and an even number of other points. Then if $T = \{x\}$, $c_T(\eta, d\zeta)$ puts a mass of δ on $\{1 - \eta(x)\}$ if $\eta(x)$ agrees with a majority of the coordinates $\{\eta(y), y - x \in N\}$, and a mass of $1 - \delta$ on $\{1 - \eta(x)\}$ otherwise. Thus at exponential times with parameter one, a coordinate $\eta(x)$ looks at its "neighborhood" $x + N$, and then flips to the majority symbol with probability $1 - \delta$ and to the minority symbol with probability δ. Note that if $\delta = \frac{1}{2}$, this process corresponds to having independent flips at all the sites in Z^d.

In this example, $\varepsilon = \min\{1, 2\delta\}$ and $M = 2n|1 - 2\delta|$ where the cardinality of N is $2n + 1$. Thus Theorem 4.1 yields ergodicity whenever

$$\frac{n}{2n + 1} < \delta < \frac{2n + 1}{4n}.$$

As will be seen in Chapters III and IV, the process is in fact ergodic for all $\delta \in (0, \frac{1}{2})$ if $d = 1$ and $N = \{-1, 0, +1\}$. It is not known whether or not it is ergodic for all $\delta \in (0, 1)$ in other cases of interest, such as $d = 1$ and $N = \{-n, -n + 1, \ldots, 0, \ldots, n - 1, n\}$ for $n \geq 2$, and $d \geq 2$ and $N = \{x \in Z^d : |x| \leq 1\}$.

The final two applications of the construction give sufficient conditions for approximate independence of the behavior of coordinates of the system which are distant from one another. The first of these does this for finite times under rather weak assumptions, while the second does it for invariant measures (i.e., at infinite times) under the assumption of exponential ergodicity of the system. The following result gives the key estimate for both applications.

Proposition 4.4. *Assume conditions* (3.3) *and* (3.8). *For* $f, g \in D(X)$,

$$\|S(t)fg - [S(t)f][S(t)g]\|$$

$$\leq \sum_{u,v} \left[\sum_{T \ni u,v} c_T \right] \int_0^t e^{-2\varepsilon s} (e^{s\Gamma}\Delta_f)(u)(e^{s\Gamma}\Delta_g)(v) \, ds.$$

Proof. For f and $g \in D(X)$, define F and G from $[0, \infty)$ to $C(X)$ by

$$F(t) = S(t)fg - [S(t)f][S(t)g]$$

and

$$G(t) = \Omega[(S(t)f)(S(t)g)] - [S(t)f][\Omega S(t)g] - [S(t)g][\Omega S(t)f].$$

Note that since $\Delta_{fg} \leq \|f\|\Delta_g + \|g\|\Delta_f$, products of elements of $D(X)$ are again in $D(X)$. Therefore by part (d) of Theorem 3.9, $F(t) \in D(X) \subset \mathcal{D}(\Omega)$ for all $t \geq 0$. By part (c) of Theorem 3.9, $S(t)f$ and $S(t)g$ are continuous functions of t relative to the norm $\|\| \cdot \|\| + \| \cdot \|$. Therefore by the bound in part (a) of Proposition 3.2, $\Omega F(t)$ and $G(t)$ are continuous functions from $[0, \infty)$ to $C(X)$. Furthermore $F(0) = 0$, and by part (c) of Theorem 2.9,

$$\frac{d}{dt} F(t) = \Omega S(t)fg - [S(t)f][\Omega S(t)g] - [S(t)g][\Omega S(t)f] = \Omega F(t) + G(t).$$

Therefore by Theorem 2.15,

$$F(t) = \int_0^t S(t-s)G(s)\,ds.$$

Since $S(t)$ is a contraction, this implies that

(4.5) $$\|F(t)\| \le \int_0^t \|G(s)\|\,ds.$$

Now by the definition of Ω on $D(X)$ in Proposition 3.2,

$$G(0)(\eta) = \sum_T \int_{W^T} c_T(\eta,\,d\zeta)[f(\eta^\zeta) - f(\eta)][g(\eta^\zeta) - g(\eta)],$$

so

$$\|G(0)\| \le \sum_T c_T \left[\sum_{u\in T} \Delta_f(u)\right]\left[\sum_{u\in T} \Delta_g(u)\right]$$

$$= \sum_{u,v} \Delta_f(u)\Delta_g(v) \sum_{T\ni u,v} c_T.$$

Replacing f and g by $S(s)f$ and $S(s)g$ respectively in this inequality and using part (c) of Theorem 3.9, this gives

$$\|G(s)\| \le e^{-2\varepsilon s} \sum_{u,v} \left[\sum_{T\ni u,v} c_T\right](e^{s\Gamma}\Delta_f)(u)(e^{s\Gamma}\Delta_g)(v).$$

The required result follows from this and (4.5). $\square$

For $R \subset S$, let $C_R(X)$ denote the set of functions in $C(X)$ which depend only on the coordinates in R. In order to motivate the next result, suppose R_1 and R_2 are disjoint subsets of S, and suppose the process on R_1 envolves independently of the process on R_2. Then of course if $f \in C_{R_1}(X)$ and $g \in C_{R_2}(X)$,

$$S(t)(fg)(\eta) = E^\eta f(\eta_t)g(\eta_t) = E^\eta f(\eta_t)E^\eta g(\eta_t) = [S(t)f(\eta)][S(t)g(\eta)].$$

Thus the next result asserts that at finite times, distant coordinates evolve almost independently,

Theorem 4.6. *In addition to conditions (3.3) and (3.8), suppose that*

(4.7) $$\sup_{x\in S} \sum_{T\ni x} |T|c_T < \infty, \quad \text{and}$$

(4.8) $$M^* = \sup_u \sum_x \gamma(x, u) < \infty.$$

Then for each $f \in C(X)$ and each $t_0 > 0$,

$$\lim_{\substack{R \uparrow S}} \sup_{\substack{g \in C_{R^c}(X) \\ \|g\| \le 1}} \sup_{0 \le t \le t_0} \|S(t)(fg) - [S(t)f][S(t)g]\| = 0.$$

Proof. Since $D(X)$ is dense in $C(X)$ and each $S(t)$ is a contraction, it suffices to prove the result for $f \in D(X)$. Let g_R be defined by

$$g_R(u) = \begin{cases} 2 & \text{if } u \notin R, \\ 0 & \text{if } u \in R. \end{cases}$$

Then if $\|g\| \le 1$ and $g \in C_{R^c}(X)$, $\Delta_g \le g_R$. Therefore by Proposition 4.4, we need to show that

$$\lim_{R \uparrow S} \sum_{u,v} \left[\sum_{T \ni u,v} c_T \right] \int_0^{t_0} (e^{s\Gamma} \Delta_f)(u)(e^{s\Gamma} g_R)(v) \, ds = 0.$$

But this is a consequence of the Dominated Convergence Theorem since by (4.8), $e^{s\Gamma} g_R(v)$ is bounded above by $2e^{sM^*}$ and tends to zero as $R \uparrow S$ for each v, while by (4.7),

$$\sum_{u,v} \left[\sum_{T \ni u,v} c_T \right] e^{s\Gamma} \Delta_f(u) = \sum_u e^{s\Gamma} \Delta_f(u) \sum_{T \ni u} |T| c_T$$

$$\le e^{sM} \|f\| \sup_u \sum_{T \ni u} |T| c_T < \infty. \quad \square$$

It should be noted that assumptions (4.7) and (4.8) above are only slight additional restrictions. In order to see that, observe that if $c_T(\eta, d\zeta) = 0$ whenever $|T| \ge k$ for some fixed k, then (4.7) follows from (3.3). On the other hand, if $S = Z^d$ and the transition mechanism is translation invariant, then $\gamma(x, u) = \gamma(0, u - x)$, so that (4.8) is equivalent to (3.8).

The main applications of Theorem 4.6 occur in the translation invariant case, so we will restrict ourselves to that context for the remainder of this section. Before turning to the applications, we need to discuss a little ergodic theory. For $x \in Z^d$, define the shift transformation τ_x on X by

$$(\tau_x \eta)(y) = \eta(y - x).$$

These induce in a natural way shift transformations on the space of all functions on X via

$$\tau_x f(\eta) = f(\tau_x \eta),$$

and then on elements of $\mathcal{P}$ via

$$\int f d(\tau_x \mu) = \int (\tau_x f) \, d\mu.$$

Let $\mathcal{S}$ be the set of all $\mu \in \mathcal{P}$ such that $\tau_x \mu = \mu$ for all $x \in Z^d$.

We will need to use the following version of the multiparameter ergodic theorem. Its proof can be found in Chapter VIII of Dunford and Schwartz (1958). The inequalities in the statement are to be interpreted component-wise, and $x \to \infty$ is interpreted as saying that each component of x tends to ∞.

Theorem 4.9 (Ergodic Theorem). *If $\mu \in \mathcal{S}$ and f is a bounded measurable function on X, then*

$$\lim_{x \to \infty} \frac{\sum_{0 \le y \le x} \tau_y f}{|\{y \in Z^d : 0 \le y \le x\}|}$$

exists a.s. and in L_1 relative to μ. If $d = 1$, it suffices that f be in $L_1(\mu)$.

Definition 4.10. A $\mu \in \mathcal{S}$ is said to be *ergodic* if whenever $\tau_x f = f$ for all $x \in Z^d$ and f is measurable on X, it follows that f is constant a.s. relative to μ.

The Ergodic Theorem provides a useful criterion for a measure in $\mathcal{S}$ to be ergodic:

Proposition 4.11. *A $\mu \in \mathcal{S}$ is ergodic if and only if for every f and $g \in C(X)$,*

$$(4.12) \qquad \lim_{x \to \infty} \frac{\sum_{0 \le y \le x} \int (\tau_y f) g \, d\mu}{|\{y \in Z^d : 0 \le y \le x\}|} = \int f \, d\mu \int g \, d\mu.$$

Proof. Suppose first that μ is ergodic. The limit in the statement of Theorem 4.9 is translation invariant, and hence constant a.s.(μ). Since $\mu \in \mathcal{S}$, $\int (\tau_y f) \, d\mu = \int f \, d\mu$ for all y. Hence that constant limit must be $\int f \, d\mu$, and (4.12) follows immediately. For the converse, suppose (4.12) holds for all f and $g \in C(X)$. Since every bounded measurable function can be approximated in $L_1(\mu)$ by a continuous function, it follows that (4.12) holds for bounded measurable f and g as well. Now suppose f is bounded measurable and $\tau_x f = f$ for all $x \in Z^d$. Then (4.12) gives $\int fg \, d\mu = \int f \, d\mu \int g \, d\mu$. Putting $g = f$, we see that f is constant a.s. relative to μ. $\square$

Proposition 4.13. *Suppose μ_1 and $\mu_2 \in \mathcal{S}$ are both ergodic. Then either $\mu_1 = \mu_2$ or μ_1 and μ_2 are mutually singular.*

Proof. Suppose that $\mu_1 \neq \mu_2$. Then there is an $f \in C(X)$ so that $\int f d\mu_1 \neq \int f d\mu_2$. Since μ_1 and μ_2 are both ergodic, Theorem 4.9 gives

$$\lim_{x \to \infty} \frac{\sum_{0 \leq y \leq x} \tau_y f}{|\{y \in Z^d : 0 \leq y \leq x\}|} = \int f d\mu_i$$

a.s. relative to μ_i. Therefore μ_1 and μ_2 are mutually singular. $\square$

For the next result, let $\mathcal{S}_e$ be the extreme points of $\mathcal{S}$, i.e. the elements of $\mathcal{S}$ which cannot be written as a nontrivial convex combination of elements of $\mathcal{S}$.

Corollary 4.14. $\mathcal{S}_e = \{\mu \in \mathcal{S}: \mu \text{ is ergodic}\}$.

Proof. Suppose first that $\mu \in \mathcal{S}$ is ergodic and that $\mu = \alpha\mu_1 + (1 - \alpha)\mu_2$ for some $\alpha \in (0, 1)$ and $\mu_1, \mu_2 \in \mathcal{S}$. Then μ_1 and μ_2 are absolutely continuous with respect to μ and hence are ergodic as well. By Proposition 4.13, it follows that $\mu_1 = \mu_2 = \mu$, so that $\mu \in \mathcal{S}_e$. For the converse, suppose that μ is not ergodic. Then there is a measurable translation invariant subset A of X so that $0 < \mu(A) < 1$. Writing

$$\mu(\cdot) = \mu(\cdot | A)\mu(A) + \mu(\cdot | A^c)\mu(A^c),$$

we see that μ is not extreme since both $\mu(\cdot | A)$ and $\mu(\cdot | A^c)$ are in $\mathcal{S}$. $\square$

Returning now to our discussion of particle systems, take $S = Z^d$ and assume for the remainder of this section that the mechanism (i.e. the measures $c_T(\eta, d\zeta)$) is translation invariant. Then the semigroup $S(t)$ commutes with the shift transformations τ_x, and hence $\mu \in \mathcal{S}$ implies that $\mu S(t) \in \mathcal{S}$ for all $t \geq 0$.

Theorem 4.15. *Under the assumptions of Theorem 4.6, $\mu \in \mathcal{S}_e$ implies that $\mu S(t) \in \mathcal{S}_e$ for all $t \geq 0$.*

Proof. We will check that $\mu S(t) \in \mathcal{S}_e$ by using the criterion from Proposition 4.11. Write for $f, g \in C(X)$

$$\int (\tau_y f)g \, d[\mu S(t)] = \int S(t)[(\tau_y f)g] \, d\mu$$

$$= \int [\tau_y S(t)f]S(t)g \, d\mu$$

$$+ \int \{S(t)[(\tau_y f)g] - [S(t)\tau_y f]S(t)g\} \, d\mu.$$

Since $\mu \in \mathscr{S}_e$ and $S(t)f, S(t)g \in C(X)$, the required result will follow from Proposition 4.11 and Corollary 4.14 as soon as we prove that

(4.16)
$$\lim_{y \to \infty} \| S(t)[(\tau_y f)g] - [S(t)\tau_y f]S(t)g \| = 0.$$

For f and g depending on finitely many coordinates, this follows from Theorem 4.6. For general f and $g \in C(X)$, (4.16) follows by approximation. □

Unfortunately, $\mathscr{S}_e$ is not a closed subset of $\mathscr{S}$, so one cannot conclude from Theorem 4.15 that $\lim_{t \to \infty} \mu S(t) \in \mathscr{S}_e$ whenever $\mu \in \mathscr{S}_e$ and the limit exists. An example which shows that $\mathscr{S}_e$ is not closed is the following: take $d = 1$ and $W = \{0, 1\}$ and let μ_δ be the element of $\mathscr{S}$ which is the distribution of the stationary two-state Markov chain with transition matrix

$$\begin{pmatrix} 1-\delta & \delta \\ \delta & 1-\delta \end{pmatrix}.$$

An alternative description of μ_δ is given by

$$\mu_\delta\{\eta : \eta = \zeta \text{ on } [m, n]\} = \tfrac{1}{2}(1-\delta)^k \delta^l,$$

where k is the number of $x \in [m, n-1]$ so that $\zeta(x) = \zeta(x+1)$ and l is the number of $x \in [m, n-1]$ so that $\zeta(x) \neq \zeta(x+1)$. It is easy to see from the convergence theorem for finite state Markov chains and Proposition 4.11 that $\mu_\delta \in \mathscr{S}_e$ for $0 < \delta < 1$. However $\lim_{\delta \downarrow 0} \mu_\delta$ is not ergodic since it is a convex combination of the pointmasses on $\eta \equiv 0$ and $\eta \equiv 1$ respectively. The voter model in one or two dimensions provides an example in which $\lim_{t \to \infty} \mu S(t) \notin \mathscr{S}_e$ even if $\mu \in \mathscr{S}_e$.

In spite of the above remarks, it is possible under sufficiently strong conditions to show that the unique element of $\mathscr{I}$ for an ergodic particle system is in $\mathscr{S}_e$. In fact the next results give sufficient conditions for a much stronger assertion—that ν has exponentially decaying correlations. This is important for a number of applications. For example, it implies in many cases that the random variables $\{\eta(x), x \in Z^d\}$ satisfy classical central limit behavior (see, for example, Malysev (1975), Leonenko (1976), Neaderhouser (1978), Newman (1980), Bolthausen (1982), and Takahata (1983)). Of course, any measure in $\mathscr{S}$ which has exponentially decaying correlations in the sense of Theorem 4.20 is in $\mathscr{S}_e$ by the criterion of Proposition 4.11. First we will give a simpler version of Proposition 4.4 for processes which are of finite range in the sense of the following definition.

Definition 4.17. The process is said to be of *finite range* if there is an N so

(a) $c_T = 0$ if T contains two points x, y with $|x - y| > N$, and
(b) $\gamma(x, y) = 0$ whenever $|x - y| > N$.

Proposition 4.18. *Suppose that the process is of finite range. Then there is a* $\delta > 0$ *and a* $K > 0$ *so that*

$$\|S(t)fg - [S(t)f][S(t)g]\| \leq K[\|f\|][\|g\|] \exp[4Mt - \delta d(R_1, R_2)],$$

where $R_1, R_2 \subset Z^d$, $d(R_1, R_2)$ *is the distance between* R_1 *and* R_2, $f \in C_{R_1}(X) \cap D(X)$, *and* $g \in C_{R_2}(X) \cap D(X)$.

Proof. First note that as a result of the translation invariance and finite range assumptions, conditions (3.3) and (3.8) are automatically satisfied. Therefore Theorem 3.9 and Proposition 4.4 apply. By the latter of these results,

$$\|S(t)fg - [S(t)f][S(t)g]\|$$

$$\leq \sum_{x,y} \Delta_f(x)\Delta_g(y) \sum_{u,v} \left[\sum_{T \ni u,v} c_T \right] \int_0^t \gamma_s(u-x)\gamma_s(v-y) \, ds,$$

where $\gamma_s(x)$ is defined by

$$e^{s\Gamma}\beta(u) = \sum_x \beta(x)\gamma_s(u-x).$$

Since $\|f\| = \sum_x \Delta_f(x)$ and $\|g\| = \sum_x \Delta_g(x)$, the result will follow provided that we show that

$$\sum_{u,v} \left[\sum_{T \ni u,v} c_T \right] \gamma_s(u-x)\gamma_s(v-y) \leq 4MK \, e^{4Ms - \delta|x-y|}.$$

This in turn would follow from

(4.19) $$\sup_s e^{-4Ms} \sum_{u,v,y} e^{\delta|y|} \left[\sum_{T \ni u,v} c_T \right] \gamma_s(u)\gamma_s(v-y) < \infty.$$

To show this, choose $\delta > 0$ so that

$$\sum_u \gamma(0, u) \, e^{\delta|u|} \leq 2M.$$

This is possible since $M = \sum_u \gamma(0, u)$ and $\gamma(0, u)$ is nonzero for only finitely many u. (We can assume here that $M > 0$, since otherwise the result is trivial.) Then

$$\sum_u \gamma_s(u) \, e^{\delta|u|} \leq e^{2sM}.$$

The finiteness of the expression in (4.19) then follows by replacing $|y|$

by $|v-y|+|v-u|+|u|$, and summing first on y, then on v, and finally on u. $\square$

Proposition 4.18 gives exponential decay of correlations for the distribution at time t if the initial distribution is deterministic. There is a nonuniformity in t which appears on the right-hand side of the estimate. If the process converges exponentially rapidly to equilibrium, this nonuniformity can be removed, as we will see in the next result. The resulting uniformity is important, since it implies that the limiting invariant measure ν has exponentially decaying correlations as well. One way the exponential convergence to equilibrium can be verified is by using Theorem 4.1. Another instance of this verification will occur in Section 3 of Chapter VI.

Theorem 4.20. *Suppose that the process is of finite range, and that for some $\eta \in X$, $\nu \in \mathcal{P}$, and $K, \delta > 0$,*

$$(4.21) \qquad \left| S(t)f(\eta) - \int f\,d\nu \right| \le K\,e^{-\delta t}\|\|f\|\|$$

for all $f \in D(X)$ and $t \ge 0$. Then there are $\bar{K}, \bar{\delta} > 0$ so that for that η and all $t \ge 0$,

$$(4.22) \quad |S(t)(fg)(\eta) - [S(t)f(\eta)][S(t)g(\eta)]| \le \bar{K}\,e^{-\bar{\delta}d(R_1, R_2)}\|\|f\|\|\|\|g\|\|],$$

where $R_1, R_2 \subset Z^d$, $d(R_1, R_2)$ is the distance between R_1 and R_2, $f \in C_{R_1}(X) \cap D(X)$, and $g \in C_{R_2}(X) \cap D(X)$.

Proof. Since adding a constant to f or g has no effect on either the left or right side of (4.22), we can assume throughout that $\int f\,d\nu = \int g\,d\nu = 0$. For such functions,

$$\|f\| \le \|\|f\|\|, \qquad \|g\| \le \|\|g\|\|, \quad \text{and}$$
$$(4.23)$$
$$\|\|fg\|\| \le 2[\|\|f\|\|][\|\|g\|\|].$$

By taking the larger of the two K's and the smaller of the two δ's, we can assume that the same K and δ are used in (4.21) and in the conclusion of Proposition 4.18. Take $0 < s < t$ and write

$$|S(t)(fg)(\eta) - [S(t)f(\eta)][S(t)g(\eta)]|$$
$$\le |S(s)(fg)(\eta) - [S(s)f(\eta)][S(s)g(\eta)]|$$
$$+ |S(t)(fg)(\eta) - S(s)(fg)(\eta)|$$
$$+ |S(t)f(\eta)|\,|S(t)g(\eta) - S(s)g(\eta)|$$
$$+ |S(s)g(\eta)|\,|S(t)f(\eta) - S(s)f(\eta)|$$

$$\leq K\{[\|\|f\|\|][\|\|g\|\|] \exp[4Ms - \delta d(R_1, R_2)]$$
$$+ 2 e^{-\delta s}\|\|fg\|\| + 2\|f\| e^{-\delta s}\|\|g\|\| + 2\|g\| e^{-\delta s}\|\|f\|\|\}$$
$$\leq K[\|\|f\|\|][\|\|g\|\|]\{\exp[4Ms - \delta d(R_1, R_2)] + 8 e^{-\delta s}\}.$$

The first inequality above is simply the triangle inequality. In the second inequality, Proposition 4.18 is used on the first term, and (4.21) is used on the other three terms. The final inequality comes from (4.23). Choosing $s = \delta d(R_1, R_2)/(\delta + 4M)$, we see that (4.22) holds with

$$\bar{K} = 9K \quad \text{and} \quad \bar{\delta} = \frac{\delta^2}{\delta + 4M},$$

provided that $t \geq \delta d(R_1, R_2)/(\delta + 4M)$. For $t \leq \delta d(R_1, R_2)/(\delta + 4M)$, (4.22) follows directly from Proposition 4.18. $\quad\square$

5. The Martingale Problem

The martingale approach to Markov processes will be described in this section. This approach was originally developed by Stroock and Varadhan for the purpose of constructing and studying diffusion processes with continuous coefficients. Earlier treatments of the construction problem for diffusion processes had required additional smoothness assumptions on the diffusion coefficients. A complete exposition of the theory in the diffusion context can be found in Stroock and Varadhan (1979).

Throughout this section, we adopt the setting and notation of Sections 1 and 2. The applications to interacting particle systems will be developed in Sections 6 and 7.

Definition 5.1. Suppose that Ω is a Markov pregenerator (see Definition 2.1) and that $\eta \in X$. A probability measure P on $D[0, \infty)$ is said to solve the *martingale problem* for Ω with initial point η if

(a) $P[\zeta. \in D[0, \infty): \zeta_0 = \eta] = 1$, and
(b) $f(\eta_t) - \int_0^t \Omega f(\eta_s) \, ds$ is a martingale relative to P and the σ-algebras $\{\mathcal{F}_t, t \geq 0\}$ for all $f \in \mathcal{D}(\Omega)$.

To motivate the definition, one can think in the following terms. The treatment of the Hille–Yosida theory in Sections 1 and 2 provides a way of relating the infinitesimal description of the process which is contained in Ω to the collection of measures $\{P^\eta, \eta \in X\}$ on $D[0, \infty)$. This program only works when $\bar{\Omega}$ is a Markov generator, and the proof of this fact requires in general that reasonably strong regularity assumptions be placed on the infinitesimal parameters. Definition 5.1 establishes this relationship between Ω and $\{P^\eta, \eta \in X\}$ under the much weaker assumption that Ω be a Markov pregenerator.

Of course, in order for this to be useful, one must establish existence of solutions to the martingale problem. For most purposes, it is important to establish uniqueness of the solution as well. As we will see in the next section, existence of solutions can be proved under minimal conditions on the infinitesimal parameters. Uniqueness is more difficult to establish, and often requires assumptions similar to those which are needed for the Hille–Yosida approach. An important contribution of the martingale approach to interacting particle systems is that it focuses attention on the uniqueness question. The following result demonstrates the close connection between Feller processes and solutions to the martingale problem.

Theorem 5.2. *Suppose that Ω is a Markov pregenerator and that its closure $\bar{\Omega}$ is a Markov generator. Let $\{P^{\eta}, \eta \in X\}$ be the unique Feller process which corresponds to $\bar{\Omega}$ via Theorems 1.5 and 2.9. Then for each $\eta \in X$, P^{η} is the unique solution to the martingale problem for Ω with initial point η.*

Proof. In order to show that P^{η} is a solution to the martingale problem, fix $f \in \mathscr{D}(\Omega)$ and write for $r < t$,

$$E^{\eta}\left[\int_r^t \Omega f(\eta_s)\, ds \middle| \mathscr{F}_r\right] = E^{\eta_r} \int_0^{t-r} \Omega f(\eta_s)\, ds$$

$$= \int_0^{t-r} E^{\eta_r} \Omega f(\eta_s)\, ds$$

$$= \int_0^{t-r} S(s)\Omega f(\eta_r)\, ds$$

$$= \int_0^{t-r} \bar{\Omega} S(s) f(\eta_r)\, ds$$

$$= S(t-r)f(\eta_r) - f(\eta_r).$$

In this computation, the first equality follows from the Markov property (part (c) of Definition 1.1), the second from Fubini's Theorem, the third from the definition of the semigroup, and the last two from part (c) of Theorem 2.9. Using the above equality, compute for $r < t$

$$E^{\eta}\left[f(\eta_t) - \int_0^t \Omega f(\eta_s) \middle| \mathscr{F}_r\right]$$

$$= E^{\eta}\left[f(\eta_t) \middle| \mathscr{F}_r\right] - \int_0^r \Omega f(\eta_s)\, ds - E^{\eta}\left[\int_r^t \Omega f(\eta_s) \middle| \mathscr{F}_r\right]$$

$$= E^{\eta_r} f(\eta_{t-r}) - \int_0^r \Omega f(\eta_s)\, ds - S(t-r)f(\eta_r) + f(\eta_r)$$

$$= f(\eta_r) - \int_0^r \Omega f(\eta_s)\, ds.$$

Therefore P^{η} is a solution to the martingale problem. For the uniqueness statement, fix $\eta \in X$ and let P be any solution to the martingale problem for Ω with initial point η. Clearly P is also a solution to the martingale problem for $\bar{\Omega}$ with initial point η. Given $g \in C(X)$ and $\lambda > 0$, there exists by part (b) of Proposition 2.8 an $f \in \mathscr{D}(\bar{\Omega})$ so that

$$(5.3) \qquad\qquad\qquad (\lambda - \bar{\Omega})f = g.$$

Since P solves the martingale problem,

$$(5.4) \qquad\qquad E\left[f(\eta_t) - \int_r^t \bar{\Omega}f(\eta_s)\, ds \,\Big|\, \mathscr{F}_r \right] = f(\eta_r)$$

for $r < t$, where E refers to the (conditional) expectation relative to P. Multiply (5.4) by $\lambda\, e^{-\lambda t}$ and integrate from r to ∞ to obtain

$$E\left[\int_r^\infty e^{-\lambda t}\lambda f(\eta_t)\, dt - \int_r^\infty e^{-\lambda s}\bar{\Omega}f(\eta_s)\, ds \,\Big|\, \mathscr{F}_r \right] = e^{-\lambda r}f(\eta_r),$$

or by (5.3),

$$(5.5) \qquad\qquad E\left[\int_r^\infty e^{-\lambda t}g(\eta_t)\, dt \,\Big|\, \mathscr{F}_r \right] = e^{-\lambda r}f(\eta_r).$$

Now take $0 < s_1 < \cdots < s_n$, $\lambda_i > 0$, and $h_i \in C(X)$. Multiply (5.5) with $r = s_n$ by

$$\exp\left[-\sum_{i=1}^n \lambda_i s_i \right] \prod_{i=1}^n h_i(\eta_{s_i}),$$

take expected values of both sides and integrate to obtain

$$(5.6)$$
$$\int \cdots \int_{s_1 < \cdots < s_n < t} \exp\left[-\lambda t - \sum_{i=1}^n \lambda_i s_i \right] E\left\{ g(\eta_t) \prod_{i=1}^n h_i(\eta_{s_i}) \right\} ds_1 \cdots ds_n\, dt$$
$$= \int \cdots \int_{s_1 < \cdots < s_n} \exp\left[-\lambda s_n - \sum_{i=1}^n \lambda_i s_i \right] E\left\{ f(\eta_{s_n}) \prod_{i=1}^n h_i(\eta_{s_i}) \right\} ds_1 \cdots ds_n.$$

Setting $r = 0$ in (5.5) and using the uniqueness theorem for Laplace transforms, we see that the one-dimensional distributions of P are the same as those of P^{η}, since P^{η} is another solution to the martingale problem. Applying the uniqueness theorem for multidimensional Laplace transforms to (5.6) and the corresponding identity where E is replaced by E^{η}, we obtain the following conclusion: the equality of the n-dimensional distributions of P and P^η implies the equality of the $(n+1)$-dimensional distributions of P and P^η. Thus P and P^η have the same finite-dimensional distributions by induction, and hence $P = P^\eta$. □

It is important to note that in the uniqueness proof above, the only properties of a Markov generator which are used are those contained in Definition 2.7 and Proposition 2.8. In particular, the Hille–Yosida theorem (Theorem 2.9) itself is not used in that proof. Therefore by using the martingale approach, one can avoid the use of the Hille–Yosida theorem if one prefers. Of course one still needs (5.3), and it was the proof of (5.3) which was the primary objective of Section 3. The following definition will simplify some of the statements in the next two sections.

Definition 5.7. Suppose that Ω is a Markov pregenerator. The martingale problem for Ω is said to be *well posed* if for each $\eta \in X$, the martingale problem for Ω with initial point η has a unique solution.

To conclude this section, we present a counterexample to a possible converse to Theorem 5.2. For an example of this type in the particle system context, see Example 7.6.

Example 5.8. Let $X = [0, 1]$, and define Ω by $\Omega f = \frac{1}{2} f''$ on

$$\mathscr{D}(\Omega) = \{f \in C(X): f'' \in C(X), f'(0) = f'(1) = 0, f'(\tfrac{1}{3}) = f'(\tfrac{2}{3})\}.$$

This pregenerator is a restriction of the generator for Brownian motion on $[0, 1]$ with reflecting barriers at 0 and 1 which was discussed in Examples 2.3(b) and 2.10(b). This pregenerator is closed, as can be seen by writing

$$f'(\tfrac{2}{3}) - f'(\tfrac{1}{3}) = \int_{1/3}^{2/3} f''(\eta) \, d\eta.$$

Therefore $\bar{\Omega} = \Omega$ is not a Markov generator, so the hypothesis of Theorem 5.2 is not satisfied. On the other hand, the martingale problem is well posed for Ω, and of course the unique solution is then reflecting Brownian motion on $[0, 1]$.

In order to show that the martingale problem for Ω is well posed it suffices to show that if P is any solution to the martingale problem for Ω with initial point η, then it is also a solution to the martingale problem for the generator of reflecting Brownian motion on $[0, 1]$, since then Theorem 5.2 can be applied to that generator. To do this suppose that $f \in C(X)$ satisfies $f'' \in C(X)$ and $f'(0) = f'(1) = 0$. Choose $g, h \in \mathscr{D}(\Omega)$ so that $f = g$ on

$[0, \frac{3}{5}]$ and $f = h$ on $[\frac{2}{5}, 1]$. Then

$$G(t) = g(\eta_t) - \frac{1}{2} \int_0^t g''(\eta_s) \, ds$$

and

$$H(t) = h(\eta_t) - \frac{1}{2} \int_0^t h''(\eta_s) \, ds$$

are P martingales. We need to show that

$$F(t) = f(\eta_t) - \frac{1}{2} \int_0^t f''(\eta_s) \, ds$$

is also a P martingale. Define an increasing sequence of hitting times τ_n by $\tau_0 = 0$ and

$$\tau_{n+1} = \begin{cases} \inf\{t > \tau_n \colon \eta_t \leq \frac{2}{5}\} & \text{if } n \text{ is odd,} \\ \inf\{t > \tau_n \colon \eta_t \geq \frac{3}{5}\} & \text{if } n \text{ is even.} \end{cases}$$

It is not difficult to use the fact that P solves the martingale problem for Ω to show that for $\tau_n > 0$,

$$\eta_{\tau_n} = \begin{cases} \frac{2}{5} & \text{if } n \text{ is even,} \\ \frac{3}{5} & \text{if } n \text{ is odd.} \end{cases}$$

For example, if $\eta < \frac{3}{5}$, choose $l \in \mathcal{D}(\Omega)$ satisfying $l \equiv 0$ on $[0, \frac{3}{5}]$ and $l > 0$ on $(\frac{3}{5}, 1]$ and use the identity

$$E\left\{ l(\eta_{\tau_1}) - \frac{1}{2} \int_0^{\tau_1} l''(\eta_s) \, ds \right\} = l(\eta) = 0$$

to show that $\eta_{\tau_1} \leq \frac{3}{5}$, and hence that $\eta_{\tau_1} = \frac{3}{5}$. Therefore for n such that $\tau_n > 0$,

(5.9) $g(\eta_{\tau_n}) = h(\eta_{\tau_n}) = f(\eta_{\tau_n})$.

To see that $F(t)$ is a martingale, use (5.9) to write

$$F(t) = f(\eta) + \sum_{k=0}^{\infty} [F(\tau_{k+1} \wedge t)] - F(\tau_k \wedge t)]$$

$$= f(\eta) + \sum_{\substack{k=0 \\ k \text{ odd}}}^{\infty} [H(\tau_{k+1} \wedge t) - H(\tau_k \wedge t)]$$

$$+ \sum_{\substack{k=0 \\ k \text{ even}}}^{\infty} [G(\tau_{k+1} \wedge t) - G(\tau_k \wedge t)],$$

and then use the fact that $H(t)$ and $G(t)$ are both martingales.

Remark 5.10. Example 5.8 shows also that in Proposition 2.13, the assumption that D is a core cannot be replaced by the assumption that the martingale problem for the restriction of Ω to D is well posed. To see this, note that in this example,

$$\int_{1/3}^{2/3} \Omega f \, d\eta = 0$$

for all $f \in \mathscr{D}(\Omega)$, yet the restriction of Lebesgue measure to $[\frac{1}{3}, \frac{2}{3}]$ is not invariant for reflecting Brownian motion on $[0, 1]$.

6. The Martingale Problem for Particle Systems

This section is devoted to a discussion of interacting particle systems from the point of view of the martingale problem. The context and notation will be that of Section 3. An immediate consequence of Theorems 3.9 and 5.2 is that the martingale problem for Ω is well posed under assumptions (3.3) and (3.8). In this section we will see that solutions to the martingale problem exist when these assumptions are replaced by a weakened form of (3.3). Under this weaker assumption, nonuniqueness can occur, however. Examples of nonuniqueness will be provided in Section 7. On the other hand, when the martingale problem for Ω is well posed, the solutions $\{P^\eta, \eta \in X\}$ to the martingale problem give a Feller process on X whose generator is an extension of Ω. The true generator of this process may be strictly larger than the closure of Ω, as will be seen in Section 7.

Let $\mathscr{D}$ be the set of all functions in $C(X)$ which depend on finitely many coordinates. By defining Ω on $\mathscr{D}$ instead of on $D(X)$ as we did in Section 3, we are able to weaken assumption (3.3) slightly and hence are able to consider the martingale problem in somewhat greater generality. $\mathscr{D}$ is a minimal class of functions on which to define Ω, if we expect the definition of Ω to be useful. The proof of the following result is identical to that of Proposition 3.2. Note that assumption (6.2) below is simply the assertion that there is a bound on the total rate at which a given coordinate can change.

Proposition 6.1. *Assume that*

(6.2) $$\sum_{T \ni x} c_T < \infty \quad \text{for each } x \in S.$$

(a) *For $f \in \mathscr{D}$, the series*

$$\Omega f(\eta) = \sum_T \int_{W^T} c_T(\eta, d\zeta)[f(\eta^\zeta) - f(\eta)]$$

converges uniformly and defines a function in $C(X)$.
(b) *Ω is a Markov pregenerator.*

The main results of this section depend on a certain compactness statement. In order to formulate it, give $D[0, \infty)$ the Skorohod topology. The Skorohod topology on $D[0, 1]$ is described in Chapter 3 of Billingsley (1968). In particular, Billingsley's Theorem 14.5 asserts that the Borel σ-algebra associated with this topology agrees with $\mathscr{F}$. For the extension of the Skorohod topology to $D[0, \infty)$, see Stone (1963) or Chapter III of Ethier and Kurtz (1985).

Proposition 6.3. *Let* γ *be any positive function on* S. *Let* $\mathscr{P}(\gamma)$ *be the set of all probability measures on* $D[0, \infty)$ *which are solutions to the martingale problem for some initial point* $\eta \in X$ *and for some* Ω *corresponding to a collection* $\{c_T(\eta, d\zeta)\}$ *satisfying* $\sum_T c_T < \infty$ *and*

$$(6.4) \qquad\qquad \sum_{T \ni x} c_T \le \gamma(x)$$

for all $x \in S$. *Then* $\mathscr{P}(\gamma)$ *is relatively compact (i.e., has compact closure in the space of probability measures on* $D[0, \infty)$ *with the topology of weak convergence).*

Proof. Let $\{P^\eta, \eta \in X\}$ be the Feller process whose generator is $\bar{\Omega}$, where $\sum_T c_T < \infty$. Under this assumption, the generator $\bar{\Omega}$ is a bounded operator on $C(X)$. Therefore the corresponding process is a pure jump process on X. Letting τ_x be the first time t that $\eta_t(x) \neq \eta_0(x)$, it then follows that

$$P^\eta\{\tau_x \le \delta\} \le 1 - \exp\left[-\delta \sum_{T \ni x} c_T\right] \le 1 - \exp[-\delta\gamma(x)],$$

provided that (6.4) holds. Now let $\{\tau_x^{(n)}, n \ge 1\}$ be the successive times at which $\eta_t(x)$ changes, and let $N_x(t) = \min\{n \ge 1 : \tau_x^{(n)} > t\}$. By the strong Markov property for the pure jump process,

$$P^\eta\{\tau_x^{(n+1)} - \tau_x^{(n)} \le \delta \text{ and } \tau_x^{(n)} \le t \text{ for some } n \ge 1\}$$

$$\le \sum_{n=1}^\infty E^\eta[P^{\eta_{\tau_x^{(n)}}}[\tau_x \le \delta], \tau_x^{(n)} \le t]$$

$$\le [1 - \exp(-\delta\gamma(x))] \sum_{n=1}^\infty P^\eta[N_x(t) \ge n]$$

$$= [1 - \exp(-\delta\gamma(x))]E^\eta N_x(t)$$

$$\le t\gamma(x)[1 - \exp(-\delta\gamma(x))].$$

Since this bound is independent of the choice of $\eta \in X$ and of the choice of $\{c_T(\eta, d\zeta)\}$ subject to (6.4), and since the bound tends to zero as $\delta \downarrow 0$, Theorem 15.3 of Billingsley (1968) implies that for each $x \in S$, the x-

coordinate marginals of the elements of $\mathcal{P}(\gamma)$ are relatively compact. Since the product of compact sets is compact, it follows that $\mathcal{P}(\gamma)$ itself is relatively compact. $\square$

Proposition 6.5. *Suppose $c_T^{(n)}(\eta, d\zeta)$ and $c_T(\eta, d\zeta)$ are transition rates, each of which satisfy (6.2) with a bound independent of n. Let Ω_n and Ω be the corresponding pregenerators defined on $\mathcal{D}$ as in Proposition 6.1. Assume that*

$$\Omega f = \lim_{n \to \infty} \Omega_n f$$

for all $f \in \mathcal{D}$. If P_n is a solution to the martingale problem for Ω_n with initial point $\eta_n \in X$ where $\eta_n \to \eta$ and if P_n converges weakly to P, then P is a solution to the martingale problem for Ω with initial point η.

Proof. Let E_n and E denote expectations with respect to P_n and P respectively. If $0 < u < t$ and G is any bounded continuous function on $D[0, \infty)$ which is measurable with respect to $\mathscr{F}_u$, then

$$(6.6) \qquad E_n[f(\eta_t) - \int_0^t \Omega_n f(\eta_s)\, ds]G = E_n[f(\eta_u) - \int_0^u \Omega_n f(\eta_s)\, ds]G$$

for any $f \in \mathcal{D}$. The function

$$\eta. \to f(\eta_t) - \int_0^t \Omega f(\eta_s)\, ds$$

on $D[0, \infty)$ is bounded, and is continuous on $\{\eta.: \eta_t = \eta_{t-}\}$. Since the bound in (6.2) is independent of n, $P\{\eta.: \eta_t \neq \eta_{t-}\} = 0$. Using this and the uniform convergence of $\Omega_n f$ to Ωf on X, we can pass to the limit in (6.6) to conclude that (6.6) holds with the n's deleted. Therefore P is a solution to the martingale problem for Ω with initial point η. $\square$

Theorem 6.7. *Assume (6.2). Then for each $\eta \in X$ there exists a solution P^η to the martingale problem for Ω with initial point η.*

Proof. Let S_n be finite sets which increase to S. Define $c_T^{(n)}(\eta, d\zeta)$ by

$$c_T^{(n)}(\eta, d\zeta) = \begin{cases} c_T(\eta, d\zeta) & \text{if } T \subset S_n, \\ 0 & \text{if } T \not\subset S_n. \end{cases}$$

Then

$$\sum_T c_T^{(n)} < \infty \quad \text{for each } n$$

and

$$\sum_{T \ni x} c_T^{(n)} \leq \sum_{T \ni x} c_T < \infty$$

for each $x \in S$. Therefore, by Proposition 6.3, the collection $\{P_n, n \geq 1\}$ is relatively compact, where P_n is the (unique) solution to the martingale problem for Ω_n with initial point η. By Proposition 6.5, any weak limit point of $\{P_n, n \geq 1\}$ provides a solution to the martingale problem for Ω with initial point η. $\square$

Theorem 6.8. *Assume* (6.2). *Suppose that for each* $\eta \in X$, *the solution* P^η *to the martingale problem for* Ω *with initial point* η *is unique. Then* $\{P^\eta, \eta \in X\}$ *is a Feller process on* X *whose generator is an extension of* Ω.

Proof. By the uniqueness assumption and Proposition 6.3, it is clear from the proof of Theorem 6.7 that $\{P^\eta, \eta \in X\}$ is relatively compact. To see this, it suffices to note that in the construction in Theorem 6.7, $\{P^\eta, \eta \in X\}$ are all limit points of a fixed relatively compact set of probability measures on $D[0, \infty)$. By Proposition 6.5 and the uniqueness assumption, $\eta_n \to \eta$ implies that P^{η_n} converges to P^η. Thus P^η is a continuous function of η. This will give the Feller property once we have shown that $\{P^\eta, \eta \in X\}$ is a Markov process (recall Definitions 1.1 and 1.2). In order to check the Markov property (part (c) of Definition 1.1), fix $\eta \in X$ and $s > 0$ and define a probability measure P on $D[0, \infty)$ in the following way:

$$P[\eta. \in A, \eta_{s+.} \in B] = \int_X P^\eta(A, \eta_s \in d\zeta) P^\zeta(B)$$

for $A \in \mathscr{F}_s$ and $B \in \mathscr{F}$. Note that the integration is well defined by the continuity of P^ζ in ζ which implies the measurability of $P^\zeta(B)$ in ζ for $B \in \mathscr{F}$. The Markov property will follow once we show that $P = P^\eta$. In view of the uniqueness assumption, this in turn will follow from the fact that P solves the martingale problem for Ω with initial point η. To see this, we need to verify that for $f \in \mathscr{D}$ and $0 < r < t$,

$$(6.9) \qquad E\left[f(\eta_t) - \int_r^t \Omega f(\eta_u)\, du \,\Big|\, \mathscr{F}_r\right] = f(\eta_r).$$

This is immediate if $t \leq s$, since P and P^η agree on $\mathscr{F}_s$. So, take $t > s$. If $r < s$, (6.9) can be obtained from the corresponding property where $r = s$ by conditioning on $\mathscr{F}_r$. Thus it suffices to check (6.9) when $s < r < t$. In order to do this, let A and B be sets in $\mathscr{F}_s$ and $\mathscr{F}_{r-s}$ respectively. Then

$$E[f(\eta_t) - \int_r^t \Omega f(\eta_u) \, du, \, A \cap \{\eta_{s+.} \in B\}]$$

$$= \int_X P^\eta(A, \, \eta_s \in d\zeta) E^\zeta[f(\eta_{t-s}) - \int_{r-s}^{t-s} \Omega f(\eta_u) \, du, \, B]$$

$$= \int_X P^\eta(A, \, \eta_s \in d\zeta) E^\zeta[f(\eta_{r-s}), \, B]$$

$$= E[f(\eta_r), \, A \cap \{\eta_{s+.} \in B\}].$$

In this computation, the definition of P is used in the first and last steps, and the fact that P^ζ solves the martingale problem for each ζ is used in the middle step. This completes the proof of the Markov property. It remains to show that the generator of the Feller process $\{P^\eta, \, \eta \in X\}$ is an extension of Ω. To do this, take $f \in \mathscr{D}$ and use the martingale property to write

$$E^\eta \left[f(\eta_t) - \int_0^t \Omega f(\eta_s) \, ds \right] = f(\eta).$$

Therefore

$$\lim_{t \downarrow 0} \frac{E^\eta f(\eta_t) - f(\eta)}{t} = \Omega f(\eta)$$

pointwise on X. To show that the convergence is uniform, use part (b) of Proposition 1.3, applied to the function $\Omega f \in C(X)$. $\square$

The final topic for this section deals with the problem of verifying that a given probability measure μ on X is invariant for the process $\{P^\eta, \, \eta \in X\}$ in case the martingale problem for Ω is well posed. According to Proposition 2.13, if the closure $\bar{\Omega}$ is a Markov generator, then in order to verify that $\mu \in \mathscr{I}$, it suffices to check that $\int \Omega f \, d\mu = 0$ for all $f \in \mathscr{D}$. If the martingale problem is well posed but the closure of Ω is not a Markov generator (or is not known to be a Markov generator), the application of Proposition 2.13 would require that $\int \tilde{\Omega} f \, d\mu = 0$ be verified for all f in a core for $\tilde{\Omega}$, where $\tilde{\Omega}$ is the generator of the Feller process $\{P^\eta, \, \eta \in X\}$. In this situation, a core for $\tilde{\Omega}$ will usually not be known explicitly. Hence it would be very useful to know that it suffices to check $\int \Omega f \, d\mu = 0$ for $f \in \mathscr{D}$ only. (Note from Remark 5.10 that this is not immediate.) We will next prove this statement in our context under a mild regularity assumption. It should be noted that the regularity assumption (b) on μ is automatically satisfied in case W is a finite set, which is true in most applications. Echeverría (1982) has shown that the statement of Proposition 6.10 is correct without assumption (b). The proof of his result is considerably more difficult than the one given here.

Proposition 6.10. *Assume* (6.2). *Suppose that the martingale problem for* Ω *is well posed, and that* μ *is a probability measure on* X *such that*

(a) $$\int \Omega f \, d\mu = 0 \quad \text{for all} f \in \mathcal{D},$$

and

(b)   *for each finite subset* T *of* S, *there is a kernel* $\mu_T(\gamma, d\eta)$ *which is weakly continuous as a function of* γ, *depends on* γ *only through the values of* γ *on* T, *and satisfies* $\mu_T(\gamma, \{\eta: \eta = \gamma \text{ on } T\}) = 1$ *and*

$$\int f(\eta) g(\eta) \mu(d\eta) = \int \int f(\eta) g(\gamma) \mu_T(\gamma, d\eta) \mu(d\gamma)$$

for all $f \in C(X)$ *and* $g \in C_T(X)$. (*In other words,* $\mu_T(\gamma, d\eta)$ *is a version of* $\mu(d\eta)$ *conditioned on* $\eta = \gamma$ *on* T.)

Then $\mu \in \mathcal{I}$.

Proof. Take a sequence S_n of finite subsets of S which increase to S. Let $c_T^{(n)}(\gamma, d\zeta)$ be the transition rates defined by

$$c_T^{(n)}(\gamma, d\zeta) = \int \mu_{S_n}(\gamma, d\eta) c_T(\eta, d\zeta)|_{W^{S_n}},$$

where $c_T(\eta, d\zeta)|_{W^{S_n}}$ is the measure on W^T which is the image of $c_T(\eta, d\zeta)$ under the map from W^T to W^T which takes ζ to the configuration which agrees with ζ on $T \cap S_n$ and with η on $T \backslash S_n$. By regularity assumption (b), $c_T^{(n)}(\gamma, d\zeta)$ is weakly continuous in γ, so we may define the (bounded) Markov generator Ω_n on $C(X)$ by

$$\Omega_n f(\gamma) = \sum_T \int_{W^T} c_T^{(n)}(\gamma, d\zeta)[f(\gamma^\zeta) - f(\gamma)].$$

The reason this is bounded is that the only nonzero terms in the sum are those corresponding to T's for which $T \cap S_n \neq \emptyset$, so

$$\|\Omega_n f\| \leq 2\|f\| \sum_{x \in S_n} \sum_{T \ni x} c_T.$$

Since $c_T(\eta, d\zeta)$ is continuous in η, and (6.2) holds, $\Omega_n f \to \Omega f$ for each $f \in \mathcal{D}$.

Therefore by Propositions 6.3 and 6.5 and the fact that the martingale problem for Ω is well posed,

$$(6.11) \qquad\qquad S(t)f = \lim_{n \to \infty} S_n(t)f$$

for every $f \in C(X)$, where $S_n(t)$ is the semigroup with generator Ω_n and $S(t)$ is the semigroup which corresponds to Ω via Theorem 6.8. On the other hand, for $f \in C_{S_n}(X)$,

$$
\begin{aligned}
\int \Omega_n f \, d\mu &= \sum_T \iint c_T^{(n)}(\gamma, d\zeta)[f(\gamma^\zeta) - f(\gamma)]\mu(d\gamma) \\
&= \sum_{T \cap S_n \neq \varnothing} \iiint \mu_{S_n}(\gamma, d\eta) c_T(\eta, d\zeta)[f(\gamma^\zeta) - f(\gamma)]\mu(d\gamma) \\
&= \sum_{T \cap S_n \neq \varnothing} \iint c_T(\eta, d\zeta)[f(\eta^\zeta) - f(\eta)]\mu(d\eta) \\
&= \int \Omega f \, d\mu = 0
\end{aligned}
$$

by assumption (a). Therefore μ is invariant for $S_n(t)$ by Proposition 2.13, since the process corresponding to $S_n(t)$ does not change coordinates off S_n. This says that $\mu S_n(t) = \mu$ for all n and all $t \geq 0$. This together with (6.11) implies that μ is invariant for $S(t)$. $\square$

7. Examples

In this section, several examples will be discussed which illustrate the possible behavior of particle systems when viewed from the perspective of the martingale problem. The first example shows what can happen when assumption (6.2) fails, and shows how even when (6.2) holds, nonuniqueness for the martingale problem for the exclusion process can occur and is associated with nonuniqueness in the theory of continuous time Markov chains. In the second example and its variants, we see that even in the context of "spin systems," in which $W = \{0, 1\}$ and only one coordinate is allowed to change at a time, the martingale problem may not be well posed. This can happen even if (6.2) is strengthened to (3.3). Thus in particular, (3.3) alone is not sufficient in Theorem 3.9. The third example shows that even when the martingale problem is well posed, the generator of the associated Feller process (see Theorem 6.8) may be strictly larger than the closure of Ω. Therefore the converse to Theorem 5.2 is not correct. In particular there are situations in which the semigroup approach of Section 3 must fail, while the martingale approach of Section 6 succeeds.

Since these examples play only a peripheral role in the theory, our discussion of them will be rather informal. Most verifications will be left to the interested reader.

Example 7.1 (Continuation of Example 3.1(d)). *The Exclusion Process.* In this example, we continue to assume that $p(x, y) \geq 0$ but not necessarily that $\sum_y p(x, y) = 1$. In Example 3.15, we saw that each of assumptions (3.3) and (3.8) is equivalent to

$$(7.2) \qquad \sup_y \sum_x [p(x, y) + p(y, x)] < \infty.$$

On the other hand, (6.2) is equivalent to

$$(7.3) \qquad \sum_x [p(x, y) + p(y, x)] < \infty \quad \text{for each } y \in S.$$

Recall that (6.2) was needed in order to formulate the martingale problem relative to a Markov pregenerator defined on $\mathcal{D}$. In the context of the exclusion process, it is fairly easy to see what goes wrong if (7.3) fails. Suppose that for a fixed $y \in S$, $\sum_x p(x, y) = \infty$. Let η be the configuration defined by

$$\eta(x) = \begin{cases} 1 & \text{if } x \neq y, \\ 0 & \text{if } x = y. \end{cases}$$

Given our intuitive understanding of how the process starting at this η should evolve, we would expect the particle at any $x \neq y$ to wait an exponential time with mean $[p(x, y)]^{-1}$ and then to jump to y, provided no other particle has preceded it to y. However, since $\sum_x p(x, y) = \infty$, the minimum of these exponential times is identically zero. Therefore any "reasonable" version of the process would satisfy $\eta_t(x) \equiv 1$ for all $x \in S$ and $t > 0$. But this process is not right continuous at $t = 0$, so it does not have paths in $D[0, \infty)$. A more formal way to make this argument is to approximate S by finite sets S_n and consider the exclusion process $\eta_t^{(n)}$ in which particles off S_n are not allowed to move. The easily verified fact is then that

$$\lim_{n \to \infty} P^\eta[\eta_t^{(n)}(x) = 1] = 1$$

for all $x \in S$ and all $t > 0$.

It is perhaps of greater interest to consider the exclusion process in situations in which (7.3) holds but (7.2) fails. Suppose X_t is a Markov chain on $S \cup \{\Delta\}$, where Δ is the usual "cemetery," such that $P^x(X_t \in S) = 1$ for all $x \in S$ and all $t > 0$, and the forward equation

$$\frac{d}{dt} P^x(X_t = y) = \sum_z P^x(X_t = z) p(z, y) - P^x(X_t = y) \sum_z p(y, z)$$

is satisfied for all $x, y \in S$ and $t \geq 0$. Define a process η_t by

$$\eta_t(x) = \begin{cases} 1 & \text{if } X_t = x, \\ 0 & \text{if } X_t \neq x. \end{cases}$$

Then, as a consequence of the forward equation, η_t (or more precisely, the measure on $D[0, \infty)$ induced by it) is a solution of the martingale problem for Ω on $\mathscr{D}$. There are many examples in which (7.3) holds but (7.2) fails and in which there is more than one solution to the forward equation. (See, for example, Chapter 7 of Freedman (1971) for a treatment of this theory.) A simple example is obtained by taking

$$S = \{(m, n) \in Z^2 : m = 0 \text{ or } n = 0\}$$

with

$$p((0, n), (0, n+1)) = n^2 + 1 \quad \text{if } n \geq 0,$$

$$p((0, n), (0, n-1)) = n^2 + 1 \quad \text{if } n \leq 0,$$

$$p((m, 0), (m+1, 0)) = m^2 \quad \text{if } m \leq -1,$$

$$p((m, 0), (m-1, 0)) = m^2 \quad \text{if } m \geq 1,$$

and $p(x, y) = 0$ otherwise. Whenever the forward equation has multiple solutions, the martingale problem for Ω is not well posed.

Example 7.4. In this example, let $W = \{0, 1\}$ and take $c(x, \eta)$ to be a nonnegative continuous function on $S \times X$. Define the transition rates $c_T(\eta, d\zeta)$ be setting $c_T(\eta, d\zeta) = 0$ if $|T| \geq 2$ and letting $c_{\{x\}}(\eta, d\zeta)$ put mass $c(x, \eta)$ on $1 - \eta(x)$. Thus $c(x, \eta)$ is the rate at which the x coordinate changes from $\eta(x)$ to $1 - \eta(x)$ when the entire configuration is η. In this context, (6.2) is automatic, (3.3) is just the statement that $c(x, \eta)$ is uniformly bounded, and (3.8) becomes

(7.5) $$\sup_x \sum_u \sup_\eta |c(x, \eta) - c(x, \eta_u)| < \infty$$

where η_u is defined by $\eta_u(u) = 1 - \eta(u)$ and $\eta_u(x) = \eta(x)$ for all $x \neq u$.

A very simple example of nonuniqueness for the martingale problem is obtained by taking

$$S = \{1, 2, \ldots\}$$

and

$$c(x, \eta) = \begin{cases} 2^x & \text{if } \eta(x) = 0 \text{ and } \eta(x+1) = 1, \\ 0 & \text{otherwise.} \end{cases}$$

One solution to the martingale problem starting from $\eta \equiv 0$ is clearly given by $\eta_t \equiv 0$ for all $t \geq 0$. Another solution η_t can be constructed in the following way: Let $\tau_1, \tau_2, \ldots$ be independent exponentially distributed random variables with $E\tau_k = 2^{-k}$. Then define η_t by

$$\eta_t(x) = 1 \quad \text{if and only if} \quad t \geq \sum_{k=x}^{\infty} \tau_k.$$

Perhaps the best way of verifying that this is in fact a solution to the martingale problem with initial point $\eta \equiv 0$ is to use Proposition 6.5. A natural approximation to Ω in that context would be to take

$$c_n(x, \eta) = \begin{cases} c(x, \eta) & \text{if } x \leq n, \\ 0 & \text{if } x > n. \end{cases}$$

and a natural choice for η_n would be $\eta_n(x) = 0$ for $x \leq n$ and $\eta_n(x) = 1$ for $x > n$. If instead of this we took $\eta_n \equiv 0$ in Proposition 6.5, then the resulting solution would be the trivial one: $\eta_t \equiv 0$. This example illustrates the observation that nonuniqueness in the martingale problem often occurs because "influence" can come in from infinity in finite time. A more general treatment of this idea can be found in Gray and Griffeath (1977).

Since the above rates are unbounded, one might reasonably ask whether or not uniform boundedness of the rates (i.e., condition (3.3)) is sufficient for uniqueness to hold. However the above example can be modified very simply so that the same type of nonuniqueness occurs even though the rates are uniformly bounded. To do this let

$$c(x, \eta) = \begin{cases} 1 & \text{if } \eta(y) = 1 \text{ for some } 2^{n+1} \leq y < 2^{n+2}, \\ 0 & \text{otherwise}, \end{cases}$$

where n is determined by $2^n \leq x < 2^{n+1}$. To make the connection with the earlier example, it suffices to identify intervals of the form $[2^n, 2^{n+1})$ in the modification with site n in the original example.

These examples have been relatively easy to analyze primarily because $c(x, \eta) = 0$ for many choices of the arguments. A more intricate example of nonuniqueness, which is based on similar ideas, but in which $c(x, \eta)$ is uniformly bounded above and away from zero, is given in Section 3 of Holley and Stroock (1976a). Their example is however spatially inhomogeneous. This suggests that the following might be true: If $S = Z^d$, the mechanism is translation invariant, and $c(x, \eta)$ is continuous and strictly positive (and hence uniformly bounded above and away from zero), then the martingale problem is well posed. Unfortunately this turns out to be false as was shown with a highly nontrivial example by Gray (1980).

Example 7.6. This example is designed to show that even if the martingale problem for Ω is well posed, the generator of the resulting Feller process may be strictly larger than the closure of Ω. (For an example of this in the Brownian motion context, see Example 5.8) Another way of saying this is that $\mathcal{D}$ may not be a core for the generator of the process. In this example, $W = \{0, 1\}$ again and $c_T(\eta, d\zeta)$ is defined in terms of $c(x, \eta)$ as in Example 7.4. Take $S = \{1, 2, \ldots\}$ and

$$c(x, \eta) = \begin{cases} 0 & \text{if } \eta(x) = 0, \\ 1 & \text{if } \eta(y) = 1 \text{ for all } 1 \le y \le x+1, \\ \delta(x) & \text{otherwise,} \end{cases}$$

where $\delta(x) > 1$ for all x. Then for $f \in \mathcal{D}$,

$$(7.7) \quad \Omega f(\eta) = \sum_{\substack{1 \le x < N(\eta)-1}} [f(\eta_x) - f(\eta)] + \sum_{\substack{x \ge N(\eta)-1 \\ \eta(x)=1}} \delta(x)[f(\eta_x) - f(\eta)]$$

where $\eta_x(x) = 1 - \eta(x)$, $\eta_x(y) = \eta(y)$ for all $y \ne x$, and

$$N(\eta) = \begin{cases} x & \text{if } \eta(x) = 0 \text{ and } \eta(y) = 1 \text{ for all } y < x, \\ \infty & \text{if } \eta(x) = 1 \text{ for all } x \in S. \end{cases}$$

Theorem 7.8. (a) *The martingale problem for Ω (with domain $\mathcal{D}$) is well posed.*
    (b)   *The closure $\bar{\Omega}$ of Ω is a Markov generator if and only if*

$$(7.9) \qquad \sum_{x=1}^{\infty} \frac{x}{\delta(x)} = \infty.$$

    (c)   *If (7.9) fails, then a core for the generator of the Feller process which corresponds to Ω through the martingale problem is given by*

$$\{f + c\varphi : f \in \mathcal{D}, c \text{ real}\},$$

where $\varphi \in C(X)$ is given by

$$\varphi(\eta) = \begin{cases} \sum_{k=N(\eta)}^{\infty} [\delta(k)]^{-1} & \text{if } N(\eta) < \infty, \\ 0 & \text{otherwise.} \end{cases}$$

Sketch of Proof. (a) Since $\Omega f(\eta) \le 0$ for $f(\eta) = \eta(x)$, $\eta_t(x)$ is a super-martingale with respect to any solution of the martingale problem. Therefore $\eta_s(x) = 0$ implies that $\eta_t(x) = 0$ for all $t \ge s$. Once $\eta_t(x) = 0$, the process breaks down into independent finite-state Markov chains $\{\eta_t(u), u \le x\}$, $\eta_t(x+1), \eta_t(x+2), \ldots$, for each of which uniqueness is immediate. Thus uniqueness holds for any initial point other than $\eta \equiv 1$. Furthermore it is not hard to see that the solutions $\{P^\eta, \eta \neq 1\}$ to the martingale problem extend continuously to $\eta \equiv 1$. Perhaps the simplest way to see this is to use the monotonicity of P^η as a function of η. (This type of monotonicity will be exploited systematically later in this book. See Section 2 of Chapter III, for example.) Thus it suffices to show that if P is any solution to the martingale problem starting from $\eta \equiv 1$, then $P[\eta_t \equiv 1] = 0$ for all $t > 0$. To do this, let $f_n(\eta) = \prod_{x=1}^n \eta(x)$ and note that since $\delta(x) \geq 1$,

$$\Omega f_n(\eta) \leq -n f_n(\eta).$$

Therefore since P solves the martingale problem,

$$Ef_n(\eta_t) - Ef_n(\eta_s) \leq -n \int_s^t Ef_n(\eta_r)\, dr$$

for $s < t$. Hence

$$P[\eta_t(x) = 1 \text{ for all } 1 \leq x \leq n] = Ef_n(\eta_t) \leq e^{-nt},$$

and so

$$P[\eta_t \equiv 1] = 0 \quad \text{for all } t > 0.$$

(b) It suffices to show (see Definition 2.7) that $\mathcal{R}(I - \lambda\Omega)$ is dense in $C(X)$ if and only if (7.9) holds. We will do this for $\lambda = 1$, since the proof is the same for all λ. Of course $\mathcal{R}(I - \Omega)$ is dense if and only if there is no nonzero signed measure μ on X of bounded variation such that

$$(7.10) \qquad\qquad \int f\, d\mu = \int \Omega f\, d\mu$$

for all $f \in \mathcal{D}$. The strategy is to show that there is a unique (up to multiplication by a constant) signed measure on the algebra of finite-dimensional subsets of X which satisfies (7.10) for all $f \in \mathcal{D}$, and that this measure extends to a measure on the Borel sets of X of bounded variation if and only if (7.9) fails. To begin to construct μ on the algebra of finite-dimensional sets, note that (7.10) when $f \equiv 1$ implies that $\mu(X) = 0$. Now apply (7.10) to $f_n(\eta) = \prod_{x=1}^n \eta(x)$ to conclude that

$$[n + \delta(n)] \int f_n\, d\mu = [\delta(n) - 1] \int f_{n+1}\, d\mu.$$

If we normalize μ by setting $\int f_1\, d\mu = 1$, then this can be solved recursively to obtain

$$(7.11) \qquad \mu\{\eta : \eta(x) = 1 \text{ for all } 1 \leq x \leq n\} = \prod_{k=1}^{n-1} \frac{\delta(k) + k}{\delta(k) - 1}.$$

It follows immediately that if (7.9) holds, then μ does not extend to a measure of bounded variation on X. Assume from now on, then, that (7.9) fails, so that $\prod_{k=1}^{\infty} (\delta(k)+k)/(\delta(k)-1) < \infty$. Let $\varepsilon(x)$ be equal to 0 or 1 for each $1 \le x < n$, $\varepsilon(x) = 0$ for some $1 \le x < n$, and $\varepsilon(n) = 1$. Apply (7.10) to

$$f(\eta) = \begin{cases} 1 & \text{if } \eta(x) = \varepsilon(x) \text{ for all } 1 \le x \le n, \\ 0 & \text{otherwise,} \end{cases}$$

to obtain

$$\mu\{\eta: \eta(x) = \varepsilon(x) \text{ for all } 1 \le x \le n\}$$

$$= \frac{\displaystyle\sum_{x=N}^{n-1} [1 - \varepsilon(x)]\mu\{\eta: \eta(y) = \varepsilon_x(y) \text{ for all } y \le n\}c(x, \varepsilon_x)}{1 + (N-2)^+ + \displaystyle\sum_{x=N-1}^{n} \delta(x)\varepsilon(x)}$$

where N is the first x for which $\varepsilon(x) = 0$ and ε_x is obtained from ε by reversing the x coordinate. Since $\varepsilon(x) = 0$ implies that $\varepsilon_x \ge \varepsilon$, this together with (7.11) permits one to compute recursively

(7.12) $\qquad \mu\{\eta: \eta(x) = \varepsilon(x) \text{ for all } 1 \le x \le n\}$

whenever $\varepsilon(n) = 1$. Note that the value obtained for (7.12) in this case in nonnegative. In order to compute (7.12) when $\varepsilon(n) = 0$, suppose $\varepsilon(k) = 1$ and $\varepsilon(x) = 0$ for $k < x \le n$. Then

$$\mu\{\eta: \eta(x) = \varepsilon(x) \text{ for all } 1 \le x \le n\}$$

(7.13)
$$= \mu\{\eta: \eta(x) = \varepsilon(x) \text{ for all } 1 \le x \le k\}$$

$$- \sum_{j=k+1}^{n} \mu\{\eta: \eta(x) = \varepsilon(x) \text{ for all } 1 \le x < j, \eta(j) = 1\},$$

where all terms on the right side have been computed previously. The final case of (7.12) which remains to be computed is that in which $\varepsilon(x) = 0$ for all $1 \le x \le n$, but this value, which is easily seen to be negative, is obtained by using $\mu(X) = 0$. By now μ has been defined (uniquely up to multiplication by a constant) on the algebra of finite-dimensional sets so that (7.10) holds for all $f \in \mathscr{D}$. It remains to show that when (7.9) fails, the total variation of the restriction of μ to sets which depend only on the coordinates $\eta(x)$ for $1 \le x \le n$ is bounded in n. In order to do this use (7.10) with $f(\eta) = \eta(n)$ to conclude that

$$\mu\{\eta: \eta(n) = 1\} = \frac{\delta(n)-1}{\delta(n)+1} \mu\{\eta: \eta(x) = 1 \text{ for all } 1 \le x \le n+1\}$$

$$= \frac{\delta(n)+n}{\delta(n)+1} \mu\{\eta: \eta(x) = 1 \text{ for all } 1 \le x \le n\}$$

by (7.11). Therefore the sum of (7.12) over all choices of $\varepsilon(x)$, $1 \le x \le n$, for which $\varepsilon(n) = 1$ and $\varepsilon(x) = 0$ for at least one $1 \le x < n$ is equal to

$$\frac{n-1}{\delta(n)+1} \prod_{k=1}^{n-1} \frac{\delta(k)+k}{\delta(k)-1}.$$

By this and (7.13), the sum of the positive parts of (7.12) over all choices of $\varepsilon(x)$, $1 \le x \le n$ is at most

(7.14) $$\sum_{m=1}^{n} \frac{m-1}{\delta(m)+1} \prod_{k=1}^{m-1} \frac{\delta(k)+k}{\delta(k)-1},$$

which is bounded in n when (7.9) fails. Since $\mu(X) = 0$, the sum of the negative parts is also at most (7.14), so the result follows.

(c) The details of the proof are left to the reader. An important observation is that the generator of the process applied to φ is given by

$$\sum_{1 \le x < N(\eta)-1} [\varphi(\eta_x) - \varphi(\eta)] + 1$$

for η such that $N(\eta) \ge 2$. In particular, the value of this function evaluated at $\eta \equiv 1$ is

$$\sum_{k=1}^{\infty} \frac{k}{\delta(k)} + 1,$$

so condition (7.9) is obviously relevant here as well. $\square$

Example 7.15. In 1958, Blackwell proposed the following as an example of a countable state Markov chain, all of whose states are instantaneous. (See Chapter 9 of Freedman (1971) for a discussion of this example.) Using the definition of $c_T(\eta, d\zeta)$ in terms of $c(x, \eta)$ as in Example 7.4, take $S = \{1, 2, \ldots\}$ and

$$c(x, \eta) = \begin{cases} \lambda(x) & \text{if } \eta(x) = 0, \\ \mu(x) & \text{if } \eta(x) = 1, \end{cases}$$

where $\lambda(x) > 0$, $\mu(x) > 0$,

(7.16) $$\sum_{x=1}^{\infty} \lambda(x) = \infty, \quad \text{and} \quad \sum_{x=1}^{\infty} \frac{\lambda(x)}{\lambda(x) + \mu(x)} < \infty.$$

These rates satisfy (6.2), but not (3.3). Since $c(x, \eta)$ depends on η only through $\eta(x)$, it is not hard to check that in the context of Proposition 6.1, the closure of Ω is a Markov generator. Therefore by Theorem 5.2, the martingale problem is well posed for Ω. (Of course, this is easy to see directly.) In the resulting process, $\{\eta_t(x), x \in S\}$ are independent two-state Markov chains. When regarded as a process on $X = \{0, 1\}^S$, it is a perfectly well-behaved Feller process. On the other hand, as a result of (7.16), if $\sum_{x=1}^{\infty} \eta(x) < \infty$, then

$$P^\eta \left[\sum_{x=1}^{\infty} \eta_t(x) < \infty \right] = 1$$

for all t, so the process may be viewed as a Markov chain on the countable set

$$\left\{ \eta \in X : \sum_{x=1}^{\infty} \eta(x) < \infty \right\}.$$

From this point of view, the process is quite pathological since all states are instantaneous. It is interesting to note that such "spin system" ideas were present already a quarter of a century ago.

8. Notes and References

Section 3. Most of the work on the construction of particle systems was carried out in the early 1970's. Dobrushin (1971a) constructed spin systems in which only one coordinate changes at a time under finite range assumptions, by obtaining them as limits of finite systems. Harris (1972) gave an explicit probabilistic construction for finite range exclusion processes. Holley (1970) used the semigroup approach to construct one-dimensional systems, and then in (1972b) gave a construction in the finite range case which avoided the use of the Hille–Yosida Theorem. Liggett (1972) proved a general existence result based on the Hille–Yosida Theorem. The proof of Theorem 3.9 given here follows that approach, as generalized and improved by Sullivan (1974, 1976a). Further generalizations of Theorem 3.9 can be found in Gray and Griffeath (1976).

Section 4. Results along the lines of Theorem 4.1 have been proved by Dobrushin (1971a), Sullivan (1974), Holley and Stroock (1976a, b), and Gray and Griffeath (1976). Theorem 4.6 is based on Sullivan (1976a). Theorem 4.15 was first proved for finite range interactions by Holley (1972b) and was extended to the present context by Sullivan (1976a). Proposition 4.18 is based on Holley and Stroock (1976b). In Example 4.3(a), the fact that Theorem 4.1 yields convergence for all $\beta \geq 0$ in the alternative version of the stochastic Ising model with $d = 1$, was pointed out by Holley. Theorem 4.20 resulted from conversations with Gray and Holley.

Sections 5 and 6. The martingale approach to interacting particle systems was introduced and first developed by Holley and Stroock (1976a). Most of the results in Sections 5 and 6 are based on that paper, as well as on the survey paper by Stroock (1978). Proposition 6.10 is based on remarks in Gray and Griffeath (1977), and through them on Higuchi and Shiga (1975). Similar ideas were used by Dobrushin (1971a) and by Frykman (1975). For a more general version of this result which applies outside of the particle system context, see Echeverría (1982). Example 5.8 comes from Kurtz via Griffeath. It shows that some conditions of the type imposed by Echeverría on the richness of the domain of Ω are in fact needed. Several authors have proved uniqueness for the martingale problem for a number of types of particle systems. Among these are Holley and Stroock (1976a), Gray and Griffeath (1976, 1977), Gray (1978), Cocozza and Kipnis (1977, 1980), and Cocozza and Roussignol (1979).

Section 7. The first version of Example 7.1 was mentioned in Liggett (1972) and then in Chapter II.1 of Liggett (1977b). The first version of Example 7.4 appeared first in Gray and Griffeath (1976) and later in Chapter IV of Griffeath (1979a). The modification of that example with bounded rates comes from Section VII of Stroock (1978). Example 7.6 is a generalization, with a new proof, of the example in Theorem 4 of Gray and Griffeath (1977).

9. Open Problems

1. Suppose the rates $c_T(\eta, d\zeta)$ satisfy the hypothesis $M < \varepsilon$ of Theorem 4.1. If $d_T(d\zeta)$ is a measure on W^T, then the rates

$$\tilde{c}_T(\eta, d\zeta) = c_T(\eta, d\zeta) + d_T(d\zeta)$$

still satisfy $M < \varepsilon$ since the addition of the "independent flipping" $d_T(d\zeta)$ increases ε but does not change M. This suggests the following conjecture: if the process with rates $c_T(\eta, d\zeta)$ is ergodic, then so is the process with rates $\tilde{c}_T(\eta, d\zeta)$. Such a result would have important consequences. For example, there are one-parameter families of systems for which adding some "independent flipping" to one of the systems yields another member of the family. In such cases, the above conjecture would imply the existence of a critical value for that parameter. A specific example where this would occur is the family of majority vote processes which are described in Example 4.3(e) and are discussed further in Example 2.12 of Chapter III. The point is that when a constant flip rate c is added to the majority vote process with parameter δ, the resulting process is, except for a time change, the same as the majority vote process with parameter $(\delta + c)/(1 + 2c)$. Note that as c ranges between 0 and ∞, this new parameter ranges between δ and $\frac{1}{2}$. This general problem was suggested a number of years ago by Holley, but so far no progress has been made on it.

2. The above question can also be formulated in the context of the martingale problem. The problem is then to prove that if the martingale problem is well posed for the rates $c_T(\eta, d\zeta)$, then it is also well posed for the rates $\tilde{c}_T(\eta, d\zeta)$. This problem was proposed by Gray.

3. Assuming (7.3) in Example 7.1, what are reasonable conditions under which the martingale problem is well posed? These conditions should be substantially weaker than (7.2). In particular, is the assumption that the forward equation for $p(x, y)$ have a unique solution sufficient for the martingale problem to be well posed?

4. Referring to Definition 1.9, what are reasonably weak conditions on a particle system generator so that (a) implies (b) in that definition? Note that by part (e) of Proposition 1.8, (a) implies the following weaker form of (b):

$$\lim_{T \to \infty} \frac{1}{T} \int_0^T \mu S(t) = \nu \quad \text{for all } \mu \in \mathcal{P}.$$

One fairly strong sufficient condition for (a) to imply (b) is given in Corollary 2.4 of Chapter III.

Chapter II

Some Basic Tools

This chapter has two objectives. The first is to introduce some tools which will be used frequently in succeeding chapters. These tools will be illustrated here with applications to countable state Markov chains. The hope is that after seeing how they are applied in this familiar setting, the reader will be better prepared to appreciate their usefulness in the context of particle systems. The second objective of this chapter is to prove some nonstandard results about Markov chains which we will need in later chapters. The material in the first five sections relates primarily to the first objective, while the latter sections relate to the second objective.

The coupling technique, which is discussed in Section 1, will be used directly or indirectly in all succeeding chapters. Additional illustrations of coupling appear in Section 2, which deals with questions of monotonicity for Markov processes. Duality will play an important role in Chapters III, V, VI, VIII, and IX. Relative entropy will be used in Chapter IV. In addition, it was used to prove one result which is mentioned in Chapter VII, and it can be used to give alternate proofs of some of the results in Chapter VIII. Reversibility will be used primarily in Chapters IV, VII, and VIII. The results in Sections 6, 7, and 8 are intended for use in Chapters VII, VIII, and IX respectively. The first time through, the reader should probably concentrate on the first five sections of this chapter. He can return to the other topics when the need arises in the later chapters.

The eight main sections are independent of one another, except that the first three sections should be read in that order, and Section 5 should be read before Section 6.

1. Coupling

Coupling is probably the most important and most generally applicable technique in the subject of interacting particle systems. While the technique greatly predates this subject, coupling did not really come into its own until its power and flexibility in the theory of particle systems was recognized during the 1970's. At its most basic level, a coupling is simply a construction of two or more stochastic processes on a common probability space. Of course, the coupling will normally only be useful if the mechanisms of the various stochastic processes are connected in a nontrivial way. In this section, we will see a number of applications of the coupling technique, primarily in the context of finite or countable state Markov chains. The objective is to help the reader learn how to use this important tool. In the next section, we will discuss several topics which will come up often in the sequel and which are intimately connected with coupling. These include the stochastic order relation between probability measures, monotone Feller processes, and the FKG inequality of statistical mechanics.

Perhaps the simplest application of coupling is the following. Suppose one wishes to prove the plausible statement that $f(\eta)$ and $g(\eta)$ are positively correlated whenever η is a real-valued random variable and f and g are any two bounded increasing functions on the real line. If one tries to do so using only the random variable η, it is not at all clear how to proceed. However if one lets η and ζ be two independent and identically distributed random variables, the proof is quite simple. One need only write

$$
\begin{aligned}
0 \leq & E[f(\eta) - f(\zeta)][g(\eta) - g(\zeta)] \\
= & Ef(\eta)g(\eta) + Ef(\zeta)g(\zeta) - Ef(\eta)g(\zeta) - Ef(\zeta)g(\eta) \\
= & 2\{Ef(\eta)g(\eta) - Ef(\eta)Eg(\eta)\} \\
= & 2 \operatorname{cov}\{f(\eta), g(\eta)\},
\end{aligned}
$$

where the monotonicity of f and g is used in the first step. In that inequality, it is important that R^1 be totally ordered, so that with probability one, either $\eta \leq \zeta$ or $\eta \geq \zeta$, and hence $f(\eta) - f(\zeta)$ and $g(\eta) - g(\zeta)$ will have the same sign. (The FKG inequality, which appears in the next section as Corollary 2.12, gives a version of the above result in case the space on which f and g are defined is partially ordered instead.) This argument is considered to be a coupling since two random variables are constructed on a common probability space in order to prove a statement about one of them.

The next example of the use of coupling is a proof of the basic convergence theorem for finite-state Markov chains. Let $p(x, y)$ be the transition probabilities for a discrete time Markov chain on a finite set S and let $p^{(k)}(x, y)$ be the corresponding k-step transition probabilities defined by $p^{(1)}(x, y) = p(x, y)$ and

(1.1)
$$
p^{(k+1)}(x, y) = \sum_{z \in S} p^{(k)}(x, z)p(z, y).
$$

Theorem 1.2. *Suppose that for some* $m \geq 1$,

$$
\min_{x, y \in S} p^{(m)}(x, y) = \varepsilon > 0.
$$

Then

(1.3) $$\pi(y) = \lim_{k \to \infty} p^{(k)}(x, y)$$

exists for $x, y \in S$, is independent of x, and satisfies

(1.4) $$\sum_x \pi(x)p(x, y) = \pi(y)$$

Proof. Let (X_n, Y_n) be the Markov chain on $S \times S$ which evolves in the following way:

(a) X_n and Y_n move independently with transition probabilities $p(\cdot, \cdot)$ until the first time $X_n = Y_n$, and

(b) X_n and Y_n move together with the same transition probabilities after that time.

An equivalent description of (X_n, Y_n) is obtained by giving its transition probabilities:

$$P^{(x, y)}[X_1 = u, Y_1 = v] = \begin{cases} p(x, u)p(y, v) & \text{if } x \neq y, \\ p(x, u) & \text{if } x = y \text{ and } u = v, \\ 0 & \text{if } x = y \text{ and } u \neq v. \end{cases}$$

This coupled Markov chain has two important properties:

(a) X_n and Y_n are separately Markovian with transition probabilities $p(\cdot, \cdot)$, and

(b) $\lim_{n \to \infty} P^{(x, y)}[X_n = Y_n] = 1$ for each $(x, y) \in S \times S$.

The first property is immediate. To check the second let τ be the first time the two coordinates of the coupled process agree. Then

$$P^{(x, y)}(\tau \leq m) \geq \sum_{z \in S} p^{(m)}(x, z)p^{(m)}(y, z)$$

$$\geq \varepsilon \sum_{z \in S} p^{(m)}(y, z) = \varepsilon$$

for any $(x, y) \in S \times S$, so by the Markov property,

$$P^{(x, y)}(\tau \leq km) \geq 1 - (1 - \varepsilon)^k,$$

and hence

$$P^{(x, y)}[X_n \neq Y_n] \leq P^{(x, y)}(\tau > n) \leq (1 - \varepsilon)^{[n/m]},$$

which tends to zero as $n \to \infty$ exponentially rapidly. By property (a) of the coupling,

$$\begin{aligned}
|p^{(k)}&(u, y) - p^{(k)}(v, y)| \\
&= |P^{(u, v)}[X_k = y] - P^{(u, v)}[Y_k = y]| \\
&= |P^{(u, v)}[X_k = y, Y_k \neq y] - P^{(u, v)}[X_k \neq y, Y_k = y]| \\
&\leq P^{(u, v)}[X_k \neq Y_k],
\end{aligned}$$

which tends to zero by property (b). Since $\max_x p^{(k)}(x, y)$ decreases and $\min_x p^{(k)}(x, y)$ increases in k for each y, the proof of (1.3) is now easy. Statement (1.4) follows by taking the limit in (1.1) as $k \to \infty$. □

The next application of coupling is to the problem of proving that certain Markov chains have no nonconstant bounded harmonic functions. While the primary purpose of this application is to illustrate the use of coupling, bounded harmonic functions for Markov chains will arise naturally in Chapters V and VIII. Suppose now that $p(x, y)$ are the transition probabilities for a Markov chain on the countable state space S. By a (Markov) coupling of two copies of this chain, we will mean a Markov chain on $S \times S$ with transition probabilities $p((x, y), (u, v))$ which satisfy

$$\sum_v p((x, y), (u, v)) = p(x, u)$$

and

$$\sum_u p((x, y), (u, v)) = p(y, v).$$

This property guarantees that the two marginals of the coupled process are Markovian and have transition probabilities $p(\cdot, \cdot)$. The coupling will be said to be successful if for any $(x, y) \in S \times S$, $P^{(x, y)}[X_n = Y_n$ for all sufficiently large $n] = 1$. The bivariate chain constructed in the proof of Theorem 1.2 is an example of a successful coupling. A bounded function h on S is said to be harmonic for $p(\cdot, \cdot)$ if

$$h(x) = \sum_y p(x, y) h(y)$$

for all $x \in S$.

Theorem 1.5. *Suppose there exists a successful coupling for the chain $p(\cdot, \cdot)$. Then every bounded harmonic function for $p(\cdot, \cdot)$ is constant.*

Proof. Let (X_n, Y_n) be a successful coupling, and let τ be a (finite-valued) random variable such that $X_n = Y_n$ for all $n \geq \tau$. If h is a bounded harmonic function for $p(\cdot, \cdot)$, then

$$h(x) = E^x h(X_n)$$

for all $x \in S$ and $n \geq 0$. Therefore

$$
\begin{aligned}
|h(x) - h(y)| &= |E^x h(X_n) - E^y h(X_n)| \\
&= |E^{(x,y)} h(X_n) - E^{(x,y)} h(Y_n)| \\
&\leq E^{(x,y)} |h(X_n) - h(Y_n)| \\
&\leq 2[\sup_u |h(u)|] P^{(x,y)}[X_n \neq Y_n] \\
&\leq 2[\sup_u |h(u)|] P^{(x,y)}[\tau > n].
\end{aligned}
$$

Since the right side tends to 0 as $n \to \infty$, $h(x) = h(y)$ for all $x, y \in S$, and hence h is constant. $\square$

The Markov chain X_n will be called a random walk if $S = Z^d$ for some $d \geq 1$ and $p(x, y) = p(0, y - x)$. It is said to be irreducible if for each $(x, y) \in S \times S$ there is an m so that $p^{(m)}(x, y) > 0$. A basic fact is that an irreducible random walk has no nonconstant bounded harmonic functions. (When Z^d is replaced by a general locally compact Abelian group, this result is known as the Choquet–Deny Theorem. See Chapter VIII of Meyer (1966) or Chapter 5 of Revuz (1975), for example.) There are a number of approaches to the proof of this fact, one of which will be given in Section 7 (see Corollary 7.2). To illustrate the use of coupling, suppose we attempt to give a proof based on Theorem 1.5.

Consider then the case of an irreducible random walk on Z^d. The simplest coupling which has a chance of succeeding, at least in some cases, is to let the coordinates X_n and Y_n evolve independently until they meet, and then to force them to move together afterwards. This was the coupling which was used in the proof of Theorem 1.2. To see when this coupling is successful, note that $X_n - Y_n$ is a symmetric random walk with transition probabilities

$$
q(x, y) = \sum_u p(x, u) p(y, u)
$$

until it hits 0, after which it remains at 0. The q random walk is irreducible and recurrent provided for example that $d = 1$ or 2 and that the p random walk is irreducible, aperiodic, and has finite second moments. In this case the coupling is successful. The finite second moment assumption is easy to remove by modifying the coupling slightly. It suffices to have the coordinates X_n and Y_n take the same jump if the jump is large, and take independent jumps if the jumps are small, until the coordinates meet. (This is a modification of a coupling used by Ornstein (1969).) The reader is encouraged to write down explicitly the transition probabilities for the coupled chain which accomplishes this objective.

It might appear from the previous paragraph that the usefulness of Theorem 1.5 is limited to $d = 1$ or $d = 2$, because it is only in that case that the q random walk has any chance of being recurrent. This is true only if one is inflexible in the choice of a coupling. To illustrate the possibilities, consider the simple random walk on Z^d, in which

$$p(x, y) = \frac{1}{2d}$$

for nearest neighbors x and y. This mechanism can be thought of as choosing a coordinate at random, each with probability d^{-1}, and then adding ± 1 with probability $\frac{1}{2}$ each to the coordinate chosen. With this in mind, a successful coupling can be constructed for the simple random walk by having X_n and Y_n choose the same coordinate to modify, and then having the choice of which of $+1$ or -1 to add to that coordinate be independent for the two processes if those coordinates do not agree, and the same choice if those coordinates do agree. This coupling is successful (for initial points x and y so that $x - y$ has all coordinates even) because the simple random walk on Z^1 is recurrent. Of course, more involved couplings will work for more general random walks on Z^d.

Sometimes it is convenient to define a coupling in which the joint mechanism depends on the initial states of the two processes. This will be illustrated by giving a coupling proof of the Choquet–Deny Theorem for a general irreducible random walk on Z^d. We may assume that $p(0, 0) > 0$, since otherwise $p(\cdot, \cdot)$ could be replaced by

$$\tilde{p}(x, y) = \begin{cases} \frac{1}{2}p(x, y) & \text{if } x \neq y, \\ \frac{1}{2} & \text{if } x = y, \end{cases}$$

using the fact that p and $\tilde{p}$ have the same harmonic functions. By irreducibility, it suffices to show that $h(x) = h(y)$ for any bounded harmonic function and any x and y satisfying $p(x, y) > 0$. We do this by using the idea of Theorem 1.5, but will construct the coupled process (X_n, Y_n) with $X_0 = x$ and $Y_0 = y$ only on $D = \{(u, v) \in Z^d \times Z^d : v - u \text{ is an integer multiple of } y - x\}$ for some fixed $x \neq y$ satisfying $p(x, y) > 0$. The coupled process has the transition probabilities given below until the first time that $X_n = Y_n$. Set $\varepsilon = \min\{p(0, 0), p(x, y)\} > 0$, and let

$$(u, v) \to \begin{cases} (u + w, v + w) & \text{with probability } p(0, w), \text{ for all } w \neq 0, y - x, \\ (u, v + y - x) & \text{with probability } \varepsilon, \\ (u + y - x, v) & \text{with probability } \varepsilon, \\ (u, v) & \text{with probability } p(0, 0) - \varepsilon, \\ (u + y - x, v + y - x) & \text{with probability } p(x, y) - \varepsilon. \end{cases}$$

Note that the chain remains on D at all times, and that the difference $Y_n - X_n$ has the following transition probabilities until the first time $X_n = Y_n$: for integer k,

$$k(y-x) \to \begin{cases} (k+1)(y-x) & \text{with probability } \varepsilon, \\ k(y-x) & \text{with probability } 1-2\varepsilon, \\ (k-1)(y-x) & \text{with probability } \varepsilon. \end{cases}$$

Therefore by the recurrence of the simple symmetric random walk, eventually the chain will satisfy $X_n = Y_n$. At that point the transition probabilities are modified so that $X_n = Y_n$ for all later times, and then the coupling is successful.

It should be noted from the proof of Theorem 1.5 that the existence of a successful coupling for a Markov process implies that the total variation distance between the distributions of the process at time n for different initial points tends to zero as n tends to infinity. (The converse to this statement is true also, provided one allows non-Markovian couplings as well—see Theorem 3 in Griffeath (1978b).) This total variation convergence essentially never occurs for particle systems. Therefore one cannot expect couplings to exist in that context which are successful in the sense that the two processes agree from some time on. For a coupling (η_t, ζ_t) of processes with state space $\{0, 1\}^S$, the most which can be hoped for is then that

(1.6) $$\lim_{t \to \infty} P[\eta_t(x) = \zeta_t(x)] = 1$$

for each $x \in S$. Fortunately, this turns out to be sufficient for many purposes, as will be seen in later chapters. Even when (1.6) does not hold, coupling can be extremely useful. The next section expands on this statement.

2. Monotonicity and Positive Correlations

As we will see later, it is nearly impossible to prove limit theorems for particle systems by developing and using estimates. It is therefore essential to take advantage of any monotonicity which may be present in a problem. There is an intimate connection between coupling and monotonicity. In the previous section, couplings were used which made two processes agree at large times with high probability. When used in connection with monotonicity arguments, the objective of a coupling is to show that if certain inequalities between the distributions of the processes of interest hold initially, then they continue to hold at later times. This section is devoted to a discussion of a number of ideas related to monotonicity and coupling which will be used often in this book. At the end, two results concerning measures with positive correlations will be presented.

Throughout this section, X will be a compact metric space on which there is defined a partial order. A typical example to keep in mind is $X = \{0, 1\}^S$ where S is a finite or countable set. The partial order should be compatible with the topology in the sense that

$$\{(\eta, \zeta) \in X \times X : \eta \leq \zeta\}$$

is a closed subset of $X \times X$ (with the product topology). One consequence of this which will be useful is that any upper semicontinuous increasing function on X is a decreasing pointwise limit of a sequence of continuous increasing functions (see Theorem 5 of the Appendix of Nachbin (1965), for example). $\mathcal{M}$ will denote the class of all continuous functions on X which are monotone in the sense that $f(\eta) \leq f(\zeta)$ whenever $\eta \leq \zeta$.

Definition 2.1. If μ_1 and μ_2 are two probability measures on X, we will say that $\mu_1 \leq \mu_2$ provided that

$$\int f \, d\mu_1 \leq \int f \, d\mu_2$$

for all $f \in \mathcal{M}$.

Note that by approximation, $\mu_1 \leq \mu_2$ implies that $\int f \, d\mu_1 \leq \int f \, d\mu_2$ for all increasing upper semicontinuous functions as well. In many cases it will be important to know that a Feller process (see Definition 1.2 of Chapter I) η_t on X has the property that its semigroup, acting on measures, preserves this ordering. The following result gives a simple necessary and sufficient condition for this to occur.

Theorem 2.2. *Suppose η_t is a Feller process on X with semigroup $S(t)$. The following two statements are equivalent:*

(a) $f \in \mathcal{M}$ *implies* $S(t)f \in \mathcal{M}$ *for all* $t \geq 0$.
(b) $\mu_1 \leq \mu_2$ *implies* $\mu_1 S(t) \leq \mu_2 S(t)$ *for all* $t \geq 0$.

Proof. Suppose (a) holds and $\mu_1 \leq \mu_2$. Given $f \in \mathcal{M}$, $S(t)f \in \mathcal{M}$ also, so

$$\int f \, d[\mu_1 S(t)] = \int S(t)f \, d\mu_1$$

$$\leq \int S(t)f \, d\mu_2$$

$$= \int f \, d[\mu_2 S(t)],$$

and hence $\mu_1 S(t) \leq \mu_2 S(t)$. For the other direction, assume that (b) holds, and take $f \in \mathcal{M}$. Then for $\eta \leq \zeta$, the pointmasses δ_η and δ_ζ satisfy $\delta_\eta \leq \delta_\zeta$, so that

$$S(t)f(\eta) = E^\eta f(\eta_t)$$

$$= \int f \, d[\delta_\eta S(t)]$$

$$\leq \int f \, d[\delta_\zeta S(t)]$$

$$= E^\zeta f(\eta_t) = S(t)f(\zeta). \quad \square$$

Definition 2.3. A Feller process is said to be *monotone* if the equivalent conditions of Theorem 2.2 are satisfied. (In the context of particle systems in later chapters, such a process will be called *attractive* rather than monotone.)

One connection between Definitions 2.1 and 2.3 on the one hand and coupling on the other, is that usually condition (a) of Theorem 2.2 is verified by constructing a coupled process (η_t, ζ_t) which has the property that $\eta \leq \zeta$ implies

$$P^{(\eta, \zeta)}[\eta_t \leq \zeta_t] = 1.$$

A particularly simple instance of this occurs when X is linearly ordered and the process has continuous paths. (For example, a diffusion process on $[0, 1]$, or a birth and death chain on $\{0, 1, \ldots, N\}$.) Then a simple coupling which has the desired property is obtained by letting the two copies of the process evolve independently until the first time (if ever) that they agree, and then letting them evolve together. Thus we see that such processes are monotone.

It might appear at first glance that the existence of a coupling which preserves the inequality $\eta_t \leq \zeta_t$ is a much stronger statement than (a) of Theorem 2.2. The following result shows that this is not the case, and at the same time gives additional insight into the content of Definition 2.1.

Theorem 2.4. *Suppose μ_1 and μ_2 are probability measures on X. A necessary and sufficient condition for $\mu_1 \leq \mu_2$ is that there exist a probability measure ν on $X \times X$ which satisfies*

(a) $\nu\{(\eta, \zeta): \eta \in A\} = \mu_1(A), \; and$

(b) $\nu\{(\eta, \zeta): \zeta \in A\} = \mu_2(A)$

for all Borel sets in X, and

(c) $\nu\{(\eta, \zeta): \eta \leq \zeta\} = 1.$

Proof. The sufficiency of the condition is clear. For the necessity, suppose that $\mu_1 \leq \mu_2$. The probability measure ν will be constructed with the help of the Hahn–Banach Theorem and the Riesz Representation Theorem. (See Theorems 4 of Chapter 10 and 8 of Chapter 14 of Royden (1968), for example.) Given $\varphi \in C(X \times X)$, define

$$\hat{\varphi}(\zeta) = \inf\{g(\zeta): g \in \mathcal{M} \text{ and } g(\eta) \geq \varphi(\eta, \zeta) \text{ for all } \eta\}$$

and

$$\rho(\varphi) = \int \hat{\varphi}(\zeta) \mu_2(d\zeta).$$

Note that $\hat{\varphi}$ is upper semicontinuous, so that the integral defining $\rho(\varphi)$ is meaningful. It is easy to check that ρ satisfies the properties

$$\rho(\varphi_1 + \varphi_2) \leq \rho(\varphi_1) + \rho(\varphi_2), \quad \text{and}$$

$$\rho(c\varphi) = c\rho(\varphi) \quad \text{for } c \geq 0.$$

Define a linear functional T on the subspace of all $\varphi \in C(X \times X)$ which are of the form

(2.5) $$\varphi(\eta, \zeta) = f_1(\eta) + f_2(\zeta)$$

for some $f_1, f_2 \in C(X)$ by

(2.6) $$T\varphi = \int f_1 \, d\mu_1 + \int f_2 \, d\mu_2.$$

For such a φ, let f be defined by

$$f(\zeta) = \inf\{g(\zeta): g \in \mathcal{M} \text{ and } g \geq f_1 \text{ on } X\}.$$

Then f is increasing, upper semicontinuous, and satisfies $f \geq f_1$. So, since $\mu_1 \leq \mu_2$, it follows that

$$T\varphi = \int f_1 \, d\mu_1 + \int f_2 \, d\mu_2$$

$$\leq \int f \, d\mu_1 + \int f_2 \, d\mu_2$$

$$\leq \int f \, d\mu_2 + \int f_2 \, d\mu_2$$

$$= \int (f + f_2) \, d\mu_2$$

$$= \int \hat{\varphi} \, d\mu_2 = \rho(\varphi).$$

By the Hahn–Banach Theorem, T can be extended to a linear functional on all of $C(X \times X)$ in such a way that $T\varphi \le \rho(\varphi)$ for all $\varphi \in C(X \times X)$. Note that if $\varphi \le 0$ on $X \times X$, then $\hat{\varphi} \le 0$ on X and hence $\rho(\varphi) \le 0$. Therefore $T\varphi \le 0$ whenever $\varphi \le 0$. Since T is linear, it follows that $T\varphi \ge 0$ whenever $\varphi \ge 0$. By the Riesz Representation Theorem, there is a unique probability measure ν on $X \times X$ such that

$$T\varphi = \int_{X \times X} \varphi \, d\nu$$

for all $\varphi \in C(X \times X)$. By (2.6), ν satisfies required properties (a) and (b). To check property (c), suppose $\varphi \in C(X \times X)$ satisfies

(2.7) $\varphi(\eta, \zeta) = 0$ whenever $\eta \le \zeta$.

For fixed $\zeta \in X$, define g by

$$g(\eta) = \begin{cases} 0 & \text{if } \eta \le \zeta, \\ \|\varphi\| & \text{otherwise.} \end{cases}$$

This function is lower semicontinuous, increasing, and satisfies $g(\eta) \ge \varphi(\eta, \zeta)$ for all $\eta \in X$. Therefore by Theorem 5 of the Appendix of Nachbin (1965), there is a $\tilde{g} \in \mathcal{M}$ so that

$$\varphi(\eta, \zeta) \le \tilde{g}(\eta) \le g(\eta)$$

for all $\eta \in X$. Hence

$$\hat{\varphi}(\zeta) \le \tilde{g}(\zeta) \le g(\zeta) = 0.$$

Thus for such a φ, $\rho(\varphi) \le 0$, and

$$\int \varphi \, d\nu = T\varphi \le \rho(\varphi) \le 0.$$

Since this is true for all $\varphi \in C(X \times X)$ which satisfy (2.7), and since $\{(\eta, \zeta): \eta \le \zeta\}$ is closed in $X \times X$, property (c) follows. $\square$

A measure ν satisfying properties (a), (b) and (c) of the statement of Theorem 2.4 will often be called a coupling measure for μ_1 and μ_2. Of course, ν is usually very far from being uniquely determined by these properties.

For those who may feel that the Hahn–Banach Theorem and the Riesz Representation Theorem are unduly heavy tools to be used in this proof, the following comment may be helpful. If X were a finite set, then the application of these theorems would be in a finite context, and would therefore be more elementary. Theorem 5 of the Appendix of Nachbin (1965) would also not be needed then. In particular, this would apply if $X = \{0, 1\}^S$, where S is finite. In most of our applications, X will be of the form $\{0, 1\}^S$ where S is countable. But to deduce Theorem 2.4 for $X = \{0, 1\}^S$ for S countable from the version in which S is finite, requires only a simple compactness argument.

The following simple consequence of Theorem 2.4 shows how strong the relation $\mu_1 \leq \mu_2$ is. Recall that a probability measure on a product space is very far from being determined by its marginals.

Corollary 2.8. *Suppose that* $X = \{0, 1\}^S$, *where S is a countable set, is given the natural partial ordering:* $\eta \leq \zeta$ *if and only if* $\eta(x) \leq \zeta(x)$ *for all* $x \in S$. *If* μ_1 *and* μ_2 *are probability measures on X such that* $\mu_1 \leq \mu_2$ *and if*

$$\mu_1\{\eta: \eta(x) = 1\} = \mu_2\{\eta: \eta(x) = 1\}$$

for all $x \in S$, *then* $\mu_1 = \mu_2$.

Proof. Let ν be a coupling measure for μ_1 and μ_2, whose existence is guaranteed by Theorem 2.4. Then

$$\nu\{(\eta, \zeta): \eta(x) = 0, \zeta(x) = 1\} = \mu_2\{\eta: \eta(x) = 1\} - \mu_1\{\eta: \eta(x) = 1\}$$

$$= 0.$$

Therefore $\nu\{(\eta, \zeta): \eta = \zeta\} = 1$, so $\mu_1 = \mu_2$. □

There are three reasons for discussing the next result. First, the proof is a simple application of the coupling technique. Second, the result gives a sufficient condition for two probability measures to satisfy $\mu_1 \leq \mu_2$. Finally, it has an important corollary which plays a role in the study of Gibbs states in the context of statistical mechanics. (See Corollary 2.12.)

Theorem 2.9. *Let* $X = \{0, 1\}^S$, *where S is a finite set and X is given the natural partial order. For* $\eta, \zeta \in X$ *define* $\eta \vee \zeta$ *and* $\eta \wedge \zeta$ *by*

$$(\eta \vee \zeta)(x) = \max\{\eta(x), \zeta(x)\}, (\eta \wedge \zeta)(x) = \min\{\eta(x), \zeta(x)\}.$$

Suppose μ_1 *and* μ_2 *are probability measures on X which assign a strictly positive probability to each point of X. If*

(2.10) $$\mu_1(\eta \wedge \zeta)\mu_2(\eta \vee \zeta) \geq \mu_1(\eta)\mu_2(\zeta)$$

for all $\eta, \zeta \in X$, *then* $\mu_1 \leq \mu_2$.

Proof. The idea of the proof is to define a continuous time Markov chain (η_t, ζ_t) on $\{(\eta, \zeta) \in X \times X : \eta \le \zeta\}$ which has the following properties:

(a) η_t is an irreducible Markov chain on X with stationary probability measure μ_1; and

(b) ζ_t is an irreducible Markov chain on X with stationary probability measure μ_2.

Once (η_t, ζ_t) with these properties is constructed, it follows that for $f \in \mathcal{M}$, $E^\eta f(\eta_t) \le E^\zeta f(\zeta_t)$ whenever $\eta \le \zeta$. Passing to the limit as $t \to \infty$, it follows that $\int f\, d\mu_1 \le \int f\, d\mu_2$ for $f \in \mathcal{M}$, and hence that $\mu_1 \le \mu_2$. For $x \in S$ and $\eta \in X$ define η_x as usual by $\eta_x(y) = \eta(y)$ for $y \ne x$ and $\eta_x(x) = 1 - \eta(x)$. In order to satisfy properties (a) and (b), the marginal processes η_t and ζ_t will be chosen to have transition rates (at each x)

$$
\left.
\begin{array}{ll}
\eta \to \eta_x \text{ at rate } 1 & \text{if } \eta(x) = 0 \\[2mm]
\eta \to \eta_x \text{ at rate } \dfrac{\mu_1(\eta_x)}{\mu_1(\eta)} & \text{if } \eta(x) = 1
\end{array}
\right\} \quad \text{for } \eta_t, \quad \text{and}
$$

$$
\left.
\begin{array}{ll}
\zeta \to \zeta_x \text{ at rate } 1 & \text{if } \zeta(x) = 0 \\[2mm]
\zeta \to \zeta_x \text{ at rate } \dfrac{\mu_2(\zeta_x)}{\mu_2(\zeta)} & \text{if } \zeta(x) = 1
\end{array}
\right\} \quad \text{for } \zeta_t.
$$

In order to check that μ_1 and μ_2 are the stationary measures for the two chains, it is easier to note that in fact these chains are reversible with respect to μ_1 and μ_2 respectively. (For more on this point, see Section 5.) It now remains to choose a transition mechanism for the coupled process which is consistent with the above transition rates for the marginal processes and which preserves the relation $\eta_t \le \zeta_t$. The most natural choice is the following, for $\eta \le \zeta$:

$$(\eta, \zeta) \to (\eta_x, \zeta) \text{ at rate } 1 \qquad\qquad \text{if } \eta(x) = 0 \text{ and } \zeta(x) = 1,$$

$$(\eta, \zeta) \to (\eta, \zeta_x) \text{ at rate } \frac{\mu_2(\zeta_x)}{\mu_2(\zeta)} \qquad\qquad \text{if } \eta(x) = 0 \text{ and } \zeta(x) = 1,$$

$$(\eta, \zeta) \to (\eta_x, \zeta_x) \text{ at rate } 1 \qquad\qquad \text{if } \eta(x) = \zeta(x) = 0,$$

$$(\eta, \zeta) \to (\eta_x, \zeta_x) \text{ at rate } \frac{\mu_2(\zeta_x)}{\mu_2(\zeta)} \qquad\qquad \text{if } \eta(x) = \zeta(x) = 1, \quad \text{and}$$

$$(\eta, \zeta) \to (\eta_x, \zeta) \text{ at rate } \frac{\mu_1(\eta_x)}{\mu_1(\eta)} - \frac{\mu_2(\zeta_x)}{\mu_2(\zeta)} \quad \text{if } \eta(x) = \zeta(x) = 1.$$

This choice has all the desired properties. For these rates to be permissible, it must be the case that

$$\frac{\mu_1(\eta_x)}{\mu_1(\eta)} \geq \frac{\mu_2(\zeta_x)}{\mu_2(\zeta)}$$

whenever $\eta \leq \zeta$ and $\eta(x) = \zeta(x) = 1$. This is where assumption (2.10) comes in, since in this case, $\eta \wedge \zeta_x = \eta_x$ and $\eta \vee \zeta_x = \zeta$. (In fact it is not hard to see that the above inequality for all such η and ζ is equivalent to (2.10).) □

Theorem 2.9 gives a sufficient condition for two measures to be stochastically ordered. This condition is rather strong. To see this, consider the two special cases (a) $\mu_1 = \mu_2$, and (b) X is linearly ordered. In the first case, the conclusion of Theorem 2.9 is trivial, while (2.10) can easily be made to fail. In the second case, (2.10) asserts that

$$\frac{\mu_1(\eta)}{\mu_2(\eta)}$$

is a decreasing function of η, while the conclusion $\mu_1 \leq \mu_2$ is just the statement that

$$\mu_1\{\eta: \eta \geq \zeta\} \leq \mu_2\{\eta: \eta \geq \zeta\}$$

for all $\zeta \in X$. In the linearly ordered case, the statement of Theorem 2.9 is quite trivial anyway, as can be seen from the identity

$$\int f \, d\mu_2 - \int f \, d\mu_1 = \tfrac{1}{2} \sum_{\eta, \zeta \in X} [f(\eta) - f(\zeta)][\mu_2(\eta)\mu_1(\zeta) - \mu_2(\zeta)\mu_1(\eta)].$$

Even though assumption (2.10) is quite strong, it is often satisfied by the measures which arise in statistical mechanics.

Theorem 2.9 can be restated in such a way that it gives a sufficient condition for a probability measure to have positive correlations. In order to do this, we need the following definition.

Definition 2.11. A probability measure μ on X is said to have *positive correlations* if

$$\int fg \, d\mu \geq \int f \, d\mu \int g \, d\mu$$

for all $f, g \in \mathcal{M}$.

If X is linearly ordered, then any μ has positive correlations, as can be seen from writing

$$\frac{1}{2} \iint\limits_{X \times X} [f(\eta) - f(\zeta)][g(\eta) - g(\zeta)]\, d\nu = \int fg\, d\mu - \int f\, d\mu \int g\, d\mu,$$

where ν is the product measure $\mu \times \mu$ on $X \times X$. The left side above is nonnegative since $f, g \in \mathcal{M}$ implies that $f(\eta) - f(\zeta)$ and $g(\eta) - g(\zeta)$ have the same sign for all $\eta,\ \zeta \in X$ in the linearly ordered case. Another simple observation is that the product of two measures with positive correlations has positive correlations as well. To see this, suppose that μ_1 on X_1 and μ_2 on X_2 have positive correlations, and let $\nu = \mu_1 \times \mu_2$ on $X = X_1 \times X_2$. Take $f, g \in \mathcal{M}$, and let

$$F(\eta) = \int f(\eta, \zeta)\mu_2(d\zeta), \qquad G(\eta) = \int g(\eta, \zeta)\mu_2(d\zeta).$$

Then $f(\eta, \cdot),\ g(\eta, \cdot) \in \mathcal{M}_2$ for each η, and $F,\ G \in \mathcal{M}_1$, so that

$$
\begin{aligned}
\int_X fg\, d\nu &= \int_{X_1} \left[\int_{X_2} fg\, d\mu_2 \right] d\mu_1 \\
&\geq \int_{X_1} FG\, d\mu_1 \\
&\geq \int_{X_1} F\, d\mu_1 \int_{X_1} G\, d\mu_1 \\
&= \int_X f\, d\nu \int_X g\, d\nu.
\end{aligned}
$$

Combining these two remarks, it follows that a product measure on a product of linearly ordered spaces has positive correlations. For other measures, it is not so easy to check the property of having positive correlations. The next result provides a sufficient condition which is often useful.

Corollary 2.12. *In the context of Theorem 2.9, suppose that μ is a probability measure on X which assigns a strictly positive probability to each point of X. If μ satisfies*

$$(2.13) \qquad\qquad \mu(\eta \wedge \zeta)\mu(\eta \vee \zeta) \geq \mu(\eta)\mu(\zeta)$$

for all $\eta,\ \zeta \in X$, then μ has positive correlations.

Proof. Take $f, g \in \mathcal{M}$. Note that by adding a constant to g, we can assume without loss of generality in verifying the inequality in Definition 2.11 that $g > 0$ on X. Define μ_1 and μ_2 by $\mu_1 = \mu$ and

$$\mu_2(\eta) = \frac{g(\eta)\mu(\eta)}{\displaystyle\int g\, d\mu}.$$

For $\eta,\ \zeta \in X$,

$$\mu_1(\eta \wedge \zeta)\mu_2(\eta \vee \zeta) = \frac{\mu(\eta \wedge \zeta)g(\eta \vee \zeta)\mu(\eta \vee \zeta)}{\displaystyle\int g\, d\mu}$$

$$\geq \frac{\mu(\eta)\mu(\zeta)g(\zeta)}{\displaystyle\int g\, d\mu}$$

$$= \mu_1(\eta)\mu_2(\zeta)$$

by (2.13) and the monotonicity of g. Therefore by Theorem 2.9, $\mu_1 \leq \mu_2$, and hence

$$\int f\, d\mu_1 \leq \int f\, d\mu_2$$

for $f \in \mathcal{M}$. This gives the desired conclusion, since $\mu_1 = \mu$ and

$$\int f\, d\mu_2 = \frac{\displaystyle\int fg\, d\mu}{\displaystyle\int g\, d\mu}. \quad \square$$

In Theorem 2.9 and Corollary 2.12, it was assumed that S is a finite set. These results can generally be used to check that $\mu_1 \leq \mu_2$ or that μ has positive correlations even if S is countable. To do this, it suffices to note that from the definitions of these two properties, it is enough to check the properties for the projections on $\{0, 1\}^T$ of the measures involved for all finite $T \subset S$.

While Corollary 2.12 gives a useful sufficient condition for a measure μ to have positive correlations, in order to verify assumption (2.13), it is necessary to have a reasonably explicit expression for μ. In many situations, it is convenient to know that the distribution at time t, or the limiting distribution as $t \to \infty$, of a Markov process has positive correlations. Such distributions will of course not usually be known explicitly. If the initial distribution is a point mass, then it automatically has positive correlations.

Thus the following result is often useful in showing that the distributions at later times have positive correlations as well. (Examples of its application can be found in the proofs of Proposition 2.16 and Theorem 3.14 of Chapter III, Theorem 2.5, Lemma 3.12, and Proposition 4.2 of Chapter V, and Problem 15 of Chapter VII.) The statement of the theorem is simplest when the generator of the process is bounded. This case suffices for most applications since limits of sequences of measures with positive correlations again have positive correlations, and most processes of interest can be obtained in a natural way as limits of similar processes with bounded generators. Before reading this theorem and its proof, it would be a good idea to review the first two sections of Chapter I.

Theorem 2.14. *Suppose that $S(t)$ and Ω are respectively the semigroup and the generator of a monotone Feller process on X. Assume further that Ω is a bounded operator. Then the following two statements are equivalent*:

(2.15) $\Omega fg \geq f\Omega g + g\Omega f$ *for all $f, g \in \mathcal{M}$.*

(2.16) $\mu S(t)$ *has positive correlations whenever μ does.*

Before proving this theorem, a few remarks should be made about (2.15). First note that it is easy to determine whether or not it holds, since it involves the generator, which is usually known explicitly, rather than the semigroup, which is not. Secondly, consider what it says when X is a finite partially ordered set and

(2.17) $\Omega f(\eta) = \sum_{\zeta \in X} \rho(\eta, \zeta)[f(\zeta) - f(\eta)],$

where $\rho(\eta, \zeta) \geq 0$ is the rate at which the chain goes from η to ζ. Then

(2.18)
$$\Omega(fg)(\eta) - f(\eta)\Omega g(\eta) - g(\eta)\Omega f(\eta)$$
$$= \sum_{\zeta \in X} \rho(\eta, \zeta)[f(\zeta) - f(\eta)][g(\zeta) - g(\eta)],$$

which is nonnegative for all $f, g \in \mathcal{M}$ if and only if ρ satisfies the property that

(2.19) $\rho(\eta, \zeta) > 0$ implies that $\eta \leq \zeta$ or $\eta \geq \zeta$.

To check this, note that (2.19) implies (2.15) trivially. For the converse, suppose $\eta, \zeta \in X$ satisfy neither $\eta \leq \zeta$ nor $\eta \geq \zeta$. Define $f, g \in \mathcal{M}$ by

$$f(\gamma) = \begin{cases} 1 & \text{if } \gamma \geq \zeta, \\ 0 & \text{otherwise, and} \end{cases}$$

$$g(\gamma) = \begin{cases} 0 & \text{if } \gamma \le \zeta, \\ 1 & \text{otherwise.} \end{cases}$$

Then by (2.18),

$$\Omega(fg)(\eta) - f(\eta)\Omega g(\eta) - g(\eta)\Omega f(\eta) = -\rho(\eta, \zeta),$$

so that (2.15) implies (2.19) as well. One can reexpress (2.19) by saying that only jumps between comparable states are allowed.

Proof of Theorem 2.14. The idea of the proof is to show that each of (2.15) and (2.16) is equivalent to the statement

(2.20) $S(t)fg \ge [S(t)f][S(t)g] \quad \text{for all } f, g \in \mathcal{M}.$

To do this, there are four implications to be proved. The last one is the most important.

(2.16) *implies* (2.20). For $\eta \in X$, the pointmass δ_η at η has positive correlations, so by (2.16), $\delta_\eta S(t)$ does too. Therefore, for any $f, g \in \mathcal{M}$,

$$\int fg \, d[\delta_\eta S(t)] \ge \int f \, d[\delta_\eta S(t)] \int g \, d[\delta_\eta S(t)],$$

which is the same as

$$S(t)(fg)(\eta) \ge [S(t)f(\eta)][S(t)g(\eta)].$$

(2.20) *implies* (2.16). Suppose μ has positive correlations and $f, g \in \mathcal{M}$. Since the process is monotone, $S(t)f$ and $S(t)g$ are in $\mathcal{M}$ also. Therefore

$$\int [S(t)f][S(t)g] \, d\mu \ge \int S(t)f \, d\mu \int S(t)g \, d\mu.$$

By (2.20),

$$\int S(t)fg \, d\mu \ge \int [S(t)f][S(t)g] \, d\mu.$$

Combining these two inequalities gives

$$\int fg \, d[\mu S(t)] \ge \int f \, d[\mu S(t)] \int g \, d[\mu S(t)]$$

as required.

(2.20) *implies* (2.15). Take f, $g \in \mathcal{M}$. By (2.20), for $t > 0$

$$\frac{S(t)fg - fg}{t} \geq S(t)f \left[\frac{S(t)g - g}{t} \right] + g \left[\frac{S(t)f - f}{t} \right].$$

Letting t tend to zero yields (2.15).

(2.15) *implies* (2.20). For f, $g \in \mathcal{M}$, let

$$F(t) = S(t)fg - [S(t)f][S(t)g].$$

Since $S(t)f$ and $S(t)g$ are in $\mathcal{M}$ also, (2.15) gives

$$\begin{aligned} F'(t) &= \Omega S(t)fg - [\Omega S(t)f][S(t)g] - [S(t)f][\Omega S(t)g] \\ &\geq \Omega\{S(t)fg - [S(t)f][S(t)g]\} \\ &= \Omega F(t). \end{aligned}$$

Therefore $G(t) = F'(t) - \Omega F(t) \geq 0$. By Theorem 2.15 of Chapter I,

$$F(t) = S(t)F(0) + \int_0^t S(t-s)G(s)\,ds,$$

which is nonnegative since $F(0) = 0$ and $G(s) \geq 0$. $\square$

Corollary 2.21. *Suppose that the assumptions of Theorem 2.14 are satisfied, and that the equivalent conditions* (2.15) *and* (2.16) *hold. Let η_t be the corresponding process, where the distribution of η_0 has positive correlations. Then for $t_1 < t_2 < \cdots < t_n$, the joint distribution of $(\eta_{t_1}, \ldots, \eta_{t_n})$, which is a probability measure on X^n, has positive correlations.*

Proof. The proof is by induction on n. When $n = 1$, this is just the statement of Theorem 2.14. Now suppose the result to be true for $n - 1$, and take f, g to be monotone functions on X^n. Put $s_k = t_{k+1} - t_1$ for $1 \leq k \leq n-1$ and define

$$F(\eta) = E^\eta f(\eta, \eta_{s_1}, \ldots, \eta_{s_{n-1}}) \quad \text{and}$$

$$G(\eta) = E^\eta g(\eta, \eta_{s_1}, \ldots, \eta_{s_{n-1}}).$$

Since f and g are monotone on X^n, it follows that F and G are monotone on X. Let μ be the distribution of η_{t_1}, which has positive correlations by Theorem 2.14. Then

$$Ef(\eta_{t_1}, \ldots, \eta_{t_n})g(\eta_{t_1}, \ldots, \eta_{t_n})$$

$$= \int E^\eta[f(\eta, \eta_{s_1}, \ldots, \eta_{s_{n-1}})g(\eta, \eta_{s_1}, \ldots, \eta_{s_{n-1}})]\,d\mu$$

$$\geq \int F(\eta) G(\eta) \, d\mu$$

$$\geq \int F(\eta) \, d\mu \int G(\eta) \, d\mu$$

$$= Ef(\eta_{t_1}, \ldots, \eta_{t_n}) Eg(\eta_{t_1}, \ldots, \eta_{t_n}).$$

In the above, the first inequality comes from the induction hypothesis and the fact that the pointmass at η has positive correlations, while the second comes from the monotonicity of F and G and the fact that μ has positive correlations. $\square$

It is interesting to note that Theorem 2.14 can be used to give a somewhat different proof of Corollary 2.12. Simply take $X = \{0, 1\}^S$, and define a Markov chain on X by letting the transition rates be (as in the proof of Theorem 2.9)

$$\eta \to \eta_x \text{ at rate } 1 \qquad \text{if } \eta(x) = 0, \quad \text{and}$$

$$\eta \to \eta_x \text{ at rate } \frac{\mu(\eta_x)}{\mu(\eta)} \quad \text{if } \eta(x) = 1.$$

Then assumption (2.13) implies that this chain is a monotone process. Clearly the chain is irreducible and has stationary measure μ. Now start the chain off in a deterministic configuration, apply Theorem 2.14, and let t tend to ∞.

The following result connecting monotonicity and positive correlations will be useful at times. It is not correct without the assumption $\mu_1 \leq \mu_2$.

Proposition 2.22. Suppose $\mu_1 \leq \mu_2$ and both μ_1 and μ_2 have positive correlations. Then so does $\mu_\lambda = \lambda \mu_1 + (1 - \lambda) \mu_2$ for any $\lambda \in [0, 1]$.

Proof. Take $f, g \in \mathcal{M}$, and write $F_i = \int f \, d\mu_i$ and $G_i = \int g \, d\mu_i$. Then

$$\int fg \, d\mu_\lambda - \int f \, d\mu_\lambda \int g \, d\mu_\lambda$$

$$= \lambda \int fg \, d\mu_1 + (1 - \lambda) \int fg \, d\mu_2 - \int f \, d\mu_\lambda \int g \, d\mu_\lambda$$

$$\geq \lambda F_1 G_1 + (1 - \lambda) F_2 G_2 - \lambda^2 F_1 G_1 - (1 - \lambda)^2 F_2 G_2$$

$$\quad - \lambda (1 - \lambda)[F_1 G_2 + F_2 G_1]$$

$$= \lambda (1 - \lambda)[F_2 - F_1][G_2 - G_1] \geq 0. \quad \square$$

3. Duality

When it is applicable, duality is an extremely useful technique in the study of interacting particle systems. Typically, the duality relation associates a dual process with the process of interest in such a way that problems involving the original process can be reformulated as problems involving the dual process. If the dual process is substantially simpler than the original process or if the nature of the problems changes in a useful way, then the reformulated problems may be more tractable than the original problems, and some progress may have been made. The particle system is usually a Markov process on the uncountable space $\{0, 1\}^S$, while the dual process, when it exists, is usually a countable state Markov chain whose state space is the collection of all finite subsets of S. In some cases, such as the contact process, the problems one is led to for the dual Markov chain may be no easier than the original problems for the particle system, but even then, duality yields interesting and important connections between these problems which add insight into what is going on.

We begin with a general definition of duality.

Definition 3.1. Suppose η_t and ζ_t are Markov processes with state spaces X and Y respectively, and let $H(\eta, \zeta)$ be a bounded measurable function on $X \times Y$. The processes η_t and ζ_t are said to be *dual* to one another with respect to H if

$$(3.2) \qquad\qquad E^\eta H(\eta_t, \zeta) = E^\zeta H(\eta, \zeta_t)$$

for all $\eta \in X$ and $\zeta \in Y$.

In each of the examples to be discussed in this section, we will take $X = Y \subset [0, \infty)$. In each case the duality function H will be

$$(3.3) \qquad\qquad H(\eta, \zeta) = \begin{cases} 1 & \text{if } \eta \le \zeta; \\ 0 & \text{otherwise.} \end{cases}$$

With this choice, (3.2) becomes

$$(3.4) \qquad\qquad P^\eta[\eta_t \le \zeta] = P^\zeta[\eta \le \zeta_t],$$

so in particular the transition mechanism of one of the processes determines that of the other. While only this H will be used in this section, other choices of H will occur naturally in later chapters.

Perhaps the first time relation (3.4) between two Markov processes was observed and exploited was in the following example (see Levy (1948)): Let η_t be Brownian motion on $[0, \infty)$ with absorption at 0 and ζ_t be Brownian motion on $[0, \infty)$ with reflection at 0. Then (3.4) holds since both sides are equal to

$$P(B_t \ge \eta - \zeta) + P(B_t \ge \eta + \zeta),$$

where B_t is ordinary Brownian motion on R^1 with $B_0 = 0$, as can be seen by direct computation (see Section 16.3 of Breiman (1968), for example). This example has been greatly generalized by Cox and Rösler (1983). Their result asserts that under mild conditions, if η_t is a diffusion process on $[0, \infty)$ with absorption at 0 and speed and scale functions $M(x)$ and $S(x)$ respectively, while ζ_t is a diffusion process on $[0, \infty)$ with reflection at 0 and speed and scale functions $S(x)$ and $M(x)$ respectively, then η_t and ζ_t are dual to one another relative to the function H in (3.3). Thus this duality relation interchanges the roles of the speed and scale functions. (For a treatment of one-dimensional diffusion processes, see Chapter 16 of Breiman (1968).) They also proved a similar result for birth and death chains on $\{0, 1, 2, \ldots\}$. Their motivation was the fact that in their context, the duality relation enables one to go back and forth between problems involving entrance laws and problems involving exit laws. The following result shows that the boundary conditions chosen in the above examples are not at all arbitrary, and also illustrates in a simple way how information about one of the processes can be used to deduce properties of the other process.

Theorem 3.5. *Suppose η_t and ζ_t are Feller processes on $[0, \infty)$ which are dual to one another with respect to the H in (3.3). (Here Feller means that the semigroup maps bounded continuous functions into themselves.) Then*

    (a) $P^0[\eta_t = 0] = 1$ *for all $t \geq 0$.*
    (b) $P^\zeta[\zeta_t = 0] = 0$ *for all $t \geq 0$ and $\zeta > 0$.*
    (c) $L = \lim_{\eta \to \infty} \lim_{t \to \infty} P^\eta[\eta_t = 0]$ *exists and is either 0 or 1.*
    (d)   *If $L = 1$, then $\zeta_t \to \infty$ in probability as $t \to \infty$ for any initial point $\zeta \geq 0$.*
    (e)   *If $L = 0$, then the distribution of ζ_t given $\zeta_0 = 0$ has a limit as $t \to \infty$. If, in addition,*

(3.6)
$$\lim_{t \to \infty} P^\eta[0 < \eta_t \leq \zeta] = 0$$

for all $\eta, \zeta \geq 0$, then the distribution of ζ_t given $\zeta_0 = \zeta$ has a limit as $t \to \infty$ which is independent of ζ.

Proof. (a) Take $\eta = \zeta = 0$ in (3.4).
    (b) Using (3.4), write for $\zeta > 0$,

$$P^\zeta[\zeta_t > 0] = \lim_{\eta \downarrow 0} P^\zeta[\zeta_t \geq \eta]$$

$$= \lim_{\eta \downarrow 0} P^\eta[\eta_t \leq \zeta]$$

$$\geq P^0[\eta_t < \zeta]$$

by the Feller property. But by (a), $P^0(\eta_t < \zeta) = 1$.

(c) By the Markov property and (a),

(3.7) $P^\eta[\eta_{t+s}=0]=P^\eta[\eta_t=0]+E^\eta\{P^{\eta_t}[\eta_s=0],\ \eta_t>0\},$

so $P^\eta[\eta_t=0]$ is an increasing function of t. By (3.4), $P^\eta[\eta_t=0]=P^0[\eta\le\zeta_t],$ which is a decreasing function of η. Therefore the limits occurring in the definition of L exist. By this monotonicity,

(3.8) $\lim_{t\to\infty} P^\eta[\eta_t=0]\ge L$

for every $\eta\ge0$. Passing to the limit as $s\to\infty$ in (3.7) using (3.8) gives

$$\lim_{s\to\infty} P^\eta[\eta_s=0]\ge P^\eta[\eta_t=0]+LP^\eta[\eta_t>0].$$

Letting t and η tend to ∞, it follows that

$$L\ge L+L(1-L),$$

or $L(1-L)\le0$, so $L=0$ or 1.

(d) Suppose $L=1$. Then

$$\lim_{t\to\infty} P^\eta[\eta_t=0]=1$$

for all $\eta\ge0$ by (3.8). By (3.4), it follows that

$$\lim_{t\to\infty} P^\zeta[\eta\le\zeta_t]=1$$

for all $\eta,\zeta\ge0$.

(e) Since $L=0$, (3.4) gives

$$\lim_{\eta\to\infty}\lim_{t\to\infty} P^0[\eta\le\zeta_t]=0,$$

so that the family $\{\zeta_t,\ t\ge0\}$ under $\zeta_0=0$ is tight. The limiting distribution then exists by the monotonicity of $P^0[\eta\le\zeta_t]$ in t which is provided by (3.4) and (3.7). By (3.4) and (3.6),

$$\lim_{t\to\infty}\{P^\zeta[\eta\le\zeta_t]-P^0[\eta\le\zeta_t]\}=0,$$

from which the final statement follows. $\square$

Remark When formulated in terms of Markov chains on $\{0,1,2,\ldots\}$ which are irreducible except for the fact that 0 is a trap for η_t, the last three parts of Theorem 3.5 would say that η_t escapes to ∞ with positive probability for initial states other than zero if and only if ζ_t is positive recurrent.

Furthermore, in this case, if π is the stationary probability measure for ζ_t and $h(\eta) = P^\eta[\eta_t = 0$ for some $t]$, it follows that

$$\pi(\eta) = h(\eta) - h(\eta + 1).$$

This exhibits already the close connection between stationary measures for one process and positive harmonic functions for its dual process. This connection plays a key role in the use of duality in Chapters V and VIII.

There is a close relationship between duality with respect to the function in (3.3) and monotonicity in the sense of Definition 2.3. This connection is given in the following theorem. This result depends heavily on the fact that the state space of the processes is totally ordered. In the particle system context, the state spaces are only partially ordered, and unfortunately the analogous result fails. As will be seen in later chapters, having a (reasonable) dual is a much more special property than being monotone, when the state space is not totally ordered. The reason for this large difference between totally ordered and partially ordered state spaces can best be seen from the following observation: If μ is a finite signed measure on $[0, \infty)$, then μ is positive if and only if the function $\mu\{\eta: \eta \leq \zeta\}$ is an increasing function of ζ. On the other hand, if μ is a finite signed measure on $\{0, 1\}^S$ for some countable set S, then the monotonicity of $\mu\{\eta: \eta \leq \zeta\}$ is a necessary condition for the positivity of μ, but it is very far from being a sufficient condition.

Theorem 3.9. *Suppose that ζ_t is a Feller process on $[0, 1]$ for which $\{1\}$ is an absorbing state and such that $P^\zeta[\zeta_t = \gamma] = 0$ for all $t > 0$, $\zeta \in [0, 1]$ and $\gamma \in [0, 1)$. In order that there be a Feller process η_t on $[0, 1]$ which is dual to ζ_t with respect to the function H in (3.3), it is necessary and sufficient that ζ_t be monotone.*

Proof. The necessity of the condition is immediate from (3.4). For the sufficiency, assume that ζ_t is monotone. Then the function $P^\zeta[\eta \leq \zeta_t]$ is increasing in ζ, decreasing and continuous in η, and upper semicontinuous in ζ. Hence it is right continuous in ζ, and since

$$P^1[\zeta_t = 1] = 1,$$

we may define probability measures $p_t(\eta, \cdot)$ on $[0, 1]$ for $t > 0$ and $\eta \in [0, 1]$ by requiring that

$$p_t(\eta, [0, \zeta]) = P^\zeta[\eta \leq \zeta_t].$$

For $f \in C[0, 1]$, put

$$S(t)f(\eta) = \int_0^1 f(\gamma) p_t(\eta, d\gamma).$$

Then $S(t)f \in C[0, 1]$, and by Theorem 1.5 of Chapter I, in order to define the required dual process η_t, it suffices to check that $S(t)$ is a Markov semigroup. We will check the semigroup property only, since the other properties of a Markov semigroup are immediate. To do this, write

$$p_{t+s}(\eta, [0, \zeta]) = P^\zeta[\eta \le \zeta_{t+s}]$$

$$= \int_0^1 P^\zeta[\zeta_t \in d\gamma] P^\gamma[\eta \le \zeta_s]$$

$$= \int_0^1 P^\zeta[\zeta_t \in d\gamma] p_s(\eta, [0, \gamma])$$

$$= \int_0^1 p_s(\eta, d\lambda) P^\zeta[\zeta_t \ge \lambda]$$

$$= \int_0^1 p_s(\eta, d\lambda) p_t(\lambda, [0, \zeta]). \quad \square$$

Theorem 3.9 explains why it should not be surprising that diffusion processes in one dimension often have duals. They are automatically monotone by the comment following Definition 2.3, so all that is required is that they have the correct boundary behavior.

4. Relative Entropy

A useful technique for studying the invariant measures and the limiting behaviour of particle systems is based on the observation that the relative entropy of the distribution of the process is monotone as a function of time. In this section, we will illustrate this technique in the simple context of finite state Markov chains. In particular, we will use it to give an alternate proof of Theorem 1.2. In this section, S will denote a finite set.

Definition 4.1. Suppose π is a probability measure on S such that $\pi(x) > 0$ for all $x \in S$. If μ is any probability measure on S, the *entropy* of μ relative to π is defined by

$$H(\mu) = \sum_x \mu(x) \log\left[\frac{\mu(x)}{\pi(x)}\right]$$

$$= \sum_x \pi(x) \varphi\left[\frac{\mu(x)}{\pi(x)}\right],$$

where $\varphi(s) = s \log s$ for $s > 0$ and $\varphi(0) = 0$.

Let now $p(x, y)$ be the transition probabilities for a discrete time Markov chain on S, and define $p^{(k)}(x, y)$ as in (1.1). For a probability measure μ

on S, let μP^k be defined by

$$\mu P^k(y) = \sum_x \mu(x) p^{(k)}(x, y).$$

Proposition 4.2. *Suppose π is a strictly positive probability measure on S which satisfies* (1.4). *Then for any probability measure μ on S,*

(4.3) $$H(\mu P) \le H(\mu)$$

Furthermore, equality holds in (4.3) *if and only if*

$$\frac{\mu(x)}{\pi(x)} = \frac{\mu(y)}{\pi(y)}$$

whenever $\sum_u p(x, u) p(y, u) > 0$.

Proof.

$$\begin{aligned}
H(\mu P) &= \sum_x \pi(x) \varphi \left[\frac{1}{\pi(x)} \sum_y \mu(y) p(y, x) \right] \\
&= \sum_x \pi(x) \varphi \left[\sum_y \frac{\mu(y)}{\pi(y)} \frac{\pi(y) p(y, x)}{\pi(x)} \right] \\
&\le \sum_x \pi(x) \sum_y \varphi \left[\frac{\mu(y)}{\pi(y)} \right] \frac{\pi(y) p(y, x)}{\pi(x)} \\
&= \sum_y \pi(y) \varphi \left[\frac{\mu(y)}{\pi(y)} \right] = H(\mu),
\end{aligned}$$

where the inequality follows from the convexity of φ and the fact that

$$\sum_y \frac{\pi(y) p(y, x)}{\pi(x)} = 1.$$

Since π is strictly positive and φ is strictly convex, equality will hold if and only if $\mu(y)/\pi(y)$ is constant on $\{y \in S: p(y, x) > 0\}$ for each x. $\square$

Alternative Proof of Theorem 1.2. Any limit π of a subsequence of

$$\frac{1}{N} \sum_{k=1}^N p^{(k)}(u, x)$$

as $N \to \infty$ for any fixed $u \in S$ satisfies (1.4). By the hypothesis of Theorem 1.2, any π which satisfies (1.4) will be strictly positive. Fix any such π, and define the entropy $H(\mu)$ relative to that π. By Proposition 4.2, $H(\mu P^n)$ is decreasing in n for any probability measure μ on S. Since H is continuous, if ν is the limit of any convergent subsequence of μP^n, then

$$H(\nu) = \lim_{n \uparrow \infty} H(\mu P^n).$$

If ν is the limit of a convergent subsequence of μP^n, then so is νP^k for any $k \geq 1$. Thus

$$H(\nu P^k) = H(\nu)$$

for all $k \geq 1$, so that by applying Proposition 4.2 to P^k,

$$\frac{\nu(x)}{\pi(x)} = \frac{\nu(y)}{\pi(y)}$$

whenever $\sum_u p^{(k)}(x, u) p^{(k)}(y, u) > 0$ for any $k \geq 1$. Using the hypothesis of Theorem 1.2 again, it follows that $\nu(x)/\pi(x)$ is constant, so that $\nu = \pi$. Since the limit of any convergent subsequence of μP^n is π, it follows that

$$\pi = \lim_{n \to \infty} \mu P^n$$

as required. $\square$

The relative entropy technique has to be used somewhat differently in the particle system context. For particle systems, the state space of the process will typically be $\{0, 1\}^S$ where $S = Z^d$. Any attempt to use a definition of relative entropy which corresponds exactly to Definition 4.1 is bound to fail, since the ratios $\mu(x)/\pi(x)$ which appear there would have to be replaced by the Radon–Nikodym derivative of μ with respect to π. But usually the measures μ of interest are singular with respect to the known invariant measure π. This difficulty will be resolved in Chapter IV by looking at the relative entropy of the projection of the distribution of the process on $\{0, 1\}^T$, where T is a finite subset of S. This leads to a new problem, since the projection of the process onto $\{0, 1\}^T$ will usually not be Markovian, and hence the relative entropy of the projection of the distribution of the process cannot be expected to be monotone. However, it fails to be monotone only because of terms which involve the interaction of the process inside T with the process outside T. In many cases these terms can be controlled, and the technique outlined above becomes effective. The details will be carried out in Chapter IV.

5. Reversibility

Reversibility is the probabilistic analogue of self-adjointness. As is the case in the context of operator theory, symmetry can greatly facilitate the treatment of a number of problems. In this section we will try to explain why this is the case, and how one can take advantage of reversibility in various situations. First we will introduce reversibility in the context of a Feller process on a compact state space, since it is there that the concept will be of greatest value to us. Then we will switch to the context of a Markov chain on a countable set in order to explain its usefulness. In the next section, we will develop a technique which further illustrates the power of reversibility, but which does not occur in standard treatments of the theory of Markov chains. It involves a comparison approach to the problem of determining whether a reversible Markov chain is transient or recurrent, and will be very useful in Chapter VII.

To begin, suppose that X is a compact metric space and that $\{S(t), t \geq 0\}$ is a Markov semigroup on $C(X)$. (For relevant definitions see Sections 1 and 2 of Chapter I.)

Definition 5.1. A probability measure $\mu \in \mathscr{P}$ is said to be *reversible* for the process with semigroup $S(t)$ if

$$\int fS(t)g \, d\mu = \int gS(t)f \, d\mu$$

for all $f, g \in C(X)$. The set of all reversible measures for the process will be denoted by $\mathscr{R}$.

Proposition 5.2. $\mathscr{R} \subset \mathscr{I}$.

Proof. Take $g \equiv 1$ in Definition 5.1, and refer to Proposition 1.8(a) of Chapter I. $\square$

Recall from Proposition 1.8(f) of Chapter I that in the present context $\mathscr{I}$ is necessarily nonempty. As is easy to see by looking at Markov chain examples, $\mathscr{R}$ may be empty. In fact in some sense, $\mathscr{R}$ is usually empty. Nevertheless, for many interesting processes $\mathscr{R}$ is not empty. Even when $\mathscr{R} \neq \varnothing$, it may or may not be the case that $\mathscr{R} = \mathscr{I}$.

The following result explains the term reversible, and gives a criterion for reversibility in terms of the generator Ω of $S(t)$. This criterion is similar to that given for invariance in Proposition 2.13 of Chapter I. Note that in case the process is a continuous time Markov chain on a finite-state space with generator $\Omega f(\eta) = \sum_\zeta q(\eta, \zeta)[f(\zeta) - f(\eta)]$, the conditions (b) and (c) below can be rewritten as $\mu(\eta)q(\eta, \zeta) = \mu(\zeta)q(\zeta, \eta)$ for all $\eta, \zeta \in X$.

Proposition 5.3. *Suppose that* $\{S(t), t \geq 0\}$ *is a Markov semigroup on* $C(X)$ *with generator* Ω, *and take* $\mu \in \mathscr{I}$. *Then the following statements are equivalent*:

(a) $\mu \in \mathscr{R}$;
(b) $\int f\Omega g \, d\mu = \int g\Omega f \, d\mu$ *for all* $f, g \in \mathscr{D}(\Omega)$;
(c) $\int f\Omega g \, d\mu = \int g\Omega f \, d\mu$ *for all* f, g *in a core for* Ω;

(d) $\{\eta_t, -\infty < t < \infty\}$ and $\{\eta_{-t}, -\infty < t < \infty\}$ have the same joint distributions, where η_t is the stationary process obtained by using μ as the
initial distribution and the transition mechanism corresponding to $S(t)$
as described immediately preceding Definition 1.9 of Chapter I.

Proof. The equivalence of (b) and (c) is an immediate consequence of the
definition of a core (Definition 2.11 of Chapter I). To prove (a)$\Rightarrow$(b),
suppose that $\mu \in \mathcal{R}$ and take $f, g \in \mathcal{D}(\Omega)$. Then by Definition 5.1,

$$\int f \frac{S(t)g - g}{t} \, d\mu = \int g \frac{S(t)f - f}{t} \, d\mu.$$

Since $f, g \in \mathcal{D}(\Omega)$, Theorem 2.9(a) of Chapter I gives

$$\int f\Omega g \, d\mu = \int g\Omega f \, d\mu.$$

For the converse, suppose that (b) holds. Then for $\lambda \geq 0$ and $f, g \in \mathcal{D}(\Omega)$,

$$\int f(g - \lambda\Omega g) \, d\mu = \int g(f - \lambda\Omega f) \, d\mu.$$

By replacing f and g by $(I - \lambda\Omega)^{-1}f$ and $(I - \lambda\Omega)^{-1}g$ respectively, we see
that for $f, g \in C(X)$,

$$\int [(I - \lambda\Omega)^{-1}f]g \, d\mu = \int [(I - \lambda\Omega)^{-1}g]f \, d\mu.$$

Iterating this yields

$$\int [(I - \lambda\Omega)^{-n}f]g \, d\mu = \int [(I - \lambda\Omega)^{-n}g]f \, d\mu$$

for $n \geq 1$, $\lambda \geq 0$ and $f, g \in C(X)$. Replacing λ by t/n and using Theorem
2.9(b) of Chapter I yields

$$\int gS(t)f \, d\mu = \int fS(t)g \, d\mu.$$

Since this holds for all $f, g \in C(X)$, it follows that $\mu \in \mathcal{R}$. Turning to the
proof of the equivalence of (a) and (d), suppose that (d) holds. Then for

$f, g \in C(X)$ and $t > 0$,

$$Ef(\eta_0)g(\eta_t) = Ef(\eta_0)g(\eta_{-t})$$
$$= Ef(\eta_t)g(\eta_0)$$

by (d) and the stationarity of $\{\eta_t, -\infty < t < \infty\}$. Rewriting this in terms of the semigroup yields

$$\int f(\eta)[S(t)g(\eta)]\mu(d\eta) = \int [S(t)f(\eta)]g(\eta)\mu(d\eta),$$

so that $\mu \in \mathcal{R}$. Conversely, suppose that $\mu \in \mathcal{R}$. Given $f_0 \equiv 1$, $f_1, \ldots, f_n$, $f_{n+1} \equiv 1 \in C(X)$ and $t_0 < \cdots < t_{n+1}$, define $g_0, \ldots, g_{n+1}$ and $h_0, \ldots, h_{n+1} \in C(X)$ by $g_0 = h_{n+1} = 1$,

$$g_l(\eta) = E^\eta \prod_{k=0}^{l-1} f_k(\eta_{t_l - t_k}) \quad \text{for } 1 \le l \le n+1$$

and

$$h_l(\eta) = E^\eta \prod_{k=l+1}^{n+1} f_k(\eta_{t_k - t_l}) \quad \text{for } 0 \le l \le n.$$

The main step in the proof of (d) is to show that

$$\alpha_l = \int_X g_l(\eta) f_l(\eta) h_l(\eta) \mu(d\eta)$$

is independent of l for $0 \le l \le n+1$. To see that this is enough, note that

$$\alpha_0 = \int h_0 \, d\mu = E \prod_{k=1}^{n} f_k(\eta_{t_k - t_0})$$

and

$$\alpha_{n+1} = \int g_{n+1} \, d\mu = E \prod_{k=1}^{n} f_k(\eta_{t_{n+1} - t_k}),$$

so that by the stationarity of $\{\eta_t\}$, it will follow that

$$E \prod_{k=1}^{n} f_k(\eta_{t_k}) = E \prod_{k=1}^{n} f_k(\eta_{-t_k}).$$

In order to show that α_l is independent of l, use $\mu \in \mathcal{R}$ as follows: for

$0 \leq l \leq n,$

$$\alpha_l = \int g_l f_l h_l \, d\mu$$

$$= \int g_l(\eta) f_l(\eta) S(t_{l+1} - t_l)[f_{l+1} h_{l+1}](\eta) \mu(d\eta)$$

$$= \int S(t_{l+1} - t_l)[g_l f_l](\eta) f_{l+1}(\eta) h_{l+1}(\eta) \mu(d\eta)$$

$$= \int g_{l+1} f_{l+1} h_{l+1} \, d\mu = \alpha_{l+1}. \quad \square$$

In order to explain the uses of reversibility, we turn now to a Markov chain setting. Throughout the remainder of this section, S will be a countable set and $p(x, y)$ will be the transition probabilities for a discrete time Markov chain on S. In this context it will be convenient to allow invariant measures and reversible measures to have infinite total mass. Thus we will use the following definition, which agrees exactly with our earlier ones in case $\sum_x \pi(x) = 1$. In particular, the probabilistic interpretation of (5.6) below is given by Proposition 5.3.

Definition 5.4. (a) An *invariant measure* for $p(\cdot, \cdot)$ is a nonnegative function π on S which is not identically zero and satisfies

(5.5) $\sum_x \pi(x) p(x, y) = \pi(y)$ for all $y \in S$.

(b)   A *reversible measure* for $p(\cdot . \cdot)$ is a nonnegative function π on S which is not identically zero and satisfies

(5.6) $\pi(x) p(x, y) = \pi(y) p(y, x)$ for all $x, y \in S$.

When such a reversible π exists, the chain is said to be reversible with respect to π.

As in the earlier context, (5.6) implies (5.5) as can be seen by summing (5.6) on $x \in S$. A simple example of a Markov chain which is irreducible and has both reversible measures and invariant measures which are not reversible is the asymmetric simple random walk in one dimension. Here $S = Z^1$, $p(x, x+1) = p$ and $p(x, x-1) = 1-p$ for all $x \in S$, and $p(x, y) = 0$ for $|x - y| \geq 2$. If, for example, $1 > p > \frac{1}{2}$, then all invariant measures for this chain are given by

$$\pi(x) = C_1 + C_2 \left(\frac{p}{1-p}\right)^x, \qquad x \in S,$$

where $C_1, C_2 \geq 0$. Such a π is reversible if and only if $C_1 = 0$.

A large class of chains which have reversible measures is the class of one-dimensional birth and death chains, in which $S = Z^1$ and $p(x, y) = 0$ for $|y - x| \geq 2$. In the irreducible case in which $p(x, y) > 0$ for $|x - y| = 1$, the reversible measures are given by

$$\pi(x) = \begin{cases} C \prod\limits_{y=0}^{x-1} \dfrac{p(y, y+1)}{p(y+1, y)}, & x \geq 1, \\[2ex] C, & x = 0, \\[2ex] C \prod\limits_{y=x}^{-1} \dfrac{p(y+1, y)}{p(y, y+1)}, & x \leq -1, \end{cases}$$

where C is a nonnegative constant. Again it is easy to see how the reversible measures are placed within the collection of all invariant measures. Equations (5.5) in this case become

$$\pi(y-1)p(y-1, y) + \pi(y+1)p(y+1, y) = \pi(y)[1 - p(y, y)],$$

which can be rewritten as

$$\pi(y-1)p(y-1, y) - \pi(y)p(y, y-1) = \pi(y)p(y, y+1) - \pi(y+1)p(y+1, y).$$

This says that

$$\gamma = \pi(y)p(y, y+1) - \pi(y+1)p(y+1, y)$$

is independent of y. This quantity has a natural interpretation as the net rate of flow of the chain to the right in "equilibrium." Of course the reversible measures are simply those for which $\gamma = 0$, so that there is no such net flow.

We have seen that many important chains have reversible measures. However (5.6) clearly imposes rather stringent constraints on the transition probabilities $p(x, y)$. For example, if the chain is irreducible and a reversible π is not identically zero, then it is strictly positive. Condition (5.6) then implies that $p(x, y) > 0$ if and only if $p(y, x) > 0$.

Reversible measures are more useful and are easier to deal with than invariant measures for two important reasons. First they are easy to compute explicitly, while invariant measures which are not reversible are notoriously difficult or impossible to compute. Secondly, reversible measures behave well under conditioning in the following sense. If a Markov chain is restricted to a subset of S by prohibiting transitions which would take it out of that subset, then in general the invariant measure of the restricted chain is not related in any simple way to the invariant measure of the original chain. In the reversible case, however, a reversible measure for the restricted chain is obtained simply by restricting the original measure to that subset of S. These remarks are made more precise in the following two simple results.

Proposition 5.7. *Suppose the Markov chain $p(\cdot, \cdot)$ is irreducible. Then it has a reversible measure if and only if*

(5.8)
$$\prod_{i=1}^{n} p(u_i, u_{i+1}) = \prod_{i=1}^{n} p(u_{i+1}, u_i)$$

for all $n \geq 1$ and all choices of $u_1, \ldots, u_n \in S$, with $u_{n+1} = u_1$. If this condition is satisfied, then the reversible measure is unique (up to constant multiples) and is given by the following construction: Fix a reference point $x_0 \in S$ and set $S_0 = \{x_0\}$ and

$$S_{n+1} = \left\{ x \in S \backslash \bigcup_{k=0}^{n} S_k : p(x, y) > 0 \text{ for some } y \in \bigcup_{k=0}^{n} S_k \right\}.$$

Then put $\pi(x_0) = 1$ and

$$\pi(x) = \frac{p(y, x)}{p(x, y)} \pi(y)$$

where $x \in S_{n+1}$, $y \in \bigcup_{k=0}^{n} S_k$, and $p(x, y) > 0$.

Proof. Suppose first that π is a reversible measure for the chain. Since π is nonzero and the chain is irreducible, $\pi(x)$ is positive for all $x \in S$. Therefore it suffices to check (5.8) after it has been multiplied by $\pi(u_1)$. But by (5.6),

$$\pi(u_j) \left[\prod_{i=j}^{n} p(u_i, u_{i+1}) \right] \left[\prod_{i=1}^{j-1} p(u_{i+1}, u_i) \right]$$

$$= \pi(u_{j+1}) \left[\prod_{i=j+1}^{n} p(u_i, u_{i+1}) \right] \left[\prod_{i=1}^{j} p(u_{i+1}, u_i) \right]$$

for $1 \leq j \leq n$. Therefore the quantity

$$\pi(u_j) \left[\prod_{i=j}^{n} p(u_i, u_{i+1}) \right] \left[\prod_{i=1}^{j-1} p(u_{i+1}, u_i) \right]$$

is independent of j for $1 \leq j \leq n+1$. Equating these quantities for $j = 1$ and $j = n+1$ gives (5.8). An argument similar to that just given shows that if π is reversible, then (5.6) holds for the iterates of $p(\cdot, \cdot)$ as well:

$$\pi(x) p^{(k)}(x, y) = \pi(y) p^{(k)}(y, x)$$

for all $x, y \in S$. The uniqueness statement follows from this observation and the irreducibility of the chain. To conclude the proof of the proposition, it now suffices to show that if (5.8) is satisfied, then the construction described in the statement of the proposition is well defined and yields a reversible measure. This will be done inductively. Suppose that the construction on $\bigcup_{k=0}^{n} S_k$ is well defined and that (5.6) holds on $\bigcup_{k=0}^{n} S_k$. Given $x \in S_{n+1}$ and $y, z \in \bigcup_{k=0}^{n} S_k$ so that $p(x, y) > 0$ and $p(x, z) > 0$, we need to check that

(5.9)
$$\frac{p(y, x)}{p(x, y)} \pi(y) = \frac{p(z, x)}{p(x, z)} \pi(z).$$

By the construction of the S_k's and the fact that (5.6) holds on $\bigcup_{k=0}^{n} S_k$, there are sequences $y = y_k, y_{k-1}, \ldots, y_0 = x_0$ and $z = z_l, z_{l-1}, \ldots, z_0 = x_0$ so that $p(y_i, y_{i-1}) > 0$, $p(z_i, z_{i-1}) > 0$,

$$\pi(y) = \pi(x_0) \prod_{i=1}^{k} \frac{p(y_{i-1}, y_i)}{p(y_i, y_{i-1})} \quad \text{and}$$

$$\pi(z) = \pi(x_0) \prod_{i=1}^{l} \frac{p(z_{i-1}, z_i)}{p(z_i, z_{i-1})}.$$

But then (5.9) is equivalent to

$$p(x, z) \prod_{i=1}^{k} \frac{p(y_{i-1}, y_i)}{p(y_i, y_{i-1})} p(y, x) = p(x, y) \prod_{i=1}^{l} \frac{p(z_{i-1}, z_i)}{p(z_i, z_{i-1})} p(z, x),$$

which in turn is equivalent to

$$p(x, z) \prod_{i=1}^{l} p(z_i, z_{i-1}) \prod_{i=1}^{k} p(y_{i-1}, y_i) p(y, x)$$

$$= p(x, y) \prod_{i=1}^{k} p(y_i, y_{i-1}) \prod_{i=1}^{l} p(z_{i-1}, z_i) p(z, x).$$

This is just assumption (5.8). Finally, we need to show that (5.6) holds for $x, y \in S_{n+1}$. But the proof of this is exactly the same as above. □

Remark. It should be noted that often (5.8) can be checked by checking it for small n only. For example, if $p(x, y) > 0$ for all $x, y \in S$, then it suffices to check it for $n = 3$.

Proposition 5.10. *Suppose that π is reversible for $p(\cdot, \cdot)$. Choose a $T \subset S$ and define a Markov chain $\tilde{p}(\cdot, \cdot)$ on T by*

$$\tilde{p}(x, y) = \begin{cases} p(x, y) & \text{if } x \neq y \quad \text{and} \\ p(x, y) + \sum_{u \notin T} p(x, u) & \text{if } x = y \end{cases}$$

for $x, y \in T$. *Then the restriction of* π *to* T *is reversible for* $\tilde{p}(\cdot, \cdot)$.

Proof. Since $\tilde{p}(x, y) = p(x, y)$ for $x \neq y$, $x, y \in T$,

$$\pi(x)\tilde{p}(x, y) = \pi(y)\tilde{p}(y, x).$$

On the other hand, this relation is automatic for $x = y$, no matter how $\tilde{p}(x, x)$ is defined. ☐

6. Recurrence and Transience of Reversible Markov Chains

Some important problems involving particle systems can be reduced to determining whether certain Markov chains with complicated, but countable, state spaces are transient or recurrent. In case the chain is reversible in the sense of Definition 5.4(b), this question of transience or recurrence can often be settled by comparing the given chain with another whose state space and transition mechanism are simpler, and whose recurrence properties may therefore already be known. Furthermore, the comparisons may yield good estimates on quantities such as the probability of escaping to infinity without returning to the initial state in the transient case. This section is devoted to the development of this comparison technique.

A seemingly simple problem may help to motivate the results of this section. As every reader of this book surely knows, the simple symmetric random walk in two dimensions is recurrent. Consider the simple random walk on a (connected) subgraph of Z^2, in which transitions to neighbors in the subgraph occur with equal probabilities. Must this necessarily be recurrent? It would appear that the random walk on the subgraph would have to be "more recurrent" than the original random walk on Z^2. But is this true, and how can this be made precise? The main results of this section will provide answers to these questions (see Corollary 6.9).

To begin, let $p(x, y)$ be the transition probabilities for an irreducible Markov chain X_n on the countable set S which is reversible with respect to $\pi(\cdot)$ in the sense of Definition 5.4(b). Fix a reference point $0 \in S$, and let

$$\tau = \min\{n \geq 1 : X_n = 0\}.$$

For $0 \in R \subset S$, let τ_R be the exit time

$$\tau_R = \min\{n \geq 0 : X_n \notin R\}, \quad \text{and}$$

$$\mathcal{H}_R = \{h : S \to [0, 1] : h(0) = 0 \text{ and } h(x) = 1 \text{ for } x \notin R\}.$$

Note that $h_R \in \mathcal{H}_R$, where h_R is defined by

$$h_R(x) = \begin{cases} P^x(\tau_R \leq \tau) & \text{if } x \neq 0, \\ 0 & \text{if } x = 0. \end{cases}$$

The weak rather than strict inequality is used in the definition of h_R in order to include the probability that $\tau_R = \tau = \infty$. For $h \in \mathcal{H}_R$, define

$$\Phi(h) = \sum_{x,y \in S} \pi(x)p(x, y)[h(x) - h(y)]^2.$$

The following is the key result which underlies the theory which will be developed in this section. It provides a variational expression for $P^0(\tau_R \leq \tau)$ which exhibits the important monotonicity property of this quantity. Note that part of the following theorem is an analogue for reversible Markov chains of the classical Dirichlet principle from potential theory. That principle states that among all continuously differentiable functions h on a smooth bounded domain D in Euclidean space with given boundary values, the integral

$$\int_D |\text{grad } h|^2$$

is minimized by the harmonic function with those boundary values.

Theorem 6.1. *Suppose that* $P^0(\tau_R < \infty) = 1$. *Then*

$$2\pi(0)P^0(\tau_R \leq \tau) = \Phi(h_R) = \min_{h \in \mathcal{H}_R} \Phi(h).$$

Proof. Suppose first that S is finite. To prove the first identity, put $g = 1 - h_R$ and compute

$$
\begin{aligned}
\Phi(h_R) = &\sum_{x,y} \pi(x)p(x, y)[g(x) - g(y)]^2 \\
= &\sum_{x \neq 0} \pi(x)g(x) \sum_y p(x, y)[g(x) - g(y)] \\
&+ \sum_{y \neq 0} \pi(y)g(y) \sum_x p(y, x)[g(y) - g(x)] \\
&+ 2\pi(0) \sum_y p(0, y)[1 - g(y)].
\end{aligned}
$$

(6.2)

By the Markov property,

$$h_R(x) = \sum_y p(x, y)h_R(y)$$

for $x \in R \backslash \{0\}$. Since $g(x) = 0$ for $x \notin R$, it follows that the first two sums on the right of (6.2) vanish. Therefore

$$\Phi(h_R) = 2\pi(0) \sum_y p(0, y)h_R(y) = 2\pi(0)P^0(\tau_R \leq \tau).$$

For the second identity, take partial derivatives of $\Phi(h)$ with respect to $h(x)$ for $x \in R\backslash\{0\}$ to obtain

$$\frac{\partial}{\partial h(x)}\Phi(h) = 4\pi(x)\sum_y p(x, y)[h(x) - h(y)].$$

Thus if h minimizes $\Phi(\cdot)$ on $\mathcal{H}_R$, it follows that

$$\sum_y p(x, y)[h(x) - h(y)] \geq 0 \quad \text{if } h(x) = 0,$$

$$\sum_y p(x, y)[h(x) - h(y)] = 0 \quad \text{if } 0 < h(x) < 1, \quad \text{and}$$

$$\sum_y p(x, y)[h(x) - h(y)] \leq 0 \quad \text{if } h(x) = 1.$$

In any of these cases this implies that

(6.3) $$h(x) = \sum_y p(x, y)h(y)$$

for $x \in R\backslash\{0\}$, and hence that $h = h_R$, thus proving the second identity in the statement of the theorem. Now let S be general and choose finite S_n which increase to S in such a way that $0 \in S_n$, $S_n \cap R^c \neq \varnothing$, and the Markov chain constrained to S_n (as in the statement of Proposition 5.10) is irreducible. Expressions with the subscript n will refer to the chain on S_n, which is reversible with respect to the restriction of π to S_n by Proposition 5.10. In particular

$$h_{n,R}(x) = P_n^x(\tau_{R_n} \leq \tau) \quad \text{for } x \in S_n\backslash\{0\} \quad \text{and}$$

$$\Phi_n(h) = \sum_{x,y \in S_n} \pi(x)p(x, y)[h(x) - h(y)]^2 \quad \text{where } R_n = R \cap S_n.$$

By the already proved finite version of the theorem,

(6.4) $$2\pi(0)P_n^0(\tau_{R_n} \leq \tau) = \Phi_n(h_{n,R}) = \min_{h \in \mathcal{H}_{R_n}} \Phi_n(h).$$

In order to pass to the limit in this statement, the first step is to show that for all $x \in S$,

$$P^x(\tau_R \leq \tau) = \lim_{n \to \infty} P_n^x(\tau_{R_n} \leq \tau).$$

The simplest way to see this is to couple together copies of the constrained and unconstrained chains so that they agree until the first time that the latter exits S_n. Since these times tend to ∞ as $n \to \infty$, the above statement is clear for those x for which $P^x(\tau_R < \infty) = 1$, and hence by the assumption, for all those $x \in R$ which the unconstrained chain can visit before τ_R starting from 0. But for all other $x \in S$,

$$P^x(\tau_R \leq \tau) = P^x_n(\tau_{R_n} \leq \tau) = 1$$

for all n such that $x \in S_n$. Passing to the limit in (6.4), we then see that

$$\Phi(h_R) \leq \liminf_{n \to \infty} \Phi_n(h_{n,R})$$

(6.5)
$$= \liminf_{n \to \infty} \min_{h \in \mathscr{H}_{R_n}} \Phi_n(h)$$

$$= \liminf_{n \to \infty} \min_{h \in \mathscr{H}_R} \Phi_n(h)$$

$$\leq \min_{h \in \mathscr{H}_R} \Phi(h).$$

The last equality above comes from the fact that each element of $\mathscr{H}_R$ is equal on S_n to an element of $\mathscr{H}_{R_n}$, and $\Phi_n(h)$ only involves the values of h on S_n. Since $h_R \in \mathscr{H}_R$, equality must hold throughout in (6.5). $\square$

The importance of Theorem 6.1 comes from the fact that the definition of $\mathscr{H}_R$ does not depend on the Markov chain, and that for fixed $h \in \mathscr{H}_R$, $\Phi(h)$ is an increasing function of the quantities $\pi(x)p(x, y)$. As a consequence, we have the following corollaries.

Corollary 6.6. *Suppose that $p_1(x, y)$ and $p_2(x, y)$ are irreducible Markov chains on S which are reversible with respect to $\pi_1(x)$ and $\pi_2(x)$ respectively. If $\pi_1(x)p_1(x, y) \leq \pi_2(x)p_2(x, y)$ for all $x \neq y \in S$, then*

$$\pi_1(0)P^0_1(\tau_R \leq \tau) \leq \pi_2(0)P^0_2(\tau_R \leq \tau)$$

for any R such that $0 \in R \subset S$. Here the subscripts on P^0_1 and P^0_2 refer to the two chains.

Proof. This follows from Theorem 6.1 provided that $P^0_k(\tau_R < \infty) = 1$ for $k = 1, 2$. To prove it for general $R \ni 0$, it suffices to choose finite $R_n \uparrow R$, since then $P^0_k(\tau_{R_n} < \infty) = 1$ for $k = 1, 2$ and all n. $\square$

Corollary 6.7. *Under the assumptions of Corollary 6.6, if p_1 is transient, then so is p_2. Furthermore,*

(6.8)
$$\pi_1(0)P^0_1(\tau = \infty) \leq \pi_2(0)P^0_2(\tau = \infty).$$

Proof. For the proof, it suffices to take finite R_n so that $0 \in R_n \subset S$ and $R_n \uparrow S$, and note that

$$P_k^0(\tau = \infty) = \lim_{n \to \infty} P_k^0(\tau_{R_n} \leq \tau)$$

for $k = 1, 2$, so (6.8) follows from Corollary 6.6. Since an irreducible Markov chain is transient if and only if $P^0(\tau = \infty) > 0$, the proof is complete. $\square$

Corollary 6.9. *Suppose $p(x, y) = p_2(x, y)$ is irreducible on S and reversible with respect to π. For $0 \in T \subset S$, define the Markov chain $p_1(x, y)$ constrained to T as in the statement of Proposition 5.10. If p_1 is irreducible and transient on T, then p_2 is transient also and*

$$P_1^0(\tau = \infty) \leq P_2^0(\tau = \infty).$$

Proof. For $0 < \varepsilon < 1$, define $p_\varepsilon(x, y)$ so that

$$p_\varepsilon(x, y) = p(x, y) \quad \text{if } x \neq y \text{ and either } x, y \in T \text{ or } x, y \notin T,$$

$$p_\varepsilon(x, y) = \varepsilon p(x, y) \quad \text{if } x \in T, y \notin T \text{ or } x \notin T, y \in T, \quad \text{and}$$

$$\sum_y p_\varepsilon(x, y) = 1 \qquad \text{for all } x \in S.$$

Then p_ε is irreducible on S and is reversible with respect to π. Since $p_\varepsilon(x, y) \leq p(x, y)$ for $x \neq y$, Corollary 6.6 applies with $\pi_1 = \pi_2 = \pi$, yielding

$$P_\varepsilon^0(\tau_R \leq \tau) \leq P_2^0(\tau_R \leq \tau)$$

for any finite R so that $0 \in R \subset S$. But

$$P_1^0(\tau_R \leq \tau) \leq \liminf_{\varepsilon \downarrow 0} P_\varepsilon^0(\tau_R \leq \tau),$$

so that

$$P_1^0(\tau_R \leq \tau) \leq P_2^0(\tau_R \leq \tau).$$

The result follows from this by letting R exhaust S as in the proof of Corollary 6.7. $\square$

In applications of these comparison results, one typically begins with a given Markov chain, and one wants to find two comparison chains which are more easily analyzed than is the original one. One of these should be "more recurrent" and the other "more transient" than the original chain so that one will give a lower bound and the other an upper bound for the probability of no return to 0. Usually the "more recurrent" chain is obtained from the original one by setting certain symmetrically chosen probabilities (i.e., if $p(x, y)$ is set to zero, then so is $p(y, x)$) equal to zero, and then modifying the diagonal terms $p(x, x)$ to make the new chain stochastic. To obtain a comparison chain on the "more transient" side, one should then set certain symmetrically chosen probabilities "equal to infinity." The final result in this section is intended to make sense of this comment. At the same time, the comparison chain will have the simple state space $\tilde{S} = \{0, 1, \ldots\}$, which can be a big advantage.

For the next result, suppose $p(x, y)$ is an irreducible Markov chain on S which is reversible with respect to π. Fix $0 \in S$ and write $S = \bigcup_{k=0}^{\infty} S_k$ where $S_0 = \{0\}$ and the S_k are disjoint. Assume that

$$\tilde{\pi}(k) = \sum_{x \in S_k} \pi(x) < \infty$$

for each $k \geq 0$ and that

$$P^0(\tau_{R_n} < \infty) = 1, \quad \text{where} \quad R_n = \bigcup_{k=0}^{n} S_k,$$

for each $n \geq 0$. Define a Markov chain $\tilde{p}(k, l)$ on $\tilde{S} = \{0, 1, \ldots\}$ by

$$\tilde{p}(k, l) = \frac{1}{\tilde{\pi}(k)} \sum_{\substack{x \in S_k \\ y \in S_l}} \pi(x) p(x, y).$$

Note that $\tilde{p}$ is reversible with respect to $\tilde{\pi}$ and is irreducible on $\tilde{S}$.

The idea behind the above construction is the following. If in the original Markov chain, one keeps track not of the actual position in S but only of the index k of the set S_k in which that position lies, then one usually loses the Markov property. To restore the Markov property, one modifies the process so that every time the chain visits an S_k, its position is immediately replaced by a position chosen at random from that S_k according to the probabilities $\pi(\cdot)/\tilde{\pi}(k)$. If the original chain had not been reversible, there would be little hope of finding a useful comparison between it and the new chain on $\tilde{S}$. In the reversible case, such a comparison is possible, and is the content of the following theorem.

Theorem 6.10. *Under the assumptions in the previous two paragraphs,*

$$P^0(\tau_{R_n} \leq \tau) \leq \tilde{P}^0(\tau_{\tilde{R}_n} \leq \tau)$$

for $n \geq 0$, where $R_n = \bigcup_{k=0}^{n} S_k$ and $\tilde{R}_n = \{0, \ldots, n\}$. In particular,

$$P^0(\tau = \infty) \leq \tilde{P}^0(\tau = \infty),$$

so that if $\tilde{p}(\cdot, \cdot)$ is recurrent, then so is $p(\cdot, \cdot)$.

Proof. For $\alpha > 0$ and $x, y \in S$, define

$$
p_\alpha(x, y) = \begin{cases} \dfrac{p(x, y) + \alpha \pi(y)}{1 + \alpha \tilde{\pi}(k)} & \text{if } x, y \in S_k \text{ for some } k \geq 1, \\[4mm] \dfrac{p(x, y)}{1 + \alpha \tilde{\pi}(k)} & \text{if } x \in S_k, y \notin S_k, \text{ for some } k \geq 1, \end{cases}
$$

and

$$
\pi_\alpha(x) = \pi(x)[1 + \alpha \tilde{\pi}(k)] \quad \text{for } x \in S_k, k \geq 1.
$$

Also let $\pi_\alpha(0) = \pi(0)$ and $p_\alpha(0, y) = p(0, y)$ for all $y \in S$. Then p_α is stochastic, irreducible, and reversible with respect to π_α. Furthermore,

$$
\pi_\alpha(x) p_\alpha(x, y) \geq \pi(x) p(x, y)
$$

for all $x, y \in S$. Therefore, by Corollary 6.6,

$$
P^0(\tau_{R_n} \leq \tau) \leq P^0_\alpha(\tau_{R_n} \leq \tau).
$$

So, it suffices to show that

$$(6.11) \qquad \tilde{P}^0(\tau_{\tilde{R}_n} \leq \tau) = \lim_{\alpha \to \infty} P^0_\alpha(\tau_{R_n} \leq \tau).$$

Fix n and let $h_\alpha(0) = 0$ and

$$
h_\alpha(x) = P^x_\alpha(\tau_{R_n} \leq \tau)
$$

for $x \neq 0$. Then for $x \in S_k$, $1 \leq k \leq n$,

$$
h_\alpha(x) = \sum_y p_\alpha(x, y) h_\alpha(y),
$$

which becomes

$$(6.12) \qquad [1 + \alpha \tilde{\pi}(k)] h_\alpha(x) = \sum_y p(x, y) h_\alpha(y) + \sum_{y \in S_k} \alpha \pi(y) h_\alpha(y)$$

when the definition of p_α is used. Since the last term above is independent of $x \in S_k$, (6.12) gives

$$
|h_\alpha(u) - h_\alpha(v)| \leq \frac{1}{1 + \alpha \tilde{\pi}(k)}
$$

for $u, v \in S_k$. Therefore, if $h(\cdot)$ is a limit of $h_\alpha(\cdot)$ as $\alpha \to \infty$ along any sequence, $h(\cdot)$ is constant on each S_k, say

$$h(x) = \tilde{h}(k)$$

for $x \in S_k$. Multiplying (6.12) by $\pi(x)$ and summing for $x \in S_k$ gives, after some cancellation

$$\sum_{x \in S_k} \pi(x) h_\alpha(x) = \sum_{l=1}^{\infty} \sum_{\substack{x \in S_k \\ y \in S_l}} \pi(x) p(x, y) h_\alpha(y).$$

By the Dominated Convergence Theorem, it then follows that

$$\tilde{h}(k) = \sum_{l=1}^{\infty} \tilde{h}(l) \tilde{p}(k, l),$$

so that for $1 \le k \le n$,

$$\tilde{h}(k) = \tilde{P}^k(\tau_{\tilde{R}_n} \le \tau).$$

To obtain (6.11), merely write

$$\tilde{P}^0(\tau_{\tilde{R}_n} \le \tau) = \sum_{k=1}^{\infty} \tilde{p}(0, k) \tilde{P}^k(\tau_{\tilde{R}_n} \le \tau)$$

and

$$P_\alpha^0(\tau_{R_n} \le \tau) = \sum_y p(0, y) h_\alpha(y),$$

and note that

$$\tilde{p}(0, k) = \sum_{y \in S_k} p(0, y). \quad \square$$

Corollary 6.13. *Suppose in addition to the assumption of Theorem 6.10 that* $p(x, y) = 0$ *whenever* $x \in S_k$, $y \in S_l$, *and* $|k - l| > 1$. *Then*

$$(6.14) \qquad \sum_{k=1}^{\infty} \left[\sum_{\substack{x \in S_k \\ y \in S_{k+1}}} \pi(x) p(x, y) \right]^{-1} = \infty$$

is a sufficient condition for the recurrence of $p(\cdot, \cdot)$.

Proof. Under the assumptions of the corollary, $\tilde{p}$ is a birth and death chain. Condition (6.14) is simply the necessary and sufficient condition for recurrence of that chain. (See page 32 of Hoel, Port, and Stone (1972), for example.) $\square$

While the most interesting applications of Theorem 6.10 will occur in Chapter VII, it is worthwhile to see at this point what happens when this theorem is applied to the simple random walk on Z^d. A natural choice for S_k is $S_k = \{(n_1, \ldots, n_d) \in Z^d : \max|n_i| = k\}$. Since then

$$\sum_{\substack{x \in S_k \\ y \in S_{k+1}}} \pi(x)p(x, y)$$

is of order k^{d-1} as $k \to \infty$, Corollary 6.13 implies the recurrence of the simple random walk if $d \leq 2$. When $d = 3$, a little computation shows that Theorem 6.10 yields

$$P^0(\tau = \infty) \leq \left[\sum_{k=1}^{\infty} \frac{1}{(2k-1)^2} \right]^{-1} \approx .81.$$

Of course, the correct value of $P^0(\tau = \infty)$ is $\approx .65$. (See page 360 of Feller (1968), for example.)

A final remark regarding Theorem 6.10 should be made at this point. The setup in this result involves a comparison chain on $\tilde{S} = \{0, 1, \ldots\}$. This is a natural choice in most applications. Since no assumptions are made on how $\tilde{p}(k, l)$ depends on (k, l), however, the theorem applies without change to situations in which it is more natural to think of $\tilde{S}$ as some other countable set.

7. Superpositions of Commuting Markov Chains

In this section we will prove a simple result which relates the bounded harmonic functions for the superposition of two commuting Markov chains to the bounded harmonic functions for the individual chains. This will then be used to deduce two important corollaries. The first is the Choquet–Deny Theorem, which asserts that all bounded harmonic functions for an irreducible random walk on Z^d are constant, and which will be used several times in the sequel. (Actually, the Choquet–Deny Theorem gives a similar statement in case Z^d is replaced by a general locally compact Abelian group, but we will only need the version of this result on Z^d. For another proof of the Choquet–Deny Theorem in this context, see Section 1.) The second corollary relates the bounded harmonic functions for the motion of finitely many independent particles to the bounded harmonic functions for the individual particles. This result will be used in Chapter VIII. Throughout this section, $l^\infty(S)$ will denote the Banach space of bounded functions on S with $\|f\| = \sup_{x \in S}|f(x)|$.

Proposition 7.1. *Suppose $p_1(x, y)$ and $p_2(x, y)$ are the transition probabilities for two Markov chains on the countable set S. Define transition operators P_1*

and P_2 on $l^\infty(S)$ by

$$(P_i f)(x) = \sum_{y \in S} p_i(x, y) f(y)$$

for $x \in S$ and $i = 1, 2$. Suppose that $P_1 P_2 = P_2 P_1$. Then $f \in l^\infty(S)$ satisfies

$$(\alpha_1 P_1 + \alpha_2 P_2) f = f$$

for some $\alpha_1, \alpha_2 > 0$ with $\alpha_1 + \alpha_2 = 1$ if and only if $P_1 f = f$ and $P_2 f = f$.

Proof. One implication is easy. To prove the other implication, fix α_1, α_2 and let

$$\mathscr{H} = \{f \in l^\infty(S): \|f\| \leq 1 \text{ and } (\alpha_1 P_1 + \alpha_2 P_2) f = f\}.$$

Suppose $f \in \mathscr{H}$. Then since P_1 and P_2 commute,

$$P_i f = P_i(\alpha_1 P_1 + \alpha_2 P_2) f = (\alpha_1 P_1 + \alpha_2 P_2) P_i f.$$

Hence P_1 and P_2 map $\mathscr{H}$ into itself. Since $\mathscr{H}$ is convex and compact in the topology of pointwise convergence, $\mathscr{H}$ is the closed convex hull of its extreme points by the Krein–Milman Theorem (see page 207 of Royden (1968), for example). Suppose f is an extreme point of $\mathscr{H}$. Then

$$f = \alpha_1(P_1 f) + \alpha_2(P_2 f)$$

exhibits f as a convex combination of $P_1 f$ and $P_2 f$, which are elements of $\mathscr{H}$. Thus, since $\alpha_1, \alpha_2 > 0$ and f is extreme, it follows that $P_1 f = f$ and $P_2 f = f$. Since this is true for each extreme point of $\mathscr{H}$ and $\mathscr{H}$ is the closed convex hull of its extreme points, $P_1 f = f$ and $P_2 f = f$ for all $f \in \mathscr{H}$, which concludes the proof of the proposition. $\square$

Corollary 7.2 (Choquet–Deny). *Suppose $S = Z^d$, and $p(x, y) = p(0, y - x)$ are the transition probabilities for an irreducible random walk on S. If f is a bounded harmonic function for $p(\cdot, \cdot)$, then f is constant. (See Section 1 for relevant definitions.)*

Proof. Write $S = S_1 \cup S_2$ where S_1 and S_2 are disjoint and

$$\alpha_i = \sum_{x \in S_i} p(0, x) > 0$$

for $i = 1, 2$. Define $p_i(x, y)$ for $i = 1, 2$ by

$$p_i(x, y) = \begin{cases} \dfrac{1}{\alpha_i} p(x, y) & \text{if } y - x \in S_i, \\ 0 & \text{if } y - x \notin S_i. \end{cases}$$

If P, P_1, P_2 are the corresponding transition operators, then $P = \alpha_1 P_1 + \alpha_2 P_2$ and $P_1 P_2 = P_2 P_1$, since

$$\alpha_1 \alpha_2 \sum_y p_1(x, y) p_2(y, z) = \sum_{\substack{u \in S_1 \\ v \in S_2 \\ u+v=z-x}} p(0, u) p(0, v)$$

and

$$\alpha_1 \alpha_2 \sum_y p_2(x, y) p_1(y, z) = \sum_{\substack{u \in S_2 \\ v \in S_1 \\ u+v=z-x}} p(0, u) p(0, v).$$

Now suppose f is a bounded harmonic function for P. Then by Proposition 7.1,

$$f(x) = P_i f(x) = \frac{1}{\alpha_i} \sum_{y-x \in S_i} p(x, y) f(y).$$

In particular, taking S_1 to be a singleton, we conclude that $f(x) = f(y)$ whenever $p(x, y) > 0$. By irreducibility, it follows that f is constant. $\square$

Corollary 7.3. *Suppose $p(x, y)$ are the transition probabilities for a Markov chain on the countable set S. For $f \in l^\infty(S^n)$ and $1 \le i \le n$, define $P_i f$ by*

$$P_i f(x_1, \ldots, x_n) = \sum_{y \in S} p(x_i, y) f(x_1, \ldots, x_{i-1}, y, x_{i+1}, \ldots, x_n).$$

Note that

$$P = \frac{1}{n} \sum_{i=1}^n P_i$$

is the transition operator for a chain on S^n which represents the motion of n particles on S. At each time, a particle is chosen at random with equal probabilities, and it then moves according to the probabilities $p(\cdot, \cdot)$. If $f \in l^\infty(S^n)$, then $Pf = f$ if and only if $P_i f = f$ for each $1 \le i \le n$.

Proof. Note that $P_i P_j = P_j P_i$ for $1 \le i, j \le n$. Proposition 7.1 can then be applied to the two transition operators P_i and

$$\frac{1}{n-1} \sum_{j \ne i} P_j$$

to obtain the desired conclusion. $\square$

8. Perturbations of Random Walks

It is an elementary fact that an irreducible random walk on Z^d can be transient or null recurrent, but that it cannot be positive recurrent. In particular, if $p^{(n)}(x, y)$ are the n-step transition probabilities for such a random walk, then

$$(8.1) \qquad \lim_{n \to \infty} p^{(n)}(x, y) = 0$$

for all $x, y \in Z^d$. In a number of places, but particularly in Chapter IX, certain perturbations of random walks will occur for which we will need to know properties such as (8.1). These perturbations are often obtained by modifying $p(x, y)$ for $x = 0$ or $y = 0$ or $x = y$, but not otherwise. This section is devoted to obtaining the properties of these perturbations which will be needed later.

Throughout this section, $p(x, y) = p(0, y - x)$ will be the transition probabilities for an irreducible random walk on Z^d. In the applications in Chapter IX, the random walks which arise will be symmetric, because they will correspond to differences of two independent and identically distributed random walks. However we will not assume symmetry in this section.

In both the statements of the main results and in their proofs, we will need to use the potential kernel $a(\cdot)$ defined by

$$a(x) = \sum_{n=0}^{\infty} [p^{(n)}(0, 0) - p^{(n)}(x, 0)]$$

for $x \in Z^d$. In case the random walk is transient, the above series converges trivially. In the recurrent case, the convergence of the series is a much deeper fact, which appears for example as Theorem 1 of Section 28 of Spitzer (1976). (Note in reading that result that a random walk which is irreducible in our sense is aperiodic in Spitzer's terminology.) We will need to use the following facts about $a(x)$:

$$(8.2) \qquad \sum_{y} p(x, y) a(y - z) - a(x - z) = \begin{cases} 1 & \text{if } x = z, \\ 0 & \text{if } x \neq z, \end{cases}$$

$$(8.3) \qquad a(x + y) \leq a(x) + a(y),$$

$$(8.4) \qquad \lim_{x \to \infty} [a(x + y) - a(x) + a(y - x) - a(-x)] = 0, \quad \text{and}$$

$$(8.5) \qquad \begin{aligned} a(x) &\geq 0 \quad \text{for all } x, \quad \text{and} \\ a(x) + a(-x) &> 0 \quad \text{for all } x \neq 0. \end{aligned}$$

In the recurrent case, these appear in Spitzer (1976) as Theorem 1 of Section 28, Proposition 4 and Theorem 1 of Section 29, and Proposition 2 of Section 30, respectively. In the transient case, (8.4) appears there as Proposition 3 of Section 24, while (8.2), (8.3), and (8.5) are elementary.

While the asymptotic behavior of $a(x)$ as $x \to \infty$ will not be explicitly used here, it is good to keep it in mind in order to assess the assumptions which appear in the main results. If the random walk is transient, then $a(x)$ is bounded. If $d = 1$,

$$\sum_x xp(0, x) = 0 \quad \text{and} \quad \sigma^2 = \sum_x x^2 p(0, x) < \infty,$$

then

$$\lim_{x \to +\infty} \frac{a(x) + a(-x)}{x} = \frac{2}{\sigma^2},$$

while in all other cases,

$$\lim_{|x| \to \infty} \frac{a(x)}{|x|} = 0.$$

For proofs of these results, see Sections 12 and 28 of Spitzer (1976).

An immediate consequence of (8.2) is that $\sum_y p(x, y) a(y) < \infty$ for all x. It will be convenient to use the symmetrized expression $a(x) + a(-x)$ in what follows. Thus we will assume in addition that

$$\sum_y p(x, y) a(-y) < \infty$$

for all x. This is of course automatically true if the random walk is symmetric.

Now let $q(x, y) \geq 0$ be defined for $x, y \in Z^d$. This is the perturbation of $p(x, y)$ which we wish to study. Note that $q(x, y)$ is not assumed to be stochastic (i.e., to satisfy $\sum_y q(x, y) = 1$ for all x). This is important, since the perturbations of $p(x, y)$ which will arise in Chapter IX will often not be stochastic.

Recall that an irreducible Markov chain is said to be positive recurrent if the mean return time to a state for the chain starting at that state is finite. An equivalent condition is that the chain have a finite stationary measure (see Corollary 4 of Section 2.5 of Hoel, Port, and Stone (1972), for example). Thus if $q(x, y)$ are the transition probabilities for an irreducible Markov chain, then the conclusion of the following result is just the statement that that chain is not positive recurrent.

Lemma 8.6. *Suppose that*

(8.7) $\sum_{x,y} |p(x, y) - q(x, y)|[a(y) + a(-y)] < 1.$

If

$$\pi(x) \ge 0 \quad and$$

$$\sum_x \pi(x)q(x, y) = \pi(y),$$

then either $\pi(x) = 0$ *for all* x *or* $\sum_x \pi(x) = \infty$.

Proof. Suppose $\sum_x \pi(x) < \infty$. Define ν by

(8.8)
$$\nu(y) = \sum_x \pi(x)[p(x, y) - q(x, y)]$$
$$= \sum_x \pi(x)p(x, y) - \pi(y).$$

By (8.7) and the first identity in (8.8),

(8.9)
$$\sum_y |\nu(y)|[a(y) + a(-y)] < \infty,$$

while by the second identity in (8.8),

(8.10)
$$\sum_y \nu(y) = 0.$$

Now define

(8.11)
$$\mu(y) = \sum_y \nu(x)a(x - y).$$

Then by (8.3), (8.5) and (8.10),

$$|\mu(y)| = |\sum_x \nu(x)[a(x - y) - a(-y)]|$$
$$\le \sum_x |\nu(x)|[a(x) + a(-x)],$$

which is finite by (8.9). Therefore

$$\sup_y |\mu(y)| < \infty.$$

By (8.2),

$$\sum_x \mu(x)p(x, y) - \mu(y) = \sum_x \sum_z \nu(z)a(z - x)\}p(x, y) - \sum_x \nu(x)a(x - y)$$
$$= \nu(y).$$

Therefore by (8.8),

$$\sum_x [\pi(x) - \mu(x)] p(x, y) = \pi(y) - \mu(y).$$

Since $\pi(y) - \mu(y)$ is bounded, Corollary 7.2 implies that

$$\pi(y) - \mu(y) = c$$

for all $y \in Z^d$, where c is a constant. Therefore by (8.10) and (8.11),

$$\pi(y) + \pi(-y) = \mu(y) + \mu(-y) + 2c$$
$$= \sum_x \nu(x)[a(x-y) - a(-y) + a(x+y) - a(y)] + 2c.$$

By (8.3),

$$|a(x-y) - a(-y) + a(x+y) - a(y)| \le 2[a(x) + a(-x)],$$

so that (8.4) and (8.9) imply that

$$\lim_{y \to \infty} [\pi(y) + \pi(-y)] = 2c.$$

By assumption, $\sum_x \pi(x) < \infty$, so that it follows that $c = 0$, and hence that $\pi(y) = \mu(y)$ for all $y \in Z^d$. Therefore by (8.8), (8.10), and (8.11),

$$\pi(y) = \mu(y) = \sum_x \nu(x)[a(x-y) - a(-y)]$$
$$= \sum_{x,z} \pi(z)[p(z, x) - q(z, x)][a(x-y) - a(-y)],$$

Taking absolute values inside the summation, and replacing $\pi(z)$ by its supremum yields

$$\sup_y \pi(y) \le [\sup_y \pi(y)] \sum_{x,z} |p(z, x) - q(z, x)|[a(x) + a(-x)].$$

Therefore by (8.7), $\pi(y) = 0$ for all $y \in Z^d$. □

Lemma 8.6 is a step in the right direction. Unfortunately, however, the assumption there is that the sum of the series appearing in (8.7) is less than one, rather than just being finite. In our later applications, this sum will automatically be finite, but there will be no reason for it to be less than one. In order to relax assumption (8.7), we will need to impose further conditions on $q(x, y)$. That some additional assumption is necessary should be clear. For example, if $p(x, y)$ is the simple symmetric random walk on

Z^1, then $q(x, y)$ can be taken to agree with $p(x, y)$ for all but two values of x, with the resulting chain having a finite closed set of states, and hence a finite stationary distribution.

The following simple Markov chain result will be needed for the proof of the desired improvement of Lemma 8.6. The reader should have no difficulty in finding an example to show that the corresponding result in which the transition probabilities are modified for finitely many x instead of finitely many y is not correct.

Lemma 8.12. *Suppose that $r_1(x, y)$ and $r_2(x, y)$ are the transition probabilities for two irreducible Markov chains on the countable set S, and that there is a finite subset R of S so that*

$$r_1(x, y) = r_2(x, y)$$

for all $x \in S$ and $y \notin R$. Then either both chains are positive recurrent, or neither is.

Proof. The expected hitting time of R is the same for both chains, no matter what the initial state is. Thus it suffices to show that an irreducible Markov chain X_n is positive recurrent if and only if for some finite set R, the expected hitting time of R is finite for any initial state in R. Note that the definition of positive recurrence is simply this statement when R is a singleton. So, letting τ_R and τ_x denote the hitting times of R and x respectively, we need to show that if $E^x(\tau_R) < \infty$ for all $x \in R$, then $E^x(\tau_x) < \infty$ for all $x \in R$. Let σ_n denote the successive hitting times of R:

$$\sigma_1 = \min\{k \geq 1 : X_k \in R\}, \quad \text{and}$$

$$\sigma_n = \min\{k > \sigma_{n-1} : X_k \in R\} \quad \text{for } n \geq 2.$$

For a fixed $u \in R$, let

$$N = \min\{n \geq 1 : X_{\sigma_n} = u\}.$$

Then $\tau_u = \sigma_N$. Now, X_{σ_n} is an irreducible finite state Markov chain with state space R. Therefore it is positive recurrent and hence satisfies $E^x N < \infty$ for all $x \in R$. Applying the stopping time theorem to the supermartingale

$$\sigma_n - n \max_{x \in R} E^x(\tau_R)$$

then yields

$$E^u \tau_u = E^u \sigma_N \leq \max_{x \in R} E^x(\tau_R) E^u N < \infty,$$

which completes the proof of the lemma. $\square$

In order to state the main result of this section, define $q^{(n)}(x, y)$, $p_t(x, y)$, and $q_t(x, y)$ by

$$q^{(n)}(x, y) = \sum_z q(x, z) q^{(n-1)}(z, y),$$

$$p_t(x, y) = e^{-t} \sum_{n=0}^{\infty} \frac{t^n}{n!} p^{(n)}(x, y), \quad \text{and}$$

$$q_t(x, y) = e^{-t} \sum_{n=0}^{\infty} \frac{t^n}{n!} q^{(n)}(x, y).$$

Thus $p_t(x, y)$ are the transition probabilities for the continuous time random walk corresponding to $p(x, y)$, and $q_t(x, y)$ is the analogous expression based on $q(x, y)$.

Theorem 8.13. *Suppose that*

(8.14) $q_t(x, y) > 0 \quad$ *for all* $x, y \in Z^d$,

(8.15) $\sum_{x,y} |p(x, y) - q(x, y)|[a(y) + a(-y)] < \infty,$

and that there is a positive function h on Z^d which is bounded away from zero so that

(8.16) $\sum_y q(x, y) h(y) = h(x) \quad$ *for all* $x \in Z^d$, *and*

(8.17) $\sum_{y \neq 0} p(x, y) h(y) \leq h(x) \quad$ *for all* $x \in Z^d$.

Then

$$\lim_{t \to \infty} q_t(x, y) = 0$$

for all $x, y \in Z^d$.

Before giving the proof of this theorem, some remarks about the hypotheses are in order, since they may appear somewhat strange at first glance. First note that if $q(x, y)$ is stochastic, then $h(x) \equiv 1$ satisfies (8.16) and (8.17), so that in particular the theorem says that (8.15) is a sufficient condition for an irreducible Markov chain to not be positive recurrent. An important observation is that no moment assumptions are imposed on the random walk or on the Markov chain. As mentioned earlier, $q(x, y)$ is often not stochastic in the applications we have in mind. In those cases, $h(x)$ arises as the mixed second moments $E W_0 W_x$ of a translation invariant collection of nonnegative random variables indexed by $x \in Z^d$. These are connected with $q(x, y)$ in a such a way that (8.16) is automatic. With the moment interpretation for $h(x)$, (8.17) is immediate when $x = 0$ by the Schwarz inequality. Furthermore, in many examples, $q(x, y)$ will satisfy

$$p(x, y) \leq q(x, y)$$

for all $x, y \neq 0$. Therefore (8.17) for $x \neq 0$ will follow from (8.16) via

$$\sum_{y \neq 0} p(x, y)h(y) \leq \sum_{y} q(x, y)h(y) = h(x).$$

Finally, (8.15) will be easy to verify, since in many cases we will have

$$|p(x, y) - q(x, y)| \leq c[p^{(0)}(x, y) + p(x, y) + p^{(2)}(x, y)] \quad \text{if } x \text{ or } y = 0,$$

$$|p(x, y) - q(x, y)| \leq cp(0, x) \quad\quad\quad\quad\quad\quad\quad\quad \text{if } x = y \neq 0,$$

and $p(x, y) = q(x, y)$ otherwise. When such bounds are available, (8.15) follows from (8.2). To check this, recall that in all our applications, $p(x, y)$ will be symmetric, and hence $a(x) = a(-x)$ for all $x \in Z^d$.

Proof of Theorem 8.13. The idea of the proof is to use Lemma 8.12 to make the transition from assumption (8.15) to assumption (8.7) of Lemma 8.6. Without loss of generality, take $h(0) = 1$. Given two finite subsets R and T of Z^d which contain 0, and two numbers α and β in $(0, 1)$, define $r(x, y)$ for $x, y \in Z^d$ by

$$r(x, y) = q(x, y) \quad \text{if } x \notin T \text{ or } y \notin R,$$

while if $x \in T$, then

$$r(x, y) = \alpha p(x, y) + \beta q(x, y) \quad \text{for } y \in R \setminus \{0\}$$

and

$$r(x, 0) = \beta q(x, 0) + (1 - \beta) \sum_{y \in R} q(x, y)h(y) - \alpha \sum_{y \in R \setminus \{0\}} p(x, y)h(y).$$

Then a simple computation shows that (8.16) implies

(8.18) $$\sum_{y} r(x, y)h(y) = h(x) \quad \text{for all } x \in Z^d.$$

In fact, it was to guarantee (8.18) that the specific expression for $r(x, 0)$

above was chosen. Now, since $a(0) = 0$,

$$\sum_{x,y} |r(x, y) - p(x, y)|[a(y) + a(-y)]$$

$$\leq \sum_{\substack{x \notin T \\ \text{or } y \notin R}} |q(x, y) - p(x, y)|[a(y) + a(-y)]$$

$$+ \sum_{\substack{x \in T \\ y \in R}} [(1-\alpha)p(x, y) + \beta q(x, y)][a(y) + a(-y)].$$

(8.19)

$$\leq \sum_{\substack{x \notin T \\ \text{or } y \notin R}} |q(x, y) - p(x, y)|[a(y) + a(-y)]$$

$$+ (1-\alpha) \sum_{\substack{x \in T \\ y \in Z^d}} p(x, y)[a(y) + a(-y)]$$

$$+ \beta \sum_{\substack{x \in T \\ y \in Z^d}} q(x, y)[a(y) + a(-y)].$$

Note that the second term in the last expression is finite by assumption, while the third term is finite by this together with (8.15). Now using (8.15), first choose R and T, and then α and β so that the right side of (8.19) is less than one and $\alpha + \beta < 1$. Then, using $\alpha + \beta < 1$, (8.16), and (8.17), choose R even larger so that for all $x \in T$,

$$(1-\beta) \sum_{y \in R} q(x, y)h(y) \geq \alpha \sum_{y \in R \setminus \{0\}} p(x, y)h(y),$$

which implies that

$$r(x, 0) \geq \beta q(x, 0) \geq 0.$$

By (8.14), (8.16), (8.18), and the fact that $r(x, y) \geq \beta q(x, y)$ for all $x, y \in Z^d$,

$$\frac{q(x, y)h(y)}{h(x)} \quad \text{and} \quad \frac{r(x, y)h(y)}{h(x)}$$

are the transition probabilities for irreducible Markov chains on Z^d which satisfy the assumption of Lemma 8.12. Therefore by that result, either both are positive recurrent or neither is. Since

$$\frac{q_t(x, y)h(y)}{h(x)}$$

gives the transition probabilities for the continuous time chain corresponding to the first of these, to prove the theorem it suffices to show that the chain with transition probabilities $r(x, y)h(y)/h(x)$ is not positive recurrent. So, suppose $\pi(x) > 0$ satisfies

$$\sum_x \pi(x) \frac{r(x, y)h(y)}{h(x)} = \pi(y).$$

Then

$$\sum_x \frac{\pi(x)}{h(x)} r(x, y) = \frac{\pi(y)}{h(y)},$$

so that $\sum_y \pi(y)/h(y) = \infty$ by (8.19) and Lemma 8.6. Since h is bounded away from zero, it follows that $\sum_x \pi(x) = \infty$, which is the desired result. $\quad\square$

The primary use of Theorem 8.13 will be to obtain the limit as $t \to \infty$ of $\sum_y q_t(x, y)f(y)$, where f is a positive definite function on Z^d. This positive definite function arises as the covariance or product moment of order two of a stationary sequence of positive random variables indexed by Z^d. Let Γ be the d-dimensional torus $[-\pi, \pi)^d$, which is to be thought of as the group dual to Z^d. Let

$$\langle x, \gamma \rangle = \sum_{k=1}^d x_k \gamma_k$$

be the natural pairing for $x = (x_1 \ldots x_d) \in Z^d$ and $\gamma = (\gamma_1, \ldots, \gamma_d) \in \Gamma$. We then have the following result.

Corollary 8.20. *In addition to the hypotheses of Theorem 8.13, assume that the function h satisfies*

(8.21)

$$\lim_{x \to \infty} h(x) = 1, \quad and \ that$$

$$\lim_{x \to \infty} \sum_y |p(x, y) - q(x, y)| = 0.$$

Suppose that v is a finite measure on Γ, and define

$$f(x) = \int_\Gamma \exp[i\langle x, \gamma \rangle] v(d\gamma).$$

Then

(8.22)

$$\lim_{t \to \infty} \sum_y q_t(x, y)f(y) = v\{0\}h(x).$$

Proof. Since h is bounded away from 0 and ∞ and since by (8.16),

$$\sum_y q_t(x, y)h(y) = h(x),$$

it follows that

(8.23) $$\sup_{x,t} \sum_y q_t(x, y) < \infty.$$

Therefore by the Dominated Convergence Theorem, in order to prove (8.22), it suffices to prove

(8.24) $$\lim_{t \to \infty} \sum_y q_t(x, y) = h(x) \quad \text{and}$$

(8.25) $$\lim_{t \to \infty} \sum_y q_t(x, y) \exp[i\langle y, \gamma \rangle] = 0 \quad \text{for } \gamma \neq 0.$$

For (8.24), write

$$\sum_y q_t(x, y) - h(x) = \sum_y q_t(x, y)[1 - h(y)],$$

which tends to zero as $t \to \infty$ by (8.21), (8.23), and Theorem 8.13. To prove (8.25), let $g(x) = \exp[i\langle x, \gamma \rangle]$ for some fixed $\gamma \neq 0$ in Γ, and let

$$w = \sum_x p(0, x)g(x).$$

Note that $|w| \leq 1$ and $\operatorname{Re} w < 1$. Then

$$\sum_y p_t(x, y)g(y) = e^{-t} \sum_{n=0}^{\infty} \sum_y \frac{p^{(n)}(x, y)t^n g(y)}{n!}$$

$$= e^{-t} \sum_{n=0}^{\infty} \frac{t^n}{n!} g(x) \sum_y p^{(n)}(0, y)g(y)$$

(8.26)

$$= e^{-t} g(x) \sum_{n=0}^{\infty} \frac{t^n}{n!} w^n$$

$$= g(x) \exp[-t(1 - w)].$$

The next step is to write

$$p_t(x, y) - q_t(x, y) = \int_0^t \frac{d}{ds}\left[\sum_u q_{t-s}(x, u)p_s(u, y) \right] ds$$

(8.27)

$$= \int_0^t \left[\sum_{u, v} q_{t-s}(x, u)[p(u, v) - q(u, v)]p_s(v, y) \right] ds.$$

Put

$$\varepsilon(u) = \sum_v |p(u, v) - q(u, v)|,$$

which tends to zero as $u \to \infty$ by hypothesis. Multiplying (8.27) by $g(y)$, summing on y, and using (8.26) yields

(8.28)
$$\left| \sum_y p_t(x, y)g(y) - \sum_y q_t(x, y)g(y) \right|$$
$$\leq \int_0^t \left[\sum_u q_{t-s}(x, u)\varepsilon(u) \right] \exp[-s(1 - \operatorname{Re} w)] \, ds.$$

The right-hand side of (8.28) tends to zero as $t \to \infty$, as is seen by using (8.23), Theorem 8.13, and the previously noted facts that $\lim_{u \to \infty} \varepsilon(u) = 0$ and $1 - \operatorname{Re} w > 0$. Using this last fact and (8.26), it follows that

(8.29)
$$\lim_{t \to \infty} \sum_y p_t(x, y)g(y) = 0.$$

Combining (8.29) with the fact that the expression in (8.28) tends to zero, we conclude that (8.25) holds, as required. □

For later applications, note that if $\{\eta(x), x \in Z^d\}$ is a stationary and ergodic collection of random variables with finite second moments and $f(x) = E\eta(x)\eta(0)$, then $\nu\{0\} = [E\eta(0)]^2$. This follows easily from Proposition 4.11 of Chapter I.

9. Notes and References

Section 1. The use of coupling techniques is usually traced back to Doeblin (1938), who gave the present proof of Theorem 1.2. Proofs along these lines have by now appeared in textbooks such as Hoel, Port, and Stone (1972), Ross (1983), and Billingsley (1979). For a rather comprehensive treatment of coupling in the theory of Markov processes, as well as for a large number of additional references, see Griffeath (1978b). Coupling techniques have been used in other parts of probability, such as renewal theory. For examples of this, see Lindvall (1977), Athreya, McDonald, and Ney (1978), and Ney (1981). Coupling has also been used extensively in the analysis of certain classes of learning models; see Kaijser (1981) and his references.

Section 2. Processes which are monotone in the sense of Definition 2.3 have arisen in a number of contexts. See Daley (1968) and his references for early examples of this. More recent references are Van Doorn (1980), Van Doorn (1981), and Aldous (1983).

The proof of Theorem 2.4 is based on the proof of a similar result in Strassen (1965). In fact Theorem 2.4 is a special case of his Theorem 11. This result also appears in a somewhat different form as Theorem 8 in the Appendix of Nachbin (1965), which is based on an earlier paper by the same author. A different proof in the case of finite X, which is based on the min–cut, max–flow theorem of graph theory can be found in Preston (1974a). For further generalizations of this result, see Chapter XI of Meyer (1966), Kamae, Krengel, and O'Brien (1977), Edwards (1978a), and Shortt (1983).

Corollary 2.12 is known as the FKG inequality, since it was first proved by Fortuin, Kasteleyn, and Ginibre (1971). Theorem 2.9 and its proof are due to Holley (1974b). This result also appears in Preston (1974b). Other references for this and related results are Preston (1974a), Batty (1976), Edwards (1978b), Thomas (1980), Batty and Bollmann (1980), and Karlin and Rinott (1980).

Theorem 2.14 and Corollary 2.21 are due to Harris (1977). The proof given here of the former is new. A similar proof has been given by Cox (1984).

Section 3. Theorem 3.9 is due to Siegmund (1976). That paper has additional references to the use of duality in the contexts of random walks and queueing theory. Clifford and Sudbury (1985) have given a sample path approach to this result. For an early paper in which duality was used in studying birth and death chains, see Karlin and McGregor (1957). A more recent paper on that topic is Van Doorn (1980).

Section 4. The first use of relative entropy techniques in the study of interacting particle systems of the type discussed in this book occurred in Holley (1971) and Holley (1972c). Subsequently it has been used in Higuchi and Shiga (1975), Moulin Ollagnier and Pinchon (1977), Holley and Stroock (1977a), and Liggett (1983a). For finite particle systems, or in a Markov chain context, the monotonicity of the relative entropy has been observed and used often. Examples of this can be found on page 98 of Kac (1959), and page 18 of Kelly (1979). The alternate proof of Theorem 1.2 given in this section can also be found in Section 9 of Chapter IX of Renyi (1970) and in Chapter I of Spitzer (1971b). Renyi credits Linnik (1959) for the idea of using relative entropy techniques to prove limit theorems in probability theory.

Section 5. Reversibility has been discussed in a number of places. Three examples are Sections 4.7 and 5.6 of Ross (1983), Chapter 2 of Spitzer (1971b) and Section 9.10 of Kemeny, Snell, and Knapp (1976). More extensive treatments of reversibility in the theory of Markov chains are given in Keilson (1979) and Kelly (1979). Propositions 5.7 and 5.10 appear in Kelly (1979) as Theorem 1.8 and Corollary 1.10 respectively. The former result is referred to there as Kolmogorov's criterion.

Section 6. The first general recurrence criterion for a reversible Markov chain is due to Nash-Williams (1959). The sufficiency part of his theorem is that given in Corollary 6.13. The treatment given here is based on Griffeath and Liggett (1982). (Note that there is a small error in the statement and proof of Theorem 2.1 there which is corrected here by defining $h_R(x)$ with a $\leq$ instead of a $<$.) For other criteria for transience and recurrence of a reversible Markov chain, see Lyons (1983) and Varopoulos (1984). A detailed exposition of this general topic is given in Doyle and Snell (1985). A similar approach to recurrence problems for reversible diffusions is carried out in Ichihara (1978). The problem mentioned at the beginning of this section involving the recurrence of simple random walks on subgraphs of Z^2 occurs on page 425 of Feller (1968).

Section 7. Corollary 7.3 comes from Lemma 3.14 of Liggett (1973a). The proof of Proposition 7.1 is based on the proof of that lemma. For a generalization of Proposition 7.1 with a different proof, see Lemma 1.1 of Chapter 5 of Revuz (1975). The motivation there as it is partially here is to prove the Choquet–Deny Theorem. For the original paper containing that theorem, see Choquet and Deny (1960). Versions of it appear in Chapter VIII of Meyer (1966) and Chapter 5 of Revuz (1975).

Section 8. Theorem 8.13 and Corollary 8.20 are new. For other criteria for the preservation of positive recurrence under perturbations, see Tweedie (1975), Tweedie (1980), Cocozza and Roussignol (1983), and Cocozza, Kipnis, and Roussignol (1983). Of these, the latter paper is the one whose criterion is most similar to that of Theorem 8.13. However, they require that $d = 1$, that the random walk have a finite second moment, and that $q(x, y)$ be stochastic. Even then, their sufficient condition is not the same as that of Theorem 8.13. Lemma 8.12 is a special case of a theorem in Tweedie (1975).

The problem of stability of recurrence, rather than positive recurrence, under perturbations, has also been considered by a number of authors. For examples, see Filonov (1980) and the above-mentioned papers.

Chapter III

Spin Systems

A spin system is an interacting particle system in which each coordinate has two possible values, and only one coordinate changes in each transition. Throughout this chapter, the state space of the system will be taken to be $X = \{0, 1\}^S$ where S is a finite or countable set. There are numerous possible interpretations of the two possible values 0 and 1. The next three chapters deal with three classes of spin systems in which the transition mechanisms have a particular form, and each of these classes corresponds to a different interpretation for 0 and 1. In the stochastic Ising model, they represent the two possible spins of an iron atom (for example). In the case of the voter model, they denote two possible positions of a "voter" on some political issue. In the case of the contact process, 0 and 1 represent healthy and infected individuals respectively.

The transition mechanism is specified by a nonnegative function $c(x, \eta)$ defined for $x \in S$ and $\eta \in X = \{0, 1\}^S$. It represents the rate at which the coordinate $\eta(x)$ flips from 0 to 1 or from 1 to 0 when the system is in state η. Thus the process η_t with state space X will satisfy

(0.1) $$P^\eta[\eta_t(x) \neq \eta(x)] = c(x, \eta)t + o(t)$$

as $t \downarrow 0$ for each $x \in S$ and $\eta \in X$. The requirement that only one coordinate change in each transition can be described by saying that

(0.2) $$P^\eta[\eta_t(x) \neq \eta(x), \eta_t(y) \neq \eta(y)] = o(t)$$

as $t \downarrow 0$ for each $x, y \in S$ with $x \neq y$ and each $\eta \in X$. The interaction among sites comes from the dependence of $c(x, \eta)$ on η.

Of course, the preceding description of the process is rather informal. For a formal description and construction of the process, we need to appeal to the results of Chapter I. Specifically we will assume throughout this chapter that $c(x, \eta)$ is a uniformly bounded nonnegative function which is continuous in η for each x, and which in addition satisfies the condition

(0.3) $$\sup_{x \in S} \sum_{u \in S} \sup_{\eta \in X} |c(x, \eta) - c(x, \eta_u)| < \infty,$$

where $\eta_u \in X$ is defined by

$$(0.4) \qquad \eta_u(v) = \begin{cases} \eta(v) & \text{if } v \neq u, \\ 1 - \eta(v) & \text{if } v = u. \end{cases}$$

Assumption (0.3) puts limits on the amount of dependence which the flip rate at one site can have on the rest of the configuration. With these assumptions, Theorem 3.9 of Chapter I yields the following statement: The closure in $C(X)$ of the operator Ω which is defined on $D(X)$ by

$$(0.5) \qquad \Omega f(\eta) = \sum_x c(x, \eta)[f(\eta_x) - f(\eta)]$$

is the generator of a Markov semigroup $S(t)$. By Theorem 1.5 of Chapter I, there is then a unique Markov process η_t on X corresponding to $S(t)$. This process is the spin system determined by the flip rates $c(x, \eta)$. By applying (0.5) to arbitrary functions f which depend on one or two coordinates and using Theorem 2.9 of Chapter I, it is easy to check that the spin system satisfies (0.1) and (0.2). Conversely, this is the only Feller process on X which satisfies (0.1) and (0.2).

The results of Section 4 of Chapter I are also available to us at this point. The most important one to keep in mind is Theorem 4.1 from that chapter. Phrased in the present context, it asserts that a sufficient condition for the spin system with rates $c(x, \eta)$ to be ergodic is that

$$(0.6) \qquad \sup_{\substack{x \in S}} \sum_{u \neq x} \sup_{\eta \in X} |c(x, \eta) - c(x, \eta_u)| < \inf_{\substack{x \in S \\ \eta \in X}} [c(x, \eta) + c(x, \eta_x)].$$

This chapter is intended as an introduction to spin systems, and particularly to the use of coupling and duality techniques in this context. Certain important special classes of spin systems in which the rates $c(x, \eta)$ take a particular form are discussed in greater detail in the next three chapters. Some of the most important couplings are described in the first section. It is here, for example, that we will see that if two spin systems are related in certain ways, then the ergodicity of one implies the ergodicity of the other. The general theory of attractive spin systems is the subject of Section 2. ("Attractive" is the name given to particle systems which are monotone in the sense of Definition 2.3 of Chapter II.) It is in the case of attractive systems that coupling and monotonicity arguments are most effective. In Section 3, more complete results are given on the possible limiting behavior of attractive spin systems in the special case that $S = Z^1$, the set of integers, $c(x, \eta)$ depends on η only through $\eta(x)$, $\eta(x-1)$, and $\eta(x+1)$, and does so in a spatially homogeneous way. Section 4 contains a duality theory for spin systems. Applications of duality, particularly to questions of ergodicity, will be presented in Section 5. Section 6 deals with the class of additive spin systems, and contains a description of the graphical representation, which can be used in their study. Additive processes include many of the more important models, such as the voter model and the contact process. This section plays a unifying role for this chapter, since there are close connections between the graphical representation for additive spin systems on the one hand, and both coupling and duality on the other. In particular, every additive spin system is attractive.

While this chapter deals exclusively with spin systems, it should be clear than many of the ideas and techniques presented here carry over to more general interacting particle systems. It would be a good idea to read Sections 1 and 2 of Chapter II before reading the first three sections of this chapter, and Section 3 of Chapter II before reading the next three sections of this chapter.

The six sections depend on one another in the following way:

$$1 \to 2 \to 3 \searrow 6$$
$$4 \to 5 \nearrow$$

1. Couplings for Spin Systems

Suppose that $c_1(x, \eta)$ and $c_2(x, \eta)$ are uniformly bounded nonnegative flip rates wich satisfy (0.3). We will begin by introducing a coupling of the corresponding spin systems η_t and ζ_t which is designed to make them agree as much as possible. This coupling is known as the basic coupling, or the Vasershtein coupling. An essentially different coupling, which is useful for certain spin systems, is described in Section 6. An informal description of the basic coupling is as follows: If $\eta_t(x) \neq \zeta_t(x)$ for some $t \geq 0$ and $x \in S$, let the two processes flip at that site independently according to their respective flip rates. If $\eta_t(x) = \zeta_t(x)$, then let the x coordinate of the two processes flip together with as large a rate as possible, consistent with the requirement that each process flip at the correct rate. This rate at which equal coordinates will flip together is of course the minimum of the rates at which they would flip in the respective marginal processes. Therefore it is natural to define

$$c(x, \eta, \zeta) = \min\{c_1(x, \eta), c_2(x, \zeta)\}$$

whenever $\eta(x) = \zeta(x)$.

We will now give a detailed formal definition of the coupled process based on this basic coupling. When other couplings are used later in this book, only the informal description will be given, and it will be understood that a similar formal construction lies in the background. The coupled process (η_t, ζ_t) will be a Feller process with state space $X \times X$. Since $X \times X$

can be identified with

$$\left\{\begin{pmatrix}0\\0\end{pmatrix},\begin{pmatrix}1\\0\end{pmatrix},\begin{pmatrix}0\\1\end{pmatrix},\begin{pmatrix}1\\1\end{pmatrix}\right\}^S,$$

the coupled process fits within the framework of Section 3 of Chapter I. So, define $\tilde{\Omega}$ on $D(X \times X)$ by

$$\tilde{\Omega}f(\eta, \zeta) = \sum_{x:\eta(x)\neq\zeta(x)} c_1(x, \eta)[f(\eta_x, \zeta) - f(\eta, \zeta)]$$

$$+ \sum_{x:\eta(x)\neq\zeta(x)} c_2(x, \zeta)[f(\eta, \zeta_x) - f(\eta, \zeta)]$$

$$+ \sum_{x:\eta(x)=\zeta(x)} c(x, \eta, \zeta)[f(\eta_x, \zeta_x) - f(\eta, \zeta)]$$

$$+ \sum_{x:\eta(x)=\zeta(x)} [c_1(x, \eta) - c(x, \eta, \zeta)][f(\eta_x, \zeta) - f(\eta, \zeta)]$$

$$+ \sum_{x:\eta(x)=\zeta(x)} [c_2(x, \zeta) - c(x, \eta, \zeta)][f(\eta, \zeta_x) - f(\eta, \zeta)].$$

While the above expression for $\tilde{\Omega}f$ clearly exhibits the various possible transitions and the corresponding rates, it is more convenient to rewrite it in the following form which is simpler, but gives less insight:

$$\tilde{\Omega}f(\eta, \zeta) = \sum_x c_1(x, \eta)[f(\eta_x, \zeta) - f(\eta, \zeta)]$$

(1.1)
$$+ \sum_x c_2(x, \zeta)[f(\eta, \zeta_x) - f(\eta, \zeta)]$$

$$+ \sum_{x:\eta(x)=\zeta(x)} c(x, \eta, \zeta)$$

$$\times [f(\eta_x, \zeta_x) - f(\eta_x, \zeta) - f(\eta, \zeta_x) + f(\eta, \zeta)].$$

Lemma 1.2. *Suppose* $g \in D(X)$.

   (a)   *If* $f(\eta, \zeta) = g(\eta)$, *then* $\tilde{\Omega}f(\eta, \zeta) = \Omega_1 g(\eta)$, *and*
   (b)   *if* $f(\eta, \zeta) = g(\zeta)$, *then* $\tilde{\Omega}f(\eta, \zeta) = \Omega_2 g(\zeta)$,

where Ω_1 *and* Ω_2 *are the generators corresponding to the rates* c_1 *and* c_2 *respectively.*

Proof. If $f(\eta, \zeta) = g(\eta)$, then the last two terms on the right of (1.1) vanish, so

$$\tilde{\Omega}f(\eta, \zeta) = \sum_x c_1(x, \eta)[g(\eta_x) - g(\eta)] = \Omega_1 g(\eta).$$

This proves part (a), and part (b) is similar.   □

Theorem 1.3. *The closure of $\tilde{\Omega}$ in $C(X \times X)$ is the generator of a Markov semigroup $\tilde{S}(t)$ on $C(X \times X)$. Given $g \in C(X)$, if $f(\eta, \zeta) = g(\eta)$, then $\tilde{S}(t)f(\eta, \zeta) = S_1(t)g(\eta)$, while if $f(\eta, \zeta) = g(\zeta)$, then $\tilde{S}(t)f(\eta, \zeta) = S_2(t)g(\zeta)$, where $S_i(t)$ is the semigroup with generator Ω_i. In particular, if (η_t, ζ_t) is the Feller process with semigroup $\tilde{S}(t)$, then η_t and ζ_t are separately Markovian with semigroups $S_1(t)$ and $S_2(t)$ respectively.*

Proof. The first statement follows from Theorem 3.9 of Chapter I, once the hypotheses of that theorem have been verified. Using the notation from that section, $c_T = 0$ and $c_T(u) = 0$ unless $|T| = 1$. If $T = \{x\}$, then

$$c_T \leq \sup_\eta c_1(x, \eta) + \sup_\zeta c_2(x, \zeta), \quad \text{and}$$

$$c_T(u) \leq 3[\sup_\eta |c_1(x, \eta) - c_1(x, \eta_u)| + \sup_\zeta |c_2(x, \zeta) - c_2(x, \zeta_u)|],$$

so that the assumptions of Theorem 3.9 of Chapter I applied to $\tilde{\Omega}$ follow from the basic assumptions made about $c_1(x, \eta)$ and $c_2(x, \zeta)$ in this section. To prove the second statement, we will agree to identify in the natural way $C(X)$ with the collection $\mathscr{C}_1$ of all functions in $C(X \times X)$ which depend only on the first coordinate. For $f \in C(X)$, set

$$h = (I - \lambda\Omega_1)^{-1}f \in \mathscr{D}(\Omega_1).$$

Then

$$h - \lambda\Omega_1 h = f,$$

so by Lemma 1.2,

$$h - \lambda\tilde{\Omega}h = f.$$

Therefore,

(1.4) $$(I - \lambda\Omega_1)^{-1}f = (I - \lambda\tilde{\Omega})^{-1}f$$

for all functions $f \in \mathscr{C}_1$ and by looking at the left side of (1.4), $(I - \lambda\Omega_1)^{-1}f$ is again in $\mathscr{C}_1$. Iterating this yields

$$\left(I - \frac{t}{n}\Omega_1\right)^{-n}f = \left(I - \frac{t}{n}\tilde{\Omega}\right)^{-n}f$$

for all $t \geq 0$, $n \geq 1$, and $f \in \mathscr{C}_1$. Passing to the limit as $n \to \infty$ and using part (b) of Theorem 2.9 of Chapter I, we see that

$$S_1(t)f = \tilde{S}(t)f$$

for all $f \in \mathscr{C}_1$. The corresponding argument for $S_2(t)$ is identical, so the proof is complete. $\square$

Let

$$K = \{(\eta, \zeta) \in X \times X : \eta \leq \zeta\}.$$

This is a closed subset of $X \times X$, and it is natural to ask under what conditions K is closed for the evolution of the coupled process, in the sense that $(\eta, \zeta) \in K$ implies that

$$P^{(\eta,\zeta)}[(\eta_t, \zeta_t) \in K] = 1$$

for all $t \geq 0$. Since K is a closed subset of $X \times X$ and the paths of the process are right continuous, this is equivalent to the statement that $(\eta, \zeta) \in K$ implies that

$$P^{(\eta,\zeta)}[(\eta_t, \zeta_t) \in K \text{ for all } t \geq 0] = 1.$$

An answer to this question is given by the following result.

Theorem 1.5. *Suppose that whenever* $\eta \leq \zeta$,

(1.6)
$$c_1(x, \eta) \leq c_2(x, \zeta) \quad \text{if } \eta(x) = \zeta(x) = 0, \quad \text{and}$$
$$c_1(x, \eta) \geq c_2(x, \zeta) \quad \text{if } \eta(x) = \zeta(x) = 1.$$

Then for all $(\eta, \zeta) \in K$ *and* $t \geq 0$,

$$P^{(\eta,\zeta)}[(\eta_t, \zeta_t) \in K] = 1.$$

Proof. Let $\mathscr{A}$ be the set of all functions f in $C(X \times X)$ such that $f \geq 0$ and $f = 0$ on K. For $\lambda > 0$ and $f \in \mathscr{A}$, define $h \in \mathscr{D}(\tilde{\Omega})$ by

$$h - \lambda \tilde{\Omega} h = f.$$

Let $(\eta, \zeta) \in K$ be a point where h achieves its maximum on K. We will show that $\tilde{\Omega} h(\eta, \zeta) \leq 0$, so that $h(\eta, \zeta) \leq f(\eta, \zeta) = 0$. Since $h \geq 0$, it will then follow that $h \in \mathscr{A}$ as well. To show $\tilde{\Omega} h(\eta, \zeta) \leq 0$, we will check that each of the terms in the sum defining $\tilde{\Omega} h$ is nonpositive. Consider the following three cases:

(a) $\eta(x) \neq \zeta(x)$, in which case $(\eta, \zeta) \in K$ implies that $(\eta_x, \zeta) \in K$ and $(\eta, \zeta_x) \in K$, so that $h(\eta_x, \zeta) \leq h(\eta, \zeta)$ and $h(\eta, \zeta_x) \leq h(\eta, \zeta)$.

(b) $\eta(x) = \zeta(x) = 0$, in which case $(\eta, \zeta) \in K$ implies that $(\eta_x, \zeta_x) \in K$ and $(\eta, \zeta_x) \in K$, so that $h(\eta_x, \zeta_x) \leq h(\eta, \zeta)$ and $h(\eta, \zeta_x) \leq h(\eta, \zeta)$.

However, in this case $(\eta_x, \zeta) \notin K$, so we need to have $c_1(x, \eta) = c(x, \eta, \zeta)$, which is the first part of assumption (1.6).

(c) $\eta(x) = \zeta(x) = 1$, which is treated like case (b), using the second part of assumption (1.6).

We now know that $(I - \lambda\tilde{\Omega})^{-1}$ maps $\mathscr{A}$ to itself. It follows then that $\tilde{S}(t)$ maps $\mathscr{A}$ to itself as required (see Theorem 2.9(b) of Chapter I). □

Corollary 1.7. *Under the assumptions of Theorem* 1.5, *if* μ_1 *and* μ_2 *are probability measures on* X *which satisfy* $\mu_1 \leq \mu_2$, *then*

$$\mu_1 S_1(t) \leq \mu_2 S_2(t)$$

for all $t \geq 0$.

Proof. Let ν be a coupling measure for μ_1 and μ_2 as in the statement of Theorem 2.4 of Chapter II. Let (η_t, ζ_t) be the coupled process constructed in Theorem 1.3, with initial distribution ν. By Theorem 1.5, $\eta_t \leq \zeta_t$ with probability one for each $t \geq 0$. By Theorem 1.3, η_t and ζ_t have distributions $\mu_1 S_1(t)$ and $\mu_2 S_2(t)$ respectively. Therefore the joint distribution of (η_t, ζ_t) provides a coupling measure for $\mu_1 S_1(t)$ and $\mu_2 S_2(t)$, thus showing that

$$\mu_1 S_1(t) \leq \mu_2 S_2(t). \quad □$$

Corollary 1.8. *Under the assumptions of Theorem* 1.5, *if the process with semigroup* $S_2(t)$ *is ergodic with invariant measure the pointmass at* $\eta \equiv 0$, *then the same is true for the process with semigroup* $S_1(t)$.

Proof. By Corollary 1.7, if μ is any probability measure on X, then

$$\mu S_1(t) \leq \mu S_2(t)$$

for all $t \geq 0$. By hypothesis, the limit of $\mu S_2(t)$ as $t \to \infty$ is the pointmass at $\eta \equiv 0$. Therefore the same is true for the limit of $\mu S_1(t)$. □

Up to this point, we have coupled together only two processes at a time. We proceed now to discuss a situation in which it is useful to couple together three processes at once. An additional illustration of this will occur in Section 3. To begin, let $c(x, \eta)$ be the flip rates for a spin system which satisfies the basic smoothness and boundedness assumptions. Define new flip rates by

$$\bar{c}(x, \xi) = \sup\{|c(x, \eta) - c(x, \zeta)|: |\eta(u) - \zeta(u)| \leq \xi(u) \text{ for all } u \in S\}$$

if $\xi(x) = 0$, and

$$\bar{c}(x, \xi) = \inf\{[c(x, \eta) + c(x, \zeta)]: \eta(x) \neq \zeta(x)$$
$$\text{and } |\eta(u) - \zeta(u)| \leq \xi(u) \text{ for all } u \in S\}$$

if $\xi(x) = 1$. Then $\bar{c}(x, \xi)$ is continuous in ξ and satisfies

$$\sup_{x, \xi} \bar{c}(x, \xi) \leq 2 \sup_{x, \eta} c(x, \eta), \quad \text{and}$$

$$|\bar{c}(x, \xi_u) - \bar{c}(x, \xi)| \leq 2 \sup_{\eta} |c(x, \eta_u) - c(x, \eta)|$$

for $u \neq x$, so $\bar{c}(x, \xi)$ is uniformly bounded and satisfies (0.3) as well. Note that $\bar{c}(x, \xi)$ has some properties not necessarily shared by the original rates which can be useful in analyzing the corresponding spin system:

(a) $\bar{c}(x, \xi) = 0$ if $\xi \equiv 0$, which makes $\xi \equiv 0$ an absorbing point for the process, and
(b) $\bar{c}(x, \xi)$ is increasing in ξ if $\xi(x) = 0$ and decreasing in ξ if $\xi(x) = 1$, which makes the process attractive in the sense of the next section.

Theorem 1.9. *If the process ξ_t corresponding to $\bar{c}(x, \xi)$ is ergodic, then so is the process η_t corresponding to $c(x, \eta)$.*

Proof. It suffices to construct a coupling (η_t, ζ_t, ξ_t) with the following properties:

(a) η_t and ζ_t are separately Markovian spin systems with flip rates given by $c(x, \cdot)$,
(b) ξ_t is a Markovian spin system with flip rates $\bar{c}(x, \cdot)$, and
(c) if $|\eta_0(u) - \zeta_0(u)| \leq \xi_0(u)$ for all $u \in S$, then $|\eta_t(u) - \zeta_t(u)| \leq \xi_t(u)$ with probability one for all $t \geq 0$ and $u \in S$.

First we will show that the existence of such a coupling is enough, and then we will construct the required coupling. Suppose ξ_t is ergodic. Since $\xi \equiv 0$ is an absorbing point for that process, it follows that

$$(1.10) \qquad\qquad \lim_{t \to \infty} P^\xi[\xi_t(u) = 1] = 0$$

for all $\xi \in X$ and all $u \in S$. Therefore, given initial configurations η and ζ for η_t and ζ_t, we can choose a $\xi \in X$ so that $|\eta(u) - \zeta(u)| \leq \xi(u)$ for all $u \in S$ and use the above coupling to conclude that for $f \in D(X)$,

$$|E^\eta f(\eta_t) - E^\zeta f(\zeta_t)| \leq \sum_x \Delta_f(x) P[\eta_t(x) \neq \zeta_t(x)]$$

$$\leq \sum_x \Delta_f(x) P[\xi_t(x) = 1],$$

which tends to 0 as $t \to \infty$ by (1.10). The second inequality follows from property (c) of the coupling. Let μ be any invariant probability measure for the process with rates $c(x, \cdot)$. (Recall $\mathcal{I} \neq \varnothing$, by Proposition 1.8(f) of

Chapter I.) By the above statement, then,

$$\int f \, d\mu - E^{\zeta} f(\zeta_t) = \int \{E^{\eta} f(\eta_t) - E^{\zeta} f(\zeta_t)\} \mu(d\eta)$$

tends to zero as $t \to \infty$, so that

$$\lim_{t \to \infty} E^{\zeta} f(\zeta_t) = \int f \, d\mu$$

for all $\zeta \in X$ and $f \in D(X)$. It follows that the process with rates $c(x, \cdot)$ is ergodic. We turn now to the construction of a coupling with properties (a), (b), and (c). We will simply describe the coupling and leave it to the reader to check that these properties are satisfied. The flip rates for the coupled system will only be defined when the configuration (η, ζ, ξ) satisfies $|\eta(u) - \zeta(u)| \le \xi(u)$ for all $u \in S$, since this class of configurations will be closed for the motion. The possible triples at each site are therefore $(a, a, 0)$, $(a, a, 1)$, or $(a, 1-a, 1)$, where $a = 0$ or 1. The flip rates at x when the system is in configuration (η, ζ, ξ) are given by the following list, where $\varepsilon = \varepsilon(x, \eta, \zeta, \xi)$ is defined by

$$\varepsilon = \begin{cases} \bar{c}(x, \xi)[c(x, \eta) + c(x, \zeta)]^{-1} & \text{if } c(x, \eta) + c(x, \zeta) > 0, \\ 0 & \text{otherwise:} \end{cases}$$

Transition	Rate
$(a, a, 0) \to (1-a, 1-a, 0)$	$\min[c(x, \eta), c(x, \zeta)]$
$\to (a, 1-a, 1)$	$c(x, \zeta) - \min[c(x, \eta), c(x, \zeta)]$
$\to (1-a, a, 1)$	$c(x, \eta) - \min[c(x, \eta), c(x, \zeta)]$
$\to (a, a, 1)$	$\bar{c}(x, \xi) - |c(x, \eta) - c(x, \zeta)|$
$(a, a, 1) \to (1-a, 1-a, 1)$	$\min[c(x, \eta), c(x, \zeta)]$
$\to (a, 1-a, 1)$	$c(x, \zeta) - \min[c(x, \eta), c(x, \zeta)]$
$\to (1-a, a, 1)$	$c(x, \eta) - \min[c(x, \eta), c(x, \zeta)]$
$\to (a, a, 0)$	$\bar{c}(x, \xi)$
$(a, 1-a, 1) \to (a, a, 1)$	$c(x, \zeta)(1 - \varepsilon)$
$\to (a, a, 0)$	$c(x, \zeta)\varepsilon$
$\to (1-a, 1-a, 1)$	$c(x, \eta)(1 - \varepsilon)$
$\to (1-a, 1-a, 0)$	$c(x, \eta)\varepsilon.$

Note that all these transition rates are nonnegative. In fact, it was in order to guarantee that property that this particular form for $\bar{c}(x, \xi)$ was chosen. The verification that this coupling is well defined and has properties (a), (b), (c) follows the same lines as the proofs of Lemma 1.2 and Theorem 1.3. The main difference is that the expression for the generator will have more terms in it now than before. The properties of the above rates which lead to (a), (b), and (c) are respectively: (a') in each case, the sum of the flip rates corresponding to a change in the first coordinates is $c(x, \eta)$ and the sum of the flip rates corresponding to a change in the second coordinate is $c(x, \zeta)$, (b') in each case, the sum of the flip rates corresponding to a change in the third coordinate is $\bar{c}(x, \xi)$ (in checking this in the third case, note that if $\eta(x) \neq \zeta(x)$, $|\eta(u) - \zeta(u)| \leq \xi(u)$ for all $u \in S$, and $c(x, \eta) + c(x, \zeta) = 0$, then $\bar{c}(x, \xi) = 0$), and (c') only transitions which lead to (a, b, c) with $|a - b| \leq c$ are permitted. □

It is important to observe that

$$\inf_{\xi}[\bar{c}(x, \xi) + \bar{c}(x, \xi_x)] \leq \inf_{\eta}[c(x, \eta) + c(x, \eta_x)]$$

for all $x \in S$ and

$$\sup_{\xi}|\bar{c}(x, \xi) - \bar{c}(x, \xi_u)| \geq \sup_{\eta}|c(x, \eta) - c(x, \eta_u)|$$

for all $x \neq u$. Therefore, Theorem 1.9 gives no improvement over previous ergodicity conditions if one verifies its hypothesis by simply applying the ergodicity criterion (0.6) to the rates $\bar{c}(x, \xi)$. However, since $\bar{c}(x, \xi)$ has a simpler structure than does $c(x, \eta)$, other techniques for proving its ergodicity become available. It should also be clear that the converse of Theorem 1.9 is false. The following provides counterexample.

Example 1.11. Consider the stochastic Ising model in one dimension (see Examples 3.1(a) and 4.3(a) of Chapter I), which in our current notation, has $S = Z^1$ and rates given by

$$c(x, \eta) = \begin{cases} e^{-2\beta} & \text{if } \eta(x-1) = \eta(x) = \eta(x+1), \\ 1 & \text{if } \eta(x-1) \neq \eta(x+1), \\ e^{2\beta} & \text{if } \eta(x) \neq \eta(x-1) = \eta(x+1), \end{cases}$$

As will be seen in Chapter IV, this spin system is ergodic for all $\beta \geq 0$. The comparison system in Theorem 1.9 has rates which satisfy

$$\bar{c}(x, \xi) \leq 2 \quad \text{if } \xi(x) = 1, \quad \text{and}$$

$$\bar{c}(x, \xi) \geq \frac{e^{2\beta} - e^{-2\beta}}{2}[\xi(x-1) + \xi(x+1)] \quad \text{if } \xi(x) = 0.$$

Therefore the hypotheses of Theorem 1.5 are satisfied if we take $c_2(x, \eta) = \bar{c}(x, \eta)$ and

$$
c_1(x, \eta) = \begin{cases} 2 & \text{if } \eta(x) = 1 \\ \dfrac{e^{2\beta} - e^{-2\beta}}{2} [\eta(x-1) + \eta(x+1)] & \text{if } \eta(x) = 0, \end{cases}
$$

with $\beta \geq 0$. Except for a factor of 2, these are the rates for the one-dimensional contact process (see Examples 3.1(c) and 4.3(c) of Chapter I). Therefore the results of Chapter VI will imply that the system with rates $c_1(x, \eta)$ is not ergodic for β so large that $e^{2\beta} - e^{-2\beta} \geq 8$. By Corollary 1.8, the same is true for the system with rates $\bar{c}(x, \eta)$.

We will conclude this section by giving an illustration of the use of coupling to prove ergodicity for a class of one-dimensional spin systems in which structural assumptions replace assumptions of the type of (0.6).

Theorem 1.12. *Suppose that $S = Z^1$ and that $c(x, \eta)$ is strictly positive and depends on η only through the coordinates $\{\eta(y), y > x\}$. Then the corresponding spin system η_t is ergodic. Its unique invariant measure ν is the product measure with $\nu\{\eta: \eta(x) = 1\} = \frac{1}{2}$ for each $x \in S$.*

Proof. Since $c(x, \eta)$ depends only on the coordinates of η which are strictly to the right of x,

(1.13)
$$
P^\eta[\eta_t(y) = \gamma(y) \text{ for } x \leq y \leq x+n]
$$
$$
= P^{\eta_x}[\eta_t(y) = \gamma_x(y) \text{ for } x \leq y \leq x+n]
$$

for all $\eta, \gamma \in X$, $x \in S$, and $n \geq 0$. For a fixed $\eta \in X$ and $x \in S$, consider the basic coupling (η_t, ζ_t) of two copies of the spin system with initial configuration (η, η_x). Then clearly $\eta_t(y) = \zeta_t(y)$ for all $t \geq 0$ and $y > x$. Applying (1.1) to $f(\eta, \zeta) = |\eta(x) - \zeta(x)|$, we have

$$
\tilde{\Omega} f(\eta, \zeta) = \begin{cases} |c(x, \eta) - c(x, \zeta)| & \text{if } \eta(x) = \zeta(x), \\ -[c(x, \eta) + c(x, \zeta)] & \text{if } \eta(x) \neq \zeta(x), \end{cases}
$$

so that if $\eta(y) = \zeta(y)$ for all $y > x$,

$$
\tilde{\Omega} f(\eta, \zeta) = -2c(x, \eta) f(\eta, \zeta).
$$

Let $\varepsilon(x) = \min_{\eta \in X} c(x, \eta) > 0$. It then follows that

$$
\frac{d}{dt} P[\eta_t(x) \neq \zeta_t(x)] \leq -2\varepsilon(x) P[\eta_t(x) \neq \zeta_t(x)],
$$

so that

$$P[\eta_t(x) \neq \zeta_t(x)] \le e^{-2\varepsilon(x)t}.$$

We therefore conclude that for any $\gamma \in X$,

$$\left| P^\eta[\eta_t(y) = \gamma(y) \text{ for } x \le y \le x+n] \right.$$
$$\left. - P^{\eta_x}[\eta_t(y) = \gamma(y) \text{ for } x \le y \le x+n] \right| \le e^{-2\varepsilon(x)t}.$$

Combining this with (1.13) yields

(1.14)
$$\left| P^\eta[\eta_t(y) = \gamma(y) \text{ for } x \le y \le x+n] \right.$$
$$\left. - P^\eta[\eta_t(y) = \gamma_x(y) \text{ for } x \le y \le x+n] \right| \le e^{-2\varepsilon(x)t}.$$

When $n = 0$, this gives

$$\left| P^\eta[\eta_t(x) = 1] - P^\eta[\eta_t(x) = 0] \right| \le e^{-2\varepsilon(x)t},$$

which can be rewritten as

$$\left| P^\eta[\eta_t(x) = 1] - \tfrac{1}{2} \right| \le \tfrac{1}{2} e^{-2\varepsilon(x)t}.$$

Since

$$2^{n+1} P^\eta[\eta_t(y) = \gamma(y) \text{ for } x \le y \le x+n] - 1$$
$$= 2^n P^\eta[\eta_t(y) = \gamma(y) \text{ for } x+1 \le y \le x+n] - 1$$
$$+ 2^n P^\eta[\eta_t(y) = \gamma(y) \text{ for } x \le y \le x+n]$$
$$- 2^n P^\eta[\eta_t(y) = \gamma_x(y) \text{ for } x \le y \le x+n],$$

estimate (1.14) implies that

$$\left| 2^{n+1} P^\eta[\eta_t(y) = \gamma(y) \text{ for } x \le y \le x+n] - 1 \right|$$
$$\le \left| 2^n P^\eta[\eta_t(y) = \gamma(y) \text{ for } x+1 \le y \le x+n] - 1 \right|$$
$$+ 2^n e^{-2\varepsilon(x)t}.$$

Therefore by induction,

$$\left| P^\eta[\eta_t(y) = \gamma(y) \text{ for } x \le y \le x+n] - \frac{1}{2^{n+1}} \right|$$

$$\le \sum_{k=0}^{n} \frac{1}{2^{k+1}} e^{-2\varepsilon(x+k)t}.$$

This gives the desired result, together with an exponential rate of convergence to equilibrium. □

2. Attractive Spin Systems

This section presents the general theory of attractive spin systems. These are
the spin systems which are monotone in the sense of Definition 2.3 of
Chapter II. The coupling techniques and results from the previous section
will play an important role here. More refined information about attractive
one-dimensional nearest-neighbor spin systems will be obtained in the next
section.

Definition 2.1. The spin system with rates $c(x, \eta)$ is said to be *attractive* if
whenever $\eta \leq \zeta$,

$$c(x, \eta) \leq c(x, \zeta) \quad \text{if } \eta(x) = \zeta(x) = 0, \quad \text{and}$$

$$c(x, \eta) \geq c(x, \zeta) \quad \text{if } \eta(x) = \zeta(x) = 1.$$

This definition asserts that a coordinate which takes a given value is more
likely to flip to the opposite value in a short period of time if it generally
disagrees with its environment than if it generally agrees with it. Thus the
evolution will tend to make a coordinate agree with neighboring coordinates.
Since coordinates tend to "attract" one another, the evolution is called
attractive. Not only is this property patently natural, but many important
spin systems are automatically attractive. Examples are the voter model,
the contact process, and the most frequently studied stochastic Ising models.

Theorem 2.2. *A spin system is attractive if and only if it is monotone.*

Proof. Suppose first that the system is monotone. Since $f(\eta) = \eta(x)$ is a
monotone function, $S(t)f \in \mathcal{M}$ for all $t \geq 0$. Since $f \in \mathcal{D}(\Omega)$,

$$\Omega f(\eta) = \lim_{t \downarrow 0} \frac{S(t)f(\eta) - f(\eta)}{t}$$

by Theorem 2.9 of Chapter I. Therefore $\eta \leq \zeta$ and $\eta(x) = \zeta(x)$ imply that

$$\Omega f(\eta) \leq \Omega f(\zeta).$$

Since

$$\Omega f(\eta) = c(x, \eta)[1 - 2\eta(x)]$$

by (0.5), it follows that the system is attractive. For the converse, note that
attractiveness implies that (1.6) is satisfied by $c_1(x, \eta) = c_2(x, \eta) = c(x, \eta)$.
Therefore attractiveness implies monotonicity by Corollary 1.7, together
with Theorem 2.2 of Chapter II. $\square$

In what follows, δ_0 and δ_1 will denote the pointmasses at $\eta \equiv 0$ and $\eta \equiv 1$ respectively. Recall (from the end of Section 4 of Chapter I) that in the translation invariant case, $S = Z^d$ for some $d \geq 1$ and the rates are translation invariant in the sense that $c(x, \eta) = c(0, \tau_{-x}\eta)$ for all $x \in S$ and $\eta \in X$. In this case, $\mathcal{S}$ denotes as usual the set of all translation invariant probability measures on X.

Theorem 2.3. *Let $S(t)$ be the semigroup for an attractive spin system. Then*

(a) $\delta_0 S(s) \leq \delta_0 S(t)$ *for* $0 \leq s \leq t$,
(b) $\delta_1 S(s) \geq \delta_1 S(t)$ *for* $0 \leq s \leq t$,
(c) $\delta_0 S(t) \leq \mu S(t) \leq \delta_1 S(t)$ *for* $t \geq 0$ *and* $\mu \in \mathcal{P}$,
(d) $\underline{\nu} = \lim_{t \to \infty} \delta_0 S(t)$ *and* $\bar{\nu} = \lim_{t \to \infty} \delta_1 S(t)$ *exist*,
(e) *if* $\mu \in \mathcal{P}$, $t_n \to \infty$, *and* $\nu = \lim_{n \to \infty} \mu S(t_n)$, *then* $\underline{\nu} \leq \nu \leq \bar{\nu}$, *and*
(f) $\underline{\nu}, \bar{\nu} \in \mathcal{S}_e$.

Proof. By definition,

$$\delta_0 \leq \delta_0 S(t - s) \quad \text{for } 0 \leq s \leq t,$$

so by attractiveness and the semigroup property,

$$\delta_0 S(s) \leq \delta_0 S(t - s) S(s) = \delta_0 S(t).$$

This gives (a), and of course the proof of (b) is similar. For (c), note again that by definition of stochastic monotonicity,

$$\delta_0 \leq \mu \leq \delta_1$$

for all $\mu \in \mathcal{P}$. Therefore by attractiveness,

$$\delta_0 S(t) \leq \mu S(t) \leq \delta_1 S(t).$$

For (d), it suffices to use (a) and (b), the compactness of $\mathcal{P}$ in the topology of weak convergence, and the fact that $\mathcal{M}$ has the following property:

$$\int f \, d\mu_1 = \int f \, d\mu_2$$

for all $f \in \mathcal{M}$ and some $\mu_1, \mu_2 \in \mathcal{P}$ implies that $\mu_1 = \mu_2$. Part (e) is an immediate consequence of (c) and (d). To prove part (f), note that $\underline{\nu}$ and $\bar{\nu} \in \mathcal{S}$ by Proposition 1.8(d) of Chapter I. To show that they are extremal in $\mathcal{S}$, suppose, for example, that

$$\bar{\nu} = \alpha \mu_1 + (1 - \alpha)\mu_2,$$

where $0 < \alpha < 1$ and $\mu_1, \mu_2 \in \mathcal{I}$. Then $\mu_1 \leq \bar{\nu}$ and $\mu_2 \leq \bar{\nu}$ by part (e) of this theorem. Therefore $f \in \mathcal{M}$ implies that

$$\int f \, d\mu_1 \leq \int f \, d\bar{\nu}, \qquad \int f \, d\mu_2 \leq \int f \, d\bar{\nu}, \quad \text{and}$$

$$\int f \, d\bar{\nu} = \alpha \int f \, d\mu_1 + (1 - \alpha) \int f \, d\mu_2,$$

and hence that

$$\int f \, d\bar{\nu} = \int f \, d\mu_1 = \int f \, d\mu_2.$$

This then implies that $\bar{\nu} = \mu_1 = \mu_2$. $\square$

Corollary 2.4. *For an attractive spin system, the following three statements are equivalent:*

(a)   *The process is ergodic.*
(b) $\mathcal{I}$ *is a singleton.*
(c) $\underline{\nu} = \bar{\nu}$, *where* $\underline{\nu}$ *and* $\bar{\nu}$ *are the measures defined in part* (d) *of Theorem 2.3.*

Proof. That (a) implies (b) follows from Definition 1.9 of Chapter I. That (b) implies (c) follows from part (f) of Theorem 2.3. Finally, consider the proof that (c) implies (a). If $\mu \in \mathcal{P}$, then the family of probability measures $\{\mu S(t), t \geq 0\}$ is relatively compact. By part (e) of Theorem 2.3, all subsequential limits of this family as $t \to \infty$ are equal to the common value of $\underline{\nu}$ and $\bar{\nu}$, and hence $\lim_{t \to \infty} \mu S(t)$ exists and equals that common value. Therefore, by Definition 1.9 of Chapter I, the process is ergodic. $\square$

The following example illustrates in a simple case how Corollary 2.4 can be applied.

Example 2.5. Suppose that $S = Z^d$, $p(x) \geq 0$ for $x \in S$, $p(0) = 0$, and

$$\sum_x p(x) = 1.$$

Consider the spin system with rates

$$c(x, \eta) = \begin{cases} \beta + \sum_u p(u - x)\eta(u) & \text{if } \eta(x) = 0, \\ \delta + \sum_u p(u - x)[1 - \eta(u)] & \text{if } \eta(x) = 1, \end{cases}$$

where β, $\delta \geq 0$ and $\beta + \delta > 0$. This system is clearly attractive and translation invariant. Therefore

$$\rho_i(t) = \delta_i S(t)\{\eta: \eta(x) = 1\}$$

is independent of x for $i = 0$ and 1. Let $f(\eta) = \eta(0)$. Then

$$\Omega f(\eta) = c(0, \eta)[1 - 2\eta(0)]$$

$$= \begin{cases} \beta + \sum_u p(u)\eta(u) & \text{if } \eta(0) = 0, \\ -\delta - \sum_u p(u)[1 - \eta(u)] & \text{if } \eta(0) = 1, \end{cases}$$

so that

$$\int \Omega f \, d\mu = \beta - (\beta + \delta) \int f \, d\mu$$

$$+ \sum_u p(u)[\mu\{\eta(u) = 1\} - \mu\{\eta(0) = 1\}]$$

for all $\mu \in \mathcal{P}$. If $\mu \in \mathcal{S}$, then the last term vanishes, and we see that

$$\int \Omega f \, d\mu = \beta - (\beta + \delta) \int f \, d\mu.$$

Therefore, since $\delta_i S(t) \in \mathcal{S}$ for all $t \geq 0$ and $i = 0$ and 1,

$$\frac{d}{dt} \rho_i(t) = \frac{d}{dt} \int f \, d[\delta_i S(t)]$$

$$= \int \Omega f \, d[\delta_i S(t)]$$

$$= \beta - (\beta + \delta)\rho_i(t).$$

So,

$$\lim_{t \to \infty} \rho_i(t) = \frac{\beta}{\beta + \delta}$$

for $i = 0$ and 1, which implies that

$$\nu\{\eta(x) = 1\} = \frac{\beta}{\beta + \delta} = \bar{\nu}\{\eta(x) = 1\}$$

for each $x \in S$. By Corollary 2.8 of Chapter II, it follows that $\nu = \bar{\nu}$, and hence by Corollary 2.4 above, the process is ergodic. This result could also have been obtained from criterion (0.6) for ergodicity. When $\beta = \delta = 0$, this process is not ergodic. Its limiting behavior will be described rather completely in Chapter V.

Corollary 2.6. *A translation invariant attractive spin system is ergodic if and only if $\mathscr{I} \cap \mathscr{S}$ is a singleton.*

Proof. This follows immediately from Corollary 2.4, since in the translation invariant case, $\delta_0 S(t)$ and $\delta_1 S(t)$ are in $\mathscr{S}$, and hence so are $\underline{\nu}$ and $\bar{\nu}$. $\quad\square$

Corollary 2.6 is useful since as we shall see, it is often easier to identify the elements of $\mathscr{I} \cap \mathscr{S}$ than to find all of $\mathscr{I}$. Of course we know now that if $\mathscr{I} \cap \mathscr{S}$ is a singleton for an attractive translation invariant spin system, then $\mathscr{I} \cap \mathscr{S} = \mathscr{I}$.

Often we will verify that $\mathscr{I}$ is a singleton by analyzing finite approximations to the spin system. In reversible situations, as will be seen in Chapters IV and VII, it is generally possible to compute explicitly the invariant measures for these finite approximations. Thus the next result is quite useful. In order to define the relevant approximations, let S_n be finite sets which increase to S. For $i = 0$ and 1, define flip rates for a spin system on S by

$$c_i^n(x, \eta) = \begin{cases} c(x, \eta^i) & \text{if } x \in S_n, \\ 0 & \text{if } x \notin S_n \text{ and } \eta(x) = i, \\ M(x) & \text{if } x \notin S_n \text{ and } \eta(x) \neq i, \end{cases}$$

where $M(x) = \sup_\eta c(x, \eta)$, $\eta^i(u) = \eta(u)$ for $u \in S_n$, and $\eta^i(u) = i$ for $u \notin S_n$. These can be thought of as spin systems obtained from the original one by keeping the coordinates outside S_n "frozen" at the value i, once such a coordinate reaches i. The rate $M(x)$ above is included only so that the process will be well defined for all initial configurations. For most purposes it will be irrelevant, since the process will usually have $\eta(x) = i$ for $x \notin S_n$ initially, and hence for all $t \geq 0$. The processes corresponding to $c_0^n(x, \eta)$ and $c_1^n(x, \eta)$ are respectively lower and upper approximations to the original system. Let $S_0^n(t)$ and $S_1^n(t)$ be the corresponding semigroups. Note that for $u \neq x$

$$\sup_\eta |c_i^n(x, \eta_u) - c_i^n(x, \eta)| \leq \sup_\eta |c(x, \eta_u) - c(x, \eta)|$$

and that

$$\lim_{n \to \infty} c_i^n(x, \eta) = c(x, \eta)$$

for $x \in S$ and $i = 0$ and 1, uniformly for $\eta \in X$. Therefore

$$\lim_{n \to \infty} S_i^n(t)f = S(t)f$$

for $i = 0$ and 1, $t \geq 0$, and $f \in C(X)$ by Theorem 2.12 of Chapter I.

Theorem 2.7. *Suppose $c(x, \eta)$ is attractive, and let $\underline{\nu}$ and $\bar{\nu}$ be defined as in part* (d) *of Theorem 2.3. Then $c_i^n(x, \eta)$ is attractive for each i and n. If $\mu_0, \mu, \mu_1 \in \mathcal{P}$ satisfy $\mu_0 \leq \mu \leq \mu_1$, then*

$$(2.8) \qquad \mu_0 S_0^n(t) \leq \mu S(t) \leq \mu_1 S_1^n(t)$$

for all $t \geq 0$. Finally, if we define $\underline{\nu}^n$ and $\bar{\nu}^n$ by $\underline{\nu}^n = \lim_{t \to \infty} \delta_0 S_0^n(t)$ and $\bar{\nu}^n = \lim_{t \to \infty} \delta_1 S_1^n(t)$, then

$$\underline{\nu} = \lim_{n \to \infty} \underline{\nu}^n \quad and \quad \bar{\nu} = \lim_{n \to \infty} \bar{\nu}^n.$$

Proof. Since $\eta \leq \zeta$ implies $\eta^i \leq \zeta^i$, the attractiveness of $c_i^n(x, \eta)$ follows from that of $c(x, \eta)$. By Corollary 1.7, in order to prove (2.8), it suffices to check that

$$c_0^n(x, \eta) \leq c(x, \eta) \leq c_1^n(x, \eta) \quad \text{if } \eta(x) = 0, \quad \text{and}$$

$$c_0^n(x, \eta) \geq c(x, \eta) \geq c_1^n(x, \eta) \quad \text{if } \eta(x) = 1.$$

This is true for $x \in S_n$ since $\eta^0 \leq \eta \leq \eta^1$ and for $x \notin S_n$ since

$$0 \leq c(x, \eta) \leq M(x).$$

Since $c_i^n(x, \eta)$ is attractive, the limits which define $\underline{\nu}^n$ and $\bar{\nu}^n$ exist by part (d) of Theorem 2.3. By taking $\mu_0 = \mu = \delta_0$ first, and $\mu_1 = \mu = \delta_1$ next in (2.8) and letting $t \to \infty$, we see that

$$(2.9) \qquad \underline{\nu}^n \leq \underline{\nu} \quad and \quad \bar{\nu} \leq \bar{\nu}^n$$

for each n. Another similar application of Corollary 1.7 shows that

$$\underline{\nu}^n \leq \underline{\nu}^{n+1} \quad and \quad \bar{\nu}^n \geq \bar{\nu}^{n+1}$$

for all n. Therefore

$$\lim_{n \to \infty} \underline{\nu}^n \quad and \quad \lim_{n \to \infty} \bar{\nu}^n$$

exist and by (2.9) satisfy

$$(2.10) \qquad \lim_{n \to \infty} \underline{\nu}^n \leq \underline{\nu} \quad \text{and} \quad \lim_{n \to \infty} \bar{\nu}^n \geq \bar{\nu}.$$

These two limits are invariant for the original process with rates $c(x, \eta)$ by Proposition 2.14 of Chapter I, so they both lie between $\underline{\nu}$ and $\bar{\nu}$ by part (e) of Theorem 2.3. So by (2.10),

$$\underline{\nu} = \lim_{n \to \infty} \underline{\nu}^n \quad \text{and} \quad \bar{\nu} = \lim_{n \to \infty} \bar{\nu}^n,$$

and the proof is complete. $\square$

Corollary 2.11. *An attractive spin system is ergodic if and only if*

$$\lim_{n \to \infty} \underline{\nu}^n = \lim_{n \to \infty} \bar{\nu}^n.$$

Proof. This follows immediately from Corollary 2.4 and Theorem 2.7. $\square$

The following example illustrates the use of Corollary 2.11. A more general treatment of these ideas will be given in Chapter IV.

Example 2.12. (The nearest-neighbor *Majority Vote Process* in one dimension; see Example 4.3(e) of Chapter I.) Let $S = Z^1$ and

$$c(x, \eta) = \begin{cases} 1 - \delta & \text{if } \eta(x+1) = \eta(x-1) \neq \eta(x), \\ \delta & \text{otherwise.} \end{cases}$$

In Example 4.3(e) of Chapter I, we saw that this process is ergodic for $\frac{1}{3} < \delta < \frac{3}{4}$. It is not ergodic if $\delta = 0$ or $\delta = 1$. It is attractive if $0 \leq \delta \leq \frac{1}{2}$. We will now use Corollary 2.11 to show that it is ergodic for $0 < \delta \leq \frac{1}{2}$. For δ in this range, let ν be the probability measure on $\{0, 1\}^Z$ induced by the stationary two-state Markov chain with transition matrix

$$P = \frac{1}{\sqrt{\delta} + \sqrt{1-\delta}} \begin{pmatrix} \sqrt{1-\delta} & \sqrt{\delta} \\ \sqrt{\delta} & \sqrt{1-\delta} \end{pmatrix}.$$

Let $S_n = \{-n, -n+1, \ldots, n-1, n\}$, and let $\underline{\nu}^n$ and $\bar{\nu}^n$ be the measures on $\{0, 1\}^Z$ defined by "conditioning" ν on the events $E_0 = \{\eta : \eta(x) = 0 \text{ for all } x \notin S_n\}$ and $E_1 = \{\eta : \eta(x) = 1 \text{ for all } x \notin S_n\}$ respectively. For $x \in S_n$,

$$c(x, \eta)\underline{\nu}^n\{\eta\} = c(x, \eta_x)\underline{\nu}^n\{\eta_x\} \quad \text{if } \eta \in E_0$$

and

$$c(x, \eta)\bar{\nu}^n\{\eta\} = c(x, \eta_x)\bar{\nu}^n\{\eta_x\} \quad \text{if } \eta \in E_1.$$

Therefore by Propositions 5.2 and 5.3 of Chapter II, $\underline{\nu}^n$ and $\bar{\nu}^n$ are invariant measures for the processes with rates $c_0^n(x, \eta)$ and $c_1^n(x, \eta)$ respectively. Since these processes are irreducible Markov chains on E_0 and E_1 respectively, the convergence theorem for finite-state Markov chains shows that $\underline{\nu}^n$ and $\bar{\nu}^n$ are the same as the measures defined in the statement of Theorem 2.7. Therefore, in order to apply Corollary 2.11 to conclude that the nearest-neighbor majority vote process in one dimension is ergodic, it suffices to show that

$$\lim_{n\to\infty} \underline{\nu}^n = \lim_{n\to\infty} \bar{\nu}^n.$$

But this is a consequence of the convergence theorem for finite-state Markov chains, applied to the chain with transition matrix P. For example,

$$\underline{\nu}^n\{\eta: \eta(0) = 1\} = \frac{p^{(n+1)}(0, 1)p^{(n+1)}(1, 0)}{p^{(2n+2)}(0, 0)}$$

where $p^{(n)}(i, j)$ are the n-step transition probabilities for that two-state chain. In fact, the common value of the two limits above is ν itself.

Since we have just introduced the approximations $S_0^n(t)$ and $S_1^n(t)$, this is a good place to state the following important consequence of Harris' theorem on positive correlations.

Theorem 2.13. *For an attractive spin system, $\underline{\nu}$ and $\bar{\nu}$ have positive correlations (see Definition 2.11 of Chapter II).*

Proof. By Theorem 2.14 of Chapter II and the remarks following its statement,

$$\delta_0 S_0^n(t) \quad \text{and} \quad \delta_1 S_1^n(t)$$

have positive correlations for all $t \geq 0$. The important property of a spin system which is used in checking (2.19) of Chapter II is that only transitions from η to some η_x are allowed, and for each x and η, either $\eta \leq \eta_x$ or $\eta_x \leq \eta$. Passing to the limit as $t \to \infty$, we see that $\underline{\nu}^n$ and $\bar{\nu}^n$ have positive correlations for each n. The result then follows by letting $n \to \infty$ and using Theorem 2.7. $\square$

So far in this section, we have seen how coupling can be used to give various sufficient conditions for ergodicity of attractive processes, and have seen how these conditions can be verified in special examples. Another important application of coupling is to the proof that critical values exist for certain one-parameter families of spin systems. The following is the main result of this type which is available via coupling techniques. Note that this result does not apply if the rates are strictly positive. Another approach to the existence of critical values will be presented in Chapter IV. (See Corollary 2.18 from that chapter.) It is based on reversibility, and the related fact that in certain cases, the measures $\underline{\nu}^n$ and $\bar{\nu}^n$ are more or less explicitly known.

Theorem 2.14. *Given an attractive spin system* $c(x, \eta)$ *such that* $c(x, \eta) = 0$ *if* $\eta \equiv 0$, *define a one-parameter family of spin systems* $c_\lambda(x, \eta)$ *for* $\lambda \geq 0$ *by*

$$c_\lambda(x, \eta) = \begin{cases} \lambda c(x, \eta) & \text{if } \eta(x) = 0, \\ c(x, \eta) & \text{if } \eta(x) = 1. \end{cases}$$

Then there is a critical value $\lambda_c \in [0, \infty]$ *such that the system with parameter* λ *is ergodic if* $\lambda < \lambda_c$ *and not ergodic if* $\lambda > \lambda_c$.

Proof. If $0 \leq \lambda_1 \leq \lambda_2$, then $c_i(x, \eta) = c_{\lambda_i}(x, \eta)$ for $i = 1, 2$ satisfies (1.6). Therefore by Corollary 1.8, if the process with $\lambda = \lambda_2$ is ergodic, so is the one with $\lambda = \lambda_1$. The result follows from this statement by defining

$$\lambda_c = \sup\{\lambda \geq 0: \text{process is ergodic}\}. \quad \square$$

Theorem 2.14 raises a number of questions for special choices of $c(x, \eta)$. Examples are (a) is $\lambda_c > 0$?; (b) is $\lambda_c < \infty$?; (c) what is λ_c exactly?; (d) is the process ergodic for $\lambda = \lambda_c$? These will be partially resolved for the contact process and the nearest-particle system in Chapters VI and VII respectively. Note that Corollary 1.7 provides a means of comparing the critical values of two different one-parameter families of spin systems. Hypothesis (1.6) becomes somewhat simpler when at least one of the processes is attractive, since then (1.6) for $\eta \leq \zeta$ follows from (1.6) for $\eta = \zeta$.

Sometimes we will obtain information about the invariant measures for attractive spin systems by first studying the invariant measures for the corresponding coupled process. Therefore it is important to establish connections between these invariant measures. One immediate connection is that the marginals of any invariant measure for the coupled process are invariant for the marginal processes. In order to state a type of converse to this assertion, let $\tilde{\mathscr{I}}$ be the set of invariant measures for the basic coupling of two copies of the attractive spin system η_t with rates $c(x, \eta)$. In the translation invariant case, $\tilde{\mathscr{I}}$ will denote the set of translation invariant measures on $X \times X$.

Theorem 2.15. (a) *If ν_1, $\nu_2 \in \mathcal{I}$, there is a $\nu \in \tilde{\mathcal{I}}$ with marginals ν_1 and ν_2.*
 (b)   *If ν_1, $\nu_2 \in \mathcal{I}_e$, then ν can be taken in $\tilde{\mathcal{I}}_e$.*
 (c)   *In each of the preceding two statements, if in addition $\nu_1 \leq \nu_2$, then ν can be taken so that*

$$\nu\{(\eta, \zeta): \eta \leq \zeta\} = 1.$$

 (d)   *In the translation invariant case, each of the preceding three statements holds if $\mathcal{I}$ and $\tilde{\mathcal{I}}$ are replaced by $\mathcal{I} \cap \mathcal{S}$ and $\tilde{\mathcal{I}} \cap \tilde{\mathcal{S}}$ respectively.*

Proof. Given ν_1, $\nu_2 \in \mathcal{I}$, let $\tilde{\nu}$ be any probability measure on $X \times X$ with marginals ν_1 and ν_2. In case (c), take $\tilde{\nu}$ to concentrate on $\{(\eta, \zeta): \eta \leq \zeta\}$, which may be done by Theorem 2.4 of Chapter II. In case (d), take $\tilde{\nu}$ to be translation invariant. Then for each t, $\tilde{\nu}S(t)$ has marginals ν_1 and ν_2, concentrates on $\{(\eta, \zeta): \eta \leq \zeta\}$ in case (c) by Theorem 1.5, and is translation invariant in case (d). The same statements hold for any weak limit along a sequence of t's tending to ∞ of

$$\frac{1}{t} \int_0^t \tilde{\nu}S(s) \, ds.$$

Such limits exist by the compactness of $X \times X$, and are in $\tilde{\mathcal{I}}$ by Proposition 1.8 of Chapter I. This proves (a), and its restatements in cases (c) and (d). We will now prove (b) only, since the corresponding statements in cases (c) and (d) are proved analogously. Let

$$\mathcal{A} = \{\nu \in \tilde{\mathcal{I}}: \nu \text{ has marginals } \nu_1 \text{ and } \nu_2\}.$$

Then $\mathcal{A}$ is compact and convex, and is nonempty by part (a). It follows that $\mathcal{A}_e \neq \varnothing$ by the Krein–Milman Theorem (see Royden (1968)). It therefore suffices to prove that $\mathcal{A}_e \subset \tilde{\mathcal{I}}_e$. In order to do so, suppose that $\nu \in \mathcal{A}_e$ and $\nu = \lambda \alpha + (1 - \lambda)\beta$ for some α, $\beta \in \tilde{\mathcal{I}}$ and $0 < \lambda < 1$. The marginals of ν are the same convex combinations of the marginals of α and β. Therefore α, $\beta \in \mathcal{A}$ since ν_1 and ν_2 are extremal in $\mathcal{I}$ and the marginals of α and β are in $\mathcal{I}$. Since $\nu \in \mathcal{A}_e$, it follows that $\nu = \alpha = \beta$. Therefore $\nu \in \tilde{\mathcal{I}}_e$. $\square$

For the sake of simplicity, the above result was formulated in the context of the basic two process coupling. Analogous results for multiprocess couplings can be proved in the same way. See the proof of Theorem 3.13 for an example of their use.

Finally, we return to the translation invariant case, to obtain a further property of the invariant measures $\underline{\nu}$ and $\bar{\nu}$.

Proposition 2.16. *For a translation invariant attractive spin system, $\underline{\nu}$ and $\bar{\nu}$ are in $\mathcal{S}_e$.*

Proof. It suffices to prove the statement for $\bar{\nu}$. By Theorem 4.15 of Chapter I, $\delta_1 S(t) \in \mathcal{S}_e$ for each $t \geq 0$. By the remarks following that theorem, we cannot simply pass to the limit as $t \uparrow \infty$ in this statement to conclude that $\bar{\nu} \in \mathcal{S}_e$. Instead, we will use the attractiveness of the process, together with the criterion for shift ergodicity in Proposition 4.11 of Chapter I. In verifying that $\bar{\nu}$ satisfies (4.12) from that chapter, it suffices to check it for nonnegative $f, g \in \mathcal{M}$. But by the attractiveness assumption, for such f and g,

$$\int (\tau_y f) g \, d\bar{\nu} \leq \int (\tau_y f) g \, d[\delta_1 S(t)],$$

so that

$$(2.17) \qquad \lim_{x \to \infty} \frac{\sum\limits_{0 \leq y \leq x} \int (\tau_y f) g \, d\bar{\nu}}{|\{y \in Z^d : 0 \leq y \leq x\}|} \leq \int f \, d[\delta_1 S(t)] \int g \, d[\delta_1 S(t)].$$

Letting t tend to ∞, we see that the left-hand side of (2.17) is bounded above by $\int f \, d\bar{\nu} \int g \, d\bar{\nu}$. But by Theorem 2.13, the left-hand side of (2.17) is bounded below by this as well, so the result follows. $\square$

3. Attractive Nearest-Neighbor Spin Systems on Z^1

In this section, we will restrict ourselves to the very nicest special case of the situation discussed in the previous section. We assume here that $S = Z^1$, that the mechanism is translation invariant and attractive, and that the rates $c(x, \eta)$ depend on η only through $\{\eta(x-1), \eta(x), \eta(x+1)\}$. There are then only eight parameters in the process—the values of $c(x, \eta)$ for the eight possible values of $(\eta(x-1), \eta(x), \eta(x+1))$. These parameters are restricted only by the requirement that they be nonnegative and by the inequalities contained in the attractiveness assumption. At this point, we only know that the process is ergodic in relatively few cases. Sufficient condition (0.6) gives ergodicity in a certain rather small open subset of the eight-dimensional parameter space. The argument of Example 2.12 applies only on a six-dimensional submanifold of this eight-dimensional space. In order to describe it, use the following notation: $c(111) = c(x, \eta)$ when $\eta(x-1) = \eta(x) = \eta(x+1) = 1$, $c(110) = c(x, \eta)$ when $\eta(x-1) = \eta(x) = 1$, $\eta(x+1) = 0$, etc. Then this manifold is defined (when $c(x, \eta) > 0$ for all η) by the two constraints

$$\frac{c(100)}{c(110)} = \frac{c(001)}{c(011)}$$

and

$$\frac{c(101)c(000)}{c(111)c(010)} = \frac{c(100)c(001)}{c(110)c(011)}.$$

The result which covers this class of cases is Theorem 3.13 of Chapter IV. The above constraints mean that the process is a stochastic Ising model.

Our objective is to determine for exactly what values of the eight parameters the process is ergodic. When it is not ergodic, we also want to find out whether there are invariant measures other than the ν and $\bar{\nu}$ which are defined in the statement of Theorem 2.3. While it is not yet possible to answer the first question completely, this section contains several results which go a long way toward giving a good picture of the situation. The size of the set of invariant measures for the process depends on the degree of positivity of the rates $c(x, \eta)$. We will begin by considering the simplest and least interesting case, in which many of the flip rates are zero. Then we will proceed in stages to the deepest and most important result along these lines which states that the process is ergodic whenever all the rates are strictly positive.

Theorem 3.1. *Suppose that*

$$(3.2) \qquad c(x, \eta) + c(x, \eta_x) = 0$$

when $\eta(x-1) = 0$ *and* $\eta(x+1) = 1$ *or when* $\eta(x-1) = 1$ *and* $\eta(x+1) = 0$. *Then* $\mathscr{I}_e$ *contains infinitely many points.*

Proof. By symmetry, it suffices to prove the result when (3.2) holds for $\eta(x-1) = 0$ and $\eta(x+1) = 1$. This, together with attractiveness, implies that

$$c(000) = c(001) = c(011) = c(111) = 0.$$

For $n \in Z^1$, define $\eta^n \in X$ by

$$(3.3) \qquad \eta^n(x) = \begin{cases} 1 & \text{if } x \geq n, \\ 0 & \text{if } x < n. \end{cases}$$

Then $c(x, \eta^n) = 0$ for all x and all n. Therefore $\Omega f(\eta^n) = 0$ for all $f \in \mathscr{D}(\Omega)$, and hence the pointmass on η^n is in $\mathscr{I}$ (and hence in $\mathscr{I}_e$) by Proposition 2.13 of Chapter I. $\square$

The next objective is to show that $\mathscr{I}_e$ consists only of the measures ν and $\bar{\nu}$ if $c(x, \eta)$ satisfies

$$(3.4) \qquad c(x, \eta) + c(x, \eta_x) > 0 \quad \text{whenever} \quad \eta(x-1) \neq \eta(x+1).$$

This will be the content of Theorem 3.13. Of course under this condition, ν and $\bar{\nu}$ may be either equal or distinct. Trivial cases in which they are distinct occur when

$$(3.5) \qquad c(000) = c(111) = 0,$$

since then $\underline{\nu}$ and $\bar{\nu}$ are the pointmasses on $\eta \equiv 0$ and $\eta \equiv 1$ respectively. A much more interesting case is the contact process, in which

$$(3.6) \qquad c(x, \eta) = \begin{cases} 1 & \text{if } \eta(x) = 1, \\ \lambda[\eta(x+1) + \eta(x-1)] & \text{if } \eta(x) = 0. \end{cases}$$

As will be seen in Chapter VI, $\underline{\nu} \neq \bar{\nu}$ in this example, provided that $\lambda \geq 2$.

The proof that $\mathscr{I}_e = \{\underline{\nu}, \bar{\nu}\}$ under (3.4) is fairly long and requires several preliminary results. The following comments should make it easier to keep in mind where we are headed. The idea of the proof is to exploit the monotonicity of a certain functional of a coupled process. The point is that if a functional of a process is always monotone in time, then it must in fact be constant if the process is in equilibrium. The fact that the functional is constant will say a lot about the possible equilibria. As will be seen, the functional which will be used here is monotone only because the interaction is nearest-neighbor. Thus the present technique will presumably not generalize beyond the nearest-neighbor case. The use of the monotonicity of a different functional was discussed in Section 4 of Chapter II, and will be used further in Section 5 of Chapter IV.

Let

$$W = \{\eta \in X: \eta(x) = \eta(x+1) \text{ for all sufficiently large positive and}$$

$$\text{negative } x\}.$$

In order to describe how the monotonicity technique will be used, consider the problem of proving that $\nu \in \mathscr{I}$ and $\nu(W) = 1$ imply that ν is a convex combination of $\underline{\nu}$ and $\bar{\nu}$ under the assumption that (3.4), (3.5), $c(010) > 0$, and $c(101) > 0$ all hold. By (3.5), W is closed for the motion of the process. For $\eta \in W$, let

$$f(\eta) = \text{number of intervals of zeros and ones in } \eta$$

$$= 1 + \sum_{x=-\infty}^{\infty} 1_{\{\eta(x) \neq \eta(x+1)\}} < \infty.$$

By (3.5), $f(\eta_t)$ is nonincreasing in t with probability one. Since $c(010) > 0$ and $c(101) > 0$, any $\eta \in W$ for which there is an x such that $\eta(x-1) = \eta(x+1) \neq \eta(x)$ has a positive rate of having f decrease strictly. Since $\nu \in \mathscr{I}$, ν therefore must concentrate on those $\eta \in W$ for which there are no intervals of zeros or ones of length one. By (3.4), any configuration which has at least two intervals of zeros and ones of finite length leads with a positive rate in a finite number of steps to a configuration with a singleton in it. Thus ν must concentrate on

$$\{\eta \in W: f(\eta) \leq 3\}.$$

By looking separately at the process on

$$\{\eta \in W: f(\eta) = 2\} \quad \text{and} \quad \{\eta \in W: f(\eta) = 3\},$$

it is easy to see that ν puts no mass on either set. For example, the process restricted to the set $\{\eta^n, n \in Z^1\}$ defined in (3.3) can be identified with a random walk on Z^1. By (3.4), this random walk is not the trivial one which does not move at all. Since such a walk cannot have a finite stationary distribution, ν can put no mass on $\{\eta \in W: f(\eta) = 2\}$.

There are basically two additional difficulties which arise in the proof of the general result. First, since (3.5) will not be assumed, it will be necessary to couple the process of interest with two other copies, starting at $\eta \equiv 0$ and at $\eta \equiv 1$ respectively. This makes the definition of the functional which plays the role of f more cumbersome. Secondly, even if (3.5) were true, we would want to consider $\nu \in \mathscr{I}$ which do not necessarily concentrate on W. If $\eta \notin W$, then $f(\eta) = \infty$, so f itself is not particularly useful. Thus we will need to consider functionals which give the number of intervals of zeros and ones in a large finite part of a configuration. Then this functional loses its monotonicity, but it is still close enough to being monotone to serve the required purpose.

In order to prepare for the proof, it is necessary to consider a three-way version of the basic coupling which was discussed at the beginning of Section 1. The coupled process $(\eta_t, \gamma_t, \zeta_t)$ will be a Feller process on

$$X_3 = \{(\eta, \gamma, \zeta) \in X^3 : \eta \le \gamma \le \zeta\}.$$

At each site, the coupled process will take one of the four values $(0, 0, 0)$, $(0, 0, 1)$, $(0, 1, 1)$, or $(1, 1, 1)$. The flip rates for the coupled process at site $x \in Z^1$ are given by the following table:

	$(0, 0, 0)$	$(0, 0, 1)$	$(0, 1, 1)$	$(1, 1, 1)$
$(0, 0, 0)$	—	$c(x, \zeta) - c(x, \gamma)$	$c(x, \gamma) - c(x, \eta)$	$c(x, \eta)$
$(0, 0, 1)$	$c(x, \zeta)$	—	$c(x, \gamma) - c(x, \eta)$	$c(x, \eta)$
$(0, 1, 1)$	$c(x, \zeta)$	$c(x, \gamma) - c(x, \zeta)$	—	$c(x, \eta)$
$(1, 1, 1)$	$c(x, \zeta)$	$c(x, \gamma) - c(x, \zeta)$	$c(x, \eta) - c(x, \gamma)$	—

For example, the third row should be interpreted in the following way: if $\eta_t(x) = 0$, $\gamma_t(x) = 1$, and $\zeta_t(x) = 1$, then $\gamma_t(x)$ and $\zeta_t(x)$ will flip together at rate $c(x, \zeta_t)$, $\gamma_t(x)$ will flip alone at rate $c(x, \gamma_t) - c(x, \zeta_t)$, and $\eta_t(x)$ will flip alone at rate $c(x, \eta_t)$. The attractiveness assumption and the fact that $(\eta, \gamma, \zeta) \in X_3$ guarantees that all the entries in the table are nonnegative. The rates have the property that the marginal processes η_t, γ_t and ζ_t are Markovian spin systems with rates $c(x, \cdot)$, and the pairs (η_t, γ_t), (η_t, ζ_t) and (γ_t, ζ_t) are separately Markovian and evolve according to the two-process basic coupling. Formal proofs of these facts would proceed along the lines of the proof of Theorem 1.3. The generator and the set of invariant measures for the coupled process will be denoted by $\tilde{\Omega}$ and $\tilde{\mathscr{I}}$ respectively.

The idea which is exploited in the proof is that as t increases, γ_t will agree with η_t or ζ_t on longer and longer blocks of sites. Therefore if one looks at a fixed block of sites at a large time, then γ_t will agree with either η_t or ζ_t over that entire block with high probability. If initially $\eta_0 \equiv 0$ and $\zeta_0 \equiv 1$, then at large times η_t and ζ_t are distributed approximately as ν and $\bar{\nu}$ respectively. Hence γ_t should be distributed approximately as a convex combination of ν and $\bar{\nu}$.

For $m \leq n$ and $l \geq 1$, define functions $f_{m,n}$ and $g^l_{m,n}$ on X_3 in the following way: Let $m \leq x_1 < x_2 < \cdots < x_k \leq n$ be all those x's between m and n for which $\zeta(x) = 1$ and $\eta(x) = 0$. Then

$$f_{m,n}(\eta, \gamma, \zeta) = \begin{cases} 0 & \text{if } k = 0, \\ 1 + \text{number of } i \text{ such that } \gamma(x_{i+1}) \neq \gamma(x_i) & \text{if } k \geq 1, \end{cases}$$

and $g^l_{m,n}(\eta, \gamma, \zeta) = $ number of i such that $i \geq 1$, $i + l + 1 \leq k$, and

$$\gamma(x_i) \neq \gamma(x_{i+1}) = \gamma(x_{i+2}) = \cdots = \gamma(x_{i+l}) \neq \gamma(x_{i+l+1}).$$

In words, if one only looks at sites at which η and ζ differ, $f_{m,n}(\eta, \gamma, \zeta)$ is the number of intervals of zeros or ones in γ between m and n, and $g^l_{m,n}(\eta, \gamma, \zeta)$ is the number of interior intervals of length l. If $f_{-\infty, +\infty}(\eta, \gamma, \zeta)$ were finite, it would be this functional whose monotonicity in time we would be exploiting in the proof. Let $K = \max_\eta c(x, \eta)$ and

$$\varepsilon = \min\{c(100) + c(110), c(001) + c(011),$$

$$c(011) + c(110), c(100) + c(001)\}.$$

Lemma 3.7. (a) $f_{m,n}(\eta, \gamma, \zeta)$ and $g^l_{m,n}(\eta, \gamma, \zeta)$ increase as n increases or as m decreases.

(b) $$f_{m,n}(\eta, \gamma, \zeta) \leq 2 + \sum_{l=1}^\infty g^l_{m,n}(\eta, \gamma, \zeta).$$

(c) $$\sum_{l=1}^\infty l g^l_{m,n}(\eta, \gamma, \zeta) \leq n - m + 1.$$

If $\nu \in \tilde{\mathscr{I}}$, then

(d) $$\varepsilon \int g^1_{m,n} \, d\nu \leq K \int [f_{m-1,n} + f_{m,n+1} - 2f_{m,n}] \, d\nu$$

and

(e)
$$\varepsilon \int g_{m,n}^{l+1} \, d\nu \leq 4Kl \int g_{m,n}^{l} \, d\nu \quad \text{for } l \geq 1.$$

Proof. The first three statements are immediate consequences of the definitions of $f_{m,n}$ and $g_{m,n}^{l}$. Parts (d) and (e) are obtained from the statements

(3.8)
$$\int \tilde{\Omega} f_{m,n} \, d\nu = 0, \quad \text{and}$$

(3.9)
$$\int \tilde{\Omega} g_{m,n}^{l} \, d\nu = 0$$

respectively, which are true by Proposition 2.13 of Chapter I since $\nu \in \tilde{\mathcal{I}}$ and $f_{m,n}$ and $g_{m,n}^{l}$ are in the domain of $\tilde{\Omega}$. In order to do so, we must compute the left sides of (3.8) and (3.9). The only way $f_{m,n}$ can increase as the result of one flip is if $f_{m-1,n} = f_{m,n} + 1$, in which case the flip must occur at m, or if $f_{m,n+1} = f_{m,n} + 1$, in which case the flip must occur at n. The maximal rate at which these flips can occur is K. Therefore the positive terms in the sum defining $\tilde{\Omega} f_{m,n}$ are bounded above by

$$K[f_{m-1,n} + f_{m,n+1} - 2f_{m,n}].$$

On the other hand, there are $g_{m,n}^{l}$ sites x such that $m < x < n$ with the property that a flip at x will decrease $f_{m,n}$ by two. At such an x, $\gamma(x) = 1$ or 0. By symmetry, we may assume that $\gamma(x) = 1$. Then $\gamma(x-1) = \eta(x-1)$ and $\gamma(x+1) = \eta(x+1)$. Therefore a flip at x will occur at one of the following four rates: $c(111) + c(101)$, $c(110) + c(100)$, $c(011) + c(001)$, $c(010) + c(000)$. Since

$$c(101) \geq \max\{c(001), c(100)\}, \quad \text{and}$$

$$c(010) \geq \max\{c(110), c(011)\}$$

by attractiveness, each of the four rates is at least $\varepsilon/2$. Therefore, the negative terms in the sum defining $\tilde{\Omega} f_{m,n}$ are bounded above by $-\varepsilon g_{m,n}^{l}$. This then gives

$$\tilde{\Omega} f_{m,n} \leq K[f_{m-1,n} + f_{m,n+1} - 2f_{m,n}] - \varepsilon g_{m,n}^{l},$$

so that part (d) of the lemma follows from (3.8). The argument for part (e) is similar, but a bit more involved. First note that $g_{m,n}^{l}$ can only decrease via flips at no more than $lg_{m,n}^{l}$ x_i's or their neighbors. The rate at which that happens is at most $2K$ for an x_i and K for a neighbor of an x_i. Such a flip results in a decrease of one in $g_{m,n}^l$ except in case $x_{i+1} = x_i + 1$ and $\gamma(x_i) \neq \gamma(x_{i+1})$, in which case $g_{m,n}^l$ may decrease by two. But then x_i and x_{i+1} are each other's neighbors. Therefore, in any case, the negative terms in the sum defining $\tilde{\Omega} g_{m,n}^l$ are bounded below by $-4Kl g_{m,n}^l$. Next, note that $g_{m,n}^l$ can increase by one at no fewer than $g_{m,n}^{l+1}$ pairs of sites. These pairs of sites are the leftmost and rightmost x_i's in an interval of length $l+1$. To see what the minimal rate is for such a flip to occur, suppose that $u < v$ is such a pair of sites and that $\gamma(u) = \gamma(v) = 1$. Then $\gamma(u-1) = \eta(u-1)$ and $\gamma(v+1) = \eta(v+1)$. There is a flip at u at rate at least $c(100)$ if $\gamma(u-1) = \eta(u-1) = 1$, and at rate at least $c(011)$ if $\gamma(u-1) = \eta(u-1) = 0$. Similarly, there is a flip at v at rate at least $c(001)$ if $\gamma(v+1) = \eta(v+1) = 1$ and at rate at least $c(110)$ if $\gamma(v+1) = \eta(v+1) = 0$. Therefore in any case, the sum of the flip rates at u and v is at least ε, so that the positive terms in the sum defining $\tilde{\Omega} g_{m,n}^l$ are bounded below by $\varepsilon g_{m,n}^{l+1}$. Hence

$$\tilde{\Omega} g_{m,n}^l \geq \varepsilon g_{m,n}^{l+1} - 4Kl g_{m,n}^l.$$

This, together with (3.9), gives part (e) of the lemma. □

Now let

$$W_1 = \{(\eta, \gamma, \zeta) \in X_3 \colon \eta \equiv \gamma\}, \quad W_2 = \{(\eta, \gamma, \zeta) \in X_3 \colon \gamma \equiv \zeta\},$$

$$W_3 = \{(\eta, \gamma, \zeta) \in X_3 \setminus W_1 \cup W_2 \colon \text{there is an } x \in Z^1 \text{ so}$$

$$\text{that } \eta(y) = \gamma(y) \text{ for } y \leq x \text{ and } \gamma(y) = \zeta(y) \text{ for } y > x\},$$

$$W_4 = \{(\eta, \gamma, \zeta) \in X_3 \setminus W_1 \cup W_2 \colon \text{there is an } x \in Z^1 \text{ so}$$

$$\text{that } \eta(y) = \gamma(y) \text{ for } y > x \text{ and } \gamma(y) = \zeta(y) \text{ for } y \leq x\}.$$

Lemma 3.10. *Suppose that* $\varepsilon > 0$. *If* $\nu \in \tilde{\mathcal{I}}$, *then*

$$\nu(W_1 \cup W_2 \cup W_3 \cup W_4) = 1.$$

If $\nu \in \tilde{\mathcal{I}}_e$, *then* $\nu(W_i) = 1$ *for some* $1 \leq i \leq 4$.

Proof. The second statement follows from the first since for each $1 \leq i \leq 4$,

$$P^{(\eta, \gamma, \zeta)}[(\eta_t, \gamma_t, \zeta_t) \in W_i] = 1$$

whenever $(\eta, \gamma, \zeta) \in W_i$. This is a consequence of the nearest-neighbor character of the interaction. The first statement is equivalent to the assertion that

(3.11) $$\int g_{m,n}^l \, d\nu = 0 \quad \text{for all } l \geq 1 \text{ and } m \leq n.$$

To prove this, note that since

$$f_{m-1,n} \leq f_{m,n} + 1 \quad \text{and} \quad f_{m,n+1} \leq f_{m,n} + 1,$$

part (d) of Lemma 3.7 implies that

$$\sup_{m \leq n} \int g^1_{m,n} \, d\nu < \infty.$$

Repeated use of part (e) gives

$$\sup_{m \leq n} \int g^l_{m,n} \, d\nu < \infty$$

for all $l \geq 1$. By this and parts (b) and (c) of Lemma 3.7,

$$(3.12) \qquad \lim_{n-m \to \infty} \frac{1}{n-m} \int f_{m,n} \, d\nu = 0.$$

By part (d), if $N \geq 1$,

$$\varepsilon \sum_{m=-N+1}^{0} \sum_{n=0}^{N-1} \int g^1_{m,n} \, d\nu \leq K \sum_{n=0}^{N-1} \int f_{-N,n} \, d\nu + K \sum_{m=-N+1}^{0} \int f_{m,N} \, d\nu.$$

Therefore by (3.12),

$$\lim_{N \to \infty} \frac{1}{N^2} \sum_{m=-N+1}^{0} \sum_{n=0}^{N-1} \int g^1_{m,n} \, d\nu = 0.$$

By part (a) of Lemma 3.7, $\int g^1_{m,n} \, d\nu = 0$ for all $m \leq n$. Therefore by part (e) again, we obtain (3.11). $\square$

In order to prove the following theorem, it is necessary to couple together four copies of the spin system in a manner analogous to the three-process coupling with which Lemmas 3.7 and 3.10 are concerned. This coupled process is a Feller process on

$$X_4 = \{(\eta, \gamma^1, \gamma^2, \zeta) \in X^4 : \eta \leq \gamma^1, \gamma^2 \leq \zeta\},$$

in which the four coordinates at a given site flip together as much as possible. Not only are the four marginal processes versions of the spin system being studied, but also the processes $(\eta_t, \gamma^1_t, \zeta_t)$ and $(\eta_t, \gamma^2_t, \zeta_t)$ evolve according to the earlier described three-process coupling. Thus the statements of Lemmas 3.7 and 3.10 apply to each of them.

Theorem 3.13. *Under assumption* (3.4), *it follows that* $\mathcal{I}_e = \{\underline{\nu}, \bar{\nu}\}$.

Proof. First note that the result is immediate if (3.4) holds and $\varepsilon = 0$, since in that case either $c(011) = c(110) = 0$ or $c(100) = c(001) = 0$. In the first of these cases, for example, $c(100) > 0$ and $c(001) > 0$ by (3.4), and $c(111) = 0$ and $c(101) > 0$ by attractiveness. Therefore if η_t ever contains two consecutive ones, then the process converges to the pointmass on $\eta \equiv 1$, while if it never contains two consecutive ones, then the process converges to the pointmass on $\eta \equiv 0$. Hence we may assume that $\varepsilon > 0$. Now, take $\mu_1 \in \mathcal{I}_e$, and define μ_2 to be the following translate of μ_1:

$$\mu_2\{\eta: \eta(x) = 1 \text{ for } x \in T\} = \mu_1\{\eta: \eta(x+1) = 1 \text{ for } x \in T\}$$

for finite $T \subset Z^1$. By part (f) of Theorem 2.3 and by the four-process version of Theorem 2.15, there is a probability measure ν on X_4 with marginals $\underline{\nu}, \mu_1, \mu_2, \bar{\nu}$ which is extremal invariant for the coupled process $(\eta_t, \gamma_t^1, \gamma_t^2, \zeta_t)$. Let ν_1 and ν_2 be the probability measures on X_3 which are obtained from ν via the projections $(\eta, \gamma^1, \gamma^2, \zeta) \to (\eta, \gamma^1, \zeta)$ and $(\eta, \gamma^1, \gamma^2, \zeta) \to (\eta, \gamma^2, \zeta)$ respectively. By Lemma 3.10, $\nu_1(W_i) = 1$ for some $1 \le i \le 4$ and $\nu_2(W_i) = 1$ for some $1 \le i \le 4$. Since μ_1 and μ_2 are translates of one another, there is one i so that $\nu_1(W_i) = \nu_2(W_i) = 1$. Then

$$\nu\left\{(\eta, \gamma^1, \gamma^2, \zeta): \sum_x |\gamma^1(x) - \gamma^2(x)| < \infty\right\} = 1.$$

But the process (γ_t^1, γ_t^2) has the property that $P^{(\gamma,\gamma)}\{\gamma_t^1 = \gamma_t^2\} = 1$ and $P^{(\gamma^1,\gamma^2)}\{\gamma_t^1 = \gamma_t^2\} > 0$ for $t > 0$ whenever $\sum_x |\gamma^1(x) - \gamma^2(x)| < \infty$. Therefore since ν is invariant,

$$\nu\{(\eta, \gamma^1, \gamma^2, \zeta): \gamma^1 = \gamma^2\} = 1,$$

so that $\mu_1 = \mu_2$. Hence μ_1 is translation invariant, and therefore $i = 1$ or 2. If $i = 1$, then $\mu_1 = \underline{\nu}$, while if $i = 2$, then $\mu_1 = \bar{\nu}$. $\square$

Theorems 3.1 and 3.13 together tell us that $\mathcal{I}_e$ consists of at most two elements if and only if (3.4) is satisfied. The next step is to determine when $\mathcal{I}_e$ is a singleton, i.e., when the process is ergodic. The following result gives an important solution to this problem. While the fundamental ideas used in its proof are rather straightforward, the proof involves a large number of verifications. If these were carried out in complete detail, the proof would become quite lengthy. Thus, we will present an outline of the proof, referring the interested reader to the original paper (Gray (1982)) for the details.

Theorem 3.14. *If* $c(x, \eta) > 0$ *for all* η, *then* $\underline{\nu} = \bar{\nu}$, *and hence the system is ergodic* (*by Corollary 2.4*).

Outline of the Proof. The main idea is to exploit a coupling among the processes η_t^0, η_t^1, $\eta_t^{L_n}$, $\eta_t^{R_n}$, l_t^n, and r_t^n for $n \in Z^1$ which are defined in the following way:

(a) η_t^0, η_t^1, $\eta_t^{L_n}$, $\eta_t^{R_n}$ are versions of the spin system with rates $c(x, \eta)$ which have initial states $\eta_0^0 \equiv 0$, $\eta_0^1 \equiv 1$,

$$\eta_0^{L_n}(x) = \begin{cases} 1 & \text{if } x \geq n, \\ 0 & \text{if } x < n, \end{cases}$$

and

$$\eta_0^{R_n}(x) = \begin{cases} 1 & \text{if } x < n, \\ 0 & \text{if } x \geq n. \end{cases}$$

These are all coupled together using the basic coupling which was described at the beginning of Section 1. In particular, this coupling has the property that $\eta_t^1 \geq \eta_t^{L_n} \geq \eta_t^{L_{n+1}} \geq \eta_t^0$ and $\eta_t^0 \leq \eta_t^{R_n} \leq \eta_t^{R_{n+1}} \leq \eta_t^1$ for all $n \in Z^1$ and all $t \geq 0$.

(b) Superimposed on these processes are the left- and right-edge processes l_t^n and r_t^n. They have values in $Z^1 + \frac{1}{2} = \{x + \frac{1}{2}, x \in Z^1\}$. Initially, $l_0^n = r_0^n = n - \frac{1}{2}$. The transitions will be defined so that at all times,

(3.15) $$\eta_t^{L_n}(l_t^n - \tfrac{1}{2}) = 0, \qquad \eta_t^{L_n}(l_t^n + \tfrac{1}{2}) = 1,$$

(3.16) $$\eta_t^{R_n}(r_t^n - \tfrac{1}{2}) = 1, \quad \text{and} \quad \eta_t^{R_n}(r_t^n + \tfrac{1}{2}) = 0.$$

A transition occurs for l_t^n exactly at those times at which $\eta_t^{L_n}$ experiences a flip at $l_t^n - \frac{1}{2}$ or $l_t^n + \frac{1}{2}$. If the flip occurs at $l_t^n - \frac{1}{2}$, then l_t^n moves to the left to the nearest position satisfying (3.15), while if it occurs at $l_t^n + \frac{1}{2}$, it moves to the right to the nearest position satisfying (3.15). The transition rules for r_t^n are similar, designed to maintain relations (3.16). The processes described above have the following properties:

(3.17) $$\{l_t^n, t \geq s\} \text{ is determined by } l_s^n \text{ and } \{\eta_t^{L_n}, t \geq s\},$$

(3.18) $$\{r_t^n, t \geq s\} \text{ is determined by } r_s^n \text{ and } \{\eta_t^{R_n}, t \geq s\},$$

(3.19) $$\eta_s^0(x) = \eta_s^{L_n}(x) \quad \text{for all } x < l_s^n,$$

(3.20) $$\eta_s^0(x) = \eta_s^{R_n}(x) \quad \text{for all } x > r_s^n,$$

(3.21) $$\eta_s^1(x) = \eta_s^{L_n}(x) \quad \text{for all } x > l_s^n,$$

(3.22) $$\eta_s^1(x) = \eta_s^{R_n}(x) \quad \text{for all } x < r_s^n,$$

(3.23) $$\text{if } m \leq n, \text{ then } l_s^m \leq l_s^n \text{ and } r_s^m \leq r_s^n,$$

(3.24)   if $l_s^m = l_s^n$, then $l_t^m = l_t^n$ for all $t \geq s$,

(3.25)   if $r_s^m = r_s^n$, then $r_t^m = r_t^n$ for all $t \geq s$,

                                if $l_t^m + 2 \leq r_t^n$ for all $t \in [s, u]$ and

(3.26)   if $\eta_s^0(x) = \eta_s^1(x)$ for $l_s^m < x < r_s^n$, then $\eta_t^0(x) = \eta_t^1(x)$

                                for all $t \in [s, u]$ and $l_t^m < x < r_t^n$.

                                if $r_t^m + 2 \leq l_t^n$ for all $t \in [s, u]$ and

(3.27)   if $\eta_s^0(x) = \eta_s^1(x)$ for $r_s^m < x < l_s^n$, then

                    $\eta_t^0(x) = \eta_t^1(x)$ for all $t \in [s, u]$ and $r_t^m < x < l_t^n$.

The verification of the above properties is somewhat tedious, but is quite
straightforward. The last two above are consequences of (3.19), (3.20),
(3.21), and (3.22). The proof of the following property is less straightforward,
but is similar to the proof of Lemma 3.10. It relies on properties (3.23),
(3.24), and (3.25). For each $m \in Z^1$,

(3.28) $\lim_{s \to \infty} P[r_s^m = r_s^{m+1}] = \lim_{s \to \infty} P[l_s^m = l_s^{m+1}] = 1$,

and hence

(3.29) $\lim_{s \to \infty} P[|l_s^n| \leq k \text{ or } |r_s^n| \leq k \text{ for some } n] = 0$

for all $k \geq 0$. Next, note that l_t^m is a decreasing function of the process
$\{\eta_s^{L_m}, s \leq t\}$ and r_t^n is an increasing function of the process $\{\eta_s^{R_n}, s \leq t\}$.
Therefore Corollary 2.21 of Chapter II applied to the process $(\eta_s^{L_m}, \eta_s^{R_n})$
implies that

(3.30) $P(l_t^m > j \text{ and } r_t^n > k) \leq P(l_t^m > j) P(r_t^n > k)$

for all integers $m, n, j,$ and k. This, together with (3.29), can be used to
identify two edges on which we will focus for the rest of the proof. In order
to do this, define

$$m(s) = \sup\{m \in Z^1 : P(r_s^m < 0) > \tfrac{1}{2}\}.$$

Then by (3.30),

(3.31) $P(r_s^{m(s)} < 0 \text{ and } l_s^{m(s)} > 0) + P(r_s^{m(s)+1} > 0 \text{ and } l_s^{m(s)} < 0)$

                $\geq \tfrac{1}{2} P(l_s^{m(s)} > 0) + \tfrac{1}{2} P(l_s^{m(s)} < 0) = \tfrac{1}{2}.$

Therefore, by (3.29),

$$\lim_{s\to\infty}\{P(r_s^{m(s)}<-k \text{ and } l_s^{m(s)}>k)+P(r_s^{m(s)}>k \text{ and } l_s^{m(s)}<-k)\}\geq\tfrac{1}{2}$$

for any $k\geq 0$. So, for each k and sufficiently large s, with substantial probability there is a pair of right and left edges which started together but which are on opposite sides of the interval $[-k, k]$ at time s. In order to show that $\eta_s^0(x)=\eta_s^1(x)$ for all x between those two edges, it is enough by (3.26) and (3.27) to show that the same is true at the last time at which those two edges were at most two units apart. But now the strict positivity of the rates implies that this is the case with a probability which is bounded away from zero, independently of k and s. In the verification of this statement, properties (3.17) and (3.18) are used in an essential way. Therefore we conclude that

$$\lim_{k\to\infty}\lim_{t\to\infty} P(\eta_t^0(x)=\eta_t^1(x) \text{ for all } |x|\leq k)>0,$$

so that ν and $\bar{\nu}$ are not mutually singular with respect to one another. Hence by Proposition 2.16 of this chapter and Proposition 4.13 of Chapter I, $\nu=\bar{\nu}$. Therefore the process is ergodic by Corollary 2.4. $\square$

In order to gain a better understanding of when $\nu\neq\bar{\nu}$, consider now the case in which $c(000)=c(111)=0$, so that $\nu\neq\bar{\nu}$ trivially. Let $c_\varepsilon(x, \eta)$ be the perturbed system defined by

$$c_\varepsilon(x, \eta) = \begin{cases} c(x, \eta) & \text{if } \eta(x)=0, \\ c(x, \eta)+\varepsilon & \text{if } \eta(x)=1. \end{cases}$$

For $\varepsilon>0$, δ_1 is no longer invariant, so it is not clear whether or not the system is ergodic. Gray and Griffeath (1982) found a necessary and sufficient condition on $c(x, \eta)$ so that $c_\varepsilon(x, \eta)$ is nonergodic for some $\varepsilon>0$. We will give the proof of necessity in order to motivate the condition. The proof of the more important and much deeper sufficiency direction is based on an intricate contour argument, and will be omitted. (The proof in Gray and Griffeath's paper could be simplified substantially by using the techniques in Bramson and Gray (1985). It would still be nontrivial, however.) In the special case $c(010)=0$, the perturbed process is essentially the contact process so the sufficiency statement follows from results in Chapter VI.

Theorem 3.32. *A necessary and sufficient condition so that* $c_\varepsilon(x, \eta)$ *is non-ergodic for some* $\varepsilon>0$ *is that*

$$c(001)+c(100)>c(011)+c(110).$$

Proof of Necessity. The idea of the proof is quite simple. A block of zeros in the configuration of length at least two decreases in length by one at rate $c_\varepsilon(001) + c_\varepsilon(100)$ and increases in length by at least one at a rate which is at least $c_\varepsilon(011) + c_\varepsilon(110)$. Therefore if

$$c_\varepsilon(001) + c_\varepsilon(100) < c_\varepsilon(011) + c_\varepsilon(110),$$

the length of the block of zeros will increase at least at a linear rate. Using translation invariance, it follows that $\delta_1 S(t)$ converges weakly to δ_0 as $t \to \infty$. In order to formalize this argument, suppose $\nu \in \mathcal{I} \cap \mathcal{S}$, and let

$$f_n(\eta) = \begin{cases} 1 & \text{if } \eta(0) = \eta(1) = \cdots = \eta(n-1) = 0; \\ 0 & \text{otherwise,} \end{cases}$$

for $n \geq 1$. Since $f_n \in \mathcal{D}(\Omega_\varepsilon)$ where Ω_ε is the generator of the perturbed process, Proposition 2.13 of Chapter I implies that

(3.33) $$\int \Omega_\varepsilon f_n \, d\nu = 0$$

for all $n \geq 1$. Writing this out for $n \geq 2$ yields

$$0 = c_\varepsilon(010)\nu\{\eta(-1) = 0, \eta(0) = 1, \eta(1) = \cdots = \eta(n-1) = 0\}$$
$$+ c_\varepsilon(110)\nu\{\eta(-1) = \eta(0) = 1, \eta(1) = \cdots = \eta(n-1) = 0\}$$
$$+ \sum_{k=1}^{n-2} c_\varepsilon(010)\nu\{\eta(0) = \cdots = \eta(k-1) = 0, \eta(k) = 1,$$
$$\eta(k+1) = \cdots = \eta(n-1) = 0\}$$
$$+ c_\varepsilon(011)\nu\{\eta(0) = \cdots = \eta(n-2) = 0, \eta(n-1) = 1, \eta(n) = 1\}$$
$$+ c_\varepsilon(010)\nu\{\eta(0) = \cdots = \eta(n-2) = 0, \eta(n-1) = 1, \eta(n) = 0\}$$
$$- c_\varepsilon(100)\nu\{\eta(-1) = 1, \eta(0) = \cdots = \eta(n-1) = 0\}$$
$$- c_\varepsilon(001)\nu\{\eta(0) = \cdots = \eta(n-1) = 0, \eta(n) = 1\}.$$

Using the definition of $c_\varepsilon(x, \eta)$, the attractiveness assumption, and the fact that $\nu \in \mathcal{S}$, this implies that

$$0 \geq [c_\varepsilon(110) + c_\varepsilon(011)][F(n-1) - F(n)]$$
$$- [c_\varepsilon(100) + c_\varepsilon(001)][F(n) - F(n+1)],$$

where F is defined by

$$F(n) = \int f_n \, d\nu.$$

Therefore if $c_\varepsilon(001) + c_\varepsilon(100) < c_\varepsilon(011) + c_\varepsilon(110)$, it follows that $F(n) = F(1)$ for all $n \geq 1$. In particular,

$$\nu\{\eta(0) = 1,\ \eta(1) = 0\} = \nu\{\eta(0) = 0,\ \eta(1) = 1\} = F(1) - F(2) = 0.$$

Since $\nu \in \mathscr{S}$, it follows that ν is a convex combination of δ_0 and δ_1. Using (3.33) with $n = 1$ implies that $\nu = \delta_0$. Thus the system is ergodic whenever

$$c(001) + c(100) \leq c(011) + c(110)$$

and $\varepsilon > 0$. $\square$

As mentioned above, the proof of the sufficiency of the condition in Theorem 3.32 is fairly difficult. The result should be plausible, however, in view of the following argument. Suppose that initially $\eta \equiv 1$. If ε is positive but small, an occasional zero will appear in the configuration. When it does, it may grow into an interval of zeros. But for small ε, the length of an interval of zeros which is surrounded by ones grows at a negative rate. Therefore the length of this interval of zeros should behave something like a positive recurrent Markov chain on $\{0, 1, \ldots\}$ which is absorbed at 0. In other words, the interval of zeros should die out after a time with finite expected value. If the rate of production of zeros due to the ε is small compared with this expected time, then one would expect the limiting distribution as $t \to \infty$ to be different from δ_0.

The following example, which satisfies all the assumptions of this section except attractiveness, is analyzed at the end of Section 1 of Chapter VIII:

$$c(x, \eta) = \begin{cases} 1 & \text{if } \eta(x-1) \neq \eta(x+1), \\ 0 & \text{if } \eta(x-1) = \eta(x+1). \end{cases}$$

By transforming this system into an exclusion process, $\mathscr{I}_e$ is described completely in this case. The example turns out to have the following interesting properties: (a) $\mathscr{I}_e$ is not closed, and (b) $(\mathscr{I} \cap \mathscr{S})_e \not\subset \mathscr{I}_e$.

4. Duality for Spin Systems

The first step in developing a duality theory for a class of Markov processes is to choose a duality function H and a state space for the dual processes (recall Definition 3.1 of Chapter II). In the case of spin systems, the state space for the dual processes is taken to be

$$Y = \{A \colon A \text{ is a finite subset of } S \cup \{\infty\}\},$$

which is countable. The dual processes will therefore be Markov chains on

$Y.$ As will be seen later, the inclusion of the point ∞ is a useful device to guarantee that the transition rates for these Markov chains are nonnegative in many cases, when otherwise they could easily be negative. In order to make the duality computations tractable, it is important to choose a duality function with a rather simple structure. In the present context, the two most useful and important choices are

$$
(4.1) \qquad H_1(\eta, A) = \begin{cases} \displaystyle\prod_{x \in A \cap S} [1 - \eta(x)] & \text{if } \infty \notin A, \\[2ex] -\displaystyle\prod_{x \in A \cap S} [1 - \eta(x)] & \text{if } \infty \in A, \end{cases}
$$

and

$$
(4.2) \qquad H_2(\eta, A) = \begin{cases} \displaystyle\prod_{x \in A \cap S} [2\eta(x) - 1] & \text{if } \infty \notin A, \\[2ex] -\displaystyle\prod_{x \in A \cap S} [2\eta(x) - 1] & \text{if } \infty \in A. \end{cases}
$$

(The product over the empty set is taken to be 1.)

Suppose that the rates $c(x, \eta)$ have been written in the form

$$
(4.3) \quad c_1(x, \eta) = c(x)\{[1 - \eta(x)] + [2\eta(x) - 1] \sum_{A \in Y} p(x, A) H_1(\eta, A)\}
$$

in the first case and

$$
(4.4) \qquad c_2(x, \eta) = \frac{c(x)}{2}\left\{1 - [2\eta(x) - 1] \sum_{A \in Y} p(x, A) H_2(\eta, A)\right\}
$$

in the second case, where

$$
(4.5) \qquad\qquad c(x) \geq 0, \qquad \sup_x c(x) < \infty,
$$

$$
(4.6) \qquad\qquad p(x, A) \geq 0, \qquad \sum_A p(x, A) = b(x) \leq 1, \quad \text{and}
$$

$$
(4.7) \qquad\qquad \sup_x c(x) \sum_A p(x, A)|A| < \infty.
$$

Here $|A|$ denotes the cardinality of A. Note that the nonnegativity assumption on $p(x, A)$ is no restriction, since for $A \not\ni \infty$,

$$
H_i(\eta, A) = -H_i(\eta, A \cup \{\infty\}).
$$

The main content of (4.6) is the requirement that $b(x) \leq 1$.

Note that

$$H_1(\eta_u, A) - H_1(\eta, A) = \begin{cases} [2\eta(u) - 1]H_1(\eta, A\backslash u) & \text{if } u \in A, \\ 0 & \text{if } u \notin A, \end{cases}$$

and

$$H_2(\eta_u, A) - H_2(\eta, A) = \begin{cases} -2H_2(\eta, A) & \text{if } u \in A, \\ 0 & \text{if } u \notin A, \end{cases}$$

so that in either case,

$$(4.8) \qquad \sup_\eta |c_i(x, \eta_u) - c_i(x, \eta)| \le c(x) \sum_{A \ni u} p(x, A)$$

for $u \ne x$. Therefore (4.7) guarantees that (0.3) is satisfied. Note that in case (4.3), if $p(x, A) = 0$ whenever $\infty \in A$, then the spin system is attractive.

The next step is to compute Ω_i applied to H_i as a function of η, where Ω_i is defined by (0.5) using the rates $c_i(x, \eta)$. (Note that since A is finite, $H_i(\cdot, A)$ is in $\mathcal{D} \subset \mathcal{D}(\Omega_i)$.) In doing so, it is useful to note that

$$H_1(\eta, A)H_1(\eta, B) = \begin{cases} H_1(\eta, A \cup B) & \text{if } \infty \notin A \cap B, \\ H_1(\eta, (A \cup B)\backslash\infty) & \text{if } \infty \in A \cap B, \end{cases}$$

and

$$H_2(\eta, A)H_2(\eta, B) = H_2(\eta, A \Delta B)$$

where Δ denotes the symmetric difference. Then

$$
\begin{aligned}
\Omega_1 H_1(\eta, A) &= \sum_{x \in S} c_1(x, \eta)[H_1(\eta_x, A) - H_1(\eta, A)] \\
(4.9) \qquad &= \sum_{x \in A \cap S} c(x)\left\{ -H_1(\eta, A) + \sum_B p(x, B)H_1(\eta, B)H_1(\eta, A\backslash x) \right\} \\
&= \sum_B q_1(A, B)[H_1(\eta, B) - H_1(\eta, A)] - V(A)H_1(\eta, A),
\end{aligned}
$$

where $V(A) = \sum_{x \in A \cap S} c(x)(1 - b(x)) \ge 0$ and

$$q_1(A, B) = \sum_{x \in A \cap S} c(x) \sum_F p(x, F) \ge 0,$$

where the sum is over all $F \in Y$ such that $(A\backslash x) \cup F = B$ if $\infty \notin A \cap F$ and $((A\backslash x) \cup F)\backslash\infty = B$ if $\infty \in A \cap F$. Note that $q_1(A, B)$ can be interpreted as the transition rates for a continuous time Markov chain on Y which evolves in the following way:

(a) each $x \in A \cap S$ is removed from A at rate $c(x)b(x)$ and is replaced by the set F with probability $[b(x)]^{-1}p(x, F)$, and

(b) when an attempt is made to put a point at a site which is already occupied, the points coalesce if the site is in S, and annihilate one another if the site is ∞.

When a similar computation is carried out in the second case, one finds that

(4.10)
$$\begin{aligned}
\Omega_2 H_2(\eta, A) &= \sum_{x \in S} c_2(x, \eta)[H_2(\eta_x, A) - H_2(\eta, A)] \\
&= \sum_{x \in A \cap S} c(x)\left\{-H_2(\eta, A) + \sum_B p(x, B)H_2(\eta, B)H_2(\eta, A\backslash x)\right\} \\
&= \sum_B q_2(A, B)[H_2(\eta, B) - H_2(\eta, A)] - V(A)H_2(\eta, A)
\end{aligned}$$

where $V(A)$ is as before and

$$q_2(A, B) = \sum_{x \in A \cap S} c(x) \sum_{F: F\Delta(A\backslash x)=B} p(x, F) \ge 0.$$

The interpretation of the Markov chain with these transition rates is the same as in the first case, except that points which end up at the same site in S annihilate one another instead of coalescing. Thus duality with respect to H_1 in case (4.3) will be referred to as coalescing duality, while duality with respect to H_2 in case (4.4) will be referred to as annihilating duality.

In order to show in both cases that the dual chains on Y are nonexplosive, note that

$$\begin{aligned}
\sum_B q_i(A, B)[|B \cap S| - |A \cap S|] \\
\le \sum_{x \in A \cap S} c(x) \sum_F p(x, F)[|((A\backslash x) \cup F) \cap S| - |A \cap S|] \\
\le \sum_{x \in A \cap S} c(x) \sum_F p(x, F)[|F \cap S| - 1] \\
\le \omega |A \cap S|
\end{aligned}$$

where

(4.11)
$$\omega = \sup_x c(x) \sum_F p(x, F)[|F \cap S| - 1],$$

which is finite by (4.7). Not only does this estimate imply that the chains are nonexplosive, but it also yields the following bound on the growth of the chain A_t:

(4.12)
$$E^A|A_t \cap S| \le |A \cap S| e^{\omega t}.$$

(For more on explosions of Markov chains, see Chapter 15 of Breiman (1968) or Chapters 7 and 8 of Freedman (1971).)

The following is the basic duality theorem. When $b(x) \equiv 1$, so $V(A) \equiv 0$, the statement relating the spin system and the dual Markov chain is exactly the duality relation which was discussed in Section 3 of Chapter II. When $V(A)$ is not identically zero, an additional "Feynman–Kac" term appears. This term can be interpreted as killing at rate $V(A_t)$.

Theorem 4.13. *Let $S_i(t)$ be the semigroup for the spin system with rates $c_i(x, \eta)$ given in* (4.3) *or* (4.4) *under assumptions* (4.5), (4.6), *and* (4.7), *and let A_t be the Markov chain on Y with transition rates $q_i(A, B)$. Then for every $\eta \in X$, $A \in Y$ and $t \geq 0$,*

$$(4.14) \quad S_i(t) H_i(\,\cdot\,, A)(\eta) = E^A \left[H_i(\eta, A_t) \exp \left\{ -\int_0^t V(A_s)\, ds \right\} \right].$$

Proof. Let

$$u_\eta(t, A) = E^\eta H_i(\eta_t, A) = S_i(t) H_i(\,\cdot\,, A)(\eta).$$

By Theorem 2.9(c) of Chapter I, together with (4.9) and (4.10),

$$\frac{d}{dt} u_\eta(t, A) = S_i(t)\Omega_i H_i(\,\cdot\,, A)(\eta)$$

$$= \sum_B q_i(A, B)[S_i(t) H_i(\,\cdot\,, B)(\eta) - S_i(t) H_i(\,\cdot\,, A)(\eta)]$$

$$- V(A) S_i(t) H_i(\,\cdot\,, A)(\eta)$$

$$= \sum_B q_i(A, B)[u_\eta(t, B) - u_\eta(t, A)] - V(A) u_\eta(t, A).$$

But since the chain A_t is nonexplosive, the right side of (4.14) is the unique bounded solution to this differential equation with initial condition $u_\eta(0, A) = H_i(\eta, A)$. $\square$

It is often convenient to use (4.14) in an integrated form, as follows.

Corollary 4.15. *Suppose the assumptions of Theorem 4.13 are satisfied. If μ is a probability measure on X, put*

$$\hat{\mu}(A) = \int H_i(\eta, A)\mu(d\eta)$$

for $A \in Y$. Let $\mu_t = \mu S_i(t)$. Then

$$\hat{\mu}_t(A) = \sum_{B \in Y} E^A \left[\exp \left\{ -\int_0^t V(A_s)\, ds \right\}, A_t = B \right] \hat{\mu}(B).$$

In particular, if $V(A) \equiv 0$ and $\mu \in \mathcal{I}$, $\hat{\mu}$ is a harmonic function for the dual chain A_t.

Proof. Integrate (4.14) with respect to μ. □

To conclude this section, we will consider a number of examples of spin systems which have duals. In some cases, the same spin system can have both a coalescing and an annihilating dual.

Example 4.16. *The Voter Model.* Let

$$c(x, \eta) = \sum_{y:\, \eta(y) \neq \eta(x)} p(x, y),$$

where $p(x, y) \geq 0$ and $\sum_y p(x, y) = 1$. Then $c(x, \eta)$ can be written in either form (4.3) or (4.4) with, in each case, $c(x) \equiv 1$, $p(x, F) = p(x, y)$ if $F = \{y\}$ for some $y \in S$, and $p(x, F) = 0$ otherwise. Thus both coalescing and annihilating duality are available for the voter model.

Example 4.17. *The Anti-Voter Model.* Let

$$c(x, \eta) = \sum_{y:\, \eta(y) = \eta(x)} p(x, y),$$

where again $p(x, y) \geq 0$ and $\sum_y p(x, y) = 1$. The interpretation here is that at exponential times with parameter one, a "voter" at x chooses a y with probability $p(x, y)$ and then changes to the position opposite to that held by the voter at y. The anti-voter model rates cannot be written in the form (4.3), so it does not have a coalescing dual (unless one is willing to give up the assumption that $b(x) \leq 1$ in (4.6)). On the other hand, they can be written in the form (4.4) with $c(x) \equiv 1$, $p(x, F) = p(x, y)$ if $F = \{y, \infty\}$ for some $y \in S$, and $p(x, F) = 0$ otherwise, so that it does have an annihilating dual. Note that in this case, the point ∞ must be used.

Example 4.18. *The Contact Process.* Here $S = Z^d$ and

$$c(x, \eta) = \begin{cases} \lambda \sum_{|y-x|=1} \eta(y) & \text{if } \eta(x) = 0, \\ 1 & \text{if } \eta(x) = 1, \end{cases}$$

where $\lambda \geq 0$. This process has a coalescing dual, as can be seen by writing $c(x, \eta)$ in the form (4.3) with

$$c(x) \equiv 1 + 2\lambda d, \qquad p(x, \varnothing) = (1 + 2\lambda d)^{-1},$$

$$p(x, \{x, y\}) = \frac{\lambda}{1 + 2\lambda d} \quad \text{if } |x - y| = 1,$$

and $p(x, A) = 0$ otherwise. Note that since $c(x)p(x, \varnothing) = 1$ and

$$c(x)p(x, \{x, y\}) = \lambda \quad \text{for } |x - y| = 1,$$

the dual chain is exactly the contact process itself, restricted to configurations having finitely many ones. Thus in this sense, the contact process is self dual.

Various applications of duality will appear in the next section, as well as later in the book. It is worthwhile noting that in the above three examples, $b(x) \equiv 1$, so $V(A) \equiv 0$.

5. Applications of Duality

This section is devoted to several applications of the duality relation expressed in Theorem 4.13. Additional applications to the voter model and the contact process will appear in Chapters V and VI respectively. Further applications of duality outside the context of spin systems will be made in Chapters VIII and IX. Unless otherwise stated, the results in this section apply to both coalescing and annihilating duality. We will always assume that the rates $c(x, \eta)$ have the form (4.3) in the coalescing case and (4.4) in the annihilating case, where $c(x)$ and $p(x, A)$ satisfy (4.5), (4.6), and (4.7). The subscript i will be omitted in $c_i(x, \eta)$, $H_i(\eta, A)$, $S_i(t)$, etc. The main applications are to determine when the spin system is ergodic. Let τ be the hitting time of $\{\varnothing, \{\infty\}\}$ for the dual process.

Theorem 5.1. *Suppose*

$$\inf_{x \in S} c(x)[1 - b(x)] = \delta > 0.$$

Then the spin system with rates $c(x, \eta)$ is ergodic. The unique stationary measure ν is given by (using the notation of Corollary 4.15)

(5.2)
$$\hat{\nu}(A) = E^A\left[\exp\left\{-\int_0^\tau V(A_s)\, ds\right\}, \tau < \infty, A_\tau = \varnothing\right]$$
$$- E^A\left[\exp\left\{-\int_0^\tau V(A_s)\, ds\right\}, \tau < \infty, A_\tau = \{\infty\}\right].$$

Furthermore, if μ is any probability measure on X and $\mu_t = \mu S(t)$, then

$$|\hat{\mu}_t(A) - \hat{\nu}(A)| \leq 2e^{-\delta t}.$$

Proof. Since $V(\varnothing) = V(\{\infty\}) = 0$, Corollary 4.15 implies that

$$\hat{\mu}_t(A) = E^A\left[\exp\left\{-\int_0^\tau V(A_s)\,ds\right\}, \tau \le t, A_\tau = \varnothing\right]$$

$$- E^A\left[\exp\left\{-\int_0^\tau V(A_s)\,ds\right\}, \tau \le t, A_\tau = \{\infty\}\right]$$

$$+ \sum_{B \in Y} E^A\left[\exp\left\{-\int_0^t V(A_s)\,ds\right\}, \tau > t, A_t = B\right]\hat{\mu}(B).$$

Therefore, using the facts that $|\hat{\mu}(B)| \le 1$ for all $B \in Y$ and $V(A) \ge \delta$ for all $A \ne \varnothing, \{\infty\}$, it follows that $\hat{\mu}_t(A)$ differs from the right side of (5.2) by at most $2\exp(-\delta t)$. Since linear combinations of $H(\cdot, A)$ are dense in $C(X)$, all three statements in the theorem are immediate. $\square$

In view of Theorem 5.1, one would expect that most other interesting applications of duality would occur when $b(x) = 1$ for all $x \in S$. Consequently we will assume this to be the case for the rest of the section. Therefore $V(A) = 0$ for all A and the conclusion of Corollary 4.15 becomes

(5.3) $\hat{\mu}_t(A) = E^A\hat{\mu}(A_t),$

where $\mu_t = \mu S(t)$ and μ is any probability measure on X.

Now let γ be the pointmass on $\eta \equiv 1$ in the coalescing case, and let it be the product measure with density $\frac{1}{2}$ in the annihilating case. Then

$$\hat{\gamma}(A) = \begin{cases} 1 & \text{if } A = \varnothing, \\ -1 & \text{if } A = \{\infty\}, \\ 0 & \text{otherwise,} \end{cases}$$

so that (5.3) becomes in this case

(5.4) $\hat{\gamma}_t(A) = P^A(A_t = \varnothing) - P^A(A_t = \{\infty\}).$

Since $\varnothing$ and $\{\infty\}$ are traps for the dual process A_t, $\nu = \lim_{t\to\infty} \gamma_t$ exists, is invariant, and

(5.5) $\hat{\nu}(A) = P^A(A_\tau = \varnothing, \tau < \infty) - P^A(A_\tau = \{\infty\}, \tau < \infty).$

Theorem 5.6. *A sufficient condition for the spin system to be ergodic is that*

(5.7) $P^A(\tau < \infty) = 1 \quad \text{for all } A \in Y.$

If $p(x, A) = 0$ whenever $\infty \in A$, then (5.7) is a necessary condition for ergodicity as well.

Proof. The first statement follows from (5.3) by writing

$$\hat{\mu}_t(A) = P^A(A_\tau = \varnothing, \tau \le t) - P^A(A_\tau = \{\infty\}, \tau \le t) + E^A[\hat{\mu}(A_t), \tau > t]$$

and taking the limit as $t \to \infty$. For the second statement, note that if $p(x, A) = 0$ whenever $\infty \in A$, then δ_0 is invariant for the spin system in the coalescing case, and δ_1 is invariant in the annihilating case. If (5.7) fails, then it is clear from (5.5) that $\nu \ne \delta_0$ in the first case and $\nu \ne \delta_1$ in the second, so that the spin system is not ergodic. $\square$

Corollary 5.8. *If*

$$\sup_x c(x) \sum_F p(x, F)[|F \cap S| - 1] < 0,$$

then the spin system is ergodic.

Proof. By (4.11) and (4.12),

$$P^A(\tau = \infty) \le \lim_{t \to \infty} E^A |A_t \cap S| = 0. \quad \square$$

Sometimes duality can be turned around so that we get information about the dual chain rather than about the original spin system. The following is an illustration of this.

Proposition 5.9. *Let $\sigma(A) = P^A(\tau = \infty)$ for $A \in Y$. In the case of coalescing duality, if $p(x, A) = 0$ whenever $\infty \in A$, then $\sigma(A)$ is submodular in the sense that*

$$\sigma(A \cup B) + \sigma(A \cap B) \le \sigma(A) + \sigma(B)$$

whenever $A, B \subset S$.

Proof. For $H(\eta, A)$ given by (4.1) and $A, B \subset S$,

$$H(\eta, A) + H(\eta, B) \le H(\eta, A \cup B) + H(\eta, A \cap B),$$

so that for such A, B,

(5.10) $\hat{\nu}(A) + \hat{\nu}(B) \le \hat{\nu}(A \cup B) + \hat{\nu}(A \cap B).$

Since $p(x, A) = 0$ whenever $\infty \in A$,

$$\hat{\nu}(A) = P^A(\tau < \infty) = 1 - \sigma(A)$$

for finite $A \subset S$, so the submodularity of σ follows from (5.10). $\square$

Example 5.11. *The Contact Process* (continuation of Example 4.18). In this case, with respect to coalescing duality,

$$\sup_x c(x) \sum_F p(x, F)[|F \cap S| - 1] = (2d\lambda - 1),$$

so Corollary 5.8 gives ergodicity of the contact process for $\lambda < (2d)^{-1}$. This is the same result which was obtained in Example 4.3(c) of Chapter I. This can be improved by using Proposition 5.9 together with Theorem 5.6 in the following way. In order to conclude ergodicity from Theorem 5.6, it is necessary to show that $\sigma(A) = 0$ for all finite $A \subset S$. By Proposition 5.9, this will follow once we have shown that $\sigma(A) = 0$ whenever $|A| = 1$. Of course σ is harmonic for the dual chain A_t, so that for $A \neq \varnothing$,

$$(5.12) \qquad \sigma(A) = \frac{1}{|A|(1 + 2\lambda d)} \sum_{x \in A} \left\{ \sigma(A \backslash x) + \lambda \sum_{|y - x| = 1} \sigma(A \cup y) \right\}.$$

Let $\sigma_1 = \sigma(\{x\})$ and $\sigma_2 = \sigma(\{x, y\})$ for $|x - y| = 1$. These are independent of x and y by the translation invariance and symmetry of the mechanism. Applying (5.12) to a singleton A gives, since $\sigma(\varnothing) = 0$,

$$(5.13) \qquad\qquad\qquad \sigma_1 = \frac{2\lambda d}{1 + 2\lambda d} \sigma_2.$$

Applying it to a doubleton A containing two nearest neighbors gives, using Proposition 5.9,

$$\sigma_2 \leq \frac{1}{1 + 2\lambda d} \{\sigma_1 + \lambda \sigma_2 + \lambda(2d - 1)[2\sigma_2 - \sigma_1]\}.$$

Simplifying this, we see that

$$[\sigma_1 - \sigma_2][1 - (2d - 1)\lambda] \geq 0.$$

Therefore, if $\lambda < (2d - 1)^{-1}$, it follows that $\sigma_1 \geq \sigma_2$. This together with (5.13) implies that $\sigma_1 = \sigma_2 = 0$, and hence that the contact process is ergodic. Thus we obtain the following improved lower bound for the critical value λ_d in d dimensions:

$$(5.14) \qquad\qquad\qquad \lambda_d \geq \frac{1}{2d - 1}.$$

By using (5.12) for more sets A before using Proposition 5.9 to close up the system of inequalities, improved lower bounds can be obtained. For example, if $d = 1$ and (5.12) is used for all $A \subset \{0, 1, 2\}$, and if $d = 2$ and

(5.12) is used for all $A \subset \{(0, 0), (0, 1), (1, 0), (1, 1)\}$, one obtains

$$\lambda_1 \geq 1.18 \quad \text{and} \quad \lambda_2 \geq .36.$$

These computations can be found in Section 7 of Griffeath (1975). An alternate derivation of the first of these will be given in Section 2 of Chapter VI.

The next example illustrates the use of (5.3) in proving ergodicity in situations in which (5.7) fails. In particular, it shows that the additional assumption $p(x, A) = 0$ whenever $\infty \in A$ is needed in the second part of Theorem 5.6.

Example 5.15. *The Anti-Voter Model* (continuation of Example 4.17). Suppose that the discrete time Markov chain on S with transition probabilities $p(x, y)$ has the property that two independent copies of that chain with arbitrary initial states will occupy the same site at some time with probability one. Under this assumption, we will use annihilating duality to show that the corresponding anti-voter model is ergodic. In order to show that μ_t converges to ν for any initial probability measure μ, it suffices by (5.3) and (5.5) to show that for $A \in Y$,

$$(5.16) \qquad \lim_{t \to \infty} \sum_{\substack{B \subset S \\ B \neq \varnothing, |B| < \infty}} |P^A(A_t = B) - P^A(A_t = B \cup \{\infty\})| = 0,$$

since $\hat{\mu}(B) = -\hat{\mu}(B \cup \{\infty\})$ for any finite $B \subset S$. By our assumption, if $|A \cap S|$ is even, $P^A(\tau < \infty) = 1$, so (5.16) is immediate. On the other hand, if $|A \cap S|$ is odd, $P^A(\tau < \infty) = 0$, but $P^A(|A_t \cap S| = 1$ for all large $t) = 1$, so that in verifying (5.16) it suffices to show it when $|A \cap S| = 1$. Since each transition adds ∞ to the configuration if it was not there before, or deletes it if it was there before, this amounts to showing that

$$(5.17) \qquad e^{-t} \sum_y \left| \sum_{n=0}^{\infty} \left(\frac{t^{2n}}{(2n)!} p^{(2n)}(x, y) - \frac{t^{2n+1}}{(2n+1)!} p^{(2n+1)}(x, y) \right) \right|$$

tends to zero as $t \to \infty$ for each $x \in S$. But (5.17) is bounded above by

$$e^{-t} \sum_{n=0}^{\infty} \frac{t^{2n}}{(2n)!} \sum_z p(x, z) \sum_y |p^{(2n)}(x, y) - p^{(2n)}(z, y)| + e^{-t} \sum_{n=0}^{\infty} \frac{t^{2n}}{(2n)!} \left| 1 - \frac{t}{2n+1} \right|.$$

The second term goes to zero as $t \to \infty$ by the Law of Large Numbers applied to the Poisson process, while the first term goes to zero because there is a successful coupling between two copies of the discrete time Markov chain with initial states x and z respectively. (See the final remarks in Section 1 of Chapter II.) Matloff (1977, 1980) obtained necessary and sufficient conditions on $p(x, y)$ for the anti-voter model to be ergodic. He then found all invariant measures in the nonergodic case, and proved the appropriate convergence theorem.

As was seen in the proof of Theorem 5.6, if $p(x, A) = 0$ whenever $\infty \in A$ and

$$P^A(\tau < \infty) < 1 \quad \text{for some } A \in Y,$$

then ν and δ_1 or δ_0 are two distinct invariant measures for the spin system. In the translation invariant case, these are both translation invariant. The following result gives sufficient conditions for these and their convex combinations to be the only invariant measures which are translation invariant (in the coalescing case).

Theorem 5.18. *Assume that the spin system is translation invariant on $S = Z^d$ and that coalescing duality holds (i.e., assume that the rates are given by (4.3) where $c(x) = c > 0$ is constant, $b(x) = 1$, and $p(x, A) = p(0, A - x)$). Suppose further that*

(5.19) $p(x, A) = 0 \quad \text{whenever } \infty \in A,$

(5.20) $p(x, \emptyset) > 0 \quad (\text{recall that this does not depend on } x),$

(5.21) $\begin{aligned} P^{\{x\}}[y \in A_t] > 0 \quad &\text{for all } x, y \in Z^d \text{ such that } y \geq x \text{ componentwise} \\ &\text{and all } t > 0, \quad \text{and} \end{aligned}$

(5.22) $P^A(\tau < \infty) < 1 \quad \text{for some } A \in Y.$

Then for any $\mu \in \mathscr{S}$,

$$\lim_{t \to \infty} \mu S(t) = \alpha \delta_0 + (1 - \alpha)\nu,$$

where $\alpha = \mu\{\eta: \eta \equiv 0\}$ and ν is the invariant measure given in (5.5). In particular, all elements of $\mathscr{I} \cap \mathscr{S}$ are convex combinations of δ_0 and ν.

Proof. Suppose that $\mu \in \mathscr{S}$ and $\mu\{\eta: \eta \equiv 0\} = 0$. Then we need to show that $\lim_{t \to \infty} \mu S(t) = \nu$. By (5.3) and (5.5), this is equivalent to

$$\lim_{t \to \infty} \sum_{B \neq \emptyset} P^A(A_t = B)\hat{\mu}(B) = 0$$

for all finite $A \subset S$. The idea is to show that

(5.23) $\lim_{t \to \infty} P^A(0 < |A_t| \leq k) = 0$

for any k and any finite $A \subset S$, and that

(5.24) $\lim_{k \to \infty} \sup\{\hat{\mu}(B) : |B| = k, B \subset S\} = 0.$

Unfortunately, (5.24) is not true for all the μ's we are considering. For example, it fails for the measure which puts mass $\frac{1}{2}$ on each of the configurations $\cdots 101010 \cdots$ and $\cdots 010101 \cdots$. Therefore we will show (5.24) for μ replaced by $\mu S(t)$ for any $t > 0$. By the semigroup property and the fact that $\mu S(t) \in \mathscr{S}$ and satisfies $\mu S(t)\{\eta : \eta \equiv 0\} = 0$, that is enough. To check (5.23), simply note that $\varnothing$ is a trap for the dual chain, and that

(5.25) $P^A(|A_t| = k - 1) \geq kct\, e^{-kct} p(x, \varnothing)$

whenever $|A| = k \geq 1$. By (5.20), the right side above is strictly positive. Using (5.25) and the Markov property,

$$P^A(|A_{s+t}| = k - 1) \geq kct\, e^{-kct} p(x, \varnothing) P^A(|A_s| = k)$$

for $k \geq 1$, so that knowing that

(5.26) $\lim_{t \to \infty} P^A(|A_t| = k) = 0$

for one $k \geq 1$ implies the same statement for $k + 1$. To show (5.26) for $k = 1$, use (5.25) again to show that

$$\frac{d}{dt} P^A(A_t = \varnothing) \geq P^A(|A_t| = 1) cp(x, \varnothing).$$

Therefore

$$\int_0^\infty P^A(|A_t| = 1)\, dt < \infty,$$

from which it follows easily that (5.26) holds for $k = 1$. It remains now to prove that (5.24) is correct when μ is replaced by $\mu S(t)$ for $t > 0$. Let $\mu_t = \mu S(t)$. Then

(5.27)
$$\hat{\mu}_t(B) = \int \left[\prod_{x \in B} [1 - \eta(x)] \right] \mu_t(d\eta)$$
$$= \int \left[E^\eta \prod_{x \in B} [1 - \eta_t(x)] \right] \mu(d\eta).$$

By Theorem 4.6 of Chapter I, for every $\varepsilon > 0$ and $t \geq 0$, there is an L

depending on ε and t so that

$$(5.28) \qquad E^{\eta} \prod_{x \in B} [1 - \eta_t(x)] \le \prod_{x \in B} E^{\eta}[1 - \eta_t(x)] + \varepsilon|B|$$

whenever

$$(5.29) \qquad \min\{|x - y|: x, y \in B, x \neq y\} \ge L.$$

By Hölder's inequality and the fact that $\mu \in \mathcal{S}$,

$$(5.30) \quad \int \left[\prod_{x \in B} E^{\eta}[1 - \eta_t(x)] \right] \mu(d\eta) \le \int [E^{\eta}[1 - \eta_t(0)]]^{|B|} \mu(d\eta).$$

Putting together (5.27), (5.28), and (5.30), we see that for any B satisfying (5.29),

$$(5.31) \qquad \hat{\mu}_t(B) \le \varepsilon|B| + \int [E^{\eta}[1 - \eta_t(0)]]^{|B|} \mu(d\eta).$$

Since $\mu \in \mathcal{S}$, $\mu\{\eta: \eta(y) = 0 \text{ for all } y \ge x\}$ is independent of x, and hence since $\mu\{\eta: \eta \equiv 0\} = 0$,

$$\mu\{\eta: \eta(y) = 0 \text{ for all } y \ge x\} = 0.$$

Therefore by (5.21) and the fact that

$$E^{\eta}[1 - \eta_t(0)] = E^{\{0\}} \prod_{x \in A_t} [1 - \eta(x)],$$

$$\mu\{\eta: E^{\eta}\eta_t(0) > 0\} = 1 \quad \text{for } t > 0.$$

It follows from the Dominated Convergence Theorem that

$$(5.32) \qquad \lim_{k \to \infty} \int [E^{\eta}[1 - \eta_t(0)]]^k \mu(d\eta) = 0$$

for $t > 0$. Now, given L and k, if $|A|$ is sufficiently large, there is a $B \subset A$ so that $|B| = k$ and B satisfies (5.29) for that L. Then (5.31) gives

$$\hat{\mu}_t(A) \le \hat{\mu}_t(B) \le \varepsilon k + \int [E^{\eta}[1 - \eta_t(0)]]^k \mu(d\eta),$$

so that

$$\limsup_{l \to \infty} \sup\{\hat{\mu}_t(A): |A| = l, A \subset S\} \le \varepsilon k + \int [E^{\eta}[1 - \eta_t(0)]]^k \mu(d\eta)$$

for every $\varepsilon > 0$ and $k \geq 1$. By (5.32), it follows that μ_t satisfies (5.24) for all $t > 0$, thus completing the proof. $\square$

We conclude this section with a few remarks and examples related to Theorem 5.18. The contact process satisfies all the hypotheses of the theorem. In one dimension, Theorem 3.13 provides in that case a conclusion which is stronger in one sense but weaker in another than that of Theorem 5.18. However in higher dimensions Theorem 5.18 provides information which we have not seen before. The voter model in dimensions greater than two shows that assumption (5.20) is needed in the theorem. The first of the next two examples shows that assumption (5.21) is needed as well. The second example shows that even when the assumptions of Theorem 5.18 are all satisfied, the spin system may have invariant measures which are not translation invariant.

Example 5.33. Suppose that $d = 1$ and

$$c(x, \eta) = \begin{cases} 1 & \text{if } \eta(x) = 1, \\ \lambda[\eta(x-2) + \eta(x+2)] & \text{if } \eta(x) = 0. \end{cases}$$

Then the odd and even coordinates are independent copies of the one-dimensional contact process. For λ larger that its critical value, the one-dimensional contact process has exactly two extremal invariant measures by Theorem 3.13. Hence for such λ, the process described above has exactly four extremal invariant measures. Two of these are not translation invariant, but the average of them is.

Example 5.34. Take $S = Z^2$, $c((k, l)) = 1 + 2\lambda$,

$$p((k, l), \{(k, l), (k, l+1)\}) = \frac{\lambda}{1 + 2\lambda},$$

$$p((k, l), \{(k, l), (k, l-1)\}) = \frac{\lambda}{1 + 2\lambda},$$

$$p((k, l), \{(k+1, l)\}) = \frac{\frac{1}{2}}{1 + 2\lambda},$$

$$p((k, l), \varnothing) = \frac{\frac{1}{2}}{1 + 2\lambda},$$

and $p((k, l), A) = 0$ otherwise. Using coalescing duality, the flip rates become

$$c((k, l), \eta) = \begin{cases} \lambda\eta(k, l+1) + \lambda\eta(k, l-1) + \frac{1}{2}\eta(k+1, l) & \text{if } \eta(k, l) = 0, \\ 1 - \frac{1}{2}\eta(k+1, l) & \text{if } \eta(k, l) = 1. \end{cases}$$

Note that if $\eta(k, l) = 0$ whenever $k \geq 1$, then

$$P^{\eta}[\eta_t(k, l) = 0] = 1$$

for all $k \geq 1$ and $t \geq 0$. Furthermore, when using such an initial configuration, $\{\eta_t(0, l), l \in Z^1\}$ evolves just like the one-dimensional contact process from Example 4.18. Also, when restricted to $\{(0, l), l \in Z^1\}$, the dual chain is the same as the one-dimensional contact process dual. Therefore, if λ is larger than the critical value for the one-dimensional contact process, all of the assumptions of Theorem 5.18 are satisfied, yet the system has an invariant measure which concentrates on $\{\eta : \eta(k, l) = 0$ for all $k \geq 1\}$ which is not the pointmass on $\eta \equiv 0$.

6. Additive Spin Systems and the Graphical Representation

The graphical representation which will be described in this section provides an important link between coupling and coalescing duality for spin systems which have a certain additivity property. The setting is that of Section 4. Let $c(x)$, $b(x)$, and $p(x, A)$ satisfy (4.5), (4.6), and (4.7). We will assume in this section that $p(x, A) = 0$ whenever $\infty \in A$, so we may as well redefine Y to be the set of all finite subsets of the countable set S.

The main idea is to construct a percolation structure on which to define the spin system of interest. Begin with $S \times [0, \infty)$, which should be thought of as giving a time line to each $x \in S$. In order to fix the ideas, think of the points of S as being laid out on a horizontal axis, with the time lines being placed vertically, above that axis. Define independent Poisson processes $N_x(t)$ with rate $c(x)[1 - b(x)]$ for each $x \in S$ and $N_x^A(t)$ with rate $c(x)p(x, A)$ for each $x \in S$ and $A \in Y$. At each event time s of $N_x(\cdot)$, place the symbol β at the point (x, s) in the percolation structure. At each event time s of $N_x^A(\cdot)$, place arrows from (y, s) to (x, s) for every $y \in A \setminus \{x\}$. If $x \notin A$, also place the symbol δ at (x, s). Of course, β and δ represent births and deaths respectively.

A typical realization of this percolation structure is given in Figure 1. In this figure, t_1, t_3, and t_5 are event times for $N_x(\cdot)$, $N_u(\cdot)$, and $N_z(\cdot)$ respectively, while t_2 is an event time for $N_v^A(\cdot)$ where $A = \{u, w\}$ and t_4 is an event time for $N_y^A(\cdot)$ where $A = \{x, y, z\}$.

For $x, y \in S$ and $0 \leq s \leq t$, say that there is an active path from (x, s) to (y, t) if (y, t) can be reached from (x, s) along an alternating sequence of upward and horizontal (traversed in the direction of the arrow) edges in such a way that no δ lies in the interior of an upward edge, and that there is no δ at (y, t) itself. For $\eta \in X$, let η_t^{η} be defined by $\eta_t^{\eta}(y) = 1$ if and only if there is an active path to (y, t) from some $(x, 0)$ such that $\eta(x) = 1$ or from some (x, s) with $s \leq t$ such that there is a β at (x, s). The interpretation is that if fluid is placed at each $(x, 0)$ such that $\eta(x) = 1$ and also at all

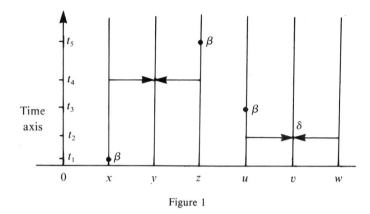

Figure 1

points marked by a β, and the fluid is allowed to move up and horizontally in the direction of the arrows, with δ's stopping the vertical flow of fluid, then $\{(x, t): \eta_t^\eta(x) = 1\}$ is the set of all points which can be reached by the fluid.

From the construction, we see that $\eta_{t-}^\eta(x) = 1$ and $\eta_t^\eta(x) = 0$ will occur if and only if t is an event time of the process $N_x^A(\cdot)$ for some A such that $\eta_{t-}^\eta \equiv 0$ on A. Thus the rate of flipping from 1 to 0 is

$$c(x) \sum_A p(x, A) H_1(\eta, A)$$

where H_1 is given by (4.1). Likewise, $\eta_{t-}^\eta(x) = 0$ and $\eta_t^\eta(x) = 1$ will occur if and only if either t is an event time for $N_x(\cdot)$ or t is an event time for $N_x^A(\cdot)$ for some A such that $\eta_{t-}^\eta \not\equiv 0$ on A. Therefore the rate of flipping from 0 to 1 is

$$c(x) \left\{ (1 - b(x)) + \sum_A p(x, A)[1 - H_1(\eta, A)] \right\}.$$

In other words η_t^η is the spin system with rates given by (4.3) and initial configuration η. Thus the above construction provides a graphical representation of the process.

The connection between the graphical representation and coupling should now be clear. The definition of the spin system with rates given by (4.3) in terms of the percolation structure provides a simultaneous coupling of copies of the spin system for all possible initial configurations. This coupling is not in general the same as the basic coupling which was discussed in Section 1 of this chapter. This point can perhaps be made clearest by considering the special cases of the voter model on Z^1 with $p(x, y) = \frac{1}{2}$ whenever $|x - y| = 1$ and the contact process on Z^1. (See Examples 4.16 and 4.18.)

In the case of the voter model,

$$p(x, \{x+1\}) = p(x, \{x-1\}) = \tfrac{1}{2}.$$

Thus the graphical representation consists entirely of arrows from (x, t) to $(x \pm 1, t)$ for some t, with δ's at the heads of all arrows. For a given nearest neighbor pair and given arrow orientation, the arrows are put down according to a Poisson process with rate $\tfrac{1}{2}$. The resulting coupling corresponds to having a given site in each version of the process use the same time to reassess his position, and then choose the same neighbor to imitate. In particular, if $\eta(x) + \zeta(x) = 1$ for all x initially, then $\eta_t(x) + \zeta_t(x) = 1$ for all x at all later times. The basic coupling certainly does not have this property. If initially $\eta(x) \leq \zeta(x)$ for all x, then the basic coupling does coincide with the one coming from the graphical representation.

In the case of the contact process, the graphical representation consists of δ's (without arrows) placed according to Poisson processes with rate 1 and arrows (without δ's) joining (x, t) to $(x \pm 1, t)$ at times determined by Poisson processes with rate λ. The difference between the two couplings occurs when one copy of the process is in configuration $\cdots 100 \cdots$ and the other is in configuration $\cdots 001 \cdots$. Using the basic coupling, the middle zero flips to a one simultaneously in the two processes. In the coupling coming from the graphical representation, they flip independently. The graphical representation will play an important role in the analysis of the contact process in Chapter VI.

A useful observation is that the graphical construction permits one to see immediately that the spin system with rates given by (4.3) is additive in the sense of the following definition.

Definition 6.1. A spin system is said to be *additive* provided that there exists processes $\{\eta_t^\eta, \eta \in X\}$ defined on a common probability space so that

(a)   for each $\eta \in X$, η_t^η is a version of that system with initial configuration η, and

(b) $$\eta_t^{\eta \vee \zeta} = \eta_t^\eta \vee \eta_t^\zeta$$

for all $\eta, \zeta \in X$.

The connection between the graphical representation and (coalescing) duality is easiest to see in the case that $b(x) = 1$ for all $x \in S$. In that case the percolation structure contains no β's. Fix a $t > 0$ and for $A \in Y$ define $A_t^A \in Y$ by letting A_t^A be the set of all $x \in S$ so that there is an active path from $(x, 0)$ to (y, t) for some $y \in A$. Then A_t^A is a version of the coalescing dual chain which was introduced in Section 4 with initial set A. A fluid flow interpretation analogous to that discussed above for the process η_t^η can be obtained by putting fluid at all points (y, t) for $y \in A$ and letting it flow down and horizontally as before, except that the arrows must now be traversed in the opposite direction. A_t^A is then the set of points x so that $(x, 0)$ is wet.

With the above definitions, the processes $\{A_t^A, A \in Y\}$ and $\{\eta_t^\eta, \eta \in X\}$ have been defined on a common probability space. Furthermore, the following relationship holds with probability one:

$$\{\eta_t^\eta = 0 \text{ on } A\} = \{\eta = 0 \text{ on } A_t^A\}$$

for all $\eta \in X$ and $A \in Y$. By equating the probabilities of these two events, the basic duality relation (4.14) is obtained.

The more general case in which $b(x)$ may be less than one can be treated in a similar way, by killing the dual process if it ever meets the symbol β. A more complex graphical representation is available for spin systems of the form (4.4). For details, see Chapter III of Griffeath (1979a).

The graphical representation was presented here as a unifying device for this chapter. More importantly, the graphical representation is often a useful tool in studying processes which can be defined in terms of it. This will be seen most clearly in Chapter VI.

7. Notes and References

Section 1. Couplings of the type described in this section have been used extensively in the study of spin systems. Examples of papers in which they have played an important role are Vasershtein (1969), Dobrushin (1971a), Holley (1972a, 1974a), Harris (1974), Griffeath (1975), Liggett (1978a), and Cocozza and Roussignol (1979, 1980). The treatment here is an expanded version of Sections 2.1, 2.2, 2.3, and 4.3 of the first part of Liggett (1977b).

Section 2. Most of this material on attractive spin systems is based on Holley (1972a). The term "attractive" was introduced in that paper. Theorem 2.14 in the case of the contact process appeared in Harris (1974). A version of Theorem 2.15 in the context of the exclusion process appeared in Liggett (1976). Proposition 2.16 was proved under somewhat different hypotheses by Durrett (1980). The proof given here is based on the proof of Corollary 2.28 in Holley (1974a). Durrett's proof does not rely on Theorem 2.13, and hence applies more generally. His proof does use attractiveness, as does the present proof.

Section 3. Theorem 3.13 is the main result in Liggett (1978a). Theorem 3.14 is due to Gray (1982). Theorem 3.32 is due to Gray and Griffeath (1982).

Section 4. The concept of duality in interacting particle systems was first observed in the case of the exclusion process by Spitzer (1970). It was then used in that context by Liggett (1973a, 1974a) and Spitzer (1974b). Vasershtein and Leontovich (1970) exploited a form of duality in a particular discrete time situation. In contexts which include continuous time spin systems, duality was discovered and first used by Holley and Liggett (1975) and by Harris (1976) (who used the term "association" rather than "duality"). Duality was then used by a number of authors, including Griffeath (1977, 1979a), Bertein and Galves (1977b), Schwartz (1976a, 1977), Matloff (1977, 1980), Harris (1978), Holley and Stroock (1979b), and Spitzer (1981). In their paper, Holley and Stroock present a general theory of duality for spin systems which includes the two versions presented in this section. Their duality can be thought of as being a mixture of coalescing and annihilating duality. Theorem 4.13 and Corollary 4.15 are special cases of their Theorem 2.15. Gray (1985) has developed a type of duality theory which applies to general attractive spin systems. The anti-voter model of Example 4.17 was introduced by Matloff (1977, 1980), who analyzed it using a different duality theory. The present duality for this model comes from Holley and Stroock (1979b). Duality has also been applied in the closely related fields of measure-valued diffusions (see, for example, Section 5.2 of Dawson and Hochberg (1982)) and infinite systems of stochastic differential equations (see, for example, Shiga (1980a, b)).

Section 5. Theorems 5.1 and 5.6 and Example 5.15 come from Holley and Stroock (1979b). Corollary 5.8 is from Liggett (1977b). Proposition 5.9 for the contact process was proved via coupling by Harris (1974), who then used it to deduce (5.14), as well as the improved value of λ_1. Theorem 5.18 is due to Harris (1976). Example 5.33 is from Liggett (1978a), while Example 5.34 is due to Holley and Stroock (1979b).

Section 6. The graphical approach to additive interacting particle systems was initiated at about the same time by Bertein and Galves (1977b), Gray and Griffeath (1977), and Harris (1978). Harris demonstrated most clearly the importance of this technique. He used it to obtain growth rates for additive processes and to prove an individual ergodic theorem for them. Some of the main ideas appeared earlier in other contexts. For examples, see Broadbent and Hammersley (1957), Toom (1968), Vasilyev (1969), and Clifford and Sudbury (1973). More recently, the graphical approach and related contour methods have been exploited by Griffeath (1979a), Gray (1982), Gray and Griffeath (1982), and Durrett and Griffeath (1983).

8. Open Problems

1. Show that critical values exist for the majority vote process described in Example 4.3(e) of Chapter I. In other words, show that if for a given N and $\delta \in (0, \frac{1}{2})$ the process is ergodic, then the process with the same N and

$\delta' \in (\delta, \frac{1}{2})$ is also ergodic. Of course the one-dimensional nearest-neighbor case is not of interest in view of Example 2.12. Something beyond coupling techniques would seem to be required to solve this problem. Note that this problem is a special case of Problem 1 in Chapter I.

2. Find ways other than that of Theorem 1.5 to compare spin systems with different rates. As a particular example of this problem, consider the stochastic Ising model and majority vote process of Example 4.3 of Chapter I, with N being the set of $2d$ nearest neighbors of the origin in Z^d. As mentioned there, the stochastic Ising model is not ergodic for large β if $d \geq 2$, but it is not known whether the majority vote process is ergodic for small δ if $d \geq 2$. One way of proving that it is not would be to find a good comparison between the stochastic Ising model with parameter β and the majority vote process with parameter δ (depending on β). A reasonable guess for this dependence is

$$\delta = [1 + e^{4\beta d}]^{-1},$$

since this would make the quantity

$$\frac{\max\limits_{\eta} c(x, \eta)}{\min\limits_{\eta} c(x, \eta)}$$

the same for both processes. One would expect that if the stochastic Ising model is not ergodic for a given β, then the corresponding majority vote process would also fail to be ergodic, since the rates for the latter are always equal to their maximum or minimum values, while the rates for the former usually lie strictly between them. Again, something more subtle than coupling is almost certainly needed to solve this problem.

3. Prove Theorem 1.12 in case $c(x, \eta)$ is allowed to depend on the coordinates $\{\eta(y), y \geq x\}$. The invariant measure will no longer be the product measure with density $\frac{1}{2}$. This problem is open even if $c(x, \eta) > 0$ depends only on $\eta(x)$ and $\eta(x+1)$. Holley and Stroock (1979b) have solved it partially, but not completely, in this case.

4. Generalize Theorem 3.13 to the finite range attractive case on Z^1. It is not clear what assumption should be used in place of (3.4). It is not enough to assume that

$$c(x, \eta) + c(x, \eta_x) > 0$$

for all η. A counterexample to this is provided by Example 5.33.

5. Prove some version of Theorem 3.13 in the nearest-neighbor attractive case on Z^2. Again (3.4) would have to be strengthened, since a number of counterexamples exist, including Example 5.34. No theorem of this type on Z^d with $d \geq 3$ is possible, since the voter model and the stochastic Ising model provide counterexamples there. As an indication of the difficulty of this problem, it should be noted that via the relationship established in Chapter IV between the stochastic Ising model and Gibbs states, such a theorem with (3.4) replaced by strict positivity of the rates would imply some deep recent results about Gibbs states in two dimensions (see Aizenmann (1980) and Higuchi (1979)).

6. Replace the nearest-neighbor assumption in Theorem 3.14 by a finite range assumption. Hence prove that a finite range, attractive, one-dimensional, translation invariant spin system with strictly positive rates is ergodic. This conjecture, with or without the attractiveness assumption, is often referred to as the positive rates conjecture, and is one of the most important open problems in this subject. Such a result is false in higher dimensions—the stochastic Ising model again provides a counterexample for dimensions greater than one. Perhaps the major reason for believing the positive rates conjecture is Theorem 3.13 of Chapter IV. See the discussion following that theorem.

7. Suppose ν is the unique invariant measure for a translation invariant ergodic spin system on Z^d. Is it true, under minimal assumptions, that $\nu \in \mathscr{S}_e$? By Proposition 2.16, this is true in the attractive case. By Theorem 4.20 and Proposition 4.11 of Chapter I, this is also true under assumption (0.6) in the finite range case. In thinking about this problem, it may be useful to keep in mind the following example of an ergodic translation invariant system on Z^1 whose unique invariant measure is not in $\mathscr{S}_e$. Given a configuration $\eta \in \{0, 1\}^{Z^1}$, entire consecutive blocks of zeros or ones in η flip together at rate 1. This system is ergodic with invariant measure $\frac{1}{2}\delta_0 + \frac{1}{2}\delta_1$. To see this, note that $\{\eta_t(x), |x| \leq n\}$ is Markovian for each n. Of course, this system is neither a spin system nor is it a monotone process.

Chapter IV

Stochastic Ising Models

Stochastic Ising models can be thought of loosely as reversible spin systems with strictly positive rates. (For a more precise version of this statement, see Theorem 2.13.) The measures with respect to which they are reversible are the Gibbs states of classical statistical mechanics. Thus stochastic Ising models can be viewed as models for nonequilibrium statistical mechanics. These were among the first spin systems to be studied, because of their close connections with physics. While no prior knowledge of statistical mechanics will be assumed in this chapter, the reader may find it useful to refer occasionally to more comprehensive treatments of that subject. Recommended for this purpose are Ruelle (1969, 1978), Griffiths (1972), Preston (1974b), Sinai (1982), and Simon (1985).

This chapter is devoted to the study of stochastic Ising models, and their relation to the Gibbs states. The first section contains the definition and some of the elementary properties of Gibbs states. The concept of phase transition is introduced there, and a number of inequalities are proved which serve to clarify that concept. Stochastic Ising models are defined in Section 2. It is shown there that in the finite range case, they coincide with the spin systems with positive rates which are reversible with respect to some probability measure. Furthermore, the reversible measures are exactly the corresponding Gibbs states. Hence phase transition for the Gibbs states is seen to imply nonergodicity for the associated stochastic Ising models. The reverse implication is correct as well in certain cases. This is the first connection between these two important concepts of phase transition and ergodicity.

The third section is devoted to results which assert the presence or absence of phase transition in special cases. In particular, we will see that phase transition does not occur in one-dimensional models with finite range, but that it does occur in higher dimensions at sufficiently low temperatures. There will be an interplay in this section between stochastic Ising models and the corresponding Gibbs states. Results developed here or earlier about one of these will yield results about the other.

The main idea in Section 4 is to use the spectral theorem to study stochastic Ising models. This tool is available to us because reversibility corresponds exactly to the self-adjointness of the generator of the spin system in the L_2 space of the corresponding Gibbs state. The principal results are that convergence in $L_2(\nu)$ for an appropriate Gibbs state ν is equivalent to ν being an extremal Gibbs state, and that exponential convergence in $L_2(\nu)$ is equivalent to there being a gap in the spectrum of $-\Omega$ to the right of the origin.

Section 5 is devoted to the use of relative entropy techniques in the study of the invariant measures of stochastic Ising models. The main result asserts under appropriate conditions that in one and two dimensions, every invariant measure is a Gibbs state. In higher dimensions, every invariant measure which is translation invariant is a Gibbs state.

Throughout this chapter, we will adopt the setting of Chapter III. In particular, the flip rates of all spin systems will be assumed to be uniformly bounded and to satisfy (0.3) of that chapter.

Sections 1, 2, and 3 should be read in that order. Following that, Sections 4 and 5 can be read in either order. Sections 5 and 4 of Chapter II should be read before reading Sections 2 and 5 of this chapter respectively.

1. Gibbs States

Before giving the definition of Gibbs states, we need to introduce the concept of a potential. A potential describes the interactions among the particles located at the sites in S.

Definition 1.1. A *potential* is a collection $\{J_R\}$ of real numbers indexed by finite subsets R of S which satisfies

$$(1.2) \qquad \sum_{R \ni x} |J_R| < \infty, \qquad x \in S.$$

For finite $R \subset S$ and $\eta \in X = \{0, 1\}^S$, let

$$(1.3) \qquad \chi_R(\eta) = \prod_{x \in R} [2\eta(x) - 1].$$

In order to lead up to the definition of the Gibbs states corresponding to a potential, consider first the case in which S is finite. The Gibbs state corresponding to the potential $\{J_R, R \subset S\}$ is then the unique probability measure ν on X which is given by

$$(1.4) \qquad \nu\{\eta\} = K \exp\left\{\sum_R J_R \chi_R(\eta)\right\},$$

where K is the normalizing constant

$$K = \left[\sum_\eta \exp\left\{\sum_R J_R \chi_R(\eta)\right\}\right]^{-1}.$$

In this case, of course, any probability measure on X which assigns strictly positive probabilities to all points is a Gibbs state for some potential. In order to see this, it suffices to express the logarithm of $\nu\{\eta\}$ as a linear combination of $\{\chi_R(\eta), R \subset S\}$. This can be done, since this set forms a basis for the $2^{|S|}$ dimensional vector space of all functions on X.

When S is countable, the definition of Gibbs states is considerably more subtle. After all, the summation in (1.4) is meaningless in general, and in fact we expect $\nu\{\eta\}$ to be zero in this case. There are two ways to overcome these difficulties. The first is to specify certain conditional probabilities of ν instead of the probabilities given in (1.4). The second is to define Gibbs states as in (1.4), but for configurations on finite subsets of S only, and then to pass to a limit. We will adopt the first of these approaches as the definition, and will then show that the second procedure yields the same class of Gibbs states.

Definition 1.5. A probability measure ν on X is said to be a *Gibbs state* relative to the potential $\{J_R\}$ provided that for all $x \in S$, a version of the conditional probability

$$(1.6) \qquad \rho_x(\zeta) = \nu\{\eta: \eta(x) = \zeta(x) | \eta(u) = \zeta(u) \text{ for all } u \neq x\}$$

is given by

$$(1.7) \qquad \frac{1}{1 + \exp\left[-2 \sum_{R \ni x} J_R \chi_R(\zeta)\right]}.$$

Note that the function in (1.7) is well defined and continuous on X for each x by (1.2). The class of all Gibbs states relative to a fixed potential $\{J_R\}$ will be denoted by $\mathscr{G}$.

Of course it is important to check that this definition coincides with (1.4) in case S is finite. This follows from the next result, which will be needed again later.

Proposition 1.8. *Suppose that S is finite and that ν is a probability measure on X. Then ν is given by (1.4) if and only if it is a Gibbs state relative to $\{J_R\}$ in the sense of Definition 1.5.*

Proof. Clearly $\nu\{\eta\} > 0$ for all $\eta \in X$ in either case. So, suppose (1.4) holds. Then

$$\nu\{\eta: \eta(x) = \zeta(x) | \eta(u) = \zeta(u) \text{ for all } u \neq x\}$$

$$= \frac{\nu\{\zeta\}}{\nu\{\zeta\} + \nu\{\zeta_x\}}$$

$$= \frac{\exp\left\{\sum_R J_R \chi_R(\zeta)\right\}}{\exp\left\{\sum_R J_R \chi_R(\zeta)\right\} + \exp\left\{\sum_R J_R \chi_R(\zeta_x)\right\}}$$

$$= \frac{1}{1+\exp\left\{\sum_R J_R[\chi_R(\zeta_x)-\chi_R(\zeta)]\right\}}$$

$$= \frac{1}{1+\exp\left\{-2\sum_{R\ni x} J_R\chi_R(\zeta)\right\}},$$

since

(1.9) $$\chi_R(\eta_x) = \begin{cases} -\chi_R(\eta) & \text{if } x\in R, \\ \chi_R(\eta) & \text{if } x\notin R. \end{cases}$$

For the converse, suppose ν is Gibbs in the sense of Definition 1.5. By the comment following (1.4), ν satisfies (1.4) for some possibly different potential $\{\tilde{J}_R\}$. But then by the first part of this proof,

$$\sum_{R\ni x} J_R\chi_R(\eta) = \sum_{R\ni x} \tilde{J}_R\chi_R(\eta).$$

Since this is true for all x, it follows that $J_R = \tilde{J}_R$ for all $R \neq \varnothing$. But changing $J_\varnothing$ simply amounts to changing the normalizing constant K, so the proof is complete. $\square$

Next we will see that any sufficiently nice measure is a Gibbs state for some potential.

Proposition 1.10. *Suppose that ν is a probability measure on X such that for all $x\in S$, there is a version of the conditional probabilities $\rho_x(\zeta)$ given in (1.6) which can be written in the form*

(1.11) $$\rho_x(\zeta) = \frac{1}{1+\exp\left[-2\sum_R J_R^x\chi_R(\zeta)\right]}$$

for some family $\{J_R^x\}$ which satisfies

$$\sum_R |J_R^x| < \infty$$

for each $x\in S$. Then ν is a Gibbs state relative to some potential $\{J_R\}$.

Proof. Comparing (1.7) and (1.11), it is clear that we need only show that

(1.12) $$J_R^x = 0 \quad \text{if } x\notin R, \quad \text{and}$$

(1.13) $$J_R^x = J_R^y \quad \text{if } x, y\in R,$$

since then we can define $\{J_R\}$ by

$$J_R = J_R^x \quad \text{for } x \in R.$$

By definition, $\rho_x(\zeta) + \rho_x(\zeta_x) = 1$ for all $x \in S$ and $\zeta \in X$. Therefore

$$\sum_R J_R^x \chi_R(\zeta) = -\sum_R J_R^x \chi_R(\zeta_x),$$

so that (1.12) follows from (1.9) and the fact that the functions χ_R are linearly independent. In order to prove (1.13), we will first show that the expression

(1.14)
$$\left[\frac{1}{\rho_x(\zeta)} - 1 \right]\left[\frac{1}{\rho_y(\zeta_x)} - 1 \right]$$

is symmetric in x and y. To do this, let T be a finite subset of S which contains neither x nor y. Note by (1.11), that ν assigns positive probability to every finite-dimensional set. Then

$$\left[\frac{1}{\nu\{\eta: \eta(x) = \zeta(x) | \eta = \zeta \text{ on } T \cup \{y\}\}} - 1 \right]$$
$$\times \left[\frac{1}{\nu\{\eta: \eta(y) = \zeta(y) | \eta = \zeta_x \text{ on } T \cup \{x\}\}} - 1 \right]$$
$$= \frac{\nu\{\eta: \eta(x) \neq \zeta(x), \eta(y) \neq \zeta(y), \eta = \zeta \text{ on } T\}}{\nu\{\eta: \eta = \zeta \text{ on } T \cup \{x, y\}\}},$$

which is symmetric in x and y, and converges to (1.14) as T increases to $S \setminus \{x, y\}$. Now, by (1.9), (1.11), (1.12), and the fact that (1.14) is symmetric in x and y, it follows that

$$\sum_{R \ni x} J_R^x \chi_R(\zeta) + \sum_{R \ni y} J_R^y \chi_R(\zeta_x) = \sum_{R \ni x, y} (J_R^x - J_R^y)\chi_R(\zeta)$$
$$+ \sum_{\substack{R \ni x \\ R \not\ni y}} J_R^x \chi_R(\zeta) + \sum_{\substack{R \ni y \\ R \not\ni x}} J_R^y \chi_R(\zeta)$$

is symmetric in x and y. Since the sum of the second and third terms on the right is symmetric, it follows that the first must be also. But this implies that

$$\sum_{R \ni x, y} (J_R^x - J_R^y)\chi_R(\zeta) = 0,$$

which gives (1.13). □

Note that the hypothesis of Proposition 1.10 is satisfied whenever the conditional probabilities in (1.6) are strictly positive and depend on ζ through only finitely many coordinates. Thus in particular, the result applies to Markov random fields (see, for example, Spitzer (1971a, b)). The hypothesis of Proposition 1.10 can be thought of as saying that ν is nearly a Markov random field. Sullivan (1973) has shown that if $S = Z^d$ and $\nu \in \mathcal{S}$, then this hypothesis can be replaced by the assumption that the conditional probabilities $\rho_x(\zeta)$ are strictly positive and continuous in ζ.

The next step is to carry out the limiting procedure described just before Definition 1.5, and to show that it leads to the same concept of a Gibbs state. For finite $T \subset S$ and $\zeta \in \{0, 1\}^{S \setminus T}$, let $\nu_{T,\zeta}$ be the probability measure on $\{0, 1\}^T$ given by

$$(1.15) \qquad \nu_{T,\zeta}\{\eta\} = K(T, \zeta) \exp\left[\sum_{R \cap T \neq \varnothing} J_R \chi_R(\eta^\zeta) \right],$$

where $K(T, \zeta)$ is again a normalizing constant and $\eta^\zeta \in X$ is defined by

$$\eta^\zeta(x) = \begin{cases} \eta(x) & \text{if } x \in T, \\ \zeta(x) & \text{if } x \notin T. \end{cases}$$

This expression is clearly analogous to (1.4). The summation converges by (1.2), since T is finite. Terms corresponding to R's which are disjoint from T are not included in the sums since their effect can be incorporated into the normalizing constant. The measure $\nu_{T,\zeta}$ can be regarded as a probability measure on X by setting

$$\nu_{T,\zeta}\{\eta \in X : \eta(x) = \zeta(x) \text{ for all } x \notin T\} = 1.$$

It is to be viewed as the Gibbs state with boundary condition ζ. Let $\mathcal{G}(T)$ be the closed convex hull of

$$\{\nu_{T,\zeta} : \zeta \in \{0, 1\}^{S \setminus T}\}.$$

Theorem 1.16. (a) $T_1 \subset T_2$ implies $\mathcal{G}(T_1) \supset \mathcal{G}(T_2)$.

(b)   If $\nu \in \mathcal{G}$, then for finite $T \subset S$ and $\zeta \in \{0, 1\}^{S \setminus T}$,

$$\nu(\cdot \,|\, \eta(u) = \zeta(u) \text{ for all } u \notin T) = \nu_{T,\zeta}(\cdot).$$

(c) $\mathcal{G} = \bigcap_T \mathcal{G}(T).$

(d) $\mathcal{G}$ is nonempty, convex, and compact.

Proof. To prove (a), suppose that $T_1 \subset T_2$, and take $\zeta \in \{0, 1\}^{S \setminus T_2}$. For $\gamma \in \{0, 1\}^{S \setminus T_1}$ such that $\gamma = \zeta$ on $S \setminus T_2$,

$$(1.17) \qquad \nu_{T_2, \zeta}(\cdot \mid \eta = \gamma \text{ on } T_2 \setminus T_1) = \nu_{T_1, \zeta}(\cdot)$$

as measures on $\{0, 1\}^{T_1}$, as can easily be seen from (1.15). Therefore

$$(1.18) \qquad \nu_{T_2, \zeta} = \sum_{\substack{\gamma: \gamma = \zeta \\ \text{on } S \setminus T_2}} \nu_{T_2, \zeta}(\eta: \eta = \gamma \text{ on } T_2 \setminus T_1) \nu_{T_1, \gamma},$$

which exhibits $\nu_{T_2, \zeta}$ as a convex combination of elements of $\mathcal{G}(T_1)$. Therefore $\mathcal{G}(T_2) \subset \mathcal{G}(T_1)$. For part (b), use (1.17) with $T_2 = T$ and $T_1 = \{x\}$ where $x \in T$ to write

$$\nu_{T, \zeta}\{\eta: \eta(x) = \zeta(x) \mid \eta(u) = \zeta(u) \text{ for all } u \in T \setminus \{x\}\}$$

$$= \nu_{\{x\}, \zeta}\{\zeta(x)\}$$

$$(1.19) \qquad = \frac{\exp\left[\sum_{R \ni x} J_R \chi_R(\zeta)\right]}{\exp\left[\sum_{R \ni x} J_R \chi_R(\zeta)\right] + \exp\left[\sum_{R \ni x} J_R \chi_R(\zeta_x)\right]}$$

$$= \frac{1}{1 + \exp\left[-2 \sum_{R \ni x} J_R \chi_R(\zeta)\right]}$$

for $\zeta \in X$ by (1.9). Comparing this with Definition 1.5, one sees that in order to prove part (b), it suffices to show that $\nu_{T, \zeta}$ is the only probability measure on $\{0, 1\}^T$ whose one point conditional probabilities are given by the right side of (1.19). But this follows from Proposition 1.8. For part (c), note that $\mathcal{G} \subset \mathcal{G}(T)$ follows from (b), and $\bigcap_T \mathcal{G}(T) \subset \mathcal{G}$ follows from (1.19) and Definition 1.5. Part (d) is an immediate consequence of parts (a) and (c) and the compactness of X. $\square$

The elements of $\mathcal{G}$ are interpreted as the possible phases of the physical system described by the potential $\{J_R\}$. By Theorem 1.16, there is always at least one such phase. One of the most important problems in classical statistical mechanics is to determine when there is more than one. We are therefore led to the following definition.

Definition 1.20. The potential $\{J_R\}$ is said to exhibit a *phase transition* if $\mathcal{G}$ contains more than one element.

In the physics literature, this concept is called a first-order phase transtion. Higher-order phase transitions refer to lack of smoothness of the Gibbs state as a function of the potential. These will not play a role in this book.

A number of partial answers to the question of when phase transition occurs will be given in Section 3. For now it suffices to say that in some sense, phase transition seldom occurs in one dimension, but frequently occurs in higher dimensions. The remainder of this section is devoted to the proofs of several inequalities and monotonicity statements which will help clarify our later discussion of phase transition. The first is a pair of correlation inequalities which we will use to show that if certain potentials exhibit phase transition, then so do many others. These inequalities have many other important applications. Some of them are described briefly in Section VIB of Griffiths (1972).

Theorem 1.21. *Suppose that S is finite, and that $\{J_R, R \subset S\}$ is a potential which is ferromagnetic in the sense that $J_R \geq 0$ for all R. Let v be the corresponding Gibbs state which is given by* (1.4). *Then*

$$(1.22) \qquad \int \chi_A \, dv \geq 0 \quad \text{for all } A \subset S, \quad \text{and}$$

$$(1.23) \qquad \frac{\partial}{\partial J_B} \int \chi_A \, dv = \int \chi_A \chi_B \, dv - \int \chi_A \, dv \int \chi_B \, dv \geq 0$$

for all $A, B \subset S$.

Proof. Since v is the product measure with density $\frac{1}{2}$ when $J_R = 0$ for all R, (1.22) for general ferromagnetic potentials is an immediate consequence of (1.23). However, the proof of (1.23) uses (1.22), so we must prove (1.22) first. In order to do so, write

$$\int \chi_A \, dv = K \sum_\eta \chi_A(\eta) \exp\left\{ \sum_R J_R \chi_R(\eta) \right\}$$

$$= K \sum_\eta \chi_A(\eta) \sum_{n=0}^{\infty} \frac{1}{n!} \left\{ \sum_R J_R \chi_R(\eta) \right\}^n$$

$$= K \sum_{n=0}^{\infty} \frac{1}{n!} \sum_{R_1,\dots,R_n} \left[\prod_{k=1}^{n} J_{R_k} \right] \sum_\eta \chi_A(\eta) \prod_{k=1}^{n} \chi_{R_k}(\eta).$$

To see that this is a sum of nonnegative terms, it suffices to note that

$$\chi_A(\eta) \prod_{k=1}^{n} \chi_{R_k}(\eta) = \chi_B(\eta)$$

where

$$B = \{x \in S : x \text{ is in an odd number of the sets } A, R_1, \dots, R_n\}$$

and that for any $B \subset S$,

$$\sum_{\eta} \chi_B(\eta) = \begin{cases} 2^{|S|} & \text{if } B = \varnothing, \\ 0 & \text{if } B \neq \varnothing. \end{cases}$$

Turning to the proof of (1.23), use the explicit expression for K to write

$$\int \chi_A \, d\nu = \frac{\sum\limits_{\eta} \chi_A(\eta) \exp\left\{\sum\limits_R J_R \chi_R(\eta)\right\}}{\sum\limits_{\eta} \exp\left\{\sum\limits_R J_R \chi_R(\eta)\right\}}.$$

Therefore the equality in (1.23) is the result of a simple differentiation. To check the inequality, write

$$
\begin{aligned}
(1.24) \quad & \int \chi_A \chi_B \, d\nu - \int \chi_A \, d\nu \int \chi_B \, d\nu \\
& = K^2 \sum_{\eta, \zeta} \left[\chi_A(\eta) \chi_B(\eta) - \chi_A(\eta) \chi_B(\zeta) \right] \exp\left\{\sum_R J_R [\chi_R(\eta) + \chi_R(\zeta)]\right\}.
\end{aligned}
$$

Let C be the symmetric difference $A \Delta B$ and let $\gamma \in X$ be defined by

$$\gamma(x) = \begin{cases} 1 & \text{if } \eta(x) = \zeta(x), \\ 0 & \text{if } \eta(x) \neq \zeta(x). \end{cases}$$

Then

$$\chi_A(\eta) \chi_B(\eta) = \chi_C(\eta),$$

$$\chi_A(\eta) \chi_B(\zeta) = \chi_B(\gamma) \chi_C(\eta), \quad \text{and}$$

$$\chi_R(\eta) + \chi_R(\zeta) = \chi_R(\eta)[1 + \chi_R(\gamma)].$$

Making these substitutions in (1.24) yields

$$
\begin{aligned}
(1.25) \quad & \int \chi_A \chi_B \, d\nu - \int \chi_A \, d\nu \int \chi_B \, d\nu \\
& = K^2 \sum_{\eta, \gamma} \chi_C(\eta)[1 - \chi_B(\gamma)] \exp\left\{\sum_R J_R \chi_R(\eta)[1 + \chi_R(\gamma)]\right\}.
\end{aligned}
$$

For fixed γ, we can define a new potential by

$$J_R^\gamma = J_R[1 + \chi_R(\gamma)],$$

which is again ferromagnetic. By (1.22) applied to this potential, we see that

$$\sum_{\eta} \chi_C(\eta) \exp\left\{\sum_R J_R^\gamma \chi_R(\eta)\right\} \geq 0.$$

Thus (1.23) follows from (1.25) by summing first on η and then on γ. $\quad\square$

For the rest of this section, we will specialize to the important case in which $S = Z^d$ for some $d \geq 1$ and

(1.26) $\quad \begin{cases} J_{\{x\}} = \beta H & \text{for each } x \in S, \\ J_{\{x,y\}} = \beta J(y-x) & \text{for distinct } x, y \in S, \quad \text{and} \\ J_R = 0 & \text{for } |R| \geq 3, \end{cases}$

where H is a real number, β is a nonnegative real number, and $J(x) \geq 0$ for all $x \neq 0$. The two parameters β and H represent the reciprocal of the temperature and the strength of the external magnetic field respectively. The famous Ising model, which gives this chapter its name, is the special case in which $J(x) = 1$ for the $2d$ neighbors of the origin, and $J(x) = 0$ otherwise. Under the above assumptions, Theorem 2.9 of Chapter II will provide additional monotonicity statements which are very useful. The theory in this case is quite analogous to the theory of attractive spin systems which was developed in Theorem 2.7 and Corollary 2.11 of Chapter III. The connection between them will be made more explicit in the next section. In particular, stochastic Ising models could be used in place of Theorem 2.9 of Chapter II in order to prove the next theorem and its corollaries.

Theorem 1.27. *Suppose that the potential is given as in* (1.26) *with* $\beta \geq 0$ *and* $J(x) \geq 0$. *Then* $\zeta_1 \leq \zeta_2$ *implies that* $\nu_{T,\zeta_1} \leq \nu_{T,\zeta_2}$ *for any finite* $T \subset S$, *where the last inequality is interpreted in the sense of Definition* 2.1 *of Chapter II.*

Proof. By Theorem 2.9 of Chapter II, it suffices to check that for $\eta_1, \eta_2 \in \{0, 1\}^T$,

$$\sum_{R \cap T \neq \varnothing} J_R[\chi_R(\eta_1^{\zeta_1} \wedge \eta_2^{\zeta_1}) + \chi_R(\eta_1^{\zeta_2} \vee \eta_2^{\zeta_2})] \geq \sum_{R \cap T \neq \varnothing} J_R[\chi_R(\eta_1^{\zeta_1}) + \chi_R(\eta_2^{\zeta_2})]$$

whenever $\zeta_1 \leq \zeta_2$. Using the special form for J_R which we have assumed, this can be rewritten as the statement that the expression

$$2\beta H \sum_{x \in T} [(\eta_1 \wedge \eta_2)(x) + (\eta_1 \vee \eta_2)(x) - \eta_1(x) - \eta_2(x)]$$

$$+ 2\beta \sum_{\substack{x,y \in T \\ x \neq y}} J(y-x)[(\eta_1 \wedge \eta_2)(x)(\eta_1 \wedge \eta_2)(y)$$

$$+ (\eta_1 \vee \eta_2)(x)(\eta_1 \vee \eta_2)(y)$$

$$- \eta_1(x)\eta_1(y) - \eta_2(x)\eta_2(y)]$$

$$+4\beta \sum_{\substack{x \in T \\ y \notin T}} J(y-x)[(\eta_1 \wedge \eta_2)(x)\zeta_1(y) + (\eta_1 \vee \eta_2)(x)\zeta_2(y)$$

$$- \eta_1(x)\zeta_1(y) - \eta_2(x)\zeta_2(y)]$$

is nonnegative. The terms in the first sum are all zero, so the sign of H is irrelevant in verifying the nonnegativity of this expression. The term in brackets in the second sum is zero unless $\eta_1(x) = \eta_2(y) = 0$ and $\eta_2(x) = \eta_1(y) = 1$ or $\eta_1(x) = \eta_2(y) = 1$ and $\eta_2(x) = \eta_1(y) = 0$, in which case it is equal to 1. The term in brackets in the third sum is zero unless $\eta_1(x) = 1$ and $\eta_2(x) = 0$, in which case it is equal to $\zeta_2(y) - \zeta_1(y)$. So, since $\beta \geq 0$ and $J(y-x) \geq 0$, the required sums are nonnegative whenever $\zeta_1 \leq \zeta_2$. □

For the next result, let $\underline{\nu}_T$ and $\bar{\nu}_T$ be defined by (1.15) with $\zeta \equiv 0$ and $\zeta \equiv 1$ respectively. By Theorem 1.27,

(1.28) $$\underline{\nu}_T \leq \nu_{T,\zeta} \leq \bar{\nu}_T$$

for all $\zeta \in \{0, 1\}^{S \setminus T}$.

Corollary 1.29. *Under the assumptions of Theorem* 1.27, $T_1 \subset T_2$ *implies that*

$$\underline{\nu}_{T_1} \leq \underline{\nu}_{T_2} \quad and \quad \bar{\nu}_{T_1} \geq \bar{\nu}_{T_2}.$$

Proof. The two statements are similar, so we will prove only the latter. By Theorem 1.27 $\nu_{T_1,\gamma} \leq \bar{\nu}_{T_1}$ for all γ. Therefore $\bar{\nu}_{T_2} \leq \bar{\nu}_{T_1}$ follows from (1.18). □

Corollary 1.30. *Under the assumptions of Theorem* 1.27,

(a) $\underline{\nu} = \lim_{T \uparrow S} \underline{\nu}_T \in \mathcal{S} \cap \mathcal{G}$ *exists*,
(b) $\bar{\nu} = \lim_{T \uparrow S} \bar{\nu}_T \in \mathcal{S} \cap \mathcal{G}$ *exists*,
(c) $\underline{\nu} \leq \nu \leq \bar{\nu}$ *for all* $\nu \in \mathcal{G}$,
(d) *phase transition occurs if and only if* $\underline{\nu} \neq \bar{\nu}$, *and*
(e) *phase transition occurs if and only if* $\underline{\nu}\{\eta: \eta(x) = 1\} \neq \bar{\nu}\{\eta: \eta(x) = 1\}$.

Proof. The fact that the limits exist in (a) and (b) and are translation invariant follows from Corollary 1.29. That these limits are in $\mathcal{G}$ follows from parts (a) and (c) of Theorem 1.16. To prove part (c), use (1.28) to show that

$$\underline{\nu}_T \leq \nu \leq \bar{\nu}_T$$

for all $\nu \in \mathcal{G}(T)$, and then pass to the limit as $T \uparrow S$ using part (c) of Theorem 1.16. Part (d) is an immediate consequence of part (c) and Definition 1.20. Part (e) follows from part (d) and Corollary 2.8 of Chapter II. □

Theorem 1.31. *In addition to the assumptions of Theorem* 1.27, *suppose that* $H = 0$. *Then*

(a) $\underline{\nu}\{\eta: \eta(x) = 1\} + \bar{\nu}\{\eta: \eta(x) = 1\} = 1$,

(b) $\bar{\nu}\{\eta\colon \eta(x)=1\}$ *is an increasing function of* β,
(c)   *there is a critical value* $0\le\beta_c\le\infty$ *such that there is no phase transition*
      if $\beta<\beta_c$ *and there is phase transition if* $\beta>\beta_c$, *and*
(d) β_c *is a decreasing function of the numbers* $J(x)$.

Proof. Since $H=0$, $\bar{\nu}_T$ is obtained from ν_T by interchanging the roles of 0 and 1. Therefore

$$\nu_T\{\eta\colon \eta(x)=1\}+\bar{\nu}_T\{\eta\colon \eta(x)=1\}=1.$$

Part (a) follows from this and Corollary 1.30. For part (b), note that if $\zeta\equiv1$,

$$\chi_R(\eta^\zeta)=\chi_{R\cap T}(\eta),$$

so that (1.15) takes the form (1.4). Therefore we can apply (1.23) with $A=\{x\}$ to conclude that

$$\bar{\nu}_T\{\eta\colon \eta(x)=1\}-\bar{\nu}_T\{\eta\colon \eta(x)=0\}=2\bar{\nu}_T\{\eta\colon \eta(x)=1\}-1$$

is an increasing function of β for each $T\ni x$. Therefore part (b) follows from part (b) of Corollary 1.30. Part (c) is an immediate consequence of parts (a) and (b), together with the characterization of phase transition in part (e) of Corollary 1.30. Part (d) follows from (1.23) in the same way that part (b) did. $\square$

It is of course an important problem to decide when $0<\beta_c<\infty$. It is relatively easy, as we will see in Section 3, to show under very weak assumptions that $\beta_c>0$. This says that phase transition does not occur at high temperatures (i.e., for small β). It is more difficult to show that phase transition does occur often at low temperatures, so that $\beta_c<\infty$. The importance of part (d) of Theorem 1.31 is that it implies that once we have shown $\beta_c<\infty$ for one choice of $\{J(x), x\in S\}$, we can conclude that $\beta_c<\infty$ for many other choices as well.

2. Reversibility of Stochastic Ising Models

We begin this section with the definition of a stochastic Ising model. While the definition may appear a bit strange at first glance, it will be motivated shortly.

Definition 2.1. Given a potential $\{J_R\}$, a spin system with strictly positive rates $c(x,\eta)$ is said to be a *stochastic Ising model* relative to that potential provided that for each $x \in S$, the function

(2.2) $$ c(x, \eta) \exp\left[\sum_{R \ni x} J_R \chi_R(\eta) \right] $$

does not depend on the coordinate $\eta(x)$.

The statement that (2.2) does not depend on $\eta(x)$ is often referred to as the condition of detailed balance. Special cases of stochastic Ising models have appeared in various guises earlier in this book. In Examples 3.1(a) and 4.3(a), of Chapter I, the potential is given by $J_R = \beta$ if R consists of a nearest-neighbor pair, and $J_R = 0$ otherwise. In Theorem 1.12 of Chapter III, the potential is identically zero. In Example 2.12 of Chapter III, $J_R = \frac{1}{4} \log(1-\delta)/\delta$ if R is a nearest-neighbor pair, and $J_R = 0$ otherwise.

In order to further motivate Definition 2.1, recall that in this chapter we are interested in finding spin systems which are reversible with respect to the Gibbs states associated with the potential $\{J_R\}$. Suppose that S is finite so that the Gibbs state ν is given by (1.4). By Proposition 5.3 of Chapter II, the spin system with rates $c(x, \eta)$ is reversible with respect to ν if and only if

(2.3) $$ c(x, \eta)\nu\{\eta\} = c(x, \eta_x)\nu\{\eta_x\} $$

for all $x \in S$ and $\eta \in X$. But that is exactly the statement that (2.2) is independent of $\eta(x)$.

There are of course infinitely many stochastic Ising models corresponding to a given potential. Two commonly used ones are

(2.4) $$ c(x, \eta) = \exp\left[- \sum_{R \ni x} J_R \chi_R(\eta) \right] \quad \text{and} $$

(2.5) $$ c(x, \eta) = \left\{ 1 + \exp\left[2 \sum_{R \ni x} J_R \chi_R(\eta) \right] \right\}^{-1}. $$

In the nearest-neighbor translation invariant case, these two versions appeared earlier in Example 4.3(a) of Chapter I. The second of these is particularly useful since then $c(x, \eta)$ is automatically bounded in x and η, and in fact satisfies

$$ c(x, \eta) + c(x, \eta_x) = 1. $$

A sufficient condition for (2.4) to be uniformly bounded and for both versions to satisfy (0.3) of Chapter III is that

(2.6) $$ \sup_{x \in S} \sum_{R \ni x} |R| |J_R| < \infty, $$

where $|R|$ denotes the cardinality of R. The computation required to check this is elementary, and will in any case be carried out for version (2.5) in the proof of Theorem 3.1. While we will not assume (2.6) explicitly in this section, recall that we are assuming throughout this chapter that $c(x, \eta)$ is uniformly bounded and satisfies (0.3) of Chapter III.

There appear to be no physical grounds for isolating one particular version of the stochastic Ising model for study. Therefore it is important that the results obtained apply to as general a class of stochastic Ising models as possible. Sometimes, though, it will be convenient to assume that the processes we are working with have certain properties such as that of attractiveness. Note in this connection that if the potential is given by (1.26) with $\beta \geq 0$ and $J(x) \geq 0$, then both versions (2.4) and (2.5) of the stochastic Ising model are attractive.

The purpose of this section is to study the relationships among reversible spin systems with strictly positive rates, stochastic Ising models, and Gibbs states. Definition 5.1 and Propositions 5.2 and 5.3 of Chapter II should be recalled at this point.

Proposition 2.7. *Suppose that ν is a probability measure on X and that $c(x, \eta)$ are the rates for a spin system. Then ν is reversible for the spin system if and only if*

$$(2.8) \qquad \int c(x, \eta)[f(\eta_x) - f(\eta)] \, d\nu = 0$$

for all $x \in S$ and $f \in C(X)$. If the rates are strictly positive, then this is equivalent to the statement that ν has the following conditional probabilities:

$$(2.9) \quad \nu\{\eta: \eta(x) = \zeta(x) | \eta(u) = \zeta(u) \text{ for all } u \neq x\} = \frac{c(x, \zeta_x)}{c(x, \zeta) + c(x, \zeta_x)}.$$

Proof. If (2.8) holds for all $f \in C(X)$, then it can be applied to the function $f(\eta_x)g(\eta)$ for $f, g \in D(X)$ to obtain

$$\int c(x, \eta) f(\eta) g(\eta_x) \, d\nu = \int c(x, \eta) f(\eta_x) g(\eta) \, d\nu,$$

or equivalently

$$\int c(x, \eta) f(\eta)[g(\eta_x) - g(\eta)] \, d\nu = \int c(x, \eta) g(\eta)[f(\eta_x) - f(\eta)] \, d\nu.$$

Summing on x and using Proposition 5.3 of Chapter II leads to the conclusion that ν is reversible for the spin system. To prove the converse, assume that ν is reversible. For a finite subset T of S and an $x \in T$, let

$f(\eta) = \prod_{y \in T} \eta(y)$ and $g(\eta) = f(\eta_x)$. Then

$$g(\eta)\Omega f(\eta) = f(\eta_x) \sum_{y \in T} c(y, \eta)[f(\eta_y) - f(\eta)] = c(x, \eta)f(\eta_x), \quad \text{and}$$

$$f(\eta)\Omega g(\eta) = f(\eta) \sum_{y \in T} c(y, \eta)[g(\eta_y) - g(\eta)] = c(x, \eta)f(\eta),$$

so that (2.8) holds for that f by Proposition 5.3 of Chapter II. By linearity, it holds for all $f \in \mathcal{D}$ (the set of functions depending on finitely many coordinates). Since $\mathcal{D}$ is dense in $C(X)$, (2.8) holds for all $f \in C(X)$. Now assume that $c(x, \eta) > 0$ for all $x \in S$ and $\eta \in X$. Fix an $x \in S$, and let $c_0(\eta)$ and $c_1(\eta)$ be the unique functions on X which do not depend on $\eta(x)$ such that

$$c(x, \eta) = \begin{cases} c_0(\eta) & \text{if } \eta(x) = 0, \\ c_1(\eta) & \text{if } \eta(x) = 1. \end{cases}$$

Then (2.9) can be rewritten as the statement that

$$\int \eta(x)f(\eta) \, d\nu = \int \frac{c_0(\eta)}{c_0(\eta) + c_1(\eta)} f(\eta) \, d\nu$$

for all functions $f \in C(X)$ which do not depend on $\eta(x)$. Since $c_0(\eta) + c_1(\eta)$ does not depend on $\eta(x)$ and is strictly positive, this is equivalent to the statement that

$$\int \eta(x)g(\eta)[c_0(\eta) + c_1(\eta)] \, d\nu = \int c_0(\eta)g(\eta) \, d\nu$$

for all $g \in C(X)$ which do not depend on $\eta(x)$. But this can be rewritten as

$$\int g(\eta)\{\eta(x)c_1(\eta) - [1 - \eta(x)]c_0(\eta)\} \, d\nu = 0, \quad \text{or}$$

(2.10) $$\int c(x, \eta)g(\eta)[2\eta(x) - 1] \, d\nu = 0.$$

On the other hand, if $f \in C(X)$ is written as

$$f(\eta) = f_0(\eta)[1 - \eta(x)] + f_1(\eta)\eta(x),$$

where $f_0(\eta)$ and $f_1(\eta)$ do not depend on $\eta(x)$, then

$$f(\eta_x) - f(\eta) = [f_0(\eta) - f_1(\eta)][2\eta(x) - 1],$$

so that (2.8) can be rewritten as

(2.11) $$\int c(x, \eta)[f_0(\eta) - f_1(\eta)][2\eta(x) - 1]\, d\nu = 0.$$

The proof of the proposition is completed by comparing (2.10) and (2.11). □

The first characterization of reversibility in Proposition 2.7 points out again how much stronger reversibility is than invariance. By Proposition 2.13 of Chapter I, μ is invariant if and only if

(2.12) $$\sum_{x \in S} \int c(x, \eta)[f(\eta_x) - f(\eta)]\, d\nu = 0$$

for all $f \in D(X)$. Proposition 2.7 asserts that reversibility is equivalent to the vanishing of each term in (2.12), rather than the vanishing of the sum.

Theorem 2.13. *Suppose that $c(x, \eta)$ is strictly positive, and that for each x, $c(x, \eta)$ depends on only finitely many coordinates. If the spin system is reversible with respect to some probability measure ν, then it is a stochastic Ising model relative to some potential $\{J_R\}$.*

Proof. By Proposition 2.7, ν has conditional probabilities given by (2.9). By the finite dependence assumption on the rates and Proposition 1.10, ν is a Gibbs state relative to some potential $\{J_R\}$. Using (2.9) again and Definition 1.5, we see that

$$\frac{c(x, \zeta_x)}{c(x, \zeta) + c(x, \zeta_x)} = \frac{1}{1 + \exp\left[-2 \sum_{R \ni x} J_R \chi_R(\zeta)\right]},$$

or equivalently,

$$\frac{c(x, \zeta)}{c(x, \zeta_x)} = \exp\left[-2 \sum_{R \ni x} J_R \chi_R(\zeta)\right].$$

Using (1.9), this can be rewritten as

$$c(x, \zeta)\exp\left[\sum_{R \ni x} J_R \chi_R(\zeta)\right] = c(x, \zeta_x) \exp\left[\sum_{R \ni x} J_R \chi_R(\zeta_x)\right],$$

which says that the spin system is a stochastic Ising model relative to the potential $\{J_R\}$. □

As a result of the previous theorem, we are justified in concentrating our attention on stochastic Ising models. Recall from Section 5 of Chapter II that $\mathcal{R}$ denotes the class of all reversible measures for a process.

Theorem 2.14. *Suppose that $c(x, \eta)$ are the rates for a stochastic Ising model relative to the potential $\{J_R\}$. Then $\mathcal{R} = \mathcal{G}$, where $\mathcal{G}$ denotes the set of all Gibbs states relative to the same potential.*

Proof. By Proposition 2.7 and Definition 1.5, it suffices to show that for a stochastic Ising model,

$$\frac{c(x, \zeta_x)}{c(x, \zeta) + c(x, \zeta_x)} = \frac{1}{1 + \exp\left[-2 \sum_{R \ni x} J_R \chi_R(\zeta)\right]}.$$

But this computation is the same (but in reverse order) as the one carried out in the proof of Theorem 2.13. $\square$

We are now in a position to present two simple results which make the connection between the concepts of ergodicity of a stochastic Ising model as treated in Chapters I and III on the one hand, and phase transition as discussed in the first section of this chapter on the other.

Theorem 2.15. *Consider a stochastic Ising model relative to the potential $\{J_R\}$, and let $\mathcal{G}$ be the corresponding Gibbs states. Then $\mathcal{G} \subset \mathcal{I}$. In particular, if the stochastic Ising model is ergodic, then there is no phase transition for that potential.*

Proof. The containment is an immediate consequence of Proposition 5.2 of Chapter II, and Theorem 2.14. If the process is ergodic, then $\mathcal{I}$ is a singleton by Definition 1.9 of Chapter I. Therefore $\mathcal{G}$ is a singleton as well, so $\{J_R\}$ does not exhibit phase transition by Definition 1.20. $\square$

Theorem 2.16. *Consider an attractive stochastic Ising model relative to the potential $\{J_R\}$, and let $\underline{\nu}, \bar{\nu}$ be defined as in Theorem 2.3 of Chapter III. Then $\underline{\nu}, \bar{\nu} \in \mathcal{G}$. In particular, the stochastic Ising model is ergodic if and only if there is no phase transition for that potential.*

Proof. Let S_n be finite sets which increase to S, and let $c_i^n(x, \eta)$ be the rates for the approximating spin systems which are defined just before the statement of Theorem 2.7 of Chapter III. By checking (2.3), or using Theorem 2.14 applied to these approximating spin systems, we see that $\nu_{S_n, \zeta}$ as defined in (1.15) is invariant for $c_0^n(x, \eta)$ if $\zeta \equiv 0$ and for $c_1^n(x, \eta)$ if $\zeta \equiv 1$. So, by the convergence theorem for irreducible finite-state Markov chains, $\underline{\nu}^n$ and $\bar{\nu}^n$ as defined in Theorem 2.7 of Chapter III are equal to $\nu_{S_n, \zeta}$ with

$\zeta \equiv 0$ and $\zeta \equiv 1$ respectively. Therefore $\underline{\nu}^n$, $\bar{\nu}^n \in \mathscr{G}(S_n)$, so that $\underline{\nu}$, $\bar{\nu} \in \mathscr{G}$ by Theorem 2.7 of Chapter III and Theorem 1.16 of this chapter. For the final statement, note that one direction follows from Theorem 2.15. The other follows from Corollary 2.4 of Chapter III, together with the fact that $\underline{\nu}$, $\bar{\nu} \in \mathscr{G}$. $\square$

Corollary 2.17. *Two attractive stochastic Ising models with respect to the same potential are either both ergodic or neither ergodic.*

For an interesting application of this corollary, recall Example 4.3(a) of Chapter I. There we took rates of the form (2.4) and (2.5) on Z^d with the potential given by $J_{\{x, y\}} = \beta \geq 0$ for nearest neighbors x and y, and $J_R = 0$ otherwise. Both versions are attractive, and Theorem 4.1 of that chapter gave ergodicity in one dimension for the second version for all β, but for the first version only for sufficiently small β. By Corollary 2.17, the first version is in fact ergodic for all β as well. Note however that this argument does not yield exponentially fast convergence to the invariant measure for the first version with large β. This is in fact the case, as was proved by Holley (1985). For more on this topic of exponentially fast convergence of stochastic Ising models, see Theorem 4.16 and Corollary 4.18.

Corollary 2.18. *Consider the stochastic Ising models with rates given by either (2.4) or (2.5) with the potential given by (1.26) with $\beta \geq 0$, $H = 0$ and $J(x) \geq 0$ for all x. Let β_c be as in Theorem 1.31. Then the process is ergodic if $\beta < \beta_c$ and not ergodic if $\beta > \beta_c$.*

3. Phase Transition

This section is devoted to a number of results which show when phase transition occurs, and when it does not. The most important ones will be proved, while some of the more refined ones will only be stated. The ones which are not proved will not be formally used in the sequel. Applications to the stochastic Ising model will be given.

The first theorem is both one of the most elementary and one of the most generally applicable of these results. It implies, for example, that there is no phase transition in the Ising model at high temperatures in any dimension. In particular, it follows that the β_c defined in Theorem 1.31 is strictly positive. The proof of this result is based on the connection between the Gibbs states and stochastic Ising models.

Theorem 3.1. *Let $\{J_R\}$ be a potential. Then there is no phase transition provided either of the following conditions is satisfied:*

$$(3.2) \qquad \sup_x \sum_{y \neq x} \sup_\eta |\rho_x(\eta) - \rho_x(\eta_y)| < 1,$$

where

$$\rho_x(\eta) = \left\{ 1 + \exp\left[-2 \sum_{R \ni x} J_R \chi_R(\eta) \right] \right\}^{-1}.$$

(3.3)
$$\sup_x \sum_{R \ni x} [|R| - 1]|J_R| < \frac{\log 2}{4}.$$

In particular, if the potential satisfies (2.6), then there is no phase transition for the potential $\{\beta J_R\}$ for sufficiently small β.

Proof. Consider the stochastic Ising model with rates given by (2.5). Then

$$c(x, \eta) = \rho_x(\eta_x), \quad \text{and}$$

$$c(x, \eta) + c(x, \eta_x) = 1$$

for all $x \in S$ and $\eta \in X$. Therefore (3.2) is exactly condition (0.6) of Chapter III. By Theorem 4.1 of Chapter I, (3.2) then implies that this stochastic Ising model is ergodic, so that by Theorem 2.15, it implies that there is no phase transition for this potential. To complete the proof, it suffices to show that (3.3) implies (3.2). In order to do so, use (1.9) to write

$$\rho_x(\eta) - \rho_x(\eta_y) = \frac{\exp\left\{ -2 \sum_{R \ni x} J_R \chi_R(\eta) \right\} \left[\exp\left\{ +4 \sum_{R \ni x,y} J_R \chi_R(\eta) \right\} - 1 \right]}{\left[1 + \exp\left\{ -2 \sum_{R \ni x} J_R \chi_R(\eta) \right\} \right]\left[1 + \exp\left\{ -2 \sum_{R \ni x} J_R \chi_R(\eta_y) \right\} \right]}$$

for $x \neq y$, so that

$$\sup_\eta |\rho_x(\eta) - \rho_x(\eta_y)| \leq \exp\left[4 \sum_{R \ni x,y} |J_R| \right] - 1$$

for $x \neq y$. Therefore

$$\sum_{y \neq x} \sup_\eta |\rho_x(\eta) - \rho_x(\eta_y)| \leq \sum_{y \neq x} \left\{ \exp[4 \sum_{R \ni x,y} |J_R|] - 1 \right\}$$
$$\leq \exp\left\{ 4 \sum_{y \neq x} \sum_{R \ni x,y} |J_R| \right\} - 1$$
$$= \exp\left\{ 4 \sum_{R \ni x} |J_R|[|R| - 1] \right\} - 1,$$

so that (3.3) implies (3.2) as required. $\square$

Other techniques give other similar sufficient conditions for the absence of phase transition. For example, Theorem 5.1 of Chapter III can be used in place of Theorem 4.1 of Chapter I in order to show that the condition

(3.4) $$\sup_{x} \sum_{R \ni x} |J_R| < \frac{\pi}{4}$$

is sufficient to conclude that there is no phase transition. The computations needed to verify the assumption of Theorem 5.1 under that condition are given in the appendix of Holley and Stroock (1976b).

The technique illustrated in Theorem 3.1 of using the stochastic Ising model to deduce properties of the Gibbs states can be carried much further. For example, Theorem 4.20 of Chapter I can be used to show immediately that for finite range potentials, condition (3.2) implies that the unique Gibbs state has exponentially decaying correlations.

The next result and its corollaries will prepare us to prove that there is no phase transition in one dimension unless the potential has very long range. They will also be useful in Section 4.

Proposition 3.5. *Let $\{J_R\}$ be a potential, and $\mathcal{G}$ be the corresponding Gibbs states.*

  (a) *Suppose μ_1, $\mu_2 \in \mathcal{G}$ and μ_1 is absolutely continuous with respect to μ_2. Let $h = d\mu_1/d\mu_2$ be the Radon–Nikodym derivative. Then*

(3.6) $$h(\eta_x) = h(\eta) \quad a.e. \ (\mu_2) \quad \text{for each } x \in S.$$

  (b) *Suppose $\mu_2 \in \mathcal{G}$, $h \geq 0$, $\int h \, d\mu_2 = 1$, and h satisfies (3.6). Then $\mu_1 = h\mu_2$ is in $\mathcal{G}$.*

Proof. To prove part (a), it suffices to show that

$$\int f(\eta)h(\eta_x) \, d\mu_2 = \int f(\eta)h(\eta) \, d\mu_2$$

for all $x \in S$ and $f \in C(X)$. The right side is just $\int f(\eta) \, d\mu_1$. To compute the left side, use Definition 1.5 as follows:

$$\int f(\eta)h(\eta_x) \, d\mu_2 = \int f(\eta_x)h(\eta) \frac{\rho_x(\eta_x)}{\rho_x(\eta)} \, d\mu_2$$

$$= \int f(\eta_x) \frac{\rho_x(\eta_x)}{\rho_x(\eta)} \, d\mu_1$$

$$= \int f(\eta) \, d\mu_1$$

as required. The change of variables $\eta \to \eta_x$ is used in the first and last steps. The proof of part (b) is similar. $\square$

Corollary 3.7. *Suppose that $\mu_1, \mu_2 \in \mathcal{G}$ implies that μ_1 is absolutely continuous with respect to μ_2. Then there is no phase transition.*

Proof. Suppose $\mu_1, \mu_2 \in \mathcal{G}$. We need to show that $\mu_1 = \mu_2$. By assumption, μ_1 and μ_2 are absolutely continuous with respect to each other. Therefore $h = d\mu_1/d\mu_2 > 0$ a.e. (μ_2). In order to show that $h = 1$, assume on the contrary that $0 < \mu_2(A) < 1$, where $A = \{\eta : h(\eta) > c\}$ for some $c > 0$. Let

$$g(\eta) = \begin{cases} \dfrac{1}{\mu_2(A)} & \text{if } \eta \in A, \\ 0 & \text{if } \eta \notin A. \end{cases}$$

By Proposition 3.5(a), h and hence g satisfies (3.6). Therefore $\mu_3 = g\mu_2$ is in $\mathcal{G}$ by Proposition 3.5(b). But μ_2 is not absolutely continuous with respect to μ_3, which contradicts the hypothesis of the corollary. $\square$

Corollary 3.8. *A $\mu \in \mathcal{G}$ is extremal in $\mathcal{G}$ if and only if the only measurable functions on X which satisfy*

$$h(\eta_x) = h(\eta) \ a.e. \ (\mu) \quad \text{for each } x \in S$$

are those which are constant a.e. (μ).

Proof. Suppose that $\mu = \alpha\mu_1 + \beta\mu_2$, where $\alpha + \beta = 1$, $0 < \alpha < 1$, and μ_1, $\mu_2 \in \mathcal{G}$. Then μ_1 is absolutely continuous with respect to μ. So by part (a) of Proposition 3.5,

$$\frac{d\mu_1}{d\mu}(\eta_x) = \frac{d\mu_1}{d\mu}(\eta) \ a.e. \ (\mu) \quad \text{for all } x \in S.$$

If $\mu_1 \neq \mu$, then of course $d\mu_1/d\mu$ is not constant. This gives one direction. For the converse, use part (b) of Proposition 3.5. $\square$

Theorem 3.9. *Suppose that $S = Z^1$, $J_{R+x} = J_R$ for all $x \in S$ and all R, and*

$$(3.10) \qquad \sum_{R \ni 0} \frac{(\text{diameter } R)|J_R|}{|R|} < \infty.$$

Then there is no phase transition for the potential $\{J_R\}$.

Proof. We will verify the hypothesis of Corollary 3.7. Let T be a finite connected subset of S and let $\zeta_1, \zeta_2 \in \{0, 1\}^{S \setminus T}$. Then

$$\frac{\exp\left[\sum_{R \cap T \neq \varnothing} J_R \chi_R(\eta^{\zeta_1})\right]}{\exp\left[\sum_{R \cap T \neq \varnothing} J_R \chi_R(\eta^{\zeta_2})\right]} = \exp\left[\sum_{\substack{R \cap T \neq \varnothing \\ R \cap T^c \neq \varnothing}} J_R [\chi_R(\eta^{\zeta_1}) - \chi_R(\eta^{\zeta_2})]\right]$$

$$\leq \exp\left[2 \sum_{\substack{R \cap T \neq \varnothing \\ R \cap T^c \neq \varnothing}} |J_R|\right]$$

$$\leq \exp\left[4 \sum_{\substack{R \cap (-\infty, 0] \neq \varnothing \\ R \cap (0, \infty) \neq \varnothing}} |J_R|\right] \leq c$$

where

$$c = \exp\left[4 \sum_{R \ni 0} \frac{(\text{diameter } R)|J_R|}{|R|}\right] < \infty.$$

The factor (diameter R) comes from the fact that for fixed R, there are that many translates of R which contain both positive and nonpositive points. The factor $|R|$ comes from the fact that for fixed R, there are that many translates of R which contain 0. By (1.15), it then follows that

$$c^{-2} \leq \frac{\nu_{T,\zeta_1}\{\eta\}}{\nu_{T,\zeta_2}\{\eta\}} \leq c^2$$

for all $\eta \in \{0, 1\}^T$. Therefore if $\mu_1, \mu_2 \in \mathscr{G}(T)$,

$$c^{-2} \leq \frac{\mu_1\{\zeta: \zeta = \eta \text{ on } T\}}{\mu_2\{\zeta: \zeta = \eta \text{ on } T\}} \leq c^2$$

for all $\eta \in \{0, 1\}^T$. Since this estimate is uniform in T, Theorem 1.16 can then be used to show that any two elements of $\mathscr{G}$ are absolutely continuous with respect to each other. In fact, c^{-2} and c^2 give bounds on the corresponding Radon–Nikodym derivatives. $\square$

Of course, hypothesis (3.10) is satisfied for any finite range potential. It is of interest to see more precisely what happens when the potential has infinite range. Suppose the potential is given in the form (1.26) with $d = 1$, $H = 0$ and $J(n) \geq 0$. Then Theorem 3.9 shows that there is no phase transition when $\sum_{n=1}^{\infty} n J(n) < \infty$. Rogers and Thompson (1981) have improved this result, showing that there is no phase transition in this case provided that

$$(3.11) \qquad \lim_{N\to\infty} \frac{1}{\sqrt{\log N}} \sum_{n=1}^{N} nJ(n) = 0.$$

On the other side of the picture, Fröhlich and Spencer (1982) showed that phase transition does occur for sufficiently large positive β if $J(n) = n^{-2}$. It then follows from Theorem 1.31 that the conclusion is valid whenever

$$(3.12) \qquad \inf_{n} n^2 J(n) > 0.$$

We can now apply Theorem 3.9 to the stochastic Ising models.

Theorem 3.13. *Every attractive translation invariant finite range stochastic Ising model in one dimension is ergodic.*

Proof. Since the process has finite range (recall Definition 4.17 of Chapter I), the corresponding potential satisfies the assumptions of Theorem 3.9. By that theorem, there is no phase transition in this case. Therefore the process is ergodic by Theorem 2.16. □

Theorem 3.13 is the main reason for believing that the positive rates conjecture is true. (See Problem 6 in Chapter III.) To make the connection clear, recall from Theorem 2.13 that in the finite range case, a stochastic Ising model is simply a spin system with positive rates which has a reversible invariant measure. While reversibility is an extremely useful tool, it is not at all clear why the loss of reversibility should make it possible for a spin system in one dimension to be nonergodic. There are those who doubt the positive rates conjecture. Reasons for this skepticism range from the fact that the proof of Theorem 3.14 of Chapter III depends so heavily on the nearest-neighbor assumption, to the fact that counterexamples to similar conjectures have been proposed in recent years. (See Gacs (1985) for an example of a discrete time system with a large but finite number of states per site which is said to be nonergodic. It appears to be very unlikely that his construction can be modified to yield a monotone example.) Of course the finite range assumption is needed in the positive rates conjecture, since without it a stochastic Ising model with a potential satisfying (3.12) would be a counterexample.

We turn now to the two-dimensional Ising model, in order to find more interesting examples of phase transition. This model is the one with potential given as in (1.26) with $J(x) = 1$ for all four neighbors of the origin, and $J(x) = 0$ otherwise. It has been shown by Ruelle (1972) and Lebowitz and Martin-Löf (1972) that phase transition for this model does not occur if $H \neq 0$. (See also Preston (1974c).) Therefore we will take $H = 0$. Note that the presence of phase transition is not very robust in this model, since the phase transition disappears when an arbitrarily small change is made in the potential. This implies an analogous statement for the stochastic Ising model. Arbitrarily small changes in the rates can change a nonergodic system into an ergodic one.

Theorem 3.14. *Suppose* $S = Z^2$, $J_R = \beta$ *if* $R = \{x, y\}$ *with* $|y - x| = 1$, *and* $J_R = 0$ *otherwise. For sufficiently large positive* β, *this potential exhibits phase transition.*

Proof. For $n \geq 1$, define ν_n as in (1.15) with $\zeta \equiv 1$ and $T = [-n, n]^2 \subset Z^2$. By Corollary 1.30(e) and Theorem 1.31(a), it suffices to prove that

$$(3.15) \qquad\qquad \lim_{\beta \to \infty} \nu_n\{\eta: \eta(0) = 0\} = 0$$

uniformly in n, since phase transition will occur for any β such that

$$\lim_{n \to \infty} \nu_n\{\eta: \eta(0) = 0\} < \tfrac{1}{2}.$$

To carry out the proof, it is important to visualize a configuration $\eta \in \{0, 1\}^T$ in a certain way. Write $+$ for 1 and $-$ for 0, and agree to draw vertical and horizontal lines of unit length between adjacent sites which have opposite signs. An illustration with a particular configuration in case $n = 3$ is given in Figure 1. The outer edge consists entirely of $+$ since the boundary condition is $\zeta \equiv 1$. Let $B(\eta)$ be the union of all these vertical and horizontal lines. Note that the configuration can be reconstructed from $B(\eta)$ since the outer edge is always $+$. Also, $B(\eta)$ is a disjoint union of contours, where a contour is a closed non-self-intersecting polygonal curve. The length $|\gamma|$ of a contour γ is the number of unit edges in γ. The sum of the lengths of all the contours which make up $B(\eta)$ will be denoted by $|B(\eta)|$. With this notation, we can proceed to prove (3.15). If $\eta(0) = 0$, then 0 is surrounded

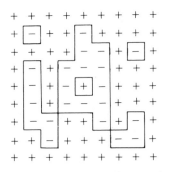

Figure 1

by at least one contour γ. Let Γ be the set of contours surrounding 0. Then

(3.16) $$\nu_n\{\eta\colon \eta(0)=0\} \leq \sum_{\gamma\in\Gamma} \nu_n\{\eta\colon \gamma\in B(\eta)\},$$

so we need to estimate $\nu_n\{\eta\colon \gamma\in B(\eta)\}$ for fixed $\gamma\in\Gamma$. To do so, use (1.15) to write

(3.17) $$\nu_n\{\eta\colon \gamma\in B(\eta)\} = \frac{\displaystyle\sum_{\eta\colon\gamma\in B(\eta)} \exp[-2\beta|B(\eta)|]}{\displaystyle\sum_{\eta} \exp[-2\beta|B(\eta)|]}.$$

If η is such that $\gamma\in B(\eta)$, define $\tilde{\eta}$ by

$$\tilde{\eta}(x) = \begin{cases} 1-\eta(x) & \text{if } \gamma \text{ surrounds } x, \\ \eta(x) & \text{otherwise.} \end{cases}$$

Then $B(\tilde{\eta})$ is obtained from $B(\eta)$ by removing γ, so that $|B(\eta)| = |\gamma| + |B(\tilde{\eta})|$. Therefore by (3.17),

$$\nu_n\{\eta\colon \gamma\in B(\eta)\} = \exp[-2\beta|\gamma|] \frac{\displaystyle\sum_{\eta\colon\gamma\in B(\eta)} \exp[-2\beta|B(\tilde{\eta})|]}{\displaystyle\sum_{\eta} \exp[-2\beta|B(\eta)|]}.$$

Since the map $\eta\to\tilde{\eta}$ is one-to-one, each term in the numerator on the right side appears also in the denominator. Therefore we can conclude that

$$\nu_n\{\eta\colon \gamma\in B(\eta)\} \leq \exp[-2\beta|\gamma|].$$

Using this in (3.16) gives

(3.18) $$\nu_n\{\eta\colon \eta(0)=0\} \leq \sum_{k=4}^{\infty} e^{-2\beta k} N(k, n),$$

where $N(k, n)$ is the number of contours $\gamma\in\Gamma$ of length k. But $N(k, n)\leq k3^k$ for all n, since each contour $\gamma\in\Gamma$ of length k must cross the positive horizontal axis at at least one of k places, and such a contour can be continued at each point in at most three ways. Thus we obtain the estimate

(3.19) $$\nu_n\{\eta\colon \eta(0)=0\} \leq \sum_{k=4}^{\infty} k3^k e^{-2\beta k},$$

from which (3.15) follows by the Dominated Convergence Theorem. $\quad\square$

We now know from Theorems 3.1 and 3.14 that for the Ising model considered in the latter theorem, $0 < \beta_c < \infty$. While it will not be of great importance to us, it is interesting to note that the exact value of β_c is known in this case: $\beta_c = \frac{1}{2}\log(1 + \sqrt{2})$. In fact, it is even known that

$$\bar{\nu}\{\eta: \eta(x) = 1\} - \tfrac{1}{2} = [1 - (\sinh 2\beta)^{-4}]^{1/8}$$

for $\beta \geq \beta_c$, so that this model does not exhibit phase transition at $\beta = \beta_c$. This formula for the "spontaneous magnetization" was obtained by Onsager in 1944. For more modern and mathematically precise versions of this result, see Abraham and Martin-Löf (1973) and Benettin, Gallavotti, and Jona-Lasinio (1973). The exact critical value for the Ising model in dimensions greater than 2 is not known. A great deal of additional information is available for the two-dimensional Ising model. For example, it was proved independently by Aizenman (1980) and Higuchi (1979) that all Gibbs states are convex combinations of ν and $\bar{\nu}$ if $\beta \geq 0$. This statement is not correct in dimensions greater than two. (See, for example, Van Beijeren (1975).)

If β is sufficiently large and negative, the two-dimensional Ising model has Gibbs states which are not translation invariant. This is an easy consequence of Theorem 3.14, since the transformation $\eta \to \tilde{\eta}$ where

$$\tilde{\eta}(x) = \begin{cases} \eta(x) & \text{if } x \text{ is an even number of steps from the origin,} \\ 1 - \eta(x) & \text{otherwise} \end{cases}$$

maps the Ising model with parameter β into the Ising model with parameter $-\beta$.

Finally, we summarize some of the main results of this section in the following way.

Corollary 3.20. *Suppose the potential satisfies the assumptions of Theorem 1.31, and let β_c be defined as in that result. Then*

(a) $\beta_c > 0$,
(b) $\beta_c = \infty$ *if $d = 1$ and $\sum_{n=1}^{\infty} nJ(n) < \infty$,*
(c) $\beta_c < \infty$ *if $d \geq 2$ and $J(x) > 0$ for all $2d$ nearest neighbors of the origin.*

Proof. Parts (a) and (b) are consequences of Theorems 3.1 and 3.9 respectively. Part (c) follows from Theorem 3.14 and part (d) of Theorem 1.31. □

Corollary 3.21. *Under the assumptions of Corollary 3.20, consider an attractive stochastic Ising model with respect to the given potential. If $\beta < \beta_c$, the system is ergodic, while if $\beta > \beta_c$, it is not.*

Proof. This result is essentially a restatement of Corollary 2.18. □

4. L_2 Theory

Throughout most of this book, the semigroup associated with a spin system is regarded as operating on $C(X)$. A stochastic Ising model is reversible with respect to appropriate Gibbs states. This implies that its semigroup is self-adjoint on the L_2 spaces of those Gibbs states. Thus it is natural to regard the semigroup as acting on these spaces as well. This section is devoted to the exploitation of this point of view. An important result will be a general L_2 convergence theorem for stochastic Ising models.

As was pointed out in Section 2, the physics of a situation dictates the use of a particular potential. On the other hand, there appear to be no physical arguments which can be used to select one of the many stochastic Ising models corresponding to that potential. Thus it would be good to know that the behavior of a stochastic Ising model depends in some sense only on the potential, and not otherwise on the choice of rates. Theorems 2.14 and 2.16 and Corollary 2.17 are results of this type. Further results of this type will be obtained in the next section (see Theorems 5.12 and 5.14). Open Problems 1 and 3 raise similar issues. In this section, we will see that the L_2 theory of the stochastic Ising model is largely independent of the choice of rates (see Corollary 4.18, for example).

In order to describe the context in which we will operate, fix a potential $\{J_R\}$, a Gibbs state ν relative to it, and a stochastic Ising model η_t relative to it with rates $c(x, \eta)$ and semigroup $S(t)$ (see Definitions 1.1, 1.5, and 2.1 respectively). The norm in $L_2(\nu)$ will be denoted by $\|\cdot\|_\nu$. The theory of contraction semigroups and their generators was described in Section 2 of Chapter I for the Banach space $C(X)$. This theory, and in particular Theorem 2.9 in that chapter, applies equally well to the space $L_2(\nu)$. The only change needed is that (c) of Definition 2.1 in that chapter should be replaced by the statement that

$$\int f\Omega f\, d\nu \le 0 \quad \text{for all } f \in \mathcal{D}(\Omega).$$

Proposition 4.1. *The semigroup $S(t)$ extends by continuity to a Markov semigroup on $L_2(\nu)$. Its generator Ω_ν is the closure of Ω in $L_2(\nu)$. The operator Ω_ν is self-adjoint on $L_2(\nu)$.*

Proof. $S(t)f(\eta) = E^\eta f(\eta_t)$ for $f \in C(X)$, so that

$$[S(t)f(\eta)]^2 \le S(t)f^2(\eta)$$

by the Schwarz inequality. By Theorems 2.14 and 2.15, $\mathcal{G} = \mathcal{R} \subset \mathcal{I}$.

Therefore, since $\nu \in \mathcal{G}$,

$$\|S(t)f\|_\nu^2 = \int [S(t)f]^2 \, d\nu$$

$$\leq \int S(t)f^2 \, d\nu$$

$$= \int f^2 \, d\nu = \|f\|_\nu^2.$$

$C(X)$ is dense in $L_2(\nu)$, so $S(t)$ extends to a Markov semigroup on $L_2(\nu)$. Let Ω_ν be its generator. Clearly Ω_ν is an extension of Ω. Since $\mathcal{R}(I - \lambda\Omega) = C(X)$, which is dense in $L_2(\nu)$, Ω_ν is the closure of Ω. Therefore by Proposition 5.3 of Chapter II,

$$(4.2) \qquad\qquad \int f\Omega_\nu g \, d\nu = \int g\Omega_\nu f \, d\nu$$

for all $f, g \in \mathscr{D}(\Omega_\nu)$. This says that Ω_ν is a symmetric operator on $L_2(\nu)$. Therefore the adjoint of Ω_ν is an extension of Ω_ν. This adjoint is also a Markov generator on $L_2(\nu)$, and hence must agree with Ω_ν. Therefore Ω_ν is self-adjoint. $\square$

The main idea of this section is to use the spectral representation of $-\Omega_\nu$ to study the convergence of $S(t)f$ in $L_2(\nu)$. In doing so, it will be important to have a more explicit expression for the symmetric bilinear form which appears in (4.2).

Lemma 4.3. *If $f, g \in \mathscr{D}(\Omega_\nu)$, then*

$$-\int f\Omega_\nu g \, d\nu = \tfrac{1}{2} \sum_x \int c(x, \eta)[f(\eta_x) - f(\eta)][g(\eta_x) - g(\eta)] \, d\nu,$$

where the series converges absolutely.

Proof. Define bilinear forms

$$L(f, g) = -\int f\Omega_\nu g \, d\nu \quad \text{for } f, g \in \mathscr{D}(\Omega_\nu)$$

and

$$R(f, g) = \tfrac{1}{2} \sum_x \int c(x, \eta)[f(\eta_x) - f(\eta)][g(\eta_x) - g(\eta)] \, d\nu$$

for all f, g such that $R(f, f) < \infty$ and $R(g, g) < \infty$. Note that the series converges absolutely by the Schwarz inequality. For $f, g \in \mathcal{D}$,

(4.4)
$$\begin{aligned} L(f, g) &= -\int f\Omega_\nu g \, d\nu \\ &= -\sum_x \int c(x, \eta) f(\eta)[g(\eta_x) - g(\eta)] \, d\nu \\ &= \sum_x \int c(x, \eta) f(\eta_x)[g(\eta_x) - g(\eta)] \, d\nu, \end{aligned}$$

where of course there are only finitely many nonzero terms in each sum. The last equality follows from Proposition 2.7, as can be seen by applying (2.8) to the function

$$\eta \to [f(\eta_x) + f(\eta)]g(\eta).$$

Taking the average of the last two terms in (4.4), we see that

(4.5)
$$\begin{aligned} L(f, g) &= -\tfrac{1}{2}\sum_x \int c(x, \eta) f(\eta)[g(\eta_x) - g(\eta)] \, d\nu \\ &\quad + \tfrac{1}{2}\sum_x \int c(x, \eta) f(\eta_x)[g(\eta_x) - g(\eta)] \, d\nu \\ &= \tfrac{1}{2}\sum_x \int c(x, \eta)[f(\eta_x) - f(\eta)][g(\eta_x) - g(\eta)] \, d\nu \\ &= R(f, g), \end{aligned}$$

which proves the result for $f, g \in \mathcal{D}$. For $f \in \mathcal{D}(\Omega_\nu)$, take $f_n \in \mathcal{D}$ so that $f_n \to f$ and $\Omega_\nu f_n \to \Omega_\nu f$ in $L_2(\nu)$. Then

(4.6)
$$\lim_{n \to \infty} L(f_n, f_n) = L(f, f), \quad \text{and}$$

$$\liminf_{n \to \infty} R(f_n, f_n) \geq R(f, f),$$

the latter coming from Fatou's Lemma. Therefore

(4.7)
$$R(f, f) \leq L(f, f),$$

so in particular, $R(f, f) < \infty$ for $f \in \mathcal{D}(\Omega_\nu)$. On the other hand,

$$0 \leq L(f - f_n, f - f_n) \leq \|f - f_n\|_\nu \|\Omega_\nu f - \Omega_\nu f_n\|_\nu$$

which tends to zero as $n \to \infty$, so that by (4.7),

$$\lim_{n \to \infty} R(f - f_n, f - f_n) = 0.$$

Therefore

(4.8) $$\lim_{n \to \infty} R(f_n, f_n) = R(f, f).$$

Putting (4.5), (4.6), and (4.8) together yields $R(f, f) = L(f, f)$ for all $f \in \mathcal{D}(\Omega_\nu)$. The desired result follows by polarization. $\square$

One consequence of Lemma 4.3 is that $-\Omega_\nu$ is not only self-adjoint, but also positive semidefinite. (This follows also from the fact that Ω_ν is a Markov generator on $L_2(\nu)$.) The spectral theorem then takes the following form:

(4.9) $$-\Omega_\nu = \int_{[0,\infty)} \lambda \, dG(\lambda),$$

where $G(\lambda)$ is a resolution of the identity. Recall that this means that $\{G(\lambda), \lambda \geq 0\}$ is a family of projections on $L_2(\nu)$ which satisfy

$$G(\lambda_1) G(\lambda_2) = G(\lambda_1 \wedge \lambda_2),$$

$$\lim_{\lambda \uparrow \infty} G(\lambda) f = f,$$

and

$$\lim_{\gamma \downarrow \lambda} G(\gamma) f = G(\lambda) f$$

for all $f \in L_2(\nu)$. Treatments of the spectral theorem can be found in Chapter 13 of Rudin (1973) and Chapter XI of Yosida (1980). The semigroup $S(t)$ on $L_2(\nu)$ can then be written as

(4.10) $$S(t) = \int_{[0,\infty)} e^{-\lambda t} \, dG(\lambda), \quad \text{or}$$

(4.11) $$S(t) = G(0) + \int_{(0,\infty)} e^{-\lambda t} \, dG(\lambda)$$

where $G(0)$ is the projection on the eigenspace corresponding to the eigenvalue 0 for Ω_ν. Since $\Omega_\nu 1 = 0$, the 0 eigenvalue always has multiplicity at least one. In particular, we obtain the following convergence result:

$$(4.12) \qquad\qquad \lim_{t \to \infty} \| S(t)f - G(0)f \|_\nu = 0$$

for all $f \in L_2(\nu)$. This is the general L_2 convergence theorem for stochastic Ising models.

As it stands, (4.12) depends on the particular choice of a stochastic Ising model for the given potential, since $G(0)$ depends on that choice. The rest of the results in this section show, however, that a number of statements can be made which are independent of that choice, subject perhaps to uniform positivity assumptions on the rates. The first deals with convergence, and the second with exponentially fast convergence. Recall that a criterion for a Gibbs state to be extremal was given in Corollary 3.8.

Theorem 4.13. *The following statements are equivalent.*

    (a) $\lim_{t \to \infty} S(t)f = \int f \, d\nu$ *in* $L_2(\nu)$ *for all* $f \in L_2(\nu)$.
    (b) *0 is a simple eigenvalue for* Ω_ν.
    (c) ν *is extremal in* $\mathcal{G}$.

Proof. By (4.12), (a) is equivalent to

$$(4.14) \qquad\qquad G(0)f = \int f \, d\nu \quad \text{for all } f \in L_2(\nu),$$

which is clearly equivalent to (b). For the equivalence of (b) and (c), note that (b) is equivalent to the statement that

$$\Omega_\nu f = 0 \quad \text{implies} \quad f = \text{constant a.e. } (\nu).$$

But by Lemma 4.3, $\Omega_\nu f = 0$ is equivalent to

$$(4.15) \qquad\qquad \int [f(\eta_x) - f(\eta)]^2 \, d\nu = 0 \quad \text{for all } x \in S.$$

Therefore (b) is equivalent to the statement that (4.15) implies $f = \text{constant}$ a.e. (ν). The equivalence to (c) then follows from Corollary 3.8. $\square$

Theorem 4.16. *Suppose that the equivalent conditions of Theorem 4.13 are satisfied. Then the following statements are equivalent.*

    (a) *There exists an* $\alpha > 0$ *so that*

$$\left\| S(t)f - \int f \, d\nu \right\|_\nu \le e^{-\alpha t} \| f \|_\nu \quad \text{for all } f \in L_2(\nu).$$

(b)   *There exists an $\alpha > 0$ so that $G(\alpha) = G(0)$.*

(c)   *There exists an $\alpha > 0$ so that*

$$\sum_x \int c(x, \eta)[f(\eta_x) - f(\eta)]^2 \, d\nu \geq \alpha \left\| f - \int f \, d\nu \right\|_\nu^2$$

for all $f \in L_2(\nu)$.

Proof. By (4.11) and (4.14),

$$S(t)f - \int f \, d\nu = \int_{(0,\infty)} e^{-\lambda t} \, dG(\lambda)f,$$

so that

$$\left\| S(t)f - \int f \, d\nu \right\|_\nu^2 = \int_{(0,\infty)} e^{-2\lambda t} \, d\|G(\lambda)f\|_\nu^2.$$

The equivalence of (a) and (b) is immediate from this. By Lemma 4.3, (c) is equivalent to the existence of an $\alpha > 0$ so that

(4.17) $$-\int f\Omega_\nu f \, d\nu \geq \alpha \left\| f - \int f \, d\nu \right\|_\nu^2$$

for all $f \in \mathcal{D}(\Omega_\nu)$. But by (4.9), (4.17) is the same as

$$\int_{(0,\infty)} \lambda \, d(G(\lambda)f, f)_\nu \geq \alpha \int_{(0,\infty)} d(G(\lambda)f, f)_\nu$$

where $(\, , \,)_\nu$ denotes the $L_2(\nu)$ inner product. The equivalence of (b) and (c) is now clear. $\square$

Corollary 4.18. *Suppose that $S_1(t)$ and $S_2(t)$ are the semigroups for two stochastic Ising models relative to a potential with rates $c_1(x, \eta)$ and $c_2(x, \eta)$, and that ν is one of the corresponding Gibbs states. Then*

$$\lim_{t \to \infty} S_1(t)f = \int f \, d\nu \quad \text{for all } f \in L_2(\nu)$$

if and only if

$$\lim_{t \to \infty} S_2(t)f = \int f \, d\nu \quad \text{for all } f \in L_2(\nu).$$

If this is the case and

(4.19) $$0 < \inf_{x,\eta} \frac{c_1(x, \eta)}{c_2(x, \eta)} \le \sup_{x,\eta} \frac{c_1(x, \eta)}{c_2(x, \eta)} < \infty,$$

then there is an $\alpha > 0$ so that

$$\left\| S_1(t)f - \int f \, d\nu \right\|_\nu \le e^{-\alpha t} \|f\|_\nu, \quad \text{for all } f \in L_2(\nu)$$

if and only if there is an $\alpha > 0$ so that

$$\left\| S_2(t)f - \int f \, d\nu \right\|_\nu \le e^{-\alpha t} \|f\|_\nu, \quad \text{for all } f \in L_2(\nu).$$

Proof. The first statement follows from Theorem 4.13 and the fact that statement (c) there depends only on ν and not on the process. The second statement follows from Theorem 4.16 in the same way, since under (4.19), statement (c) of that theorem depends only on ν and not on the process. $\square$

Corollary 4.18 shows that to a large extent, the L_2 behavior of the stochastic Ising model is independent of the particular choice of rates. Therefore one way of showing L_2 exponential convergence for one model is to verify (0.6) of Chapter III for another model with the same potential, and then use Corollary 4.18. Even in one dimension, this works only in special cases. One such case is given in Example 4.3(a) of Chapter I. For a one-dimensional finite range stochastic Ising model, the statements in Theorem 4.13 all hold since $\mathscr{G}$ is a singleton by Theorem 3.5. Holley (1985) has shown that the statements in Theorem 4.16 hold as well in this case. In the same paper, he also showed that if the process is attractive, the $L_2(\nu)$ exponential convergence can be replaced by exponential convergence in the $C(X)$ norm.

Theorem 4.13 has other interesting consequences. For the following one, let $\mathscr{G}_e$ be the set of extreme points of $\mathscr{G}$.

Corollary 4.20. $\mathscr{G}_e \subset \mathscr{I}_e.$

Proof. Suppose $\nu \in \mathscr{G}_e$ and $\nu = \alpha \nu_1 + (1 - \alpha) \nu_2$, where $0 < \alpha < 1$ and $\nu_1, \nu_2 \in \mathscr{I}$. By Theorem 4.13,

$$\lim_{t \to \infty} S(t)f = \int f \, d\nu \quad \text{in } L_2(\nu) \quad \text{for all } f \in L_2(\nu).$$

Therefore

$$\lim_{t \to \infty} S(t)f = \int f \, d\nu \quad \text{in } L_2(\nu_1) \quad \text{for all } f \in L_2(\nu).$$

Since $\nu_1 \in \mathcal{I}$,

$$\int S(t)f \, d\nu_1 = \int f \, d\nu_1$$

for all $f \in C(X)$ and all $t \geq 0$. Therefore

$$\int f \, d\nu = \int f \, d\nu_1$$

for all $f \in C(X)$, and hence $\nu_1 = \nu$. $\square$

Corollary 4.21. *Suppose that $c(x, \eta_x) = c(x, \eta)$ for all $x \in S$ and $\eta \in X$, and that*

(4.22) $$0 < \inf_{x,\eta} c(x, \eta).$$

Let ν be the product measure on X with $\nu\{\eta: \eta(x) = 1\} = \frac{1}{2}$ for all $x \in S$. Then there is an $\alpha > 0$ so that

$$\left\| S(t)f - \int f \, d\nu \right\|_\nu \leq e^{-\alpha t} \|f\|_\nu$$

for all $f \in L_2(\nu)$.

Proof. Since $c(x, \eta_x) = c(x, \eta)$, this spin system is a stochastic Ising model relative to the potential $J_R \equiv 0$. Of course, $\mathcal{G} = \{\nu\}$ in this case. By (4.22) and Theorem 4.16, it suffices to check that for some $\alpha > 0$,

(4.23) $$\sum_x \int [f(\eta_x) - f(\eta)]^2 \, d\nu \geq \alpha \left\| f - \int f \, d\nu \right\|_\nu^2$$

for all $f \in L_2(\nu)$. The collection $\{\chi_R\}$, where R ranges over the finite subsets of S is an orthonormal basis in $L_2(\nu)$. Therefore, we can write

$$f = \sum_R c_R \chi_R,$$

where the series converges in $L_2(\nu)$. Now

$$f(\eta_x) - f(\eta) = -2 \sum_{R \ni x} c_R \chi_R,$$

so that

$$\int [f(\eta_x) - f(\eta)]^2 \, d\nu = 4 \sum_{R \ni x} c_R^2.$$

Since

$$\left\| f - \int f \, d\nu \right\|_{\nu}^{2} = \sum_{R \neq \varnothing} c_{R}^{2},$$

(4.23) holds with $\alpha = 4$. $\quad\square$

5. Characterization of Invariant Measures

For a stochastic Ising model, every Gibbs state is invariant by Theorem 2.15. Since there are no known examples in which a stochastic Ising model has invariant measures which are not Gibbs, it is reasonable to conjecture that $\mathscr{I} = \mathscr{G}$, at least under some regularity conditions. At this point, we only know this to be the case if the process is attractive and $\mathscr{G}$ is a singleton (Theorem 2.16). In this section, we will use the relative entropy technique to obtain further results of this type.

The underlying idea is to define the relative entropies of the finite-dimensional distributions of a probability measure ν on X, and then to find a convenient expression for the time derivatives of these quantities corresponding to the distribution of the process at time t. If the measure is invariant, these time derivatives must be zero. This observation leads to a collection of identities which can be used to try to show that the measure is Gibbs.

Of course this technique could be used with functionals other than the relative entropy. The advantage of using the relative entropy is that it has certain monotonicity properties which are described in a Markov chain context in Section 4 of Chapter II. In the present context, this monotonicity is evidenced by the fact that many of the terms in the basic identities have a fixed sign. The relative entropy technique works extremely well in situations in which both the process and the measure are translation invariant. While it is less effective in the absence of translation invariance, it still yields important and useful conclusions which apparently cannot be obtained otherwise.

Throughout this section, $\{J_{R}\}$ will be a potential, and $c(x, \eta)$ will be the rates for a stochastic Ising model relative to that potential (see Definitions 1.1 and 2.1). In the first part of the section, T will be a fixed finite subset of S. For $\zeta \in \{0, 1\}^{T}$, define $f_{\zeta} \in \mathscr{D}$ by

$$f_{\zeta}(\eta) = \begin{cases} 1 & \text{if } \eta = \zeta \text{ on } T, \\ 0 & \text{otherwise.} \end{cases}$$

If ν is a probability measure on X, define

$$\nu(\zeta) = \int f_{\zeta} \, d\nu.$$

For $x \in T$ and $\zeta \in \{0, 1\}^T$, put

$$\Gamma(x, \zeta) = \int c(x, \eta) f_\zeta(\eta) \nu(d\eta).$$

An important consequence of Proposition 2.7 and Theorem 2.14 which shows how we will prove that certain measures are Gibbs states is the following: $\nu \in \mathcal{G}$ if and only if

$$\Gamma(x, \zeta) = \Gamma(x, \zeta_x)$$

for all T and all $x \in T$ and $\zeta \in \{0, 1\}^T$.

The entropy of ν on T relative to the potential $\{J_R\}$ is defined as

$$H_T(\nu) = \sum_\zeta \nu(\zeta) \log \nu(\zeta) - \int \left[\sum_{R \subset T} J_R \chi_R \right] d\nu,$$

where the first sum is over all $\zeta \in \{0, 1\}^T$. Note that this depends only on ν through its finite-dimensional distribution corresponding to the coordinates in T. The definition of $H_T(\nu)$ is best motivated by considering the case of a finite S with $T = S$. Then this coincides with Definition 4.1 of Chapter II if the π there is taken to be the Gibbs state given in (1.4). The first step is to obtain a useful expression for the time derivative of the relative entropy of $\nu S(t)$ on T. Note that the terms in the first sum of (5.2) below are all nonnegative.

Lemma 5.1. *Suppose that $\nu \in \mathcal{P}$ has the property that $\nu(\zeta) > 0$ for all $\zeta \in \{0, 1\}^T$. Then*

$$2 \frac{d}{dt} H_T(\nu S(t)) \bigg|_{t=0} = - \sum_{x \in T} \sum_\zeta [\Gamma(x, \zeta) - \Gamma(x, \zeta_x)] \log \frac{\Gamma(x, \zeta)}{\Gamma(x, \zeta_x)}$$

(5.2)
$$+ \sum_{x \in T} \sum_\zeta [\Gamma(x, \zeta) - \Gamma(x, \zeta_x)]$$

$$\times \left\{ V(x, \zeta) + \log \frac{\Gamma(x, \zeta)}{\nu(\zeta)} - \log \frac{\Gamma(x, \zeta_x)}{\nu(\zeta_x)} \right\},$$

where

$$V(x, \zeta) = 2 \sum_{x \in R \subset T} J_R \chi_R(\zeta).$$

Proof. Set $\nu_t = \nu S(t)$. We will compute the derivative of the two terms in $H_T(\nu_t)$ separately. Sums on ζ will always be over $\zeta \in \{0, 1\}^T$, and sums on

x will be over $x \in T$. Since $\nu(\zeta) > 0$ for all $\zeta \in \{0, 1\}^T$ and since

$$\sum_\zeta \int \Omega f_\zeta \, d\nu = \int \Omega 1 \, d\nu = 0,$$

$$\left. \frac{d}{dt} \sum_\zeta \nu_t(\zeta) \log \nu_t(\zeta) \right|_{t=0} = \sum_\zeta [1 + \log \nu(\zeta)] \frac{d}{dt} \nu_t(\zeta) \bigg|_{t=0}$$

$$= \sum_\zeta [1 + \log \nu(\zeta)] \int \Omega f_\zeta \, d\nu$$

$$= \sum_\zeta \log \nu(\zeta) \int \sum_x c(x, \eta)[f_\zeta(\eta_x) - f_\zeta(\eta)] \, d\nu$$

$$= \sum_{x, \zeta} \log \nu(\zeta)[\Gamma(x, \zeta_x) - \Gamma(x, \zeta)].$$

Making the change of variables $\zeta \to \zeta_x$ in the sum on ζ of the first term, we obtain

(5.3)
$$\left. \frac{d}{dt} \sum_\zeta \nu_t(\zeta) \log \nu_t(\zeta) \right|_{t=0} = \sum_{x, \zeta} \Gamma(x, \zeta) \log \frac{\nu(\zeta_x)}{\nu(\zeta)}.$$

For the second term in $H_T(\nu_t)$,

$$\left. \frac{d}{dt} \int \left[\sum_{R \subset T} J_R \chi_R \right] d\nu_t \right|_{t=0} = \int \left[\sum_{R \subset T} J_R \Omega \chi_R \right] d\nu$$

$$= -2 \sum_x \sum_{x \in R \subset T} J_R \int c(x, \eta) \chi_R(\eta) \, d\nu$$

since

$$\chi_R(\eta_x) - \chi_R(\eta) = \begin{cases} -2\chi_R(\eta) & \text{if } x \in R, \\ 0 & \text{if } x \notin R. \end{cases}$$

But for $R \subset T$,

$$\int c(x, \eta) \chi_R(\eta) \, d\nu = \sum_\zeta \int c(x, \eta) f_\zeta(\eta) \chi_R(\eta) \, d\nu$$

$$= \sum_\zeta \chi_R(\zeta) \Gamma(x, \zeta),$$

so that

(5.4)
$$\left. \frac{d}{dt} \int \left[\sum_{R \subset T} J_R \chi_R \right] d\nu_t \right|_{t=0} = -2 \sum_{x, \zeta} \sum_{x \in R \subset T} J_R \chi_R(\zeta) \Gamma(x, \zeta)$$

$$= -\sum_{x, \zeta} \Gamma(x, \zeta) V(x, \zeta).$$

Combining (5.3) and (5.4) gives

$$\frac{d}{dt} H_T(\nu_t)\Big|_{t=0} = \sum_{x,\zeta} \Gamma(x, \zeta)\left\{ \log \frac{\nu(\zeta_x)}{\nu(\zeta)} + V(x, \zeta)\right\}.$$

Since $V(x, \zeta_x) = -V(x, \zeta)$ and

$$\log \frac{\nu(\zeta_x)}{\nu(\zeta)} = -\log \frac{\nu(\zeta)}{\nu(\zeta_x)},$$

this can be rewritten as

$$2\frac{d}{dt} H_T(\nu_t)\Big|_{t=0} = \sum_{x,\zeta} [\Gamma(x, \zeta) - \Gamma(x, \zeta_x)]\left\{ V(x, \zeta) - \log \frac{\nu(\zeta)}{\nu(\zeta_x)}\right\},$$

from which (5.2) follows by adding and subtracting appropriate terms. □

Lemma 5.5. *Suppose that* $\nu \in \mathscr{I}$. *Then* $\nu(\zeta) > 0$ *for all* $\zeta \in \{0, 1\}^T$.

Proof. Since $\nu \in \mathscr{I}$ and $f_\zeta \in \mathscr{D}$,

$$0 = \int \Omega f_\zeta \, d\nu = \int \sum_{x \in T} c(x, \eta)[f_\zeta(\eta_x) - f_\zeta(\eta)] \, d\nu$$

$$= \sum_{x \in T} \int c(x, \eta)[f_{\zeta_x}(\eta) - f_\zeta(\eta)] \, d\nu.$$

Therefore, since $c(x, \eta) > 0$ for all x and η, $\nu(\zeta) = 0$ implies that $\nu(\zeta_x) = 0$ for all $x \in T$. Iterating this statement, we see that if $\nu(\zeta) = 0$ for some $\zeta \in \{0, 1\}^T$, then $\nu(\zeta) = 0$ for all $\zeta \in \{0, 1\}^T$. Since $\sum_\zeta \nu(\zeta) = 1$, this cannot occur. □

Theorem 5.6. *Suppose that* $\nu \in \mathscr{I}$. *Then*

$$\sum_{x \in T} \sum_{\zeta \in \{0, 1\}^T} [\Gamma(x, \zeta) - \Gamma(x, \zeta_x)] \log \frac{\Gamma(x, \zeta)}{\Gamma(x, \zeta_x)}$$

$$= \sum_{x \in T} \sum_{\zeta \in \{0, 1\}^T} [\Gamma(x, \zeta) - \Gamma(x, \zeta_x)]$$

$$\times \left\{ V(x, \zeta) + \log \frac{\Gamma(x, \zeta)}{\nu(\zeta)} - \log \frac{\Gamma(x, \zeta_x)}{\nu(\zeta_x)}\right\}.$$

Proof. By Lemma 5.5, $\nu(\zeta) > 0$ for all $\zeta \in \{0, 1\}^T$. Since $\nu \in \mathscr{I}$, $\nu S(t) = \nu$ for all $t \geq 0$, so that $(d/dt)H_T(\nu S(t)) = 0$ for all $t \geq 0$. The result then follows from Lemma 5.1. □

For $x \in T$, define

$$\alpha_T(x) = \sum_\zeta [\Gamma(x, \zeta) - \Gamma(x, \zeta_x)] \log \frac{\Gamma(x, \zeta)}{\Gamma(x, \zeta_x)},$$

$$\beta_T(x) = \sum_\zeta |\Gamma(x, \zeta) - \Gamma(x, \zeta_x)|, \quad \text{and}$$

$$\rho_T(x) = 2 \sum_{x \in R \not\subset T} |J_R| + 2 \sum_{u \notin T} \sup_\eta \frac{|c(x, \eta_u) - c(x, \eta)|}{c(x, \eta)}.$$

If $x \notin T$, they are defined to be zero.

Corollary 5.7. *Suppose that* $\nu \in \mathcal{I}$. *Then*

$$\sum_{x \in T} \alpha_T(x) \le \sum_{x \in T} \rho_T(x) \beta_T(x).$$

Proof. By Theorem 5.6, it suffices to show that

$$\left| V(x, \zeta) + \log \frac{\Gamma(x, \zeta)}{\nu(\zeta)} - \log \frac{\Gamma(x, \zeta_x)}{\nu(\zeta_x)} \right| \le \rho_T(x)$$

for all $\zeta \in \{0, 1\}^T$. In order to do this, let

$$c_T(x, \zeta) = c(x, \zeta_T),$$

where for $\zeta \in \{0, 1\}^T$, $\zeta_T \in X$ is defined by

$$\zeta_T(u) = \begin{cases} \zeta(u) & \text{if } u \in T, \\ 1 & \text{if } u \notin T. \end{cases}$$

Since the process is a stochastic Ising model, the function

$$c_T(x, \zeta) \exp\left[\sum_{R \ni x} J_R \chi_{R \cap T}(\zeta) \right] = c(x, \zeta_T) \exp\left[\sum_{R \ni x} J_R \chi_R(\zeta_T) \right]$$

does not depend on the coordinate $\zeta(x)$. Therefore by adding and subtracting the log of this function and using the definitions of $V(x, \zeta)$ and $\Gamma(x, \zeta)$, we see that

$$
V(x, \zeta) + \log \frac{\Gamma(x, \zeta)}{\nu(\zeta)} - \log \frac{\Gamma(x, \zeta_x)}{\nu(\zeta_x)}
$$

$$
= \log \frac{\displaystyle\int \frac{c(x, \eta)}{c_T(x, \zeta)} f_\zeta(\eta) \exp\left[-\sum_{x \in R \not\subset T} J_R \chi_{R \cap T}(\zeta) \right] d\nu}{\nu(\zeta)}
$$

$$
- \log \frac{\displaystyle\int \frac{c(x, \eta)}{c_T(x, \zeta_x)} f_{\zeta_x}(\eta) \exp\left[-\sum_{x \in R \not\subset T} J_R \chi_{R \cap T}(\zeta_x) \right] d\nu}{\nu(\zeta_x)}.
$$

Since $\nu(\zeta) = \int f_\zeta \, d\nu$ and $\nu(\zeta_x) = \int f_{\zeta_x} \, d\nu$, the absolute value of this expression is bounded by

$$
2 \sum_{x \in R \not\subset T} |J_R| + 2 \sup\left\{ \log \frac{c(x, \eta_1)}{c(x, \eta_2)} : \eta_1 = \eta_2 \text{ on } T \right\}.
$$

The desired bound then follows by using the inequality $\log a \le a - 1$. $\square$

In order to keep in mind where we are headed, a few observations should be made at this point. First, $\nu \in \mathcal{G}$ if and only if $\alpha_T(x) = 0$ for all T and all $x \in T$. Secondly, $\alpha_T(x) = 0$ implies that $\beta_T(x) = 0$, since all the summands in the definition of $\alpha_T(x)$ are automatically nonnegative. Therefore one should be able to control the size of $\beta_T(x)$ in terms of that of $\alpha_T(x)$. Once that is done, it should be possible to use Corollary 5.7 to conclude that $\alpha_T(x) = 0$ provided that $\rho_T(x)$ is small. But we automatically have

$$
\lim_{T \uparrow S} \rho_T(x) = 0
$$

for all $x \in S$ by the definition of the potential and by the assumption that the rates satisfy (0.3) of Chapter III. In order to make effective use of these ideas, we need the following inequalities.

Lemma 5.8. (a) $\alpha_{T_1}(x) \le \alpha_{T_2}(x)$ if $x \in T_1 \subset T_2$.

(b) $$\beta_T^2(x) \le 2[\sup_\eta c(x, \eta)] \alpha_T(x).$$

Proof. For part (a), take $x \in T_1 \subset T_2$. Let $\Gamma_1(x, \zeta^1)$ and $\Gamma_2(x, \zeta^2)$ be defined relative to T_1 and T_2 respectively. Then

$$
\alpha_{T_1}(x) = \sum_{\zeta_1} \phi[\Gamma_1(x, \zeta^1), \Gamma_1(x, \zeta_x^1)], \quad \text{and}
$$

$$\alpha_{T_2}(x) = \sum_{\zeta_2} \phi[\Gamma_2(x, \zeta^2), \Gamma_2(x, \zeta_x^2)],$$

where the sums are over $\zeta_1 \in \{0, 1\}^{T_1}$ and $\zeta_2 \in \{0, 1\}^{T_2}$ respectively and

$$\phi(u, v) = (u - v) \log \frac{u}{v}$$

for $u, v > 0$. Since ϕ is convex and homogeneous of degree one, it is subadditive. Therefore since

$$\Gamma_1(x, \zeta_1) = \sum_{\zeta_2 = \zeta_1 \text{ on } T_1} \Gamma_2(x, \zeta_2),$$

it follows that $\alpha_{T_1}(x) \leq \alpha_{T_2}(x)$. For part (b), use the symmetry and subadditivity of $\phi(u, v)$ to show that

$$\alpha_T(x) = \sum_{\zeta} \phi[\Gamma(x, \zeta), \Gamma(x, \zeta_x)]$$

$$\geq \phi(M, m),$$

where

$$M = \sum_{\zeta} \max\{\Gamma(x, \zeta), \Gamma(x, \zeta_x)\}$$

and

$$m = \sum_{\zeta} \min\{\Gamma(x, \zeta), \Gamma(x, \zeta_x)\}.$$

Since

$$\beta_T(x) = M - m$$

and

$$M \leq 2 \sup_{\eta} c(x, \eta),$$

the desired result follows from the elementary inequality

$$M - m \leq M \log \frac{M}{m} \quad \text{for } 0 < m \leq M. \quad \square$$

For $x, u \in S$, define

(5.9) $$\gamma(x, u) = \sum_{R \ni x, u} |J_R| + \sup_{\eta} |c(x, \eta_u) - c(x, \eta)|.$$

In view of (0.3) of Chapter III,

(5.10)
$$\sup_{x} \sum_{u} \gamma(x, u) < \infty$$

is equivalent to (2.6). Of course (5.10) is automatic if both the potential and flip rates are translation invariant on Z^d and of finite range. We will assume from now on that

$$\inf_{x, \eta} c(x, \eta) > 0.$$

This is automatic in the translation invariant case. By the definition of $\rho_T(x)$, we will then have the bound

(5.11)
$$\rho_T(x) \leq L \sum_{u \notin T} \gamma(x, u)$$

for some constant L.

Theorem 5.12. *Suppose that $S = Z^d$, that both the potential and the flip rates are translation invariant, and that (5.10) is satisfied. Then $\mathscr{I} \cap \mathscr{S} \subset \mathscr{G}$.*

Proof. Take $\nu \in \mathscr{I} \cap \mathscr{S}$. If T_n is a cube in Z^d of side length n, then T_{kn} is the union of k^d translates of T_n. Since $\nu \in \mathscr{S}$, it follows from part (a) of Lemma 5.8 that

(5.13)
$$\frac{1}{(kn)^d} \sum_{x \in T_{kn}} \alpha_{T_{kn}}(x) \geq \frac{1}{n^d} \sum_{x \in T_n} \alpha_{T_n}(x).$$

On the other hand, by (5.11),

$$\frac{1}{n^d} \sum_{x \in T_n} \rho_{T_n}(x) \leq \frac{L}{n^d} \sum_{x \in T_n} \sum_{u \notin T_n} \gamma(0, u - x)$$

$$= L \sum_{v} \gamma(0, v) \frac{|\{x \in T_n : x + v \notin T_n\}|}{n^d}.$$

This tends to zero as $n \to \infty$ by (5.10) and the Dominated Convergence Theorem. By Corollary 5.7 and the fact that $\beta_T(x)$ is uniformly bounded, it follows that

$$\lim_{n \to \infty} \frac{1}{n^d} \sum_{x \in T_n} \alpha_{T_n}(x) = 0.$$

Therefore by (5.13), $\alpha_{T_n}(x) = 0$ for all $x \in T_n$, and hence by part (a) of Lemma 5.8, $\alpha_T(x) = 0$ for all T and all $x \in T$. Therefore $\nu \in \mathscr{G}$. $\quad \square$

Holley (1971) in the finite range case, and Higuchi and Shiga (1975) under assumption (2.6), proved somewhat more than the conclusion of Theorem 5.12. (They considered only the choice (2.4) of flip rates.) They showed for $\nu \in \mathcal{S}$ that any limit of $\nu S(t)$ along a sequence of t's tending to ∞ must be in $\mathcal{G}$.

Theorem 5.14. *Suppose that* $S = Z^d$ *for* $d = 1$ *or* $d = 2$, *and assume that*

(5.15)
$$\sum_v |v| \sup_x \gamma(x, v+x) < \infty$$

where $|\cdot|$ *is the* l_∞ *norm on* Z^d. *Then* $\mathcal{I} = \mathcal{G}$.

Proof. Suppose $\nu \in \mathcal{I}$. By Corollary 5.7 and part (b) of Lemma 5.8,

$$\sum_{x \in T} \alpha_T(x) \le M \sum_{x \in T} \rho_T(x) \sqrt{\alpha_T(x)}$$

for some constant M. By the Schwarz inequality, this implies that

(5.16)
$$\left[\sum_{x \in T} \alpha_T(x) \right]^2 \le M^2 \left[\sum_{x \in T} \rho_T(x) \right] \left[\sum_{x \in T} \rho_T(x) \alpha_T(x) \right].$$

Let T_n be the cube $[-n, n]^d$ in Z^d. By (5.11),

$$\sum_{n=1}^{\infty} \rho_{T_n}(x) \le L \sum_{n:\, T_n \ni x} \sum_{u \notin T_n} \gamma(x, u)$$

$$\le L \sum_u \gamma(x, u) |\{n \ge 1 : x \in T_n,\ u \notin T_n\}|$$

$$\le L \sum_u \gamma(x, u) |u - x|, \quad \text{and}$$

$$\sum_{x \in T_n} \rho_{T_n}(x) \le L \sum_{x \in T_n} \sum_{u \notin T_n} \gamma(x, u)$$

$$\le L \sum_v \sum_{\substack{x \in T_n \\ v+x \notin T_n}} \gamma(x, v+x)$$

$$\le L d (2n+1)^{d-1} \sum_v |v| \sup_x \gamma(x, v+x).$$

Therefore by (5.15),

(5.17)
$$\sup_x \sum_{n=1}^{\infty} \rho_{T_n}(x) < \infty, \quad \text{and}$$

(5.18)
$$\sup_n \frac{1}{n^{d-1}} \sum_{x \in T_n} \rho_{T_n}(x) < \infty.$$

It remains to show that if $d = 1$ or $d = 2$, (5.16), (5.17), and (5.18) imply that $\alpha_{T_n}(x) = 0$ for all $x \in T_n$ and all n. In order to do this, use part (a) of Lemma 5.8 and (5.17) to show that

(5.19)
$$\sum_{x \in T_n} \alpha_{T_n}(x) \geq \varepsilon \sum_{x \in T_n} \alpha_{T_n}(x) \sum_{k=1}^{n} \rho_{T_k}(x)$$

$$\geq \varepsilon \sum_{k=1}^{n} \sum_{x \in T_k} \alpha_{T_k}(x) \rho_{T_k}(x),$$

where ε is the reciprocal of the quantity in (5.17). Let

$$\delta_k = \sum_{x \in T_k} \alpha_{T_k}(x) \rho_{T_k}(x).$$

By (5.16) and (5.19),

$$\varepsilon^2 \left(\sum_{k=1}^{n} \delta_k \right)^2 \leq M^2 \delta_n \sum_{x \in T_n} \rho_{T_n}(x),$$

so that by (5.18),

$$\left(\sum_{k=1}^{n} \delta_k \right)^2 \leq N \delta_n n^{d-1}$$

for some constant N. If for some n, $\delta_n > 0$, then for all larger n it follows that

$$\frac{1}{n^{d-1}} \leq N \left\{ \frac{1}{\sum\limits_{k=1}^{n-1} \delta_k} - \frac{1}{\sum\limits_{k=1}^{n} \delta_k} \right\}.$$

The series on the right converges, so that if $d \leq 2$, we have reached a contradiction. Therefore in this case, $\delta_n = 0$ for all n, so $\alpha_{T_n}(x) = 0$ for all $x \in T_n$ and all n by (5.16). $\square$

6. Notes and References

Section 1. Definitions of infinite Gibbs states were given at about the same time by Minlos (1967), Dobrushin (1968a), and Lanford and Ruelle (1969). Definition 1.5 is Dobrushin's version, while Minlos' definition is the equivalent form given in Theorem 1.16(c). The equivalence was proved by Dobrushin. Proposition 1.10 and its proof are in the same spirit as those results which state that every Markov random field is a Gibbs state. For more on this, see Chapter 1 of Preston (1974b) and the references given there. The conclusions of Theorem 1.21 are known as the Griffiths inequalities. They were proved by Griffiths (1967) for pair potentials and by Kelly and Sherman (1968) for many body potentials. Ginibre (1970) gave a further generalization of these inequalities. The proof of (1.23) given here is due to Ginibre (1969). The treatment of Theorem 1.27 and its corollaries follows Holley (1974a).

Section 2. Stochastic Ising models were first proposed in a special case by Glauber (1963), and then generalized and studied by Dobrushin (1971a, b). Versions of Theorem 2.13 and 2.14 were proved by Spitzer (1971b) in the case of a finite S, and by Dobrushin (1971b) and Logan (1974) for general S. Other papers containing results of this type are Higuchi and Shiga (1975), Glötzl (1981), and Ding and Chen (1981). Corresponding results for exclusion processes with speed change were proved by Logan (1974), Georgii (1979), Glötzl (1982), and Yan, Chen, and Ding (1982a, b). Theorem 2.16 is due to Holley (1974a). That paper is a survey of the stochastic Ising model as of 1974. A survey of interacting particle systems which emphasizes the role of the stochastic Ising model is given by Durrett (1981).

Section 3. Theorem 3.1 with sufficient condition (3.2) is due to Dobrushin (1968b). Although his proof did not use the connection with the stochastic Ising model, it was based on a similar type of contraction principle. The general technique of using the stochastic Ising model to prove results about the Gibbs states such as absence of phase transition, exponential decay of correlations, and analyticity of correlation functions is due to Holley and Stroock (1976b). Corollary 3.8 is due to Lanford and Ruelle (1969). Theorem 3.9 was proved independently by Ruelle (1968) and by Dobrushin (1969). The proof given here is due independently to Lanford and Preston, and comes via Spitzer. The very precise necessary condition (3.11) and sufficient condition (3.12) for the occurrence of phase transition in one-dimensional systems were only obtained after a number of weaker results had been proved. Most important was the work of Dyson (1969a, b). His necessary condition was

$$\lim_{N \to \infty} \frac{\sum_{n=1}^{N} nJ(n)}{\log(\log N)} = 0$$

and his sufficient condition was

$$\sum_{n} \frac{\log(\log n)}{n^3 J(n)} < \infty.$$

The proof of Theorem 3.14 is based on the well-known Peierls (1936) argument. This approach was further developed by Griffiths (1964) and Dobrushin (1965).

Section 4. This section is based on Holley and Stroock (1976c). Other papers which contain similar results are Higuchi and Shiga (1975) and Sullivan (1975c). One difference in the treatment given here is that Lemma 4.3 is proved without appealing to the spectral theorem.

Section 5. The results in this section are taken from Holley and Stroock (1977a). This paper is the culmination of the development of the relative entropy technique in this context which was initiated by Holley (1971) and was continued by Higuchi and Shiga (1975) and Moulin Ollagnier and Pinchon (1977). Holley and Stroock proved Theorem 5.14 in the finite range case only, but they indicated that generalizations of the type given here were possible. The relative entropy technique has been used in other contexts by Holley (1972c), Holley and Stroock (1981), Fritz (1982), and Liggett (1983a).

7. Open Problems

1. Is the conclusion of Theorem 5.14 true in dimensions greater than two?

2. Is the final statement in Theorem 2.16 true without the attractiveness assumption?

3. Is it true that any two uniformly positive stochastic Ising models with respect to the same potential are either both or neither exponentially ergodic in the uniform norm? The L_2 version of this statement is proved in Corollary 4.18. Holley (1985) has made some progress on this problem.

4. A spin system is a stochastic Ising model with respect to the potential $J_R \equiv 0$ if and only if $c(x, \eta_x) = c(x, \eta) > 0$ for all $x \in S$ and $\eta \in X$. The unique Gibbs state corresponding to this potential is the product measure ν with density $\frac{1}{2}$. Note that if $c(x, \eta_x) = c(x, \eta)$ and the process is attractive, then $c(x, \eta)$ is independent of η, so there are no interesting attractive examples in this class. In how great a generality can it be proved that $\mathscr{I} = \{\nu\}$? Some results in this direction are Corollary 4.21 and Theorem 5.14, as well as Theorem 1.12 of Chapter III.

5. Is it true that every translation invariant strictly positive spin system on Z^d with finite range has an invariant measure which is a Gibbs state? This is plausible in view of Proposition 1.10 and the fact that the strict positivity of the rates should imply that an invariant measure is somewhat smooth. Using techniques similar to those of Section 5, Künsch (1984b) has shown that if a translation invariant spin system on Z^d has an invariant measure which is translation invariant and Gibbs with respect to some potential, then every invariant measure which is translation invariant is again Gibbs with respect to the same potential. Thus an affirmative solution to this problem, together with some estimate on the size of the potential which would permit the application of Theorem 3.9, would give an affirmative solution to the positive rates conjecture for attractive spin systems (Problem 6 in Chapter III).

6. The proof of nonergodicity for appropriate stochastic Ising models is based on Theorem 3.14. Can a proof of nonergodicity be constructed which is based on a direct analysis of the time evolution itself rather than on its invariant measures? If so, that would provide an alternative proof of Theorem 3.14. More importantly, it would potentially generalize to spin systems which are not reversible.

7. In the context of Corollary 2.18, is it the case that

$$\lim_{t \to \infty} \nu_\rho S(t) = \bar{\nu} \quad \text{if } \rho > \tfrac{1}{2}, \quad \text{and}$$

$$\lim_{t \to \infty} \nu_\rho S(t) = \underline{\nu} \quad \text{if } \rho < \tfrac{1}{2},$$

where ν_ρ is the product measure on X with

$$\nu_\rho\{\eta: \eta(x) = 1\} = \rho \quad \text{for all } x \in Z^d\,?$$

What can be said about convergence for more general initial distributions?

Chapter V

The Voter Model

The voter model is the spin system with rates $c(x, \eta)$ given by

$$(0.1) \qquad c(x, \eta) = \begin{cases} \sum_y p(x, y)\eta(y) & \text{if } \eta(x) = 0, \\ \sum_y p(x, y)[1 - \eta(y)] & \text{if } \eta(x) = 1, \end{cases}$$

where $p(x, y) \geq 0$ for $x, y \in S$ and

$$\sum_y p(x, y) = 1 \quad \text{for } x \in S.$$

We will assume throughout this chapter that $p(x, y)$ is such that the Markov chain with those transition probabilities is irreducible. An equivalent way of describing the rates of the voter model is to say that a site x waits an exponential time with parameter one, at which time it flips to the value it sees at that time at a site y which is chosen with probability $p(x, y)$. The voter interpretation which gives this process its name views S as a collection of individuals, each of which has one of two possible positions on a political issue. These possible positions are denoted by 0 and 1. Periodically (i.e., at independent exponential times), an individual reassesses his view in a rather simple way: he chooses a "friend" at random with certain probabilities and then adopts his position. For an alternative interpretation, $\{x \in S: \eta(x) = 0\}$ and $\{x \in S: \eta(x) = 1\}$ can be thought of as the territories occupied by each of two populations. A site x which belongs to one population will be invaded by the other at a rate which is the sum of $p(x, y)$ over those sites y which belong to the second population.

In addition to the interest coming from the interpretations given above, there are good mathematical reasons for studying the voter model. As we saw in Example 4.16 of Chapter III, the voter model has a rich duality theory. Coalescing duality will in fact be used very heavily in this chapter. Of great technical importance is the fact that the duals of the voter model do not increase in size as time increases. This property will enable us to carry out a much more complete analysis of the voter model than is possible for other classes of spin systems. Problems involving the voter model will often be recast in terms of the dual system of coalescing Markov chains. Frequently, these problems will then be reduced to others involving independent Markov chains.

It is clear from the form of the rates given in (0.1) that the pointmasses on the configurations $\eta \equiv 0$ and $\eta \equiv 1$ are invariant for the process. (Using the voter interpretation, these configurations can be thought of as representing a consensus.) Thus the voter model is never ergodic. The natural question is then whether or not there are other extremal invariant measures. This problem will be completely resolved in Section 1. At the same time, we will prove appropriate limit theorems for the process. The most interesting special case of the voter model is that in which $S = Z^d$ and $p(x, y) = 1/2d$ for $|x - y| = 1$. In this case, the main results are that the process has only the two trivial extremal invariant measures if $d \leq 2$, but has a full one-parameter family of them if $d \geq 3$. This dichotomy is closely related to the fact that the simple random walk on Z^d is recurrent if $d \leq 2$ and transient if $d \geq 3$. We already saw the one-dimensional part of this result as a very special case of Theorem 3.13 of Chapter III. The proof in the case of the voter model is much simpler than the proof of that theorem, however.

Properties of the nontrivial extremal invariant measures for $d \geq 3$ are investigated in Section 2. In particular, we will see that in equilibrium, the coordinates of the voter model have rather high correlations. The fact that there are no nontrivial invariant measures for $d \leq 2$ indicates that clustering is occurring. The results in Section 3 give a precise description of this clustering in one dimension. The finite system, which is the subject of Section 4, is the voter model in which initially (and hence at all times), $\sum_x \eta_t(x) < \infty$. The main objective is to determine the asymptotic behavior of $P[\eta_t \neq 0]$ as $t \uparrow \infty$ in this case.

Other types of results for the voter model which will not be discussed here have been obtained by Presutti and Spohn (1983) and by Cox and Griffeath (1983). Closely related models which have been studied in some detail are the stepping stone model from genetics (see Sawyer (1977, 1979) and Shiga (1980a, b)), and the Williams–Bjerknes tumor growth model (see Bramson and Griffeath (1980c, 1981)).

Section 1 should be read first, after which the next three sections can be read in any order. Coupling and duality will both be used extensively in this chapter. Of particular relevance here are Sections 2, 4, and 6 of Chapter III.

1. Ergodic Theorems

In this section, we will find all the extremal invariant measures for the voter model, and will find the domain of attraction of each of them. Throughout the chapter, we will assume that the flip rates are given by (0.1), where

$p(x, y)$ are the transition probabilities for an irreducible Markov chain on S. Note that condition (0.3) of Chapter III is automatically satisfied, so that the spin system with these rates is well defined. Of course, the voter model is an attractive spin system in the sense of Section 2 of Chapter III.

The invariant measures for the voter model are closely related to the bounded harmonic functions for $p(x, y)$. Thus, let

$$\mathcal{H} = \left\{ \alpha : S \to [0, 1] \text{ such that } \sum_y p(x, y)\alpha(y) = \alpha(x) \text{ for all } x \right\}.$$

A certain subset of $\mathcal{H}$ will play a particularly important role in the identification of the extremal invariant measures. In order to describe it, let $X(t)$ and $Y(t)$ be independent copies of the continuous time Markov chain with transition probabilities

(1.1)
$$p_t(x, y) = e^{-t} \sum_{n=0}^{\infty} \frac{t^n}{n!} p^{(n)}(x, y),$$

where $p^{(n)}(x, y)$ are the n-step transition probabilities associated with $p(x, y)$. Let

(1.2)
$$g(x, y) = P^{(x,y)}(X(t) = Y(t) \text{ for some } t \geq 0)$$

for $x, y \in S$, where $(X(0), Y(0)) = (x, y)$. This should be thought of as a measure of how far apart x and y are. Since

$$
\begin{aligned}
E^{(x,y)}g(X(t), Y(t)) \\
&= E^{(x,y)} P^{(X(t), Y(t))}(X(s) = Y(s) \text{ for some } s \geq 0) \\
&= P^{(x,y)}(X(s) = Y(s) \text{ for some } s \geq t) \\
&\leq g(x, y)
\end{aligned}
$$

by the Markov property, $g(X(t), Y(t))$ is a (nonnegative) supermartingale. Therefore

(1.3)
$$L = \lim_{t \to \infty} g(X(t), Y(t))$$

exists with probability one. In fact, L is the indicator of the event

$$\mathscr{E} = \{\text{there exists } t_n \uparrow \infty \text{ such that } X(t_n) = Y(t_n)\}.$$

(For a discrete time version of this result, see Theorem 9.5.1 of Chung (1974).) For $\alpha \in \mathcal{H}$, $\alpha(X(t))$ is a bounded martingale, so $\lim_{t \to \infty} \alpha(X(t))$

exists with probability one. Now let $\mathcal{H}^*$ be the set of all $\alpha \in \mathcal{H}$ such that

$$\lim_{t \to \infty} \alpha(X(t)) = 0 \text{ or } 1 \text{ a.s. on } \mathcal{E}.$$

It would be a good idea to keep in mind the following three examples in what follows.

Example 1.4. $S = Z^d$, $p(x, y) = p(0, y - x)$, and the random walk $X(t) - Y(t)$ is recurrent. Then $\mathcal{H}$ consists of the constants in $[0, 1]$ by Corollary 7.2 of Chapter I. By the recurrence assumption $P(\mathcal{E}) = 1$, so that $\mathcal{H}^*$ consists only of the constants $\alpha \equiv 0$ and $\alpha \equiv 1$.

Example 1.5. $S = Z^d$, $p(x, y) = p(0, y - x)$, and the random walk $X(t) - Y(t)$ is transient. Again, $\mathcal{H}$ consists of the constants in $[0, 1]$. In this case, though, the transience assumption implies that $P(\mathcal{E}) = 0$, so that $\mathcal{H}^* = \mathcal{H}$.

Example 1.6. $S = Z^1$, and

$$p(x, y) = \begin{cases} p_l(y - x) & \text{if } x < 0, \\ p_r(y - x) & \text{if } x \geq 0, \end{cases}$$

where

$$\sum_x |x| p_l(x) < \infty, \quad \sum_x |x| p_r(x) < \infty, \quad \sum_x x p_l(x) < 0, \quad \text{and} \quad \sum_x x p_r(x) > 0.$$

Then

$$P(\lim_{t \to \infty} X(t) = +\infty) + P(\lim_{t \to \infty} X(t) = -\infty) = 1,$$

and

$$\mathcal{E} = \{\lim_{t \to \infty} X(t) = \lim_{t \to \infty} Y(t)\}.$$

In this case,

$$\mathcal{H} = \{\lambda_1 h(x) + \lambda_2 [1 - h(x)]: \lambda_1, \lambda_2 \in [0, 1]\}$$

where

$$h(x) = P^x(\lim_{t \to \infty} X(t) = +\infty),$$

and $\mathcal{H}^*$ consists of the four elements $0, 1, h, 1 - h$.

In this section, we will make heavy use of coalescing duality for the voter model. It is therefore convenient at this point to summarize the relevant results from Section 4 of Chapter III as they apply to the voter model. In doing so, we will interchange the roles of 0 and 1, which of course does not change the voter model. If μ is a probability measure on X, define

$$\hat{\mu}(A) = \mu\{\eta: \eta(x) = 1 \text{ for all } x \in A\}$$

for $A \in Y$, the class of finite subsets of S. Let A_t be the Markov chain on Y in which points move independently on S according to the continuous time Markov chain with transition probabilities $p_t(x, y)$, except that two points which occupy the same site coalesce. This is then the system of coalescing Markov chains on S which we found to be the dual of the voter model in Example 4.16 of Chapter III. The basic duality relation which is given in Corollary 4.15 of that chapter then becomes

$$(1.7) \qquad\qquad \hat{\mu}_t(A) = E^A \hat{\mu}(A_t),$$

where $\mu_t = \mu S(t)$.

For $\alpha \in \mathcal{H}$, define ν_α to be the product measure on X with marginals

$$\nu_\alpha\{\eta: \eta(x) = 1\} = \alpha(x).$$

We can now state the main result of this section.

Theorem 1.8. (a) $\mu_\alpha = \lim_{t \to \infty} \nu_\alpha S(t)$ *exists for every* $\alpha \in \mathcal{H}$, *and* $\mu_\alpha \in \mathcal{I}$.

(b) $\qquad\qquad \mu_\alpha\{\eta: \eta(x) = 1\} = \alpha(x) \quad$ *for all* $x \in S$.

(c) $\qquad\qquad\qquad \mathcal{I}_e = \{\mu_\alpha: \alpha \in \mathcal{H}^*\}$.

When applied to Example 1.4, Theorem 1.8 implies that $\mathcal{I}_e = \{\delta_0, \delta_1\}$. In the case of Example 1.5, on the other hand, there is an extremal invariant measure for each particle density between zero and one. In Example 1.6, there are four extremal invariant measures: δ_0, δ_1, a measure satisfying $\mu\{\eta: \eta(x) = 1\} = h(x)$, and one satisfying $\mu\{\eta: \eta(x) = 1\} = 1 - h(x)$. If $p_l(\cdot)$ and $p_r(\cdot)$ have finite second moments, then $\sum_{x \leq 0} h(x) < \infty$ and $\sum_{x \geq 0} [1 - h(x)] < \infty$ (see page 209 of Spitzer (1976)), so that these latter two measures have a particularly simple property: the first concentrates on

$$\left\{ \eta: \sum_{x \leq 0} \eta(x) < \infty, \sum_{x \geq 0} [1 - \eta(x)] < \infty \right\},$$

and the second concentrates on

$$\left\{ \eta: \sum_{x \leq 0} [1 - \eta(x)] < \infty, \sum_{x \geq 0} \eta(x) < \infty \right\}.$$

The following result is the basic convergence theorem.

Theorem 1.9. (a) *Suppose* $\mu \in \mathcal{P}$ *and* $\alpha \in \mathcal{H}^*$. *Then* $\lim_{t \to \infty} \mu S(t) = \mu_\alpha$ *if and only if*

(1.10)
$$\lim_{t \to \infty} \sum_y p_t(x, y) \hat{\mu}(\{y\}) = \alpha(x) \quad and$$

(1.11)
$$\lim_{t \to \infty} \sum_{u,v} p_t(x, u) p_t(x, v) \hat{\mu}(\{u, v\}) = \alpha^2(x)$$

for all $x \in S$.

(b) *Suppose* $g(x, y) = 1$ *for all* $x, y \in S$, *so that* $\mathcal{H}^* = \{0, 1\}$. *If* $\lambda \in [0, 1]$ *and* $\mu \in \mathcal{P}$, *then* $\lim_{t \to \infty} \mu S(t) = \lambda \delta_1 + (1 - \lambda) \delta_0$ *if and only if*

(1.12)
$$\lim_{t \to \infty} \sum_y p_t(x, y) \hat{\mu}(\{y\}) = \lambda \quad for \ all \ x \in S.$$

Using Corollary 8.20 of Chapter II, we can obtain the following restatement of Theorem 1.9 in the translation invariant setting.

Corollary 1.13. *Suppose* $S = Z^d$ *and* $p(x, y) = p(0, y - x)$.

(a) *If* $X(t) - Y(t)$ *is recurrent and* $\mu \in \mathcal{S}$, *then* $\lim_{t \to \infty} \mu S(t) = \alpha \delta_1 + (1 - \alpha) \delta_0$, *where* $\alpha = \mu \{\eta : \eta(x) = 1\}$.
(b) *If* $X(t) - Y(t)$ *is transient and* $\mu \in \mathcal{S}_e$, *then* $\lim_{t \to \infty} \mu S(t) = \mu_\alpha$, *where* $\alpha = \mu \{\eta : \eta(x) = 1\}$.

The remainder of this section is devoted to the proofs of Theorems 1.8 and 1.9. They will be carried out through a series of lemmas. Let $X_1(t), X_2(t), \ldots$ be independent Markov chains with transition probabilities $p_t(x, y)$, and define g on Y by

(1.14)
$$g(A) = P^A(|A_t| < A \text{ for some } t \geq 0).$$

Note that this is consistent with the definition in (1.2) in the sense that $g(\{x, y\}) = g(x, y)$ for $x \neq y$.

Lemma 1.15. *Suppose* $\alpha \in \mathcal{H}$. *Then*

(a) $\mu_\alpha = \lim_{t \to \infty} \nu_\alpha S(t)$ *exists*,
(b) $\mu_\alpha \in \mathcal{I}$,
(c) $0 \leq \hat{\mu}_\alpha(A) - \hat{\nu}_\alpha(A) \leq g(A)$ *for all* $A \in Y$, *and*
(d) $\mu_\alpha \{\eta : \eta(x) = 1\} = \alpha(x)$ *for all* $x \in S$.

Proof. If $A = \{x_1, \ldots, x_n\}$ with distinct x_i's, we can couple together A_t and $\{X_1(t), \ldots, X_n(t)\}$ with $X_i(0) = x_i$ in such a way that $A_t \subset \{X_1(t), \ldots, X_n(t)\}$ for all t. Since $\hat{\mu}(A)$ is a decreasing function of A for any $\mu \in \mathcal{P}$, and since

$\alpha \in \mathcal{H}$,

$$E^A \hat{\nu}_\alpha(A_t) \geq E^{(x_1,\ldots,x_n)} \hat{\nu}_\alpha(\{X_1(t),\ldots,X_n(t)\})$$

$$\geq \sum_{y_1,\ldots,y_n} \prod_{i=1}^n p_t(x_i, y_i)\alpha(y_i)$$

$$= \prod_{i=1}^n \alpha(x_i) = \hat{\nu}_\alpha(A).$$

By the Markov property, it follows from this that $E^A \hat{\nu}_\alpha(A_t)$ is increasing in t, and therefore has a limit which is greater than our equal to $\hat{\nu}_\alpha(A)$. Part (a) follows from this and (1.7). In fact, we have shown that

$$(1.16) \qquad\qquad \hat{\mu}_\alpha(A) = \lim_{t\to\infty} E^A \hat{\nu}_\alpha(A_t) \geq \hat{\nu}_\alpha(A).$$

Part (b) follows from part (d) of Proposition 1.8 of Chapter I. The first inequality in part (c) has already been proved in (1.16). For the second inequality, use the same coupling as above to write

$$\hat{\mu}_\alpha(A) - \hat{\nu}_\alpha(A) = \lim_{t\to\infty}\left[E^A \hat{\nu}_\alpha(A_t) - E^{(x_1,\ldots,x_n)} \prod_{i=1}^n \alpha(X_i(t)) \right]$$

$$\leq \lim_{t\to\infty} P^A(|A_t| < |A|) = g(A),$$

since $A_t = \{X_1(t),\ldots,X_n(t)\}$ on $\{|A_t| = |A|\}$. Part (d) follows from part (c), since $g(A) = 0$ when $|A| = 1$. $\square$

Lemma 1.17. *Suppose* $\alpha \in \mathcal{H}$ *and* $\mu \in \mathcal{P}$ *satisfy* (1.10) *and* (1.11). *Then*

$$\lim_{t\to\infty} \mu S(t) = \mu_\alpha.$$

Proof. By (1.7) and the definition of μ_α, it suffices to show that for each $A \in Y$,

$$(1.18) \qquad\qquad \lim_{t\to\infty}\left[E^A \hat{\mu}(A_t) - E^A \prod_{x\in A_t} \alpha(x) \right] = 0.$$

By (1.10) and (1.11), for each $x \in S$

$$\sum_y p_t(x, y)\eta(y)$$

converges in probability to $\alpha(x)$ with respect to μ. Therefore for $x_1, \ldots,$

$x_n \in S$,

$$\lim_{t \to \infty} \sum_{y_1,\ldots,y_n} \left[\prod_{i=1}^{n} p_t(x_i, y_i) \right] \hat{\mu}(\{y_1, \ldots, y_n\}) = \prod_{i=1}^{n} \alpha(x_i).$$

This can be restated as

(1.19) $$\lim_{t \to \infty} E^{(x_1,\ldots,x_n)} \hat{\mu}(\{X_1(t), \ldots, X_n(t)\}) = \prod_{i=1}^{n} \alpha(x_i).$$

Now write $A = \{x_1, \ldots, x_n\}$ with distinct $x_1, \ldots, x_n$. Let τ be the first time that $|A_t| < n$, which with the natural coupling is the first time that $X_i(t) = X_j(t)$ for some $1 \le i \ne j \le n$. Using (1.19) twice, and the Strong Markov Property, we see that

$$\lim_{t \to \infty} E^A[\hat{\mu}(A_t), \tau = \infty]$$

$$= \lim_{t \to \infty} E^{(x_1,\ldots,x_n)}[\hat{\mu}(\{X_1(t), \ldots, X_n(t)\}), \tau = \infty]$$

$$= \prod_{i=1}^{n} \alpha(x_i) - \lim_{t \to \infty} E^{(x_1,\ldots,x_n)}[\hat{\mu}(\{X_1(t), \ldots, X_n(t)\}), \tau < \infty]$$

(1.20)

$$= \prod_{i=1}^{n} \alpha(x_i) - E^{(x_1,\ldots,x_n)}\left[\prod_{i=1}^{n} \alpha(X_i(\tau)), \tau < \infty \right]$$

$$= \lim_{t \to \infty} E^{(x_1,\ldots,x_n)}\left[\prod_{i=1}^{n} \alpha(X_i(t)), \tau = \infty \right]$$

$$= \lim_{t \to \infty} E^A\left[\prod_{x \in A_t} \alpha(x), \tau = \infty \right].$$

Now we are in a position to prove (1.18) by induction on the cardinality of A as follows. The case $|A| = 1$ is just assumption (1.10). So, assume that (1.18) is true for any A with $|A| < n$, and take an A with $|A| = n$. Then

$$\lim_{t \to \infty} E^A[\hat{\mu}(A_t) - \prod_{x \in A_t} \alpha(x)] = \lim_{t \to \infty} E^A\left[(\hat{\mu}(A_t) - \prod_{x \in A_t} \alpha(x)), \tau = \infty \right]$$

$$+ \lim_{t \to \infty} E^A\left[E^{A_\tau}(\hat{\mu}(A_t) - \prod_{x \in A_t} \alpha(x)), \tau < \infty \right].$$

The first term is zero by (1.20), and the second term is zero by the induction hypothesis, since $|A_\tau| < n$ on $\{\tau < \infty\}$. $\square$

Proof of Theorem 1.9. To prove part (b), note that by (1.7), $\lim_{t\to\infty}\mu S(t)=\lambda\delta_1+(1-\lambda)\delta_0$ is equivalent to

(1.21) $$\lim_{t\to\infty} E^A\hat{\mu}(A_t)=\lambda \quad \text{for } A\neq\emptyset.$$

Applying this to $A=\{x\}$ gives (1.12). For the converse, suppose (1.12) holds for all $x\in S$. Let τ be the first time that $|A_t|=1$, which is finite with probability one by the assumption that $g(x,y)\equiv 1$. Then

$$\lim_{t\to\infty} E^A\hat{\mu}(A_t)=E^A[\lim_{t\to\infty} E^{A_\tau}\hat{\mu}(A_t)],$$

so that (1.21) follows from (1.12). Turning to the proof of part (a), note that the sufficiency of (1.10) and (1.11) for $\lim_{t\to\infty}\mu S(t)=\mu_\alpha$ has been obtained in Lemma 1.17. The necessity of (1.10) is clear from (1.7) applied to singletons. To prove the necessity of (1.11) note first that

$$0\leq \int \left[\sum_y p_t(x,y)\eta(y)-\alpha(x)\right]^2 d\mu$$

$$= \sum_{u,v} p_t(x,u)p_t(x,v)\hat{\mu}(\{u,v\})+\alpha^2(x)$$

$$-2\alpha(x)\sum_y p_t(x,y)\hat{\mu}(\{y\}).$$

Therefore by (1.10),

$$\liminf_{t\to\infty}\sum_{u,v} p_t(x,u)p_t(x,v)\hat{\mu}(\{u,v\})\geq \alpha^2(x).$$

So, it suffices to show that

(1.22) $$\limsup_{t\to\infty} E^{(x,x)}[\hat{\mu}(\{X(t),Y(t)\})-\alpha(X(t))\alpha(Y(t))]\leq 0.$$

This will be done by considering the expected value over the events $\mathscr{E}$ and $\mathscr{E}^c$ separately. Let τ be the first time that $X(t)=Y(t)$. By (1.10) and the Strong Markov Property,

$$\lim_{t\to\infty} E^{(x,y)}[\hat{\mu}(\{X(t)\})-\alpha(X(t)), \tau<\infty]$$

(1.23) $$= E^{(x,y)}\left[\lim_{t\to\infty} E^{X(\tau)}[\hat{\mu}(\{X(t)\})-\alpha(X(t))], \tau<\infty\right]$$

$$= 0.$$

Repeating this idea, we see that

$$\lim_{t\to\infty} E^{(x,y)}[\hat{\mu}(\{X(t)\}) - \alpha(X(t)), \mathcal{E}^c] = 0,$$

so that by (1.10),

$$\lim_{t\to\infty} E^{(x,y)}[\hat{\mu}(\{X(t)\}) - \alpha(X(t)), \mathcal{E}] = 0.$$

Since $\hat{\mu}(\{u, v\}) \leq \hat{\mu}(\{u\})$ and $\alpha \in \mathcal{H}^*$, it follows that

(1.24) $\lim\sup\limits_{t\to\infty} E^{(x,y)}[\hat{\mu}(\{X(t), Y(t)\}) - \alpha(X(t))\alpha(Y(t)), \mathcal{E}] \leq 0.$

On the other hand, applying (1.18) to $A = \{x, y\}$ with $x \neq y$, we see that

$$\lim_{t\to\infty} E^{\{x,y\}}\left[\hat{\mu}(A_t) - \prod_{u\in A_t} \alpha(u)\right] = 0.$$

Therefore by (1.23),

$$\lim_{t\to\infty} E^{(x,y)}[[\hat{\mu}(\{X(t), Y(t)\}) - \alpha(X(t))\alpha(Y(t))], \tau = \infty] = 0.$$

By the Strong Markov Property, we then have that

(1.25) $\lim\limits_{t\to\infty} E^{(x,y)}[\hat{\mu}(\{X(t), Y(t)\}) - \alpha(X(t))\alpha(Y(t)), \mathcal{E}^c] = 0.$

Combining (1.24) and (1.25) gives (1.22). $\square$

We have now proved Theorem 1.9, and in Lemma 1.15, we proved the first two parts of Theorem 1.8. It remains for us to prove part (c) of Theorem 1.8. In a sense, this is the most interesting result, because it rules out the possibility that invariant measures exist which are not related to the ones we already know about. In order to prepare for its proof, we begin with the following simple lemma, which is useful in other contexts as well. For example, it can be used to give another proof of Proposition 7.1 of Chapter II.

Lemma 1.26. *Let $\{q(x, y), x, y \in S\}$ be the Q matrix for a continuous time Markov chain with transition probabilities $q_t(x, y)$. Suppose that $c = \sup_x |q(x, x)| < \infty$. If for uniformly bounded functions f_n on S and some sequence $t_n \to \infty$,*

$$g(x) = \lim_{n\to\infty} \sum_y q_{t_n}(x, y)f_n(y)$$

exists for all $x \in S$, then

$$\sum_y q_t(x, y) g(y) = g(x)$$

for all $x \in S$ and $t \geq 0$. In other words, g is harmonic for the chain.

Proof. Let $r(x, y)$ be defined by

$$r(x, y) = \begin{cases} c^{-1} q(x, y) & \text{for } x \neq y, \\ 1 + c^{-1} q(x, x) & \text{for } x = y. \end{cases}$$

These are the transition probabilities for a discrete time Markov chain on S, and

(1.27) $$q_t(x, y) = e^{-ct} \sum_{n=0}^{\infty} \frac{(ct)^n}{n!} r^{(n)}(x, y).$$

To prove the lemma, it suffices to show that

$$\lim_{t \to \infty} \sum_y |q_{s+t}(x, y) - q_t(x, y)| = 0$$

for all $s \geq 0$ and $x \in S$. Using (1.27), it suffices to show that

$$\lim_{t \to \infty} e^{-ct} \sum_{n=0}^{\infty} \frac{(ct)^n}{n!} \left| e^{-cs} \left(1 + \frac{s}{t} \right)^n - 1 \right| = 0.$$

But this follows from the Law of Large Numbers, applied to the Poisson process. □

Since the dual of the voter model does not grow in size, Lemma 1.26 yields the following strengthening of part (d) of Proposition 1.8 of Chapter I in this case.

Corollary 1.28. *Suppose $\nu \in \mathcal{P}$, $t_n \to \infty$, and $\mu = \lim_{n \to \infty} \nu S(t_n)$ exists. Then $\mu \in \mathcal{I}$.*

Proof. By (1.7), it suffices to show that

(1.29) $$\hat{\mu}(A) = E^A \hat{\mu}(A_t)$$

for all $A \in Y$ and $t \geq 0$. Since

$$\hat{\mu}(A) = \lim_{n \to \infty} E^A \hat{\nu}(A_{t_n}),$$

(1.29) follows from Lemma 1.26 applied to the dual chain on $\{A: |A| \le k\}$ for each k. $\square$

The next two results are the key to identifying the elements of $\mathcal{I}_e$. They assert that under $\mu \in \mathcal{I}_e$, the coordinates $\eta(x)$ are asymptotically independent in a certain sense.

Lemma 1.30. *Suppose* $\mu \in \mathcal{I}_e$. *Then*

$$\lim_{t \to \infty} \sum_y p_t(x, y) \hat{\mu}(\{y, z\}) = \hat{\mu}(\{x\}) \hat{\mu}(\{z\}).$$

Proof. We can of course assume that $\mu \neq \delta_0, \delta_1$, since the result is immediate in those cases. Since

$$(1.31) \qquad \hat{\mu}(\{x\}) = \sum_y p_t(x, y) \hat{\mu}(\{y\})$$

by (1.7) and $p_t(x, y) > 0$ for all $x, y \in S$ and $t > 0$, it then follows that $0 < \hat{\mu}(\{x\}) < 1$ for all $x \in S$. Fix $z \in S$ and let $\lambda = \hat{\mu}(\{z\})$, $\mu_0(\cdot) = \mu(\cdot \,|\, \eta(z) = 0)$, and $\mu_1(\cdot) = \mu(\cdot \,|\, \eta(z) = 1)$. Then

$$\mu = \lambda \mu_1 + (1 - \lambda) \mu_0,$$

and hence

$$\mu = \mu S(t) = \lambda \mu_1 S(t) + (1 - \lambda) \mu_0 S(t).$$

By Corollary 1.28 and the fact that μ is extremal in $\mathcal{I}$, it follows that

$$\mu = \lim_{t \to \infty} \mu_1 S(t) = \lim_{t \to \infty} \mu_0 S(t).$$

Therefore by (1.7) and the definition of μ_1,

$$\hat{\mu}(\{x\}) = \lim_{t \to \infty} E^{\{x\}} \hat{\mu}_1(A_t)$$

$$= \lim_{t \to \infty} \sum_y p_t(x, y) \hat{\mu}_1(\{y\})$$

$$= \lim_{t \to \infty} \sum_y p_t(x, y) \frac{\hat{\mu}(\{y, z\})}{\hat{\mu}(\{z\})}. \quad \square$$

Lemma 1.32. *Suppose* $\mu \in \mathcal{I}_e$. *Then*

$$E^{(x,y)} \hat{\mu}(\{X(t), Y(t)\}) \downarrow \alpha(x) \alpha(y) \quad \text{as } t \uparrow \infty,$$

where $\alpha(x) = \hat{\mu}(\{x\})$.

Proof. First note that since $\hat{\mu}$ is a harmonic function for A_t, and A_t and $(X(t), Y(t))$ can be coupled to preserve the relation

$$A_t \subset \{X(t), Y(t)\},$$

we have the inequality

$$E^{(x,y)}\hat{\mu}(\{X(t), Y(t)\}) \leq E^{\{x,y\}}\hat{\mu}(A_t) = \hat{\mu}(\{x, y\}).$$

This gives the monotonicity in the statement of the lemma. Let

$$h(x, y) = \lim_{t\to\infty} E^{(x,y)}\hat{\mu}(\{X(t), Y(t)\}) \leq \hat{\mu}(\{x, y\}).$$

By Corollary 7.3 of Chapter II,

$$\sum_y p_t(x, y)h(y, z) = h(x, z)$$

for all $x, z \in S$. Therefore by Lemma 1.30,

(1.33) $h(x, z) \leq \alpha(x)\alpha(z).$

For the opposite inequality, apply the Schwarz inequality to the random variables $\sum_u p_t(x, u)\eta(u)$ and $\sum_u p_t(x, u)\eta(u) - \alpha(x)$ to obtain

(1.34) $\alpha^2(x) \leq h(x, x)$

and

(1.35) $[h(x, y) - \alpha(x)\alpha(y)]^2 \leq [h(x, x) - \alpha^2(x)][h(y, y) - \alpha^2(y)].$

Now (1.33) and (1.34) imply that $h(x, x) = \alpha^2(x)$, and then (1.35) implies that $h(x, y) = \alpha(x)\alpha(y)$, which completes the proof. $\square$

Proof of Theorem 1.8. Parts (a) and (b) have been proved already in Lemma 1.15. For part (c), take $\alpha \in \mathcal{H}^*$, and write

(1.36) $\mu_\alpha = \lambda\mu_1 + (1-\lambda)\mu_2,$

where $0 < \lambda < 1$ and $\mu_1, \mu_2 \in \mathcal{I}$. By Theorem 1.9, $\sum_y p_t(x, y)\eta(y)$ converges in probability to $\alpha(x)$ relative to μ_α. Therefore by (1.36), the same statement holds relative to μ_1 and μ_2. By Theorem 1.9 again,

$$\lim_{t\to\infty} \mu_1 S(t) = \lim_{t\to\infty} \mu_2 S(t) = \mu_\alpha,$$

so that $\mu_1 = \mu_2 = \mu_\alpha$. Hence $\mu_\alpha \in \mathcal{I}_e$. We have now shown that

$$\mathcal{I}_e \supset \{\mu_\alpha : \alpha \in \mathcal{H}^*\}.$$

For the reverse containment, take $\mu \in \mathcal{I}_e$ and let $\alpha(x) = \hat{\mu}(\{x\})$. By (1.31), $\alpha \in \mathcal{H}$. To show that $\alpha \in \mathcal{H}^*$, note first that by Lemma 1.32 and the convergence theorem for supermartingales,

(1.37) $$\hat{\mu}(\{X(t), Y(t)\}) - \alpha(X(t))\alpha(Y(t)) \to 0$$

with probability one. On the other hand, by (1.7) and the natural coupling,

$$\begin{aligned} \alpha(x) - \hat{\mu}(\{x, y\}) &= E^{\{x\}}\alpha(A_t) - E^{\{x,y\}}\hat{\mu}(A_t) \\ &\leq P^{\{x,y\}}(|A_t| = 2). \end{aligned}$$

Passing to the limit as $t \to \infty$, this gives

$$0 \leq \alpha(x) - \hat{\mu}(\{x, y\}) \leq 1 - g(x, y).$$

Therefore

(1.38) $$\alpha(X(t)) - \hat{\mu}(\{X(t), Y(t)\}) \to 0$$

almost surely on $\mathcal{E}$. Combining (1.37) and (1.38) gives

$$\alpha(X(t))[1 - \alpha(Y(t))] \to 0$$

almost surely on $\mathcal{E}$. Since

$$\lim_{t \to \infty} \alpha(X(t)) = \lim_{t \to \infty} \alpha(Y(t))$$

on $\mathcal{E}$, it follows that

$$\alpha(X(t))[1 - \alpha(X(t))] \to 0$$

almost surely on $\mathcal{E}$, so that $\alpha \in \mathcal{H}^*$. It remains to show that $\mu = \mu_\alpha$. But by Lemma 1.32 and Theorem 1.9, $\lim_{t \to \infty} \mu S(t) = \mu_\alpha$. Since $\mu \in \mathcal{I}$, we conclude that $\mu = \mu_\alpha$. $\square$

2. Properties of the Invariant Measures

In this section, we will obtain several properties of the invariant measures μ_α for $\alpha \in \mathcal{H}$.

Theorem 2.1. μ_α *has positive correlations for each* $\alpha \in \mathcal{H}$. *If* $\mathcal{H}$ *consists only of constants, then every* $\mu \in \mathcal{I}$ *has positive correlations.*

Proof. For each $\alpha \in \mathcal{H}$, ν_α has positive correlations by Corollary 2.12 of Chapter II. Therefore $\nu_\alpha S(t)$ has positive correlations for each t by Theorem 2.14 of the same chapter. Passing to the limit as $t \to \infty$, and using the definition of μ_α in Theorem 1.8, it follows that μ_α has positive correlations as well. Turning to the second statement, note that by part (c) of Proposition 1.8 of Chapter I and part (c) of Theorem 1.8, it suffices to show that any convex combination of μ_α's for $\alpha \in \mathcal{H}$ has positive correlations. Since $\mathcal{H}$ consists only of constants, $\mathcal{H}$ is linearly ordered. Therefore $\{\nu_\alpha, \alpha \in \mathcal{H}\}$ is linearly ordered, and hence so is $\{\mu_\alpha, \alpha \in \mathcal{H}\}$ by Theorem 2.2 of Chapter II and Theorem 2.2 of Chapter III. Thus convex combinations of μ_α's for $\alpha \in \mathcal{H}$ have positive correlations by Proposition 2.22 of Chapter II. $\square$

One consequence of Lemma 1.15 is that μ_α lies above ν_α in the sense that $\hat{\mu}_\alpha(A) \geq \hat{\nu}_\alpha(A)$ for all $A \in Y$. The point of the next result is to show that it is not the case that μ_α lies above ν_α in the stronger sense discussed in Section 2 of Chapter II.

Theorem 2.2. *If* $\alpha \in \mathcal{H}$ *is not* $\equiv 0$ *or* $\equiv 1$, *then* $\nu_\alpha \nleq \mu_\alpha$.

Proof. By Corollary 2.8 of Chapter II and Theorem 1.8(b), it suffices to show that $\mu_\alpha \neq \nu_\alpha$ if $\alpha \not\equiv 0$ and $\alpha \not\equiv 1$. To do this, take $x \neq y$ and let $X(t)$ and $Y(t)$ be independent copies of the Markov chain with transition probabilities $p_t(\cdot, \cdot)$ with initial states x and y respectively. Let τ be the first time that $X(t) = Y(t)$. Then

$$\hat{\mu}_\alpha(\{x, y\}) - \nu_\alpha(\{x, y\})$$

$$= \lim_{t \to \infty} E^{\{x, y\}} \hat{\nu}_\alpha(A_t) - \alpha(x)\alpha(y)$$

$$= E[\lim_{t \to \infty} \alpha(X_t)\alpha(Y_t), \tau = \infty]$$

$$\quad + E[\alpha(X_\tau), \tau < \infty] - \alpha(x)\alpha(y)$$

$$= E[\alpha(X_\tau), \tau < \infty] - E[\lim_{t \to \infty} \alpha(X_t)\alpha(Y_t), \tau < \infty]$$

$$= E[\alpha(X_\tau)[1 - \alpha(X_\tau)], \tau < \infty] > 0$$

if $\alpha \not\equiv 0$ and $\alpha \not\equiv 1$. $\square$

For the remainder of this section, we will assume that $S = Z^d$ and that $p(x, y) = p(0, y - x)$, so that we are in a translation invariant setting. Furthermore, we will consider only the case in which $X(t) - Y(t)$ is transient, so that $\mathcal{H} = \mathcal{H}^* = [0, 1]$. In the recurrent case, of course, there are no interesting invariant measures to discuss.

Lemma 2.3. (a) *For $\alpha \in [0, 1]$ and B, $C \in Y$,*

$$0 \le \hat{\mu}_\alpha(B \cup C) - \hat{\mu}_\alpha(B)\hat{\mu}_\alpha(C) \le \sum_{\substack{x \in B \\ y \in C}} g(x, y).$$

(b) $$\lim_{x \to \infty} g(x, y) = 0 \quad \textit{for each } y \in S.$$

Proof. The first inequality in (a) follows from Theorem 2.1, since the function $\prod_{x \in B} \eta(x)$ is a monotone function of η. For the second inequality, we can assume that $B \cap C = \varnothing$, since otherwise it is trivial. Let $A = B \cup C$, and construct in the natural way three coupled copies of the dual chain A_t, B_t and C_t such that:

(a) $A_0 = A$, $B_0 = B$, $C_0 = C$;
(b) B_t and C_t are independent; and
(c) $A_t = B_t \cup C_t$ for $t < \tau$, where τ is the first time that $B_t \cap C_t \ne 0$.

Then

$$\hat{\mu}_\alpha(B \cup C) - \hat{\mu}_\alpha(B)\hat{\mu}_\alpha(C) = \lim_{t \to \infty}[E\alpha^{|A_t|} - E\alpha^{|B_t|+|C_t|}]$$

$$\le P(\tau < \infty).$$

On the other hand, by coupling B_t and C_t with independent copies of the random walk so that if $|B| = m$ and $|C| = n$, $B_t \subset \{X_1(t), \ldots, X_m(t)\}$ and $C_t \subset \{X_{m+1}(t), \ldots, X_{m+n}(t)\}$, we see that

$$P(\tau < \infty) \le P(X_i(t) = X_j(t) \text{ for some } 1 \le i \le m < j \le m+n \text{ and some } t > 0)$$

$$\le \sum_{\substack{x \in B \\ y \in C}} g(x, y).$$

For part (b), let $Z(t) = X(t) - Y(t)$, which is a symmetric transient random walk. Then

(2.4) $$g(x, y) = P^{y-x}(Z(t) = 0 \text{ for some } t),$$

so the result follows from Proposition 3 of Section 25 of Spitzer (1976). □

Theorem 2.5. *For every $\alpha \in [0, 1]$, $\mu_\alpha \in \mathscr{S}_e$.*

Proof. It is immediate that $\mu_\alpha \in \mathscr{S}$, since $\nu_\alpha S(t) \in \mathscr{S}$ for each t. By the two parts of Lemma 2.3,

$$\lim_{x \to \infty} \hat{\mu}_\alpha(B \cup (C + x)) = \hat{\mu}_\alpha(B)\hat{\mu}_\alpha(C)$$

for all B, $C \in Y$. Therefore $\mu_\alpha \in \mathscr{S}_e$ by Proposition 4.11 and Corollary 4.14 of Chapter I. □

The final result in this section is designed to show that under μ_α, the coordinates $\eta(x)$ have rather large correlations. In particular, central limit theorems for these random variables require nonclassical norming. The first step in this direction is to compute the covariance of $\eta(x)$ and $\eta(y)$ under μ_α. By (1.7) and the definition of μ_α in Theorem 1.8,

$$\hat{\mu}_\alpha(A) = E^A \alpha^N,$$

where $N = \lim_{t\to\infty} |A_t|$. Therefore

$$\int \eta(x)\, d\mu_\alpha = \alpha, \quad \text{and}$$

$$\int \eta(x)\eta(y)\, d\mu_\alpha = \alpha g(x, y) + \alpha^2[1 - g(x, y)],$$

so that

(2.6) $$\text{cov}(\eta(x), \eta(y)) = \alpha(1 - \alpha)g(x, y).$$

It is convenient to rewrite this in terms of the Green function

$$G(x, y) = \int_0^\infty \sum_u p_t(x, u)p_t(y, u)\, dt$$

for the random walk $X(t) - Y(t)$. Then

$$g(x, y) = \frac{G(x, y)}{G(0, 0)},$$

so that (2.6) becomes

(2.7) $$\text{cov}(\eta(x), \eta(y)) = \frac{\alpha(1 - \alpha)}{G(0, 0)} G(x, y).$$

Suppose ϕ is an infinitely differentiable function on R^d with compact support, and for $r > 0$, define

$$\Phi_r = \sum_x \phi\left(\frac{x}{r}\right)\eta(x).$$

Under ν_α

$$\lim_{r\to\infty} r^{-d}\, \text{var}\, \Phi_r = \alpha(1 - \alpha)\lim_{r\to\infty} r^{-d} \sum_x \phi^2\left(\frac{x}{r}\right)$$

$$= \alpha(1 - \alpha)\int_{R^d} \phi^2\, dm,$$

where m is Lebesgue measure on R^d. The high correlations in μ_α will be seen from the fact that under μ_α, $\text{var}(\Phi_r)$ grows more rapidly than r^d as $r \uparrow \infty$. In order to obtain a precise result, we must make some assumptions about the random walk $X(t) - Y(t)$.

Theorem 2.8. *Suppose that $X(t) - Y(t)$ is in the domain of normal attraction of a (necessarily symmetric) genuinely d-dimensional stable law of index $\beta \in (0, 2]$ with characteristic function $e^{-\psi(\gamma)}$, in the sense that $\psi(\gamma) > 0$ for $\gamma \neq 0$ and the characteristic function of $[X(t) - Y(t)]/t^{1/\beta}$ converges to $e^{-\psi(\gamma)}$ as $t \uparrow \infty$. Under μ_α,*

$$\lim_{r \to \infty} r^{-(d+\beta)} \text{var } \Phi_r = \frac{\alpha(1-\alpha)}{G(0,0)(2\pi)^d} \int_{R^d} \frac{|\hat\phi(\gamma)|^2}{\psi(\gamma)} \, dm,$$

where

$$\hat\phi(\gamma) = \int_{R^d} \phi(x) \exp[i\langle x, \gamma\rangle] \, dm.$$

Proof. Let $\rho(\gamma)$ be the characteristic function of $p(0, \cdot)$, which is defined on $\Gamma = [-\pi, \pi)^d$ by

$$\rho(\gamma) = \sum_x p(0, x) \exp[i\langle x, \gamma\rangle].$$

The Fourier inversion formula then gives

$$p(0, x) = \frac{1}{(2\pi)^d} \int_\Gamma \exp[-i\langle x, \gamma\rangle]\rho(\gamma) \, dm.$$

Therefore

$$p_t(x, y) = \frac{1}{(2\pi)^d} \int_\Gamma \exp[i\langle x - y, \gamma\rangle - t[1 - \rho(\gamma)]] \, dm,$$

so that

$$\sum_u p_t(x, u)p_t(y, u) = \frac{1}{(2\pi)^d} \int_\Gamma \exp[i\langle x - y, \gamma\rangle - 2t[1 - \text{Re } \rho(\gamma)]] \, dm.$$

Hence

(2.9) $$G(x, y) = \frac{1}{(2\pi)^d} \int_\Gamma \frac{\exp[i\langle x - y, \gamma\rangle]}{2[1 - \text{Re } \rho(\gamma)]} \, dm,$$

where the integral is finite by the transience assumption. By (2.7) and (2.9),

(2.10) $$\text{var } \Phi_r = \frac{\alpha(1-\alpha)}{G(0,0)(2\pi)^d} \int_\Gamma \frac{|\hat\phi_r(\gamma)|^2}{2[1 - \text{Re } \rho(\gamma)]} \, dm,$$

where

$$\hat{\phi}_r(\gamma) = \sum_x \phi\left(\frac{x}{r}\right) \exp[i\langle x, \gamma \rangle].$$

In order to pass to the limit in (2.10), it is convenient to make a change of variable in the integral to obtain

(2.11)
$$\operatorname{var} \Phi_r = \frac{\alpha(1-\alpha)r^{-d}}{G(0,0)(2\pi)^d} \int_{r\Gamma} \frac{|\hat{\phi}_r(\gamma/r)|^2}{2[1 - \operatorname{Re} \rho(\gamma/r)]} \, dm.$$

The domain of attraction assumption can be restated as

(2.12)
$$\lim_{r\to\infty} 2r^\beta \left[1 - \operatorname{Re} \rho\left(\frac{\gamma}{r}\right) \right] = \psi(\gamma)$$

uniformly for compact sets of γ's. By using the convergence of Riemann sums to the integral defining $\hat{\phi}$, we see that

(2.13)
$$\lim_{r\to\infty} r^{-d}\hat{\phi}_r\left(\frac{\gamma}{r}\right) = \hat{\phi}(\gamma).$$

The desired result now follows from (2.11), (2.12), and (2.13), provided that we can justify the interchange of the limit with the integral. In order to do so, first use summation by parts in the definition of $\hat{\phi}_r$ to show that $|\hat{\phi}_r(\gamma)|$ is bounded above by a constant multiple of

$$\prod_{j=1}^d \min(r, |1 - e^{i\gamma_j}|^{-1}).$$

Next, to control the denominator in the integral in (2.11), note that from the uniform convergence on compact sets in (2.12) and the strict positivity of ψ, it follows that there is some r_0 and $\varepsilon > 0$ so that $r \geq r_0$ and $\psi(\gamma) = 1$ imply that

$$r^\beta \left[1 - \operatorname{Re} \rho\left(\frac{\gamma}{r}\right) \right] \geq \varepsilon.$$

Since ψ is homogeneous of degree β, it follows that

$$1 - \operatorname{Re} \rho(\gamma) \geq \varepsilon \psi(\gamma)$$

whenever $\psi(\gamma) \leq r_0^{-1/\beta}$. Furthermore, $1 - \operatorname{Re} \rho(\gamma)$ is bounded away from zero when $\gamma \in \Gamma$ is bounded away from zero by the irreducibility assumption on $p(x, y)$. Therefore for the required domination, it suffices to show that

$[\psi(\gamma)]^{-1}$ is integrable on

$$\{\gamma: \psi(\gamma) \leq r_0^{-1/\beta}\}$$

and that

$$\prod_{j=1}^{d} \min(1, |\gamma_j|^{-2})$$

is integrable on R^d. The first of these integrability statements is easily proved using the fact that $\beta < d$ (which is true by the transience assumption), while the second is clear. $\square$

A natural question raised by Theorem 2.8 is the following. Does

$$\frac{\Phi_r - E\Phi_r}{r^{(d+\beta)/2}}$$

have a limiting distribution, and if so, is it normal? An affirmative answer was given by Bramson and Griffeath (1979b) for a discrete time version of the voter model in three dimensions when $\beta = 2$. An alternate proof of their result was then given by Major (1980). See also Section 1 of Holley and Stroock (1978b).

The fact that the limiting distribution exists should not be surprising. In view of the large correlations which develop, it may be more surprising that the limit is normal. To explain why this is the case, we will give an informal account of the main idea in Major's proof. It shows that the problem can be reduced to the case of sums of independent random variables.

The first step is to reexpress the duality relation (1.7) in a more convenient form. Let $\{X_x(t), x \in S\}$ be a collection of coalescing random walks with $X_x(0) = x$ which have mean one exponential holding times and transition probabilities $p(x, y)$. Whenever two paths meet, they coalesce into one. Of course, $A_t = \{X_x(t), x \in A\}$ is a version of the dual chain starting at A. Next let $\{B(x), x \in S\}$ be independent random variables with distribution given by

$$P(B(x) = 1) = \alpha, \qquad P(B(x) = 0) = 1 - \alpha.$$

Then (1.7) can be restated in the following way. If $\{\eta_0(x), x \in S\}$ is distributed according to ν_α, then

$$\{\eta_t(x), x \in S\} \quad \text{and} \quad \{B(X_x(t)), x \in S\}$$

have the same joint distributions. In order to pass to the limit as $t \to \infty$ in this statement, define a (random) partition Π of S by identifying points x and y such that $X_x(t) = X_y(t)$ for all large t. Then under μ_α, $\Phi_r - E\Phi_r$ has the same distribution as

$$(2.14) \qquad \sum_{k \geq 1} (B_k - \alpha) \sum_{x \in S_k} \phi\left(\frac{x}{r}\right),$$

where $\Pi = \{S_k, k \geq 1\}$, and $\{B_k, k \geq 1\}$ are independent random variables with distribution

$$P(B_k = 1) = \alpha, \qquad P(B_k = 0) = 1 - \alpha,$$

which are also independent of Π. So, conditional on Π, (2.14) is a sum of independent mean zero random variables with variance

$$\alpha(1 - \alpha) \sum_{k \geq 1} \left[\sum_{x \in S_k} \phi\left(\frac{x}{r}\right) \right]^2.$$

This conditional distribution is asymptotically normal as $r \to \infty$ by the Central Limit Theorem for triangular arrays. What remains to be proved is that the asymptotic variance is essentially independent of Π. This requires certain estimates which are carried out in Proposition 3 of Bramson and Griffeath (1979b) and in Lemma 1 of Major (1980).

3. Clustering in One Dimension

Throughout this section, we will consider the basic one-dimensional voter model in which $S = Z^1$ and $p(x, x+1) = p(x, x-1) = \frac{1}{2}$ for each x. By Corollary 1.13,

$$\lim_{t \to \infty} \mu S(t) = \alpha \delta_1 + (1 - \alpha) \delta_0,$$

where $\alpha = \mu\{\eta : \eta(x) = 1\}$, provided that $\mu \in \mathcal{S}$. This result indicates that clustering occurs in this case. The aim of this section is to give a more precise description of this clustering.

The clusters of an $\eta \in X$ are defined to be the connected components of $\{x : \eta(x) = 0\}$ or of $\{x : \eta(x) = 1\}$. The mean cluster size for η is defined to be

$$C(\eta) = \lim_{n \to \infty} \frac{2n}{\text{number of clusters of } \eta \text{ in } [-n, n]},$$

provided that this limit exists.

Proposition 3.1. *Consider the voter model with initial distribution* $\mu \in \mathcal{S}_e$. *Then*

$$C(\eta_t) = \{P[\eta_t(0) \neq \eta_t(1)]\}^{-1}$$

with probability one.

Proof. By Theorem 4.15 of Chapter I, $\mu S(t) \in \mathcal{S}_e$ for all $t \geq 0$. Therefore, by Theorem 4.9 of Chapter I,

$$\lim_{n \to \infty} \frac{\text{number of } -n \leq x < n \text{ so that } \eta(x) \neq \eta(x+1)}{2n} = P[\eta_t(0) \neq \eta_t(1)]$$

a.s. relative to $\mu S(t)$. Since the number of clusters of η in $[-n, n]$ differs from the number of $-n \leq x < n$ so that $\eta(x) \neq \eta(x+1)$ by at most one, the result follows. $\square$

Theorem 3.2. *If the initial distribution μ of the voter model is in $\mathcal{S}_e$, and $\alpha = \mu\{\eta: \eta(0) = 1\} \in (0, 1)$, then*

$$\frac{C(\eta_t)}{\sqrt{t}} \to \frac{\sqrt{\pi}}{2\alpha(1-\alpha)}$$

in probability as $t \uparrow \infty$.

Proof. By Proposition 3.1, it suffices to show that

$$(3.3) \qquad \lim_{t \to \infty} \sqrt{t}\, P[\eta_t(0) \neq \eta_t(1)] = \frac{2\alpha(1-\alpha)}{\sqrt{\pi}}.$$

In order to do so, write

$$P[\eta_t(0) \neq \eta_t(1)] = P[\eta_t(0) = 1] + P[\eta_t(1) = 1] - 2P[\eta_t(0) = \eta_t(1) = 1].$$

By (1.7), it then follows that

$$P[\eta_t(0) \neq \eta_t(1)] = 2\alpha - 2E^{\{0,1\}}\hat{\mu}(A_t)$$

$$= 2\alpha P^1(\tau > t) - 2 \sum_{x=1}^{\infty} \hat{\mu}(\{0, x\}) P^1(Z(t) = x, \tau > t),$$

where $Z(t)$ is a one-dimensional simple random walk with mean $\frac{1}{2}$ exponential holding times, and

$$\tau = \inf\{t \geq 0: Z(t) = 0\}.$$

By the reflection principle (this is the discrete analogue of the reflection principle for Brownian motion which is discussed in Section 3 of Chapter 7 of Karlin and Taylor (1975), for example),

$$P^1(Z(t) = x, \tau \leq t) = P^1(Z(t) = -x)$$

for $x \geq 1$, so that

(3.4) $P^1(Z(t) = x, \tau > t) = p_t(x-1) - p_t(x+1)$

for $x \geq 1$, where $p_t(x) = P^0(Z(t) = x)$. Therefore

$$P[\eta_t(0) \neq \eta_t(1)] = 2 \sum_{x=1}^{\infty} [\alpha - \hat{\mu}(\{0, x\})][p_t(x-1) - p_t(x+1)].$$

In order to determine the asymptotic behavior of this quantity as $t \uparrow \infty$, it is convenient to use Bochner's Theorem to write

$$\hat{\mu}(\{0, x\}) = \alpha \int_{\Gamma} e^{ix\gamma} \nu(d\gamma),$$

where ν is a probability measure on $\Gamma = [-\pi, \pi)$. Since $\mu \in \mathcal{S}_e$, $\nu\{0\} = \alpha$. (See Proposition 4.11 of Chapter I.) Therefore, writing

$$P[\eta_t(0) \neq \eta_t(1)] = 2\alpha \sum_{x=1}^{\infty} [p_t(x-1) - p_t(x+1)] \int_{\Gamma} [1 - e^{ix\gamma}] \nu(d\gamma),$$

we see that in order to prove (3.3), it suffices by the Bounded Convergence Theorem to show that

(3.5) $\lim_{t \to \infty} \sqrt{\pi t} \sum_{x=1}^{\infty} [p_t(x-1) - p_t(x+1)] e^{ix\gamma} = \begin{cases} 1 & \text{if } \gamma = 0, \\ 0 & \text{if } 0 < |\gamma| \leq \pi. \end{cases}$

(Recall that $p_t(x-1) - p_t(x+1) \geq 0$ for $x \geq 1$ by (3.4).) This could be done by using a sufficiently strong form of the local Central Limit Theorem. (See, for example, Theorem 16 of Chapter VII of Petrov (1975).) However, we will proceed directly as follows. First use the Fourier inversion formula to write

(3.6) $p_t(x) = \frac{1}{2\pi} \int_{\Gamma} e^{-ix\sigma} \exp[-2t(1 - \cos \sigma)] \, d\sigma.$

Making the change of variables $\lambda = \sqrt{2t}\sigma$ and using the Dominated Convergence Theorem, we see that

(3.7) $\lim_{t \to \infty} \sqrt{2t} p_t(x) = \frac{1}{2\pi} \int_{-\infty}^{\infty} \exp(-\tfrac{1}{2}\lambda^2) \, d\lambda = \frac{1}{\sqrt{2\pi}}$

for each x. Therefore

$$\lim_{t \to \infty} \sqrt{\pi t} \sum_{x=1}^{\infty} [p_t(x-1) - p_t(x+1)] = \lim_{t \to \infty} \sqrt{\pi t}[p_t(0) + p_t(1)] = 1,$$

giving (3.5) when $\gamma = 0$. Assume now that $\gamma \neq 0$, which is the more delicate case. Integrate by parts in (3.6) to obtain

$$p_t(x-1) - p_t(x+1)$$

(3.8)
$$= \frac{i}{\pi} \int_{\Gamma} e^{-ix\sigma} \sin \sigma \exp[-2t(1 - \cos \sigma)] \, d\sigma$$

$$= \frac{1}{\pi} \int_{\Gamma} \frac{e^{-ix\sigma}}{x} [\cos \sigma - 2t \sin^2 \sigma] \exp[-2t(1 - \cos \sigma)] \, d\sigma.$$

We will need to use the facts that for $0 < \gamma \leq \pi$,

(3.9)
$$\sum_{x=1}^{\infty} \frac{e^{ix\gamma}}{x} \text{ converges, and}$$

(3.10)
$$\sup_{N \geq 1} \left| \sum_{x=1}^{N} \frac{e^{ix\gamma}}{x} \right| \leq c_1 + c_2 |\log \gamma|$$

for some constants c_1 and c_2. To check (3.9), write

$$\sum_{x=N}^{M} \frac{e^{ix\gamma}}{x} = \sum_{x=N}^{M} e^{ix\gamma} \sum_{y=x}^{\infty} \frac{1}{y(y+1)}$$

$$= \sum_{y=N}^{\infty} \frac{1}{y(y+1)} \sum_{x=N}^{y \wedge M} e^{ix\gamma},$$

which is bounded above in absolute value by

$$\frac{2}{N|1 - e^{i\gamma}|}.$$

For (3.10), use the fact that

$$\sum_{x=1}^{N} \frac{e^{ix\pi}}{x} = \sum_{x=1}^{N} \frac{(-1)^x}{x}$$

is bounded in N, and the estimate

$$\left| \frac{d}{d\gamma} \sum_{x=1}^{N} \frac{e^{ix\gamma}}{x} \right| = \left| \sum_{x=1}^{N} e^{ix\gamma} \right| \leq \frac{2}{|1 - e^{i\gamma}|}.$$

By (3.8), (3.9), (3.10), and the Dominated Convergence Theorem,

$$\sqrt{2t} \sum_{x=1}^{\infty} [p_t(x-1) - p_t(x+1)] e^{ix\gamma}$$

$$= \frac{\sqrt{2t}}{\pi} \int_{\Gamma} \sum_{x=1}^{\infty} \frac{e^{ix(\gamma-\sigma)}}{x} [\cos \sigma - 2t \sin^2 \sigma] \exp[-2t(1-\cos \sigma)] \, d\sigma.$$

By (3.10), the absolute value of the above expression, integrated only over $\{\sigma \in \Gamma: |\sigma| > |\gamma|/2\}$, tends to zero as t tends to infinity. For the above expression integrated over $\{\sigma \in \Gamma: |\sigma| \leq |\gamma|/2\}$, make the change of variable $\lambda = \sigma\sqrt{2t}$ again to conclude using the Dominated Convergence Theorem that

$$\lim_{t\to\infty} \sqrt{2t} \sum_{x=1}^{\infty} [p_t(x-1) - p_t(x+1)] e^{ix\gamma}$$

$$= \frac{1}{\pi} \sum_{x=1}^{\infty} \frac{e^{ix\gamma}}{x} \int_{-\infty}^{\infty} (1-\lambda^2) e^{-\lambda^2/2} \, d\lambda = 0. \quad \square$$

Theorem 3.2 gives some information concerning the degree of clustering in the basic one-dimensional voter model. More insight is given by limit theorems for the distribution of the size of the cluster containing a fixed point, say the origin. For $\eta \in X$, let

$$r(\eta) = \min\{x \geq 1: \eta(x) \neq \eta(0)\}, \quad \text{and}$$

$$l(\eta) = \min\{x \geq 1: \eta(-x) \neq \eta(0)\}.$$

Note that if $\mu \in \mathcal{S}$, then

$$\mu\{\eta: r(\eta) > r \text{ and } l(\eta) > l\}$$

$$= \mu\{\eta: \eta(x) = \eta(0) \text{ for all } -l \leq x \leq r\}$$

$$= \mu\{\eta: r(\eta) > r + l\}.$$

Therefore the existence of a limiting distribution for $r(\eta_t)/\sqrt{t}$ as $t \uparrow \infty$ implies the existence of a limiting joint distribution for

$$\left(\frac{r(\eta_t)}{\sqrt{t}}, \frac{l(\eta_t)}{\sqrt{t}} \right),$$

provided that the initial distribution of the process is translation invariant. Hence in the remainder of this section, we will focus on the distribution of $r(\eta_t)$.

We begin with a series of simple lemmas concerning the behavior of the dual process.

Lemma 3.11. *There is a constant c so that*

$$P^{\{0,x\}}(|A_t| = 2) \le c \frac{x}{\sqrt{t}}$$

for all $x \ge 1$ and $t > 0$.

Proof. With $Z(t)$, $p_t(x)$, and τ as in the proof of Theorem 3.2, note that

$$P^{\{0,x\}}(|A_t| = 2) = P^x(\tau > t).$$

By the reflection principle,

$$P^x(\tau \le t) = 2P^x(Z(t) \le 0)$$
$$= P^0(|Z(t)| \ge x),$$

so that

$$P^{\{0,x\}}(|A_t| = 2) = P^0(|Z(t)| < x)$$
$$\le (2x - 1)p_t(0)$$

by (3.6). The result now follows from (3.7). □

Lemma 3.12. *There is a constant c so that*

$$P^{\{-x,0,y\}}(|A_t| = 3) \le c \frac{xy}{t}$$

for all $x, y \ge 1$ and $t > 0$.

Proof. Let $X_1(t)$, $X_2(t)$ and $X_3(t)$ be independent copies of one-dimensional simple random walks with mean 1 exponential holding times, and let

$$U(t) = X_1(t) - X_2(t), \quad \text{and}$$

$$V(t) = X_3(t) - X_2(t).$$

Then $(U(t), V(t))$ is a random walk on Z^2, all of whose transitions are between states which are comparable in the natural partial order on Z^2. Therefore by Corollary 2.21 of Chapter II, if $U(0) = -x$ and $V(0) = y$ for $x, y \ge 1$, then the events $\{U(s) \ge 0$ for some $s \le t\}$ and $\{V(s) > 0$ for all $s \le t\}$ are positively correlated. Using the natural coupling between A_t and $(X_1(t)$, $X_2(t), X_3(t))$, it follows that

$$P^{\{-x,0,y\}}(|A_t| = 3) \le P^{\{-x,0\}}(|A_t| = 2) P^{\{0,y\}}(|A_t| = 2),$$

so that the required estimate follows from Lemma 3.11. □

Lemma 3.13. *There is a constant c so that*

$$P^{[0,x]}(|A_t| \geq 3) \leq c \frac{x^2}{t}$$

for all $x \geq 1$ and $t > 0$.

Proof. For $x \geq 1$,

$$P^{[0,2x]}(|A_t| \geq 3) \leq P^{\{0,x,2x\}}(|A_t| = 3) + 2P^{[0,x]}(|A_t| \geq 3)$$

$$\leq c \frac{x^2}{t} + 2P^{[0,x]}(|A_t| \geq 3)$$

by Lemma 3.12. Interating this gives

$$P^{[0,2^n]}(|A_t| \geq 3) \leq \frac{c}{t} 2^{2n-2} \sum_{k=0}^{n-1} 2^{-k} \leq \frac{c}{2t} 2^{2n}.$$

If $2^{n-1} \leq x \leq 2^n$, then

$$P^{[0,x]}(|A_t| \geq 3) \leq P^{[0,2^n]}(|A_t| \geq 3) \leq \frac{c}{2t} 2^{2n} \leq \frac{c}{2t} (2x)^2,$$

so that the required result holds with a constant which is twice that in Lemma 3.12. $\square$

The following result is a simple consequence of the invariance principle and well-known properties of Brownian paths. (See Billingsley (1968).)

Lemma 3.14. *Let B_t be finite subsets of Z^1 and A be a finite subset of R^1, all of the same cardinality k. Suppose*

$$\frac{B_t}{\sqrt{t}} \to A$$

in the natural sense. Then the distribution of $A_t/\sqrt{t}$ with $A_0 = B_t$ converges weakly to the distribution at time one of k coalescing Brownian motions which start at the points of A.

Theorem 3.15. *There is a one-parameter family $\{F_\alpha(u), 0 < \alpha < 1\}$ of continuous distribution functions with $F_\alpha(0) = 0$ such that if $\mu \in \mathcal{S}$ satisfies*

$$\lim_{\min\{x_{i+1}-x_i\} \to \infty} \hat{\mu}(\{x_1, \ldots, x_n\}) = \alpha^n$$

for every $n \geq 1$, then

$$\lim_{t \to \infty} \mu S(t) \left\{ \eta : \frac{r(\eta)}{\sqrt{t}} \leq u \right\} = F_\alpha(u)$$

for all $u \in [0, \infty)$.

Proof. By (1.7),

$$\mu S(t) \left\{ \eta : \frac{r(\eta)}{\sqrt{t}} > u \right\}$$

$$= \mu S(t) \{ \eta : \eta(x) = \eta(0) \quad \text{for all } 0 \leq x \leq u\sqrt{t} \}$$

$$= E^{[0,[u\sqrt{t}]]} [\hat{\mu}(A_t) + \hat{\bar{\mu}}(A_t)],$$

where $[\cdot]$ is the greatest integer function and $\bar{\mu}$ is the measure obtained from μ by interchanging the roles of zeros and ones. Therefore, in order to prove the theorem, we need to show that

(3.16) $$\rho_k(u) = \lim_{t \to \infty} P^{[0,[u\sqrt{t}]]}(|A_t| \leq k) \quad \text{for } k \geq 1, u \geq 0$$

exists and satisfies

(3.17) $$\lim_{k \to \infty} \rho_k(u) = 1 \quad \text{for each } u > 0,$$

(3.18) $$\rho_k(u) \text{ is continuous in } u \text{ on } [0, \infty), \text{ and}$$

(3.19) $$\lim_{u \to \infty} \rho_k(u) = 0 \quad \text{for each } k \geq 1.$$

Furthermore, we need to know that

(3.20) $$\lim_{t \to \infty} P^{[0,[u\sqrt{t}]]}(d(A_t) < n, |A_t| > 1) = 0$$

for $u > 0$ and $n \geq 1$, where

$$d(\{x_1, \ldots, x_n\}) = \min_{1 \leq i < n} |x_{i+1} - x_i|.$$

In other words, we need to show that starting with $A_0 = [0, [u\sqrt{t}]]$, the distribution of $|A_t|$ has a limit, and the points in A_t get arbitrarily far away from one another as $t \uparrow \infty$. Once (3.16), (3.17), (3.18), (3.19), and (3.20) have been shown, the statement of the theorem will follow with $F_\alpha(u)$

defined by

$$1 - F_\alpha(u) = \alpha(1-\alpha) \sum_{k=0}^{\infty} \rho_{k+1}(u)[\alpha^k + (1-\alpha)^k].$$

To prove the existence of the limit in (3.16), let

$$B_t^n = \left\{ \left[k \frac{[u\sqrt{t}]}{2^n} \right], k = 0, \ldots, 2^n - 1 \right\} \cup \{[u\sqrt{t}]\}.$$

For fixed n,

$$\frac{B_t^n}{\sqrt{t}} \to A^n = \left\{ \frac{ku}{2^n}, k = 0, 1, \ldots, 2^n \right\},$$

so that by Lemma 3.14, the distribution of $A_t/\sqrt{t}$ with $A_0 = B_t^n$ converges as $t \uparrow \infty$ to the distribution at time one of $2^n + 1$ coalescing Brownian motions which start at the points of A^n. We then need to bound the difference between the distribution of $A_t/\sqrt{t}$ under $A_0 = B_t^n$ and its distribution under $A_0 = [0, [u\sqrt{t}]]$. Coupling together the processes $A_t/\sqrt{t}$ with these two initial states in the natural way, it is clear that the processes will differ at time t only if for some consecutive points $x, y \in B_t^n$, the process starting with initial state $[x, y]$ has cardinality at least 3 at time t. For fixed x and y, this probability is at most $c(y-x)^2/t$ by Lemma 3.13. Therefore

$$\sum_D |P^{B_t^n}(A_t = D) - P^{[0, [u\sqrt{t}]]}(A_t = D)| \le \frac{2^n}{t} \left(\frac{u\sqrt{t}}{2^n} \right)^2 \le \frac{u^2}{2^n},$$

which tends to zero as $n \to \infty$ uniformly in t. This establishes the existence of the limit in (3.16), as well as (3.17) and (3.20). For (3.18), note that by Lemma 3.11,

$$|\rho_k(u) - \rho_k(v)| \le \lim_{t \to \infty} P^{\{[u\sqrt{t}], [v\sqrt{t}]\}}(|A_t| = 2) \le c|u - v|,$$

so that in fact $\rho_k(u)$ is Lipschitz continuous. Finally, (3.19) follows from the fact that coalescing Brownian motions starting from points which are far away from one another will still be far apart at time one. $\square$

4. The Finite System

This section is devoted to the finite voter model. It is the version in which $\sum_x \eta_t(x) < \infty$ for all $t \ge 0$. In order to guarantee that the system is finite for all t whenever the initial configuration is finite, we will assume throughout this section that $p(x, y)$ is doubly stochastic:

$$\sum_x p(x, y) = 1 \quad \text{for all } y \in S.$$

One consequence of the first result is that this assumption is sufficient for this purpose. Let $|\eta| = \sum_x \eta(x)$, and

$$\tau = \inf\{t \geq 0 : |\eta_t| = 0\}.$$

Proposition 4.1. (a) $E^\eta |\eta_t| = |\eta|$ for all $\eta \in X$.

(b) $\qquad\qquad P^\eta(\tau < \infty) = 1 \quad \text{whenever} \quad |\eta| < \infty.$

Proof. By (1.7),

$$E^\eta \eta_t(x) = \sum_y p_t(x, y) \eta(y).$$

Since $p(x, y)$ is doubly stochastic, so is $p_t(x, y)$. Therefore

$$E^\eta |\eta_t| = \sum_{x,y} p_t(x, y) \eta(y) = |\eta|.$$

For part (b), suppose that $|\eta| < \infty$. By part (a), $|\eta_t|$ is a martingale under P^η. Therefore

$$\lim_{t \to \infty} |\eta_t|$$

exists by the Martingale Convergence Theorem. By the irreducibility assumption on $p(x, y)$, the only possible value of this limit is zero. Since $|\eta_t|$ is integer valued, it follows that $|\eta_t| = 0$ eventually with probability one, and hence that $\tau < \infty$. $\square$

Proposition 4.1 suggests some natural problems. The first is to determine the rate of convergence to zero of $P^\eta[\tau > t]$ as $t \uparrow \infty$. Another is to describe the asymptotic size or shape of $\{x : \eta_t(x) = 1\}$, conditioned on the event $\{\tau > t\}$. Similar problems have been raised and resolved in the simpler context of branching processes (see Athreya and Ney (1972), for example). For the voter model, weak answers to the first question can be obtained in fairly general settings. Rather precise results are available in the case of a simple random walk on Z^d. The proofs of these are quite easy in one dimension, but are much harder in higher dimensions.

The next result gives some information about the behavior of the extinction probability as a function of the initial configuration. For finite subsets A of S, it will be convenient to write η_t^A for the voter model which initially satisfies $\eta = 1$ on A and $\eta = 0$ on A^c. These processes will be constructed jointly using the graphical representation described in Section 6 of Chapter III. It corresponds to the coupling in which voters at a given site in different versions of the system use the same times to reevaluate their positions, and choose the same other voter to imitate.

Proposition 4.2. $P(\eta_t^{A \cup B} \equiv 0) \leq P(\eta_t^A \equiv 0) \, P(\eta_t^B \equiv 0)$ *whenever* $A \cap B = \varnothing$.

Proof. Since the graphical representation is being used, $\eta_t^{A \cup B} = \eta_t^A \vee \eta_t^B$ for all A, B (see Definition 6.1 of Chapter III). Therefore the statement to be proved is simply the assertion that the events $\{\eta_t^A \equiv 0\}$ and $\{\eta_t^B \equiv 0\}$ have negative correlations whenever $A \cap B = \varnothing$. This will be shown as an application of Theorem 2.14 of Chapter II. The set-up is as follows. Let

$$E = \{(A, B): A, B \text{ are disjoint finite subsets of } S\}$$

with the partial order defined by $(A_1, B_1) \leq (A_2, B_2)$ if and only if $A_1 \subset A_2$ and $B_1 \supset B_2$. The direction of the containment in the second component is a device to change a statement about positive correlations into one about negative correlations. Then

$$(\{x: \eta_t^A(x) = 1\}, \{x: \eta_t^B(x) = 1\})$$

defines a Markov chain on E which is a monotone process which jumps only among sites which are comparable in the partial order. In verifying these properties, it is important to keep in mind that $(A, B) \in E$ implies that $A \cap B = \varnothing$. By the definition of the partial order, the functions

$$f(A, B) = \begin{cases} 1 & \text{if } A \neq \varnothing, \\ 0 & \text{if } A = \varnothing, \end{cases}$$

and

$$g(A, B) = \begin{cases} 1 & \text{if } B = \varnothing, \\ 0 & \text{if } B \neq \varnothing, \end{cases}$$

are increasing functions on E. Therefore by Theorem 2.14 of Chapter II,

$$P[\eta_t^A \neq 0, \, \eta_t^B \equiv 0] \geq P[\eta_t^A \neq 0] P[\eta_t^B \equiv 0].$$

Replacing $\{\eta_t^A \neq 0\}$ by its complement, we see that

$$P[\eta_t^A \equiv 0, \, \eta_t^B \equiv 0] \leq P[\eta_t^A \equiv 0] P[\eta_t^B \equiv 0]$$

as required. $\square$

Theorem 4.3. *For any* $t > 0$,

$$P^{\eta}[\tau > t] \geq 1 - \left(\frac{t}{1+t}\right)^{|\eta|}.$$

Proof. By Proposition 4.2, it suffices to prove the result in case $|\eta| = 1$. Let

$$u(x, t) = P(\eta_t^{\{x\}} \equiv 0).$$

The backward equation for the Markov chain $\eta_t^{\{x\}}$ gives

$$\frac{d}{dt} u(x, t) = [1 - p(x, x)] + \sum_{y \neq x} p(y, x) P(\eta_t^{\{x,y\}} \equiv 0) - 2[1 - p(x, x)] u(x, t).$$

By Proposition 4.2, it follows that

$$\frac{d}{dt} u(x, t) \leq [1 - p(x, x)][1 - 2u(x, t)] + u(x, t) \sum_{y \neq x} p(y, x) u(y, t)$$

$$\leq 1 - 2u(x, t) + u(x, t) \sum_y p(y, x) u(y, t).$$

Multiplying this by e^{2t}, and integrating from 0 to T, using the initial condition $u(x, 0) = 0$, we see that

$$u(x, T) \leq \int_0^T e^{-2(T-t)} \left[1 + u(x, t) \sum_y p(y, x) u(y, t) \right] dt.$$

Now let $u(t) = \max_{x \in S} u(x, t)$, and conclude that

$$u(T) \leq \int_0^T e^{-2(T-t)} [1 + u^2(t)] \, dt.$$

Therefore $u(t) \leq v(t)$, where v is the solution of

$$v(T) = \int_0^T e^{-2(T-t)} [1 + v^2(t)] \, dt.$$

Turning this integral equation back into a differential equation, we have $v(0) = 0$ and

$$v'(t) = [1 - v(t)]^2,$$

so that $v(t) = t/(1+t)$ as required. $\quad\square$

Theorem 4.4. *Suppose that there is a* $c > 0$ *and a* $\gamma \in [0, 1)$ *so that*

(4.5)
$$\sum_{\substack{\eta(x)=1 \\ \eta(y)=0}} p(x, y) \geq c|\eta|^{\gamma}$$

when $0 < |\eta| < \infty$. *Then*

$$P^{\eta}[\tau > t] \leq \frac{2|\eta|}{[[ct(1-\gamma)]^{1/(2-\gamma)}]},$$

where $[\cdot]$ *denotes the integer part.*

Proof. Let τ_n be the successive times at which η_t changes. Then since $|\eta_t|$ decreases or increases by one at the same rate $q(\eta_t)$, where

$$q(\eta) = \sum_{\substack{\eta(x)=1 \\ \eta(y)=0}} p(x, y) = \sum_{\substack{\eta(x)=0 \\ \eta(y)=1}} p(x, y),$$

it follows that $W_n = |\eta_{\tau_n}|$ is a simple symmetric random walk on $\{0, 1, \ldots\}$. Let

$$N_k = \text{number of visits of } W_n \text{ to } k \text{ before hitting } 0.$$

Then write τ in the form

$$\tau = \sum_{k=1}^{\infty} \sum_{j=1}^{N_k} T_{k, j},$$

where $T_{k, j}$ is the time spent by η_t in the state which is the jth one visited of cardinality k. Let $M = \max\{k \geq 1 : N_k \geq 1\}$. Then for any integer m,

(4.6)
$$P^{\eta}[\tau > t] \leq P^{\eta}[M > m] + P^{\eta}\left[\sum_{k=1}^{m} \sum_{j=1}^{N_k} T_{k, j} > t\right].$$

By the solution to the Gambler's Ruin Problem (see, for example, Hoel, Port, and Stone (1972)),

(4.7)
$$P^{\eta}[M > m] \leq \frac{|\eta|}{m}, \quad \text{and}$$

(4.8)
$$E^{\eta} N_k = 2 \min(|\eta|, k).$$

To estimate the second term on the right of (4.6), note that by (4.5),

$$E^{\eta} T_{k, j} \leq \frac{1}{2ck^{\gamma}}.$$

Therefore by Wald's identity and (4.8),

$$E^\eta \sum_{j=1}^{N_k} T_{k,j} \le \frac{E^\eta N_k}{2ck^\gamma}$$

$$= \frac{\min(|\eta|, k)}{ck^\gamma}.$$

So, by (4.6), (4.7) and Chebyshev's inequality,

$$P^\eta[\tau > t] \le \frac{|\eta|}{m} + \frac{|\eta|}{ct} \sum_{k=1}^m k^{-\gamma}$$

$$\le |\eta| \left[\frac{1}{m} + \frac{m^{1-\gamma}}{ct(1-\gamma)} \right].$$

The result follows by letting m be the integer part of $(ct(1-\gamma))^{1/(2-\gamma)}$. $\square$

In the case of a simple random walk on Z^d, Theorems 4.3 and 4.4 give the following bounds for $|\eta| = 1$:

$$(1+t)^{-1} \le P^\eta[\tau > t] \le c_d t^{-d/(d+1)}$$

for some constant c_d depending on d. The following gives a much more precise statement in one dimension.

Theorem 4.9. *If $S = Z^1$ and $p(x, x+1) = p(x, x-1) = \frac{1}{2}$, then*

(a) $$\lim_{t \to \infty} \sqrt{t} P^*[N_t > 0] = \frac{1}{\sqrt{\pi}}, \quad and$$

(b) $$\lim_{t \to \infty} P^* \left[\frac{N_t}{\sqrt{t}} \le u \,\middle|\, N_t > 0 \right] = 1 - e^{-u^2/4},$$

*where $N_t = |\eta_t|$ and * represents the configuration η with $\eta(0) = 1$ and $\eta(x) = 0$ for $x \ne 0$.*

Proof. With initial state *, $\{x: \eta_t(x) = 1\}$ is an interval of integers or is empty for all t. The boundaries of this interval perform independent simple random walks until they hit, at which time η_t is identically zero. Therefore the length of the interval performs a simple random walk with rate 2 until η_t is identically zero. So, with the notation in the proof of Theorem 3.2,

(4.10) $$P^*[N_t = x] = P^1[Z(t) = x, \tau > t]$$

for $x \geq 1$. Using (3.4) and (3.7), we then see that

$$\lim_{t \to \infty} \sqrt{t} P^*[N_t > 0] = \lim_{t \to \infty} \sqrt{t}[p_t(0) + p_t(1)]$$

$$= \frac{1}{\sqrt{\pi}},$$

which gives (a). For (b), use (4.10) and (3.4) to write

$$
P^*\left[\frac{N_t}{\sqrt{t}} \leq u \,\middle|\, N_t > 0\right] = \frac{P^1[Z(t)/\sqrt{t} \leq u, \tau > t]}{P^1[\tau > t]}
$$
(4.11)
$$
= \frac{p_t(0) + p_t(1) - p_t([u\sqrt{t}]) - p_t([u\sqrt{t}] + 1)}{p_t(0) + p_t(1)}.
$$

Using (3.6) as we did to prove (3.7), we see that

$$\lim_{t \to \infty} \sqrt{2t} p_t([u\sqrt{t}]) = \frac{1}{2\pi} \int_{-\infty}^{\infty} e^{iu(\lambda/\sqrt{2})} e^{-\lambda^2/2} \, d\lambda$$
(4.12)
$$= \frac{1}{\sqrt{2\pi}} e^{-u^2/4}.$$

Part (b) of the theorem follows from (4.11) and (4.12).   □

In dimensions greater than one, the analysis of N_t is much more difficult. Using deep results of Sawyer (1979), Bramson and Griffeath (1980a) have shown that in the case of simple random walk on Z^d,

$$\lim_{t \to \infty} \frac{t}{\log t} P^*[N_t > 0] = \frac{1}{\pi} \quad \text{if } d = 2, \quad \text{and}$$

$$\lim_{t \to \infty} t P^*[N_t > 0] = \frac{1}{\gamma_d} \quad \text{if } d \geq 3,$$

where γ_d is the probability that the simple random walk never returns to its initial position. Furthermore, they showed that for $d \geq 2$,

$$\lim_{t \to \infty} P^*\left[\frac{N_t}{P^*[N_t > 0]} \leq u \,\middle|\, N_t > 0\right] = 1 - e^{-u}$$

for $u \geq 0$.

The following comments provide the basis for their proof, and should help to put these asymptotics in the proper context. Consider a "multitype" voter model, in which initially all sites are of distinct types. Sites wait independent exponential times, at which times they adopt the type of a randomly chosen neighbor. Let M_t be the number of sites which at time t have the same type as the origin. Sawyer's (1979) result is that if $d \geq 2$,

$$(4.13) \qquad \lim_{t \to \infty} P\left(\frac{M_t}{EM_t} \leq u\right) = \int_0^u 4v\, e^{-2v}\, dv.$$

Kelly (1977) noticed the following connection between M_t and N_t (where initially $N_0 = 1$):

$$P(M_t = k) = kP(N_t = k) \quad \text{for all } k \geq 0.$$

Therefore

$$(4.14) \qquad P(N_t > 0) = \sum_{k=1}^{\infty} \frac{1}{k} P(M_t = k) = E\frac{1}{M_t}.$$

In the presence of appropriate domination, (4.13) and (4.14) would imply

$$(4.15) \qquad \begin{aligned} \lim_{t \to \infty} EM_t P(N_t > 0) &= \lim_{t \to \infty} E\left[\frac{EM_t}{M_t}\right] \\ &= \int_0^{\infty} 4\, e^{-2v}\, dv = 2. \end{aligned}$$

In fact, the justification of this step occupies the major portion of Bramson and Griffeath (1980a). Once we have (4.15), in order to get the asymptotics for $P(N_t > 0)$, it is sufficient to have them for EM_t. But Sudbury (1976) had observed that $EM_t = ER_{2t}$, where R_t is the number of sites visited by time t by a continuous time mean one simple random walk on Z^d. This is easy to see, since

$$EM_t = \sum_x P(x \text{ and } 0 \text{ have the same type at time } t)$$

$$= \sum_x P(\text{two random walks starting at } 0 \text{ and } x \text{ have hit by time } t)$$

$$= \sum_x P(\text{one random walk starting at } 0 \text{ has hit } x \text{ by time } 2t)$$

$$= ER_{2t}.$$

The remaining step is to have the asymptotics for ER_t. But these were obtained by Dvoretsky and Erdös (1951).

Note that while the above comments can be summarized by saying that $E(N_t | N_t > 0)$ and ER_t have the same asymptotics as $t \to \infty$, the distributions of $(A_t | A_t \neq \varnothing)$ and the set of sites visited by a simple random walk are not at all similar. One obvious difference is that the latter concentrates on connected sets, while the former does not.

5. Notes and References

Section 1. The voter model was introduced independently by Clifford and Sudbury (1973) and by Holley and Liggett (1975). The first of these papers contains some special cases of Theorems 1.8 and 1.9. The second paper contains a statement without proof of Theorem 1.8, and proofs of both theorems under the simplifying assumption that $\mathcal{H}$ consists only of constants. Analogous results in a similar context were proved by Shiga (1980a, b). In fact, the idea of proving Theorem 1.9 first, and then using it to prove Theorem 1.8(c), is due to him. Other stochastic models in which the dichotomy described in Examples 1.4 and 1.5 occurs have been investigated by Sawyer (1976b), Kallenberg (1977), Dawson (1977), Matthes, Kerstan, and Mecke (1978), and Durrett (1979). Lemma 1.26 is due to Matloff (1977).

Cox and Griffeath (1983) have discovered a more elaborate dimension dependence for the voter model by considering the behavior of occupation times. They let

$$T_t = \int_0^t \eta_s(0) \, ds,$$

where η_s is the configuration at time s of the voter model on Z^d with $p(x, y) = 1/2d$ for nearest neighbor x, y, in which the initial distribution is taken to be the product measure ν_α for some constant $\alpha \in (0, 1)$. They show that the variance of T_t behaves asymptotically as $t \uparrow \infty$ like a constant multiple of

$$t^2 \qquad \text{if } d = 1,$$

$$t^2/\log t \quad \text{if } d = 2,$$

$$t^{3/2} \qquad \text{if } d = 3,$$

$$t \log t \quad \text{if } d = 4, \quad \text{and}$$

$$t \qquad \text{if } t \geq 5.$$

They then prove a limit theorem for $(T_t - \alpha t)/\sqrt{\text{var } T_t}$. The limiting distribution turns out to be Gaussian if $d \geq 2$. It is of course not Gaussian if $d = 1$, since $0 \leq T_t \leq t$.

We used coalescing duality to analyze the voter model in this section. As we saw in Example 4.16 of Chapter III, the voter model also has an annihilating dual. It could also be used in proving ergodic theorems in this context, as was done for example by Spitzer (1981).

Systems of coalescing and annihilating random walks, which arise here as duals of the voter model, have been studied in their own right. Papers on this subject are Erdos and Ney (1974), Schwartz (1976b), Griffeath (1978a), Bramson and Griffeath (1980a, b), and Arratia (1981, 1983b).

Section 2. Theorem 2.5 is due to Holley and Liggett (1975). Dynkin (1984) (and the references given there) has made a study of random fields whose covariances are given by the Green function of a symmetric Markov process as in (2.7). Theorem 2.8 was proved for a discrete time version of the voter model in three dimensions with $\beta = 2$ by Bramson and Griffeath (1979b). For similar results in other contexts, see Dawson (1977), Holley and Stroock (1978a), and Siegmund–Schultze (1981).

Section 3. The results in this section are based on Bramson and Griffeath (1980b). They proved Theorem 3.2 under the somewhat stronger assumption that the initial distribution is mixing. In their paper, Bramson and Griffeath obtained additional information about the distribution functions $F_\alpha(u)$. They showed that F_α has a monotone decreasing continuous density function, and that it has an exponential tail. They also obtained bounds on the mean of F_α. Their paper treats coalescing and annihilating random walks, as well as the voter model. The idea of using Harris' theorem on positive correlations (Corollary 2.21 of Chapter II) to prove Lemma 3.12 is due to Arratia.

Section 4. Proposition 4.2 is due to Arratia (1981). Theorem 4.3 is Lemma 1 of Griffeath (1978a). The proof given here is somewhat different from his. Theorem 4.4 is based on the main result in Bramson and Griffeath (1979a). Theorem 4.9 comes from Bramson and Griffeath (1980a). Earlier work on this problem can be found in Sudbury (1976) and Kelly (1977). Donnelly and Welsh (1983) analyze the transient behavior of voter models on finite graphs.

Chapter VI

The Contact Process

The contact process is the spin system in which $S = Z^d$ and

$$(0.1) \qquad c(x, \eta) = \begin{cases} \lambda \sum_{|y-x|=1} \eta(y) & \text{if } \eta(x) = 0, \\ 1 & \text{if } \eta(x) = 1, \end{cases}$$

where λ is a nonnegative parameter. One interpretation of this process is as a model for the spread of an infection. An individual at $x \in S$ is infected if $\eta(x) = 1$ and healthy if $\eta(x) = 0$. Healthy individuals become infected at a rate which is proportional to the number of infected neighbors. Infected individuals recover at a constant rate, which is normalized to be 1.

The contact process has appeared independently in the high-energy physics literature. It arose there because it is equivalent to the reggeon spin model, which is itself a simplification of reggeon field theory.

Results about the contact process have been useful in studying other systems. For example, Durrett and Liggett (1981) used the linear growth of a discrete time version of the contact process to obtain information about the limiting "shape" of a growth model which was studied by Richardson (1973).

The above remarks give some of the reasons for studying this process. However, interest in the contact process comes from a number of other sources as well. It is one of the simplest particle systems which exhibits a phase transition. (This means that the behavior of the system undergoes an abrupt change at a certain value of the parameter. We have already seen important examples of this is Chapter IV.) Furthermore, practically all of the tools which have been developed in interacting particle systems apply to it. The contact process is both attractive and self-dual, so that most of the results and techniques in Chapter III are applicable. In addition, the interaction is nearest neighbor. This fact is very important, particularly in the one-dimensional case. One basic tool which is not available in this context is reversibility. In Chapter VII, we will encounter some reversible analogues of the contact process. While many of these are attractive, they fail to have duals and they are not nearest neighbor.

The first three sections of this chapter deal with the one-dimensional case. The theory here is rather complete, although several important open problems do remain. The main objective of the first section is to obtain an upper bound on the critical value λ_c. (Recall that lower bounds were given in Chapter III.) In addition, a number of simple results are presented which are used later.

The second section contains the main convergence theorems. The preliminary study of "edges" which occurs there is the main tool used in proving those theorems. Edges also provide a means of obtaining improved lower bounds on the critical value. Rates of convergence are discussed in Section 3. In particular, a number of exponential estimates are proved for $\lambda \neq \lambda_c$. The theory in dimensions greater than one is far less satisfactory than that in one dimension. Much of what is known is described in Section 4.

The four sections depend on one another in the following way:

$$1 \rightarrow 2 \nearrow^{3} \searrow_{4}$$

1. The Critical Value

Let ν_λ be the upper invariant measure for the one-dimensional contact process with parameter $\lambda \geq 0$:

$$(1.1) \qquad \nu_\lambda = \lim_{t \to \infty} \delta_1 S(t).$$

The limit exists by Theorem 2.3 of Chapter III, since the contact process is attractive. Define

$$(1.2) \qquad \rho(\lambda) = \nu_\lambda\{\eta: \eta(x) = 1\},$$

which is independent of $x \in Z^1$, since $\nu_\lambda \in \mathcal{S}$. By Corollary 2.4 of Chapter III, the process is ergodic for a given λ if and only if $\rho(\lambda) = 0$. By Corollary 1.7 of Chapter III,

$$(1.3) \qquad \lambda_1 \leq \lambda_2 \quad \text{implies} \quad \nu_{\lambda_1} \leq \nu_{\lambda_2}.$$

Therefore, $\rho(\lambda)$ is a nondecreasing function of λ.

Define the critical value λ_c as in the statement and proof of Theorem 2.14 of Chapter III:

$$(1.4) \qquad \begin{aligned} \lambda_c &= \sup\{\lambda \geq 0: \text{process is ergodic}\} \\ &= \inf\{\lambda \geq 0: \rho(\lambda) > 0\}. \end{aligned}$$

In Example 5.11 of Chapter III, we saw that $\lambda_c \geq 1$. As was pointed out there, using the same ideas, it can be shown that $\lambda_c \geq 1.18$. An alternate proof of this lower bound will be given in the next section. At this point, we do not yet know that $\lambda_c < \infty$. The main purpose of this section is to prove this, and in so doing to obtain a good upper bound for λ_c. Before doing this, we will record several facts, most of which come from specializing results of Chapter III to the contact process.

Theorem 1.5. *For any* $\lambda \geq 0$, $\mathcal{I}_e = \{\delta_0, \nu_\lambda\}$.

Proof. This is a special case of Theorem 3.13 of Chapter III. To check assumption (3.4) there, simply note that for the contact process, $c(x, \eta) = 1$ if $\eta(x) = 1$, so that $c(x, \eta) + c(x, \eta_x) \geq 1$ for all η and all x. $\quad\square$

Theorem 1.6. (a) $\rho(\lambda) = 0$ *for* $\lambda < \lambda_c$.
    (b) $\rho(\lambda) > 0$ *for* $\lambda > \lambda_c$.
    (c) *If* $\rho(\lambda) > 0$, *then* ν_λ *puts no mass on* $\eta \equiv 0$.
    (d) $\rho(\lambda)$ *is continuous on* $[\lambda_c, \infty)$.

Proof. The first two statements are just a rephrasing of the definition of λ_c. Part (c) follows from the fact that the conditional measure $\nu_\lambda(\cdot \,|\, \eta \neq 0)$ is invariant as well, and hence must be equal to ν_λ. To prove the final statement, we will check right and left continuity separately. By monotonicity, $\nu = \lim_{\gamma \downarrow \lambda} \nu_\gamma$ exists and satisfies $\nu \geq \nu_\lambda$. By Proposition 2.14 of Chapter I, ν is invariant for the contact process with parameter λ, so that $\nu \leq \nu_\lambda$. Therefore $\nu = \nu_\lambda$, so that ρ is right continuous on $[0, \infty)$. For the left continuity, note again that $\nu = \lim_{\gamma \uparrow \lambda} \nu_\lambda$ exists, is invariant for the contact process with parameter λ, and satisfies $\nu \geq \nu_\gamma$ for all $\gamma < \lambda$. If $\lambda > \lambda_c$, it then follows from part (c) above that ν puts no mass on $\eta \equiv 0$. Therefore $\nu = \nu_\lambda$ by Theorem 1.5 as required. $\quad\square$

Let Y be the collection of all finite subsets of Z^1. The finite contact process is the one for which initially, and hence at all times, $\sum_x \eta_t(x) < \infty$. With the identification

$$A = \{x: \eta(x) = 1\},$$

the finite contact process can be viewed as a Markov chain A_t on Y. Then A_t is the (coalescing) dual of η_t by Theorem 4.13 of Chapter III, as applied to the contact process in Example 4.18 of that chapter. The formal statement of this is as follows.

Theorem 1.7. *For* $\eta \in X$ *and* $A \in Y$,

$$P^\eta[\eta_t(x) = 0 \text{ for all } x \in A] = P^A[\eta(x) = 0 \text{ for all } x \in A_t].$$

In particular,

(1.8) $P^1[\eta_t(x) = 0 \ for \ all \ x \in A] = P^A[A_t = \varnothing],$

where 1 *is the configuration* $\eta \equiv 1$.

Proof. The first statement is simply Theorem 4.13 of Chapter III in this case. The second statement is obtained from the first by taking $\eta \equiv 1$. □

Let $\tau = \inf\{t \geq 0: A_t = \varnothing\}$, and define

$$\sigma(A) = P^A(\tau = \infty)$$

for all $A \in Y$. The next result gives information about how σ depends on A.

Theorem 1.9. (a) $A \subset B$ *implies* $\sigma(A) \leq \sigma(B)$.
    (b) $\sigma(A \cup B) + \sigma(A \cap B) \leq \sigma(A) + \sigma(B)$ *for all* $A, B \in Y$.
    (c) $\sigma(\{x_1, \ldots, x_n\}) \leq \sigma(\{y_1, \ldots, y_n\})$ *provided that* $x_1 < x_2 < \cdots < x_n$, $y_1 < y_2 < \cdots < y_n$, *and* $x_{i+1} - x_i \leq y_{i+1} - y_i$ *for each i.*

Proof. Part (a) comes from attractiveness, since A_t and B_t can be coupled in such a way as to preserve the relation $A_t \subset B_t$, provided that $A_0 \subset B_0$. Part (b) is obtained by specializing Proposition 5.9 of Chapter III to the contact process. For part (c), we will use a somewhat different coupling. We will couple together three processes A_t, B_t, and C_t so that A_t and C_t are copies of the contact process, $A_0 = \{x_1, \ldots, x_n\}$, $B_0 = C_0 = \{y_1, \ldots, y_n\}$, and at all times, $|A_t| = |B_t|$, B_t is more spread out than A_t in the sense described in (c), and $B_t \subset C_t$. The process B_t alone will not even be Markovian, and hence certainly will not be a copy of the contact process. Points in $C_t \backslash B_t$ die independently. Other deaths in the three processes are coupled as follows: points in A_t and B_t are paired in increasing order (i.e., the leftmost points in A_t and B_t are paired, etc.), while points in B_t are paired with the same points in C_t, which is possible since $B_t \subset C_t$. Deaths at paired points in the three processes occur simultaneously. Births at unoccupied neighbors of paired points also occur simultaneously. Since $|A_t| = |B_t| \leq |C_t|$, $A_t \neq \varnothing$ implies $C_t \neq \varnothing$. □

Next, we will use the previous two theorems to connect up the limiting and critical behavior of the finite and infinite contact processes.

Theorem 1.10. (a) $\sigma(A) = \nu_\lambda\{\eta: \eta(x) = 1 \ for \ some \ x \in A\}$, *so in particular,*
    $\rho(\lambda) = \sigma(\{x\})$ *for any* $x \in Z^1$.
    (b) $\lambda_c = \inf\{\lambda \geq 0: \sigma(A) > 0\}$ *for any* $A \in Y$.
    (c) $\rho(\lambda_c) = 0$ *if and only if* $\sigma(A) = 0$ *for all A when* $\lambda = \lambda_c$.
    (d)   *If* $\rho(\lambda) > 0$, *then* $\lim_{|A| \to \infty} \sigma(A) = 1$.

Proof. Part (a) is obtained by letting $t \to \infty$ in (1.8). Parts (b) and (c) follow from part (a) and parts (a) and (b) of Theorem 1.9, since they combine to give

$$\rho(\lambda) \leq \sigma(A) \leq |A|\rho(\lambda)$$

for $A \neq \varnothing$. For part (d), use part (c) of Theorem 1.9 to show that $|A| = n$ implies that $\sigma(A) \geq \sigma(\{1, \ldots, n\})$. To see that $\lim_{n \to \infty} \sigma(\{1, \ldots, n\}) = 1$, it suffices to use part (a) above and part (c) of Theorem 1.6. □

The remainder of this section is devoted to the proof that $\lambda_c \leq 2$. The idea of the proof is quite simple. For $\lambda \geq 2$, we wish to find a function h on Y which satisfies

(1.11) $h(\varnothing) = 0$ and $0 < h(A) \leq 1$ for $A \neq \varnothing$,

(1.12) $$\lim_{|A| \to \infty} h(A) = 1, \quad \text{and}$$

(1.13) $E^A h(A_t) \geq h(A)$ for $A \in Y$ and $t \geq 0$.

If we can find such a function, then by (1.11) and (1.12) it will follow that

$$\sigma(A) = \lim_{t \to \infty} E^A h(A_t).$$

Therefore we will have $\sigma(A) \geq h(A)$ by (1.13), so that $\rho(\lambda) > 0$ for that λ. The main difficulty is to find a good choice of h, and then to prove (1.13) for that choice. Of course, (1.13) is equivalent to

$$\frac{d}{dt} E^A h(A_t)\big|_{t=0} \geq 0$$

for all $A \in Y$. This in turn is equivalent to

(1.14)
$$\sum_{x \in A} [h(A) - h(A \backslash x)]$$
$$\leq \lambda \sum_{x \notin A} (1_A(x-1) + 1_A(x+1))[h(A \cup x) - h(A)]$$

for all $A \in Y$, where $1_A(x)$ is the indicator function of A.

In searching for a good h, a natural first step is to try functions which depend on A only through its cardinality or diameter. Not surprisingly, no choice of this type works. In order to see this, suppose first that $h(A) = g(n)$ whenever $|A| = n$ for some g on $Z_+ = \{0, 1, 2, \ldots\}$. Applying (1.14) to $A = \{1, 2, 3, \ldots, n\}$ gives

$$n[g(n) - g(n-1)] \leq 2\lambda[g(n+1) - g(n)],$$

so that g is superharmonic for the Markov chain on Z_+ which moves from n to $n+1$ at rate 2λ and from n to $n-1$ at rate n. This chain is recurrent for all λ, so that g must be constant. Similarly, if $h(A) = g(n)$ whenever A has diameter n, then applying (1.14) to $A = \{1, n\}$ gives

$$g(n) - g(1) \leq \lambda[g(n+1) - g(n)],$$

so that g is superharmonic for the Markov chain on Z_+ which moves from n to $n+1$ at rate λ and from n to 1 at rate 1. This chain is recurrent for all λ, so again g must be constant.

From the above considerations, it should be clear that a successful choice of h must depend on the structure of A in a nontrivial way. The h which works results from the following three ideas, each of which may appear to require an unreasonable leap of faith:

(a) Choose h of the form

(1.15) $$h(A) = \mu\{\eta: \eta(x) = 1 \text{ for some } x \in A\}$$

for some probability measure μ on X (with $h(\varnothing) = 0$).

(b) Choose μ to be the stationary renewal measure corresponding to some probability density $f(n)$ of finite mean on $\{1, 2, \ldots\}$. This means that $\mu \in \mathcal{S}$ and if $A = \{x_1, x_2, \ldots, x_n\}$ with $x_1 < x_2 < \cdots < x_n$, then

$$\mu\{\eta: \eta(x_i) = 1 \text{ for } 1 \leq i \leq n \text{ and } \eta(x) = 0$$
$$\text{for all } x \notin A \text{ such that } x_1 < x < x_n\}$$

(1.16)
$$= \frac{\prod\limits_{i=1}^{n-1} f(x_{i+1} - x_i)}{\sum\limits_{k=1}^{\infty} kf(k)}.$$

In other words, the distances between successive ones in η under μ are independent and identically distributed with density f.

(c) Choose the density f so that (1.14) holds with equality for all A of the form $\{1, 2, \ldots, n\}$.

The motivation for these choices is as follows. Since $h(A)$ is supposed to be a bound for $\sigma(A)$ and $\sigma(A)$ is of the form (1.15) by Theorem 1.10(a), it is not too unreasonable to choose h of that form. The choice of μ to be a renewal measure is suggested by the fact to be seen in Chapter VII that the upper invariant measure for a reversible nearest-particle system is always a renewal measure. (Recall that the contact process is a nonreversible nearest-particle system.) Another reason for choosing a renewal measure is that in order to carry out the computations involved in checking (1.14), a reasonably explicit expression for h will be needed, and hence a reasonably simple μ will be needed. Exchangeable μ's are ruled out by our earlier comments, since if μ were exchangeable, then h would depend on A only through its cardinality. Finally, idea (c) is admittedly ad hoc. It is at least a way of picking out a particular f. Furthermore, it is not hard to check that for no choice of f will μ satisfy (1.14) if $\lambda < 2$. The above ideas will now be implemented in a series of lemmas.

Lemma 1.17. *If* $\lambda \geq 2$, *the unique probability density* f *on* $\{1, 2, \ldots\}$ *with finite mean such that* h *defined on* Y *by* (1.15) *and* (1.16) *satisfies* (1.14) *with equality for all* A *of the form* $A = \{1, \ldots, n\}$ *is given by* $f(n) = F(n) - F(n+1)$ *where*

$$(1.18) \qquad F(n+1) = \frac{(2n)!}{n!\,(n+1)!}\,\frac{1}{(2\lambda)^n} \quad \text{for } n \geq 0.$$

If $0 < \lambda < 2$, *there is no such density.*

Proof. Suppose f is a probability density with finite mean, let $F(n) = \sum_{k=n}^{\infty} f(k)$, and define h and μ by (1.15) and (1.16). Then

$$F(1) = 1, \qquad \sum_{n=1}^{\infty} F(n) = \sum_{n=1}^{\infty} nf(n) < \infty,$$

and for $x \in A$,

$$(1.19) \qquad h(A) - h(A \backslash x) = \mu\{\eta : \eta(x) = 1 \text{ and } \eta(y) = 0 \text{ for all } y \in A \backslash x\}.$$

Therefore (1.14) with equality for all A of the form $\{1, \ldots, n\}$ becomes

$$(1.20) \qquad \sum_{k=1}^{n} F(k)F(n+1-k) = 2\lambda F(n+1), \qquad n \geq 1.$$

This clearly has a unique solution satisfying $F(1) = 1$ for any $\lambda > 0$. We need to show that this solution is decreasing and summable if and only if $\lambda \geq 2$. To solve (1.20) with $F(1) = 1$ explicitly, introduce the generating function

$$(1.21) \qquad \phi(u) = \sum_{n=1}^{\infty} F(n)u^n.$$

Multiply (1.20) by u^{n+1} and sum for $n \geq 1$ to get

$$\phi^2(u) = 2\lambda[\phi(u) - u],$$

or

$$\phi(u) = \lambda \pm \sqrt{\lambda^2 - 2u\lambda}.$$

Since $\phi'(0) = 1$, the appropriate sign is negative. Thus

(1.22) $$\phi(u) = \lambda - \sqrt{\lambda^2 - 2u\lambda}.$$

This is real analytic exactly for $u < \lambda/2$, so that the radius of convergence of (1.21) is $\lambda/2$. Therefore $F(n)$ is not summable if $\lambda < 2$. If $\lambda \geq 2$, expand (1.22) in a power series in u to obtain (1.18). Of course,

(1.23) $$\sum_{n=1}^{\infty} F(n) = \phi(1) = \lambda - \sqrt{\lambda^2 - 2\lambda}$$

in this case. To check that $F(n)$ is decreasing when $\lambda \geq 2$, use (1.18) to show that $F(n+1) \geq F(n+2)$ is equivalent to

$$\lambda \geq \frac{2n+1}{n+2},$$

which is true if $\lambda \geq 2$ and $n \geq 0$. $\square$

A simple proof of the next lemma could be given by using the results of Section 4 of Chapter VII. The idea is that the stationary renewal measure with positive density f is a reversible measure for the nearest-particle system with birth rates

$$\beta(l, r) = \frac{f(l)f(r)}{f(l+r)}.$$

The monotonicity of $f(n)/f(n+1)$ which is assumed below translates into the statement that this system is attractive. The conclusion of the lemma is then an immediate consequence of this attractiveness. In order not to jump ahead, we will instead give a somewhat more obscure proof based on results in Chapter II. A third proof of the part of Lemma 1.24 which we will need is given in Section 2 of Chapter VII.

Lemma 1.24. *Suppose that f is a strictly positive probability density on $\{1, 2, \ldots\}$ with finite mean, and let μ be the corresponding renewal measure defined in (1.16). For $n \geq 1$, let μ_n be the measure on $\{0, 1\}^{Z_+}$ defined by*

$$\mu_n(\cdot) = \mu(\cdot \mid \eta(-n) = 1)$$

(i.e., μ_n is obtained from μ by conditioning on $\eta(-n) = 1$). If $f(n)/f(n+1)$ is nonincreasing in n, then $\mu_{n+1} \leq \mu_n$ for all $n \geq 1$.

Proof. By the translation invariance of μ, it suffices to prove that $\mu_2 \leq \mu_1$. Since μ is a renewal measure,

$$\mu(\cdot \,|\, \eta(-2) = 1, \eta(-1) = 1) = \mu(\cdot \,|\, \eta(-1) = 1),$$

so that it suffices to show that

$$\mu(\cdot \,|\, \eta(-2) = 1, \eta(-1) = 0) \leq \mu(\cdot \,|\, \eta(-1) = 1)$$

as measures on $\{0, 1\}^{Z_+}$. By Theorem 2.9 of Chapter II, it will be enough to show that

$$\frac{\mu\{\eta:\; \eta(-2) = 1,\; \eta(-1) = 0,\; \eta(u) = \zeta(u) \text{ for } u \in A \backslash \{x, y\},\; \eta(x) = 0,\; \eta(y) = 0\}}{\mu\{\eta:\; \eta(-2) = 1,\; \eta(-1) = 0,\; \eta(u) = \zeta(u) \text{ for } u \in A \backslash \{x, y\},\; \eta(x) = 0,\; \eta(y) = 1\}}$$

$$\geq \frac{\mu\{\eta:\; \eta(-1) = 1,\; \eta(u) = \zeta(u) \text{ for } u \in A \backslash \{x, y\},\; \eta(x) = 1,\; \eta(y) = 0\}}{\mu\{\eta:\; \eta(-1) = 1,\; \eta(u) = \zeta(u) \text{ for } u \in A \backslash \{x, y\},\; \eta(x) = 1,\; \eta(y) = 1\}},$$

whenever $x \neq y$, $x, y \in A = \{0, 1, \dots, n\}$, and $\zeta \in \{0, 1\}^{Z_+}$. By (1.16) and the monotonicity assumption on $f(n)/f(n+1)$, the left side above is greater than or equal to

$$\frac{\mu\{\eta:\; \eta(-1) = 1,\; \eta(u) = \zeta(u) \text{ for } u \in A \backslash \{x, y\},\; \eta(x) = 0,\; \eta(y) = 0\}}{\mu\{\eta:\; \eta(-1) = 1,\; \eta(u) = \zeta(u) \text{ for } u \in A \backslash \{x, y\},\; \eta(x) = 0,\; \eta(y) = 1\}}.$$

Then using (1.16) and the monotonicity of $f(n)/f(n+1)$ again, it follows that this in turn is greater than or equal to

$$\frac{\mu\{\eta:\; \eta(-1) = 1,\; \eta(u) = \zeta(u) \text{ for } u \in A \backslash \{x, y\},\; \eta(x) = 1,\; \eta(y) = 0\}}{\mu\{\eta:\; \eta(-1) = 1,\; \eta(u) = \zeta(u) \text{ for } u \in A \backslash \{x, y\},\; \eta(x) = 1,\; \eta(y) = 1\}}$$

as required. $\square$

The hardest part of the program is contained in the next result.

Lemma 1.25. *Suppose that* $\lambda \geq 2$, *and let* $f(n)$ *be the probability density on* $\{1, 2, \dots\}$ *with finite mean given by* $f(n) = F(n) - F(n+1)$ *where* F *is as in* (1.18). *Let* h *and* μ *be the corresponding function on* Y *and renewal measure defined by* (1.15) *and* (1.16). *Then* (1.14) *holds for all* $A \in Y$.

Proof. Fix $A \in Y$, and write $A = \bigcup_{i=1}^{k} A_i$, where $A_i = [l_i + 1, r_i - 1]$ are the ordered maximal connected components of A. Then $r_i \leq l_{i+1} < r_{i+1} - 1$ for all i. For $x \in Z^1$, define

$$R(x) = \mu\{\eta: \eta = 0 \text{ on } A \cap (x, \infty) | \eta(x) = 1\}, \quad \text{and}$$

$$L(x) = \mu\{\eta: \eta = 0 \text{ on } A \cap (-\infty, x) | \eta(x) = 1\}.$$

By (1.19), if $x \in A$, then

$$h(A) - h(A\backslash x) = L(x) R(x) \mu\{\eta: \eta(x) = 1\}.$$

Therefore (1.14) can be rewritten as

(1.26) $$\sum_{x \in A} L(x) R(x) \leq \lambda \sum_{i=1}^{k} [L(l_i) R(l_i) + L(r_i) R(r_i)].$$

The main difficulty in proving this inequality is that the left side involves L and R on A, while the right side involves L and R on A^c. We will obtain some identities which relate the values of these functions on A to those on A^c, and then use monotonicity properties of L and R on A^c which come from Lemma 1.24. To begin, write

$$\{\eta: \eta(x) = 1, \eta = 0 \text{ on } A \cap (x, \infty)\}$$
$$= \bigcup_{\substack{y > x \\ y \in A^c}} \{\eta: \eta(x) = \eta(y) = 1, \eta = 0 \text{ on } [A \cup (x, y)] \cap (x, \infty)\},$$

so that

(1.27) $$R(x) = \sum_{\substack{y > x \\ y \in A^c}} f(y - x) R(y).$$

By (1.20),

(1.28) $$2\lambda f(n) = \sum_{k=1}^{n-1} F(k) f(n - k) - F(n)$$

for $n \geq 2$. By (1.27) and (1.28),

(1.29) $$2\lambda R(l_i) = \sum_{\substack{y > l_i \\ y \in A^c}} R(y) \left[\sum_{k=1}^{y - l_i - 1} F(k) f(y - l_i - k) - F(y - l_i) \right].$$

By (1.27) again,

(1.30) $$\sum_{\substack{y > l_i \\ y \in A^c}} R(y) F(y - l_i) = \sum_{\substack{y > z > l_i \\ y, z \in A^c}} R(y) F(z - l_i) f(y - z).$$

Adding (1.29) and (1.30) and using (1.27) gives

(1.31)
$$2\lambda R(l_i) = \sum_{\substack{y>z>l_i \\ y \in A^c, z \in A}} R(y)F(z-l_i)f(y-z)$$

$$= \sum_{\substack{z>l_i \\ z \in A}} F(z-l_i)R(z).$$

Similarly, we have

(1.32)
$$2\lambda L(r_i) = \sum_{\substack{z<r_i \\ z \in A}} F(r_i-z)L(z).$$

Therefore, the right side of (1.26) can be written as

$$\frac{1}{2} \sum_{i=1}^{k} \left\{ L(l_i) \sum_{\substack{z>l_i \\ z \in A}} F(z-l_i)R(z) + R(r_i) \sum_{\substack{z<r_i \\ z \in A}} F(r_i-z)L(z) \right\}.$$

By (1.27), the left side of (1.26) can be written as

$$\frac{1}{2} \sum_{x \in A} \left\{ L(x) \sum_{\substack{z>x \\ z \in A^c}} f(z-x)R(z) + R(x) \sum_{\substack{z<x \\ z \in A^c}} f(x-z)L(z) \right\}.$$

Hence (1.26) will follow once we have shown that $i \leq j \leq m$ and $u \in A_j$ imply

$$\sum_{z=r_m}^{l_{m+1}} R(z)f(z-u) \leq R(r_m)F(r_m-u), \quad \text{and}$$

$$\sum_{z=r_{i-1}}^{l_i} L(z)f(u-z) \leq L(l_i)F(u-l_i).$$

For these, it is sufficient to check that

$$R(z) \leq R(r_m) \quad \text{for } r_m \leq z \leq l_{m+1}, \quad \text{and}$$

$$L(z) \leq L(l_i) \quad \text{for } r_{i-1} \leq z \leq l_i.$$

These inequalities follow from Lemma 1.24. The assumption there that $f(n)/f(n+1)$ is nonincreasing in n is verified by direct computation, using (1.18). $\square$

Theorem 1.33. (a) $\lambda_c \leq 2$.

(b)   *For $\lambda \geq 2$,*

$$\sigma(A) = \nu_\lambda\{\eta: \eta(x) = 1 \text{ for some } x \in A\}$$

$$\geq \mu\{\eta: \eta(x) = 1 \text{ for some } x \in A\},$$

where μ is the renewal measure corresponding to (1.18).

(c) $\rho(\lambda) \geq \frac{1}{2} + \sqrt{\frac{1}{4} - 1/2\lambda}$ *for $\lambda \geq 2$.*

Proof. Parts (a) and (b) are consequences of Lemmas 1.17 and 1.25, together with the discussion leading up to (1.14). Part (c) is a special case of part (b). The explicit right side of (c) comes from (1.23). $\square$

It is interesting to note that $\rho(2) \geq \frac{1}{2}$, so that even when $\lambda = 2$, the upper invariant measure has a substantial density. At this point, we known that $1.18 \leq \lambda_c \leq 2$. Further information about λ_c will be given following the discussion of edges in the next section.

Using an ingenious (but unfortunately nonrigorous) technique, Brower, Furman, and Moshe (1978) have obtained an approximation for λ_c. Their value is $\lambda_c \approx 1.6494$. In the same paper, they perform analogous computations which result in estimates for the critical value of the two-dimensional contact process, as well as for several critical exponents in one and two dimensions.

Since critical exponents will occur from time to time in this chapter and the next, we will take this opportunity to give a very brief description of this important idea. Suppose one is considering a one-parameter family of models of some type indexed by λ (such as the contact process or the stochastic Ising model). Suppose also that there is a critical value λ_c at which the behavior of the model changes abruptly (as is the case for the contact process in all dimensions and for the stochastic Ising model in dimensions greater than one). There will then usually be naturally defined functions $f(\lambda)$ defined for $\lambda > \lambda_c$ or $\lambda < \lambda_c$ which tend to zero or infinity as λ approaches λ_c. An example of this for the contact process is $\rho(\lambda)$ for $\lambda > \lambda_c$. (Provided that, as expected, the critical contact process dies out.) Then one would expect in many cases that

$$f(\lambda) \sim C|\lambda - \lambda_c|^\beta$$

or that

$$\log f(\lambda) \sim \beta \log|\lambda - \lambda_c|$$

as λ approaches λ_c. The number β which appears above would then be called the critical exponent for f. The importance of critical exponents is based largely on what is known as the universality principle, which plays an important role in mathematical physics. It asserts that while the value of λ_c will usually depend on the details of the definition of the model, the value of β will be the same for large classes of models, which are then called universality classes. This principle has been verified in only very few cases. It remains an important source of problems in this general area. Some critical exponents for reversible nearest-particle systems will be computed in Chapter VII.

2. Convergence Theorems

If $\lambda > \lambda_c$, all invariant measures for the one-dimensional contact process are convex combinations of δ_0 and ν_λ by Theorem 1.5. By Theorem 5.18 of Chapter III,

$$(2.1) \qquad \lim_{t \to \infty} \mu S(t) = \gamma \delta_0 + (1 - \gamma) \nu_\lambda,$$

where $\gamma = \mu\{\eta: \eta \equiv 0\}$, provided that $\mu \in \mathcal{S}$. This section is devoted to proving an analogue of (2.1) for arbitrary $\mu \in \mathcal{P}$. This will of course provide an alternate proof of those two results in this case. The techniques developed here will be useful also in the next section.

The main tool here will be the analysis of the edge processes corresponding to systems with one-sided initial configurations. The usefulness of these systems can be seen from the first result, which connects them up with the finite system starting from a singleton. The result depends heavily on the nearest-neighbor character of the interaction. Let η_t^+ and η_t^- be the versions of the contact process with initial states

$$\eta_0^+(x) = 1 \quad \text{if and only if } x \geq 0, \quad \text{and}$$

$$\eta_0^-(x) = 1 \quad \text{if and only if } x \leq 0.$$

Whenever several copies of the contact process are used in this section, they are to be constructed simultaneously on one probability space via the graphical representation described in Section 6 of Chapter III. Define

$$l_t = \min\{x: \eta_t^+(x) = 1\}, \quad \text{and}$$

$$r_t = \max\{x: \eta_t^-(x) = 1\},$$

so that $l_0 = r_0 = 0$.

Theorem 2.2. *Let A_t be the contact process with $A_0 = \{0\}$, and η_t be the contact process with $\eta_0 \equiv 1$. Then*

$$(2.3) \qquad \{A_t \neq \varnothing\} = \{l_s \leq r_s \text{ for all } s \leq t\}.$$

On this event,

(2.4) $\eta_t(x) = \eta_t^+(x) = \eta_t^-(x)$ *for all* $l_t \leq x \leq r_t,$ *and*

(2.5) $A_t = \{x: l_t \leq x \leq r_t, \eta_t(x) = 1\}.$

Proof. To prove the theorem, one simply checks that in the graphical representation, no transition can destroy the properties (2.3), (2.4), and (2.5). The three properties hold trivially at $t = 0$.   □

The main idea behind the next few results is that survival of the contact process should correspond to r_t drifting to $+\infty$ and l_t drifting to $-\infty$. Thus, we would like to understand the asymptotic behavior of r_t and l_t as $t \uparrow \infty$. By symmetry, the process l_t has the same law as the process $-r_t$. The analysis of r_t is based on the following general result, which is known as the Subadditive Ergodic Theorem.

Theorem 2.6. *Suppose* $\{X_{m,n}, m \leq n\}$ *are random variables which satisfy the following properties:*

(a) $X_{0,0} = 0,$ $X_{0,n} \leq X_{0,m} + X_{m,n}$ *for* $0 \leq m \leq n.$
(b) $\{X_{(n-1)k,nk}, n \geq 1\}$ *is a stationary process for each* $k \geq 1.$
(c) $\{X_{m,m+k}, k \geq 0\} = \{X_{m+1,m+k+1}, k \geq 0\}$ *in distribution for each* $m.$
(d) $EX_{0,1}^+ < \infty.$

Let $\alpha_n = EX_{0,n} < \infty,$ *which is well defined by* (a), (b), *and* (d). *Then*

(2.7) $$\alpha = \lim_{n \to \infty} \frac{\alpha_n}{n} = \inf_{n \geq 1} \frac{\alpha_n}{n} \in [-\infty, \infty), \quad and$$

(2.8) $$X_\infty = \lim_{n \to \infty} \frac{X_{0,n}}{n} \quad exists \ a.s., \ with \ -\infty \leq X_\infty < \infty.$$

Furthermore, $EX_\infty = \alpha.$ *If* $\alpha > -\infty,$ *then*

$$\lim_{n \to \infty} E \left| \frac{X_{0,n}}{n} - X_\infty \right| = 0.$$

If the stationary processes in (b) *are ergodic, then* $X_\infty = \alpha$ *a.s.*

Proof. The proof is broken up into several steps.
    Step 1. To prove (2.7), take expectations in (a) and use (c) to obtain

(2.9) $$\alpha_{m+n} \leq \alpha_m + \alpha_n.$$

Define α by

$$\alpha = \inf_{n \geq 1} \frac{\alpha_n}{n} \in [-\infty, \infty).$$

Fix an $m \geq 1$ and write $n = km + l$, where $0 \leq l < m$. By (2.9),

$$\alpha_n \leq k\alpha_m + \alpha_l.$$

As $n \to \infty$, $n/k \to m$, so that

$$\limsup_{n \to \infty} \frac{\alpha_n}{n} \leq \frac{\alpha_m}{m}.$$

Since m is arbitrary,

$$\limsup_{n \to \infty} \frac{\alpha_n}{n} \leq \alpha,$$

which completes the proof of Step 1, since

$$\liminf_{n \to \infty} \frac{\alpha_n}{n} \geq \alpha$$

is automatic.

Step 2. Let $\bar{X} = \limsup_{n \to \infty}(1/n)X_{0,n}$. Assume $\alpha > -\infty$, and fix $k \geq 1$. Using (a) repeatedly, write

(2.10) $$X_{0,kn} \leq \sum_{j=1}^{n} X_{k(j-1),kj}.$$

By the Ergodic Theorem (Theorem 4.9 of Chapter I) and (b)

(2.11) $$\frac{1}{n} \sum_{j-1}^{n} X_{k(j-1),kj}$$

converges a.s. and in L_1 to a random variable with mean α_k. Therefore

(2.12) $$E \limsup_{n \to \infty} \frac{X_{0,kn}}{kn} \leq \frac{\alpha_k}{k}.$$

By (a),

(2.13) $$X_{0,kn+j} \leq X_{0,kn} + X_{kn,kn+j}.$$

By (c), the distribution of $X_{kn,kn+j}$ depends only on j, so that using (d),

$$\limsup_{n \to \infty} \frac{X_{kn,kn+j}}{n} \leq 0 \quad \text{a.s.}$$

Therefore by (2.7), (2.12), and (2.13),

(2.14) $$E\bar{X} \leq \alpha.$$

If the processes in (b) are ergodic, then the a.s. limit of (2.11) is α_k, so that

(2.15) $$\bar{X} \leq \alpha \quad \text{a.s.}$$

Step 3. Let $\underline{X} = \lim \inf_{n \to \infty} (1/n) X_{0,n}$. Assume $\alpha > -\infty$. Let U_n be a random variable which is independent of the $X_{k,l}$'s and is uniformly distributed on $\{1, 2, \ldots, n\}$. Define

$$Y_k^n = X_{0,k+U_n} - X_{0,k+U_n-1}.$$

Then

(2.16)
$$\begin{aligned}
EY_k^n &= \frac{1}{n} \sum_{l=1}^{n} E[X_{0,k+l} - X_{0,k+l-1}] \\
&= \frac{1}{n} E[X_{0,k+n} - X_{0,k}] \\
&\geq \frac{n+k}{n} \alpha - \frac{\alpha_k}{n} > -\infty
\end{aligned}$$

by (2.7). Also,

$$\begin{aligned}
E(Y_k^n)^+ &= \frac{1}{n} \sum_{l=1}^{n} E[X_{0,k+l} - X_{0,k+l-1}]^+ \\
&\leq EX_{0,1}^+ < \infty
\end{aligned}$$

by (a), (c), and (d). Therefore

$$\sup_n E|Y_k^n| < \infty$$

for each $k \geq 0$. By the tightness which this provides, there is a subsequence n_i so that the joint distributions of $\{Y_k^{n_i}, k \geq 0\}$ converge to those of some collection $\{Y_k, k \geq 0\}$. (See, for example, Theorem 6.1 of Billingsley (1968).) As a result of the averaging which was done in introducing the random variables U_n, $\{Y_k, k \geq 0\}$ is a stationary process. By (a), (c), and (d), $\{(Y_1^n)^+, n \geq 1\}$ is uniformly integrable. So, by Fatou's Lemma and (2.16),

(2.17) $$EY_1 \geq \lim_{n \to \infty} \inf EY_1^n \geq \alpha.$$

Now, by (a)

$$(X_{0,k+1} - X_{0,k}, \ldots, X_{0,k+l} - X_{0,k}) \leq (X_{k,k+1}, \ldots, X_{k,k+l}).$$

By (c)

$$(X_{k,k+1}, \ldots, X_{k,k+l}) \quad \text{and} \quad (X_{0,1}, \ldots, X_{0,l})$$

have the same distribution. Therefore, if f is any bounded increasing continuous function on R^l,

$$Ef(X_{0,k+1} - X_{0,k}, \ldots, X_{0,k+l} - X_{0,k}) \leq Ef(X_{0,1}, \ldots, X_{0,l})$$

for all k. Hence

$$Ef(Y_0, Y_0 + Y_1, \ldots, Y_0 + Y_1 + \cdots + Y_{l-1}) \leq Ef(X_{0,1}, \ldots, X_{0,l}).$$

Therefore the distribution of

$$\lim_{n \to \infty} \frac{1}{n} \sum_{k=1}^{n} Y_k,$$

which exists a.s. and has mean EY_1 by the Ergodic Theorem, is stochastically smaller than the distribution of $\underline{X}$. By (2.17) then,

(2.18) $E\underline{X} \geq \alpha.$

Step 4. Here we will complete the proof of the theorem under the assumption that $\alpha > -\infty$. By (2.14) and (2.18), (2.8) is correct, since

$$\underline{X} \leq \bar{X} \quad \text{a.s.}$$

It also follows that $EX_\infty = \alpha$. For the L_1 convergence note that $\{X_{0,n}^+/n, \, n \geq 1\}$ is uniformly integrable by (b), (d), and (2.10) with $k = 1$. Therefore

$$\lim_{n \to \infty} E\left(\frac{X_{0,n}}{n} - X_\infty\right)^+ = 0.$$

But now use

$$E\left|\frac{X_{0,n}}{n} - X_\infty\right| = 2E\left(\frac{X_{0,n}}{n} - X_\infty\right)^+ - \left(E\frac{X_{0,n}}{n} - \alpha\right)$$

and (2.7). The final statement of the theorem follows from (2.15).

Step 5. Finally consider the case $\alpha = -\infty$. Truncate $X_{m,n}$ in the following manner:

$$X_{m,n}^N = \max(X_{m,n} - N(n-m)).$$

Then $X_{m,n}^N$ satisfies the hypotheses of the theorem, with

$$\alpha^N = \inf_{n \geq 1} \frac{EX_{0,n}^N}{n} \geq -N > -\infty.$$

Since

$$\frac{1}{n} X_{0,n}^N = \max \left(\frac{1}{n} X_{0,n}, -N \right),$$

the result follows from the version of the theorem for $\alpha > -\infty$. $\square$

Next we will apply Theorem 2.6 to the edge process for the contact process.

Theorem 2.19. *Let $\alpha_t = Er_t$. Then*

(a) $$\alpha = \lim_{t \to \infty} \frac{\alpha_t}{t} = \inf_{t > 0} \frac{\alpha_t}{t} \in [-\infty, \infty),$$

(b) $\lim_{t \to \infty} (r_t / t) = \alpha$ *a. s., and*
(c) *if $\alpha > -\infty$, then*

$$\lim_{t \to \infty} E \left| \frac{r_t}{t} - \alpha \right| = 0.$$

Proof. Using the graphical representation, define

$$r_{s,t} = \max \{ x \in Z^1 : \text{there is an active path from } (y, s) \text{ to}$$
$$(x, t) \text{ for some } y \leq r_s \} - r_s$$

for $0 \leq s \leq t$, and

$$N_t = \max \{ x \in Z^1 : \text{there is a path from } (y, 0) \text{ to } (x, t) \text{ for some}$$
$$y \leq 0, \text{ where } \delta\text{'s are ignored in the definition of the path} \}.$$

Then $r_{0,t} = r_t$, and $X_{m,n} = r_{m,n}$ satisfies (a), (b) and (c) of Theorem 2.6, by the independence and stationarity properties of the Poisson processes which are used in the graphical construction. In fact, the stationary processes in (b) consist of independent random variables, so they are clearly ergodic. To check assumption (d), note that

(2.20) $r_t \leq N_t$ a.s.,

and N_t is a Poisson process with rate λ. Thus to deduce this theorem from

Theorem 2.6, we need only make the transition from convergence along integer times to convergence along real times. But this is easily done by using the Borel–Cantelli Lemma and the bounds

$$P(\max_{n \le t \le n+1} (r_t - r_n) \ge \varepsilon n) \le P(N_1 \ge \varepsilon n), \quad \text{and}$$

$$P(\max_{n-1 \le t \le n} (r_n - r_t) \ge \varepsilon n) \le P(N_1 \ge \varepsilon n). \quad \square$$

Theorem 2.19 is an important first step in proving the basic convergence theorems for the contact process. In its application, however, we will need to know that $\alpha > 0$ if $\lambda > \lambda_c$, since we want to use the theorem to show that $r_t \to \infty$ a.s. in that case. The next results are intended to prove this positivity. From now on, write $\alpha = \alpha(\lambda)$ and $\alpha_t = \alpha_t(\lambda)$ to indicate the dependence on λ. For any $B \subset Z^1$ containing only finitely many positive points, let

$$r_t^B = \max\{x : \text{there is an active path from } (y, 0) \text{ to } (x, t) \text{ for some } y \in B\},$$

where $r_t^B = -\infty$ if there is no such path.

Lemma 2.21. *Suppose $B \subset A$, where A has only finitely many positive points, and let C be any finite set. Then*

(2.22) $$0 \le r_t^{A \cup C} - r_t^A \le r_t^{B \cup C} - r_t^B.$$

In particular, for $B \subset (-\infty, -1]$,

(2.23) $$E(r_t^{B \cup \{0\}} - r_t^B) \ge E(r_t^{(-\infty, 0]} - r_t^{(-\infty, -1]}) = 1.$$

Proof. From the definition, it is clear that

$$r_t^{A \cup D} = \max(r_t^A, r_t^D)$$

for any A, D. (This is a manifestation of the additivity property which was observed in Definition 6.1 of Chapter III.) Applying this to the pairs $\{A, C\}$, $\{A, B\}$, and $\{B, C\}$, we obtain

$$r_t^{A \cup C} - r_t^A = (r_t^C - r_t^A)^+$$

$$r_t^A \ge r_t^B, \quad \text{and}$$

$$(r_t^C - r_t^B)^+ = r_t^{B \cup C} - r_t^B.$$

The inequalities in (2.22) follow from these three relations, since the function

$z \to (r_t^C - z)^+$ is nonincreasing. The equality in (2.23) comes from the translation invariance of the system. □

Theorem 2.24. (a) $\alpha_t(\lambda + \delta) \geq \alpha_t(\lambda) + \delta t$ for all $t \geq 0$ and $\delta \geq 0$.
  (b) $\alpha(\lambda + \delta) \geq \alpha(\lambda) + \delta$ for all $\delta \geq 0$. (Here $-\infty + \delta = -\infty$.)

Proof. Part (b), clearly follows from part (a) and Theorem 2.19(a). To prove part (a), construct the contact processes with parameters λ and $\lambda + \delta$ on the same probability space by using a common graphical representation. This is done by regarding the Poisson flows with parameter $\lambda + \delta$ which are the basis for the construction as superpositions of independent Poisson flows with parameters λ and δ respectively. The superscripts 1 and 2 will denote quantities relating to the processes with parameters λ and $\lambda + \delta$ respectively. Clearly $r_s^1 \leq r_s^2$ for all s. Let

$$\tau = \inf\{s > 0: r_s^1 < r_s^2\}.$$

Define an intermediate process which uses all the arrows up to time τ, but only the arrows coming from Poisson flows with parameter λ after time τ. Use the superscript 3 to denote this process. Then $r_t^2 = r_t^1$ for $t < \tau$, and $r_t^2 \geq r_t^3$ for $t > \tau$. Therefore

$$(2.25) \qquad\qquad E(r_t^2 - r_t^1) \geq E(r_t^3 - r_t^1, \tau \leq t).$$

By the Strong Markov Property and (2.23) applied to the process with parameter λ and initial configuration η_τ^1, which lies below $\eta_\tau^2 = \eta_\tau^3$,

$$(2.26) \qquad\qquad E(r_t^3 - r_t^1, \tau \leq t) \geq P(\tau \leq t) = 1 - e^{-\delta t}.$$

Combining (2.25) and (2.26), we have

$$\alpha_t(\lambda + \delta) - \alpha_t(\lambda) \geq 1 - e^{-\delta t}.$$

The final step is to write

$$\alpha_t(\lambda + \delta) - \alpha_t(\lambda) = \sum_{k=1}^{n} \left[\alpha_t\left(\lambda + \frac{k\delta}{n}\right) - \alpha_t\left(\lambda + \frac{(k-1)\delta}{n}\right) \right]$$
$$\geq n(1 - e^{-\delta t/n}),$$

and then let $n \to \infty$. □

Theorem 2.27. (a) If $\alpha(\lambda) < 0$, then $\rho(\lambda) = 0$, so $\lambda \leq \lambda_c$.
  (b) $\alpha(\lambda) \geq \lambda - \lambda_c$ for $\lambda \geq \lambda_c$.
  (c) $r_t \to \infty$ and $l_t \to -\infty$ a.s. if $\lambda > \lambda_c$.
  (d) If $\lambda < \lambda_c$, then $\alpha(\lambda) < 0$.

Proof. Suppose $\alpha(\lambda) < 0$. Then $\lim_{t \to \infty} r_t = -\infty$ a.s. by Theorem 2.19. By symmetry, $\lim_{t \to \infty} l_t = +\infty$ a.s. Therefore

$$\lim_{t \to \infty} P(A_t \neq \varnothing) = 0$$

if $A_0 = \{0\}$ by (2.3). Finally, $\rho(\lambda) = 0$ by part (a) of Theorem 1.10. Turning to part (b), note that $\alpha(\lambda) \geq 0$ for $\lambda > \lambda_c$ by part (a). By part (a) of Theorem 2.19 and the fact that $\alpha_t(\lambda)$ is continuous in λ for fixed t, $\alpha(\lambda)$ is upper semicontinuous in λ. Therefore $\alpha(\lambda_c) \geq 0$ as well. Now (b) follows from part (b) of Theorem 2.24. Part (c) follows from part (b) and Theorem 2.19. For part (d), suppose first that $\alpha(\lambda) > 0$. Then $\lim_{t \to \infty} r_t = +\infty$ and $\lim_{t \to \infty} l_t = -\infty$ a.s. by Theorem 2.19. Therefore for sufficiently large N,

$$P[r_t < -N \text{ for some } t] = P[l_t > N \text{ for some } t] < \tfrac{1}{2},$$

so that

$$P[r_t^N \geq 0 \text{ and } l_t^N \leq 0 \text{ for all } t] > 0,$$

where r_t^N and l_t^N are the right and left edges for the processes with initial configurations $(-\infty, N]$ and $[-N, \infty)$ respectively. But then by the natural extension of Theorem 2.2 to the case in which $A_0 = \{0\}$ is replaced by $A_0 = \{-N, -N+1, \ldots, N\}$,

$$P^A(A_t \neq \varnothing \text{ for all } t) > 0.$$

Therefore $\alpha(\lambda) > 0$ implies that $\lambda \geq \lambda_c$, and hence $\lambda < \lambda_c$ implies that $\alpha(\lambda) \leq 0$. To complete the proof of (d), simply use part (b) of Theorem 2.24. $\square$

In Corollary 3.8, part (d) of Theorem 2.27 will be strengthened to say that $\alpha(\lambda) = -\infty$ for $\lambda < \lambda_c$. Part (b) of Theorem 2.27 implies that $\alpha(\lambda) > 0$ for $\lambda > \lambda_c$, and that $\alpha(\lambda_c) \geq 0$. This latter statement will be improved to $\alpha(\lambda_c) = 0$ in Corollary 3.20. In Theorem 3.36, we will show that $\alpha(\lambda)$ is continuous on $[\lambda_c, \infty)$. The information we have at this point is sufficient, however, to prove the main convergence theorem.

Theorem 2.28. *Suppose $\lambda > \lambda_c$. Then for any initial distribution μ,*

$$\lim_{t \to \infty} \mu S(t) = \gamma \delta_0 + (1 - \gamma) \nu_\lambda,$$

where $\gamma = \int P^\eta[\tau < \infty] \mu(d\eta)$ and τ is the hitting time of $\varnothing$ (which is identified with $\eta \equiv 0$).

Proof. By monotonicity and part (d) of Theorem 1.10, it suffices to show that for any $A \in Y$ and $f \in \mathcal{D}$,

$$(2.29) \qquad \lim_{t \to \infty} E^A f(A_t) = f(\varnothing) P^A(\tau < \infty) + P^A(\tau = \infty) \int f \, d\nu_\lambda.$$

We will prove this first in case $A = \{0\}$. Then

$$E^A f(A_t) = f(\varnothing) P^A(\tau \le t) + E^A[f(A_t), \tau > t].$$

Suppose that f depends on η only through coordinates $\eta(x)$ with $l \le x \le r$. By Theorem 2.2,

$$f(A_t) = f(\eta_t) \quad \text{a.s.}$$

on the event $\{\tau > t, \, l_t \le l, \, r_t \ge r\}$. By part (c) of Theorem 2.27, this event converges as $t \uparrow \infty$ to the event $\{\tau = \infty\}$. Therefore, we need to know that the conditional distribution of η_t (with $\eta_0 \equiv 1$) conditional on $\{\tau = \infty\}$ (where τ is defined in terms of A_t with $A_0 = \{0\}$) converges to ν_λ. To do so, take $s < t$ and $f \in \mathcal{M}$, and use the Markov property and the attractiveness of the contact process to see that

$$E[f(\eta_t), \tau > s] = E[E^{\eta_s} f(\eta_{t-s}), \tau > s]$$
$$\le Ef(\eta_{t-s}) P(\tau > s).$$

Similarly,

$$E[f(\eta_t), \tau \le s] \le Ef(\eta_{t-s}) P(\tau \le s).$$

Since the distribution of η_t converges to ν_λ, the last two inequalities imply respectively that

$$\limsup_{t \to \infty} E[f(\eta_t), \tau > s] \le \int f \, d\nu_\lambda P(\tau > s)$$

and

$$\limsup_{t \to \infty} E[f(\eta_t), \tau \le s] \le \int f \, d\nu_\lambda P(\tau \le s).$$

Since

$$\lim_{t \to \infty} \{ E[f(\eta_t), \tau > s] + E[f(\eta_t), \tau \le s] \} = \lim_{t \to \infty} Ef(\eta_t) = \int f \, d\nu_\lambda,$$

it then follows that

$$\lim_{t \to \infty} E[f(\eta_t), \tau > s] = \int f \, d\nu_\lambda P(\tau > s).$$

Since $\lim_{s \to \infty} P(\tau > s) = P(\tau = \infty)$ and

$$|E[f(\eta_t), \tau > s] - E[f(\eta_t), \tau = \infty]| \leq \|f\| P(s < \tau < \infty),$$

we can conclude from this that

$$\lim_{t \to \infty} E[f(\eta_t), \tau = \infty] = \int f \, d\nu_\lambda P(\tau = \infty)$$

as required. Having proved (2.29) for $A = \{0\}$, we now turn to its proof for general $A \in Y$. Since every member of $\mathcal{D}$ is a finite linear combination of elements of $\mathcal{M}$, we can take $f \in \mathcal{M}$. For $t > 0$, and $A \in Y$, let

$$g_t(A) = E^A[f(A_t), A_t \neq \varnothing], \quad \text{and}$$

$$g(A) = P^A(\tau = \infty) \int f \, d\nu_\lambda.$$

Then by the Markov property,

(2.30) $g_{t+s}(\{0\}) = E^{\{0\}} g_t(A_s), \quad \text{and}$

(2.31) $g(\{0\}) = E^{\{0\}} g(A_s).$

By the special case of the theorem which we have already proved,

(2.32) $\lim_{t \to \infty} g_t(\{0\}) = g(\{0\}).$

On the other hand, by the Markov property and the attractiveness of the contact process,

$$E^A[f(A_{t+s}), \tau \geq s] = E^A[E^{A_s} f(A_t), \tau \geq s]$$
$$\leq E^1 f(\eta_t) P^A(\tau \geq s),$$

so that

$$\limsup_{t \to \infty} g_t(A) \leq g(A).$$

Therefore

$$E^{\{0\}}|g_t(A_s) - g(A_s)| = 2E^{\{0\}}(g_t(A_s) - g(A_s))^+$$
$$- E^{\{0\}}(g_t(A_s) - g(A_s))$$

tends to zero as $t \to \infty$ by the Bounded Convergence Theorem, (2.30), (2.31), and (2.32). Hence

$$\lim_{t\to\infty} g_t(A) = g(A)$$

for any A such that $P^{\{0\}}(A_s = A) > 0$. This is true for all $A \in Y$ if $s > 0$, so the proof is complete. $\square$

Techniques similar to those used in the proof of Theorem 2.28 can be used to prove the following pointwise ergodic theorem.

Theorem 2.33. *Suppose* $\lambda > \lambda_c$. *For any initial configuration* η *and any* $f \in C(X)$,

$$\lim_{t\to\infty} \frac{1}{t} \int_0^t f(\eta_s)\, ds = \begin{cases} f(\varnothing) & \text{on } \{\tau < \infty\}, \\ \\ \int f\, d\nu_\lambda & \text{on } \{\tau = \infty\}, \end{cases}$$

with probability one relative to P^η.

Proof. First let η_t be the contact process with initial distribution ν_λ. Then η_t is a stationary process. By Theorem 2.28 and part (c) of Theorem 1.6,

$$\lim_{t\to\infty} Ef(\eta_t)g(\eta_0) = \lim_{t\to\infty} \int g(\eta)E^\eta f(\eta_t)\, d\nu_\lambda$$

$$= \int g\, d\nu_\lambda \int f\, d\nu_\lambda.$$

for any $f, g \in C(X)$. Therefore by Proposition 4.11 of Chapter I, η_t is a stationary ergodic process. By the Ergodic Theorem (Theorem 4.9 of Chapter I),

$$\lim_{t\to\infty} \frac{1}{t} \int_0^t f(\eta_s)\, ds = \int f\, d\nu_\lambda \quad \text{a.s.}$$

By Fubini's Theorem, it then follows that the conclusion of the theorem holds for almost every η with respect to ν_λ. The extension to arbitrary $\eta \in X$ follows the lines of the proof of Theorem 2.28, and will be omitted. □

By approximating the edge process by a more manageable system, it is possible to obtain improved lower bounds for λ_c. The idea is the following. Modify the one-sided system η_t^- defined at the beginning of this section by keeping all coordinates which lie strictly to left of $r_t - k$ identically equal to one, where k is a fixed nonnegative integer. This process can be constructed in such a way that it always lies above the original contact process. Therefore, if r_t^k is the position of the rightmost one in the modified process, it follows that

(2.34) $r_t^k \geq r_t.$

Now let $\alpha_k(\lambda) = \lim_{t\to\infty}(Er_t^k/t)$, and

$$\lambda_k = \inf\{\lambda > 0: \alpha_k(\lambda) > 0\}.$$

By (2.34) and part (b) of Theorem 2.27,

$$\lambda_k \leq \lambda_c$$

for all $k \geq 0$. For small k it is not difficult to compute $\alpha_k(\lambda)$ and hence λ_k explicitly. For $k = 0$, the modified process can be identified with a simple random walk on Z^1 which moves to the right one unit at rate λ and to the left one unit at rate 1. Thus $\alpha_0(\lambda) = \lambda - 1$ and $\lambda_0 = 1$. This is the lower bound which was obtained in Example 5.11 of Chapter III.

When $k = 1$, we must keep track of two types of configurations η_n and ζ_n:

$$\eta_n(x) = \begin{cases} 1 & \text{if } x \leq n, \\ 0 & \text{if } x > n, \end{cases} \quad \text{and}$$

$$\zeta_n(x) = \begin{cases} 1 & \text{if } x \leq n - 2 \text{ or } x = n, \\ 0 & \text{if } x = n - 1 \text{ or } x > n. \end{cases}$$

The possible transitions and corresponding rates for the process are then given by

Transition	Rate
$\eta_n \to \eta_{n+1}$	λ
$\eta_n \to \eta_{n-1}$	1
$\eta_n \to \zeta_n$	1
$\zeta_n \to \eta_n$	2λ
$\zeta_n \to \eta_{n-2}$	1
$\zeta_n \to \eta_{n+1}$	λ

Suppressing the subscript, we obtain a Markov chain with two states called η and ζ and rates

$$\eta \to \zeta \quad \text{at rate } 1,$$

$$\zeta \to \eta \quad \text{at rate } 1+3\lambda.$$

This chain has stationary measure

$$\pi(\eta) = \frac{1+3\lambda}{2+3\lambda}, \qquad \pi(\zeta) = \frac{1}{2+3\lambda}.$$

Therefore, returning to the role of n, we see that

$$\alpha_1(\lambda) = \frac{1+3\lambda}{2+3\lambda}(\lambda - 1) + \frac{1}{2+3\lambda}(\lambda - 2)$$

$$= \frac{3\lambda^2 - \lambda - 3}{2+3\lambda}.$$

Hence

$$\lambda_1 = \frac{1+\sqrt{37}}{6} \approx 1.18.$$

Ziezold and Grillenberger (1985) have used a computer to obtain the values of λ_k for $k \leq 14$. Their results are given in the following table:

k	2	4	6	8	10	12	14
λ_k	1.280	1.387	1.445	1.482	1.507	1.525	1.539

For much larger k, it is not practical to compute the value of λ_k. However they have run a simulation of the modified processes for $k = 100$ and $k = 1000$. The results of the simulation give rather convincing evidence that λ_{100} lies between 1.60 and 1.67, and that λ_{1000} lies between 1.62 and 1.67. These lower bounds are consistent with the estimate of $\lambda_c \approx 1.6494$ which was mentioned near the end of Section 1.

In addition to yielding lower bounds for λ_c, the above computations yield upper bounds for $\alpha(\lambda)$ by (2.34). The case $k = 0$ gives

$$\alpha(\lambda) \leq \lambda - 1,$$

while the case $k = 1$ gives

$$\alpha(\lambda) \leq \frac{3\lambda^2 - \lambda - 3}{2+3\lambda} = \lambda - 1 - \frac{1}{2+3\lambda}.$$

By combining Theorems 1.33 and 2.27(b), we have the following lower bound for $\alpha(\lambda)$:

$$\alpha(\lambda) \geq \lambda - 2.$$

3. Rates of Convergence

We now know, among other things, that if $\lambda < \lambda_c$, then

(3.1) $$\lim_{t \to \infty} P^A(A_t \neq \varnothing) = 0 \quad \text{for all } A \in Y, \quad \text{and}$$

(3.2) $$\lim_{t \to \infty} \mu S(t) = \delta_0 \quad \text{for all } \mu \in \mathscr{P},$$

while if $\lambda > \lambda_c$,

(3.3) $$\lim_{t \to \infty} \delta_1 S(t) = \nu_\lambda.$$

The purpose of this section is to prove that the convergence in these limiting statements is exponentially rapid. As we do so, we will obtain other exponential estimates as well, and will even get some information about the critical contact process, in which $\lambda = \lambda_c$.

We will begin with the subcritical case, since that is the simplest one. The first result gives exponential rates for (3.1) and (3.2).

Theorem 3.4. *Suppose* $\lambda < \lambda_c$. *Then there are positive constants K and ε depending only on λ so that*

$$P^A(A_t \neq \varnothing) \leq K|A| \, e^{-\varepsilon t} \quad and$$

$$\mu S(t)\{\eta \colon \eta(x) = 1 \text{ for some } x \in A\} \leq K|A| \, e^{-\varepsilon t}$$

for all $A \in Y$, $\mu \in \mathscr{P}$ and $t \geq 0$.

Proof. The two statements are equivalent by (1.8). By looking at the second statement (or by looking at the first statement and using the additivity of the contact process), it is clear that it suffices to prove that

(3.5) $$P^{\{0\}}(A_s \neq \varnothing) \leq K e^{-\varepsilon s}.$$

By part (d) of Theorem 2.27, $\alpha(\lambda) < 0$. Hence by part (a) of Theorem 2.19, $\alpha_t(\lambda) < 0$ for some $t > 0$. Using the notation in the proof of that theorem, recall that

$$\{r_{(n-1)t,nt}, n \geq 1\}$$

are independent and identically distributed random variables with mean $\alpha_t(\lambda)$. Furthermore,

(3.6) $$r_{nt} = r_{0,nt} \leq \sum_{k=1}^{n} r_{(k-1)t,kt}.$$

Since $\alpha_t(\lambda) < 0$ and (2.20) holds,

$$\gamma = E \, e^{\theta r_t} < 1$$

for sufficiently small positive θ. Therefore by (3.6),

$$P(r_{nt} \geq 0) \leq E \, e^{\theta r_{nt}} \leq \gamma^n \quad \text{for all } n \geq 1.$$

By Theorem 2.2,

$$\begin{aligned}
P^{\{0\}}(A_{nt} \neq \varnothing) &= P(l_s \leq r_s \text{ for all } s \leq nt) \\
&\leq P(l_{nt} \leq r_{nt}) \\
&\leq P(l_{nt} \leq 0) + P(r_{nt} \geq 0) \\
&\leq 2\gamma^n.
\end{aligned}$$

Since $P^{\{0\}}(A_s \neq \varnothing)$ is monotone in s and $\gamma < 1$, it follows that (3.5) holds with $K = 2\gamma^{-1}$ and $\varepsilon = t^{-1} \log \gamma^{-1}$. $\square$

Corollary 3.7. *Suppose* $\lambda < \lambda_c$. *For any* $\eta \in X$ *and* $x \in S$,

$$P^\eta[\eta_t(x) = 0 \text{ for all sufficiently large } t] = 1.$$

Proof. By Theorem 5.2 of Chapter I,

$$M_t = f(\eta_t) - \int_0^t \Omega f(\eta_s) \, ds$$

is a martingale for each $f \in \mathcal{D}(\Omega)$. Apply this to $f(\eta) = \eta(x)$ for a fixed $x \in Z^1$. By Theorem 3.4, $E^\eta |M_t|$ is bounded in t. Therefore by the Martingale Convergence Theorem, $\lim_{t \to \infty} M_t$ exists a.s. By Theorem 3.4 again,

$$\lim_{t \to \infty} \int_0^t \Omega f(\eta_s) \, ds = \int_0^\infty \Omega f(\eta_s) \, ds$$

is finite a.s., so it follows that $\lim_{t\to\infty} f(\eta_t)$ exists a.s. The only possible limit is of course zero. $\square$

Corollary 3.8. *Suppose* $\lambda < \lambda_c$. *Then there are positive constants* K *and* ε *so that*

$$(3.9) \qquad\qquad P(r_t > -e^{\varepsilon t}) \le K e^{-\varepsilon t} \quad \text{for all } t \ge 0.$$

In particular, $\alpha(\lambda) = -\infty$.

Proof. Write the initial configuration $(-\infty, 0]$ which is used in defining r_t as

$$(-\infty, -2e^{\varepsilon t}] \cup (-2e^{\varepsilon t}, 0].$$

By the additivity property and translation invariance of the contact process, it follows that

$$P(r_t > -e^{\varepsilon t}) \le P(r_t > e^{\varepsilon t}) + P^{(-2e^{\varepsilon t}, 0]}(A_t \ne \varnothing).$$

By (2.20),

$$P(r_t > -e^{\varepsilon t}) \le P(N_t > e^{\varepsilon t}) + 2\bar K e^{\varepsilon t} e^{-\bar\varepsilon t},$$

where $\bar K$ and $\bar\varepsilon$ are the constants in Theorem 3.4. By Chebyshev's inequality,

$$P(N_t > e^{\varepsilon t}) \le \lambda t (\lambda t + 1) e^{-2\varepsilon t}.$$

Combining the last two inequalities gives (3.9) with $\varepsilon = \bar\varepsilon/2$ and appropriate K. $\square$

Not very much is known about the critical contact process, in which $\lambda = \lambda_c$. One fact which we will prove later in this section is that $\alpha(\lambda_c) = 0$. However, it is not even known whether or not the critical contact process dies out. A discussion of this question is given in the formulation of Problem 2. The next result says that if it does die out, it cannot do so very rapidly. This is in sharp contrast with the subcritical case discussed in Theorem 3.4 and Corollary 3.7.

Theorem 3.10. *Suppose* $\lambda = \lambda_c$. *Then*

$$(3.11) \qquad\qquad \lim_{t\to\infty} t P^1[\eta_t(x) = 1] = \infty.$$

Furthermore,

$$P^1[\eta_t(x) = 1 \text{ for arbitrarily large } t] = 1.$$

Here 1 *denotes the configuration* $\eta \equiv 1$.

Proof. The proof is similar to that of part (d) of Theorem 2.27. By part (b) of that theorem, $\alpha(\lambda_c) \geq 0$. Therefore by Theorem 2.19,

$$\lim_{t \to \infty} \frac{r_t}{t} \geq 0 \quad \text{a.s.}$$

So, if $\varepsilon > 0$, there is a large N so that

$$P[r_t < -\varepsilon t - N \text{ for some } t] = P[l_t > \varepsilon t + N \text{ for some } t] \leq \tfrac{1}{3}.$$

Therefore using the construction in Theorem 2.2 as before, we see that

(3.12) $\qquad\qquad\qquad P^{[-N-k, N+k]}[A_{(k/\varepsilon)} \neq \varnothing] \geq \tfrac{1}{3}$

for all integers $k \geq 1$. By the additivity property of the contact process and translation invariance,

(3.13)
$$P^{[-N-k, N+k]}[A_t \neq \varnothing] \leq \sum_{x=-N-k}^{N+k} P^{\{x\}}[A_t \neq \varnothing]$$
$$= [2(N+k)+1]P^{\{0\}}[A_t \neq \varnothing].$$

Combining (3.12) and (3.13) gives

$$P^{\{0\}}[A_{(k/\varepsilon)} \neq \varnothing] \geq \frac{1}{6(N+k)+3},$$

so that

$$P^{\{0\}}[A_t \neq \varnothing] \geq \frac{1}{6(N+[\varepsilon t])+3},$$

and hence

$$\liminf_{t \to \infty} t P^{\{0\}}[A_t \neq \varnothing] \geq \frac{1}{6\varepsilon}.$$

Since ε is arbitrary, the first statement in the theorem follows from this and (1.8). For the second statement, let

$$T_t = E^1 \int_0^t \eta_s(0) \, ds.$$

By (3.11),

(3.14) $\qquad\qquad\qquad T_t \uparrow \infty \quad \text{as} \quad t \uparrow \infty.$

Let

$$\sigma_t = \min\{s \ge t: \eta_s(0) = 1\}.$$

Then by the Strong Markov Property,

$$T_t - T_s = E^1 \int_s^t \eta_r(0) \, dr$$

$$= E^1 \left[\int_{\sigma_s}^t \eta_r(0) \, dr, \, \sigma_s \le t \right]$$

$$\le T_t P^1(\sigma_s \le t).$$

Therefore

$$P^1(\sigma_s \le t) \ge 1 - \frac{T_s}{T_t},$$

so $P^1(\sigma_s < \infty) = 1$ for each $s > 0$ by (3.14). This proves the second part of the theorem. $\square$

The remainder of this section is devoted to the proof that $\alpha(\lambda_c) = 0$, and to the derivation of exponential estimates in the supercritical case $\lambda > \lambda_c$. These facts will all follow from a construction which relates the contact process to one-dependent oriented percolation. The graphical representation for the contact process plays an essential role in this construction.
    Let

$$I = \{(j, k) \in Z^2: k \ge 0 \text{ and } j + k \text{ is even}\}.$$

For $(j, k) \in I$, put $\|(j, k)\| = \frac{1}{2}(|j| + |k|)$. A one-dependent oriented percolation model on I with parameter p is a collection of random variables $U_{j,k}$ indexed by I which satisfy the following conditions:

(a) $P(U_{j,k} = 1) = p$ and $P(U_{j,k} = 0) = 1 - p$ for all $(j, k) \in I$, and
(b) if $J \subset I$ contains no nearest-neighbor pairs (relative to the norm defined above), then $\{U_{j,k}: (j, k) \in J\}$ are mutually independent.

A path in I is a sequence $(j_1, k_1), \ldots, (j_n, k_n)$ from I which satisfies $k_{l+1} = k_l + 1$ and $j_{l+1} = j_l \pm 1$ for all $1 \le l \le n - 1$. The path is said to be active if $U_{j_l, k_l} = 1$ for each $1 \le l \le n$. Percolation from $(0, 0)$ is said to occur if there is an infinite active path starting at $(0, 0)$.
    To carry out the construction, fix $0 < \beta < \alpha/3$ and $M > 0$ so that $M\beta/2$ and $M\alpha$ are integers. For $(j, k) \in I$, define parallelograms in $Z^1 \times [0, \infty)$ by

$$L_{j,k} = \left\{ (x, t) \in Z^1 \times \left[kM, M\left(k + 1 + \frac{\beta}{\alpha}\right) \right] : \right.$$

$$\left. 0 \le x + \alpha t - M\left(j\alpha + k\alpha - j\beta + \frac{\beta}{2} \right) \le \beta M \right\}, \quad \text{and}$$

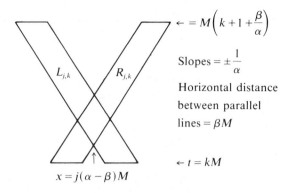

Figure 1

$$R_{j,k} = \left\{ (x, t) \in Z^1 \times \left[kM, M\left(k+1+\frac{\beta}{\alpha}\right) \right] : \right.$$

$$\left. -\beta M \leq x - \alpha t - M\left(j\alpha - k\alpha - j\beta - \frac{\beta}{2}\right) \leq 0 \right\}.$$

These are shown in Figure 1, and will be useful in establishing the connection between the contact process and the percolation model.

Let $\mathcal{E}_{j,k}$ be the event that there is an active path in the graphical representation of the contact process which goes from the bottom edge of $R_{j,k}$ to the top edge, always staying entirely within $R_{j,k}$, and also that there is an active path from the bottom edge of $L_{j,k}$ to the top edge, always staying entirely within $L_{j,k}$. Note that $P(\mathcal{E}_{j,k})$ does not depend on (j, k).

Lemma 3.15. *Suppose $J \subset I$ contains no nearest-neighbor pairs. Then the events*

$$\{\mathcal{E}_{j,k} : (j, k) \in J\}$$

are mutually independent.

Proof. This follows immediately from the fact that $\mathcal{E}_{j,k}$ depends only on the events in the Poisson processes used in the graphical representation which lie in the rectangle

$$(M\alpha(j-1) - M\beta(j+1), M\alpha(j+1) - M\beta(j-1)) \times (kM, M(k+2)).$$

These are disjoint for $(j, k) \in J$ since $\beta < \alpha/3$. $\square$

Lemma 3.16. *Consider the percolation model in which $U_{j,k}$ is the indicator of the event $\mathcal{E}_{j,k}$. Suppose that $(j_1, k_1), \ldots, (j_n, k_n)$ is an active path in I. Then* *there is an active path in the graphical representation from the bottom edge of R_{j_1,k_1} to the top edge of R_{j_n,k_n}.*

Proof. The proof consists of checking that the short active paths in the graphical representation whose existence is guaranteed by the occurrence of the events $\mathscr{E}_{j_l,k_l}$ link together in such a way that they form the desired long active paths. This is most easily seen from Figure 2, which shows how several copies of Figure 1 fit together in the plane. Pictured there are the paths whose existence is guaranteed by the occurrence of $\mathscr{E}_{j,k}$ and $\mathscr{E}_{j+1,k+1}$. □

We now know that if we let $U_{j,k}$ be the indicator of $\mathscr{E}_{j,k}$, then they form an oriented one-dependent percolation model with parameter $p = P(\mathscr{E}_{j,k})$. Furthermore, by Lemma 3.16, if percolation from $(0,0)$ occurs in the percolation model, then the contact process with initial configuration $[-2\beta M, 2\beta M]$ does not die out. The next step is to show that p can be taken arbitrarily close to one simply by taking M large (provided that $\lambda > \lambda_c$). One of the very nice aspects of the present construction is that problems concerning the contact process even near its critical value are transformed into percolation problems for p near one. There is no need to study percolation near its critical value.

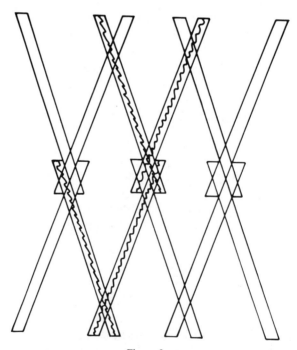

Figure 2

Lemma 3.17. *Suppose that* $\alpha(\lambda) = \alpha$ *and* $0 < \beta < \alpha/3$. *Then*

$$\lim_{M \to \infty} P(\mathscr{E}_{j,k}) = 1.$$

Proof. Let r_t^N be the right edge process for the contact process with initial configuration η^N defined by

$$\eta^N(x) = \begin{cases} 1 & \text{if } x \le N, \\ 0 & \text{if } x > N. \end{cases}$$

Let $\mathscr{E}$ be the event that there is an active path which goes from the bottom edge of $R_{0,0}$ to the top edge, always staying entirely within $R_{0,0}$. For $-3M\beta/2 \le n \le M\alpha - M\beta/2$, let F_n be the event that there is an active path from $(n, (n+3\beta M/2)/\alpha)$ to $(m, (M\alpha + M\beta)/\alpha)$ for some $m \ge M\alpha - M\beta/2 + N$, where $0 < N < M\beta/2$. Then $\mathscr{E}$ contains the intersection of the events

$$\left\{ r_t^{-M\beta/2 - N} < \alpha t - \frac{\beta}{2} M \text{ for all } t \le t_0 \right\},$$

(3.18)
$$\left\{ r_{t_0}^{-M\beta/2 - N} > M\alpha - \frac{M\beta}{2} + N \right\}, \quad \text{and}$$

$$\left\{ \bigcup_{-3M\beta/2 \le n \le M\alpha - M\beta/2} F_n \right\}^c,$$

where $t_0 = M(1 + \beta/\alpha)$. Let M and N tend to ∞ in such a way that $N \sim M\beta/3$. Then the probability of each of the first two events in (3.18) tends to one by Theorem 2.19(b) and translation invariance. To show that the probability of the third event in (3.18) tends to one as well, it is necessary to use exponential bounds on r_t as in the proof of Theorem 3.4. First write

$$P(F_n) = P\left(r_{(M\alpha - n - \beta M/2)/\alpha} \ge M\alpha - \frac{M\beta}{2} + N - n \right).$$

If $\theta > 0$, (3.6) gives

$$E(e^{\theta r_{kt}}) \le [E(e^{\theta r_t})]^k.$$

The last two relations and Chebyshev's inequality then give the following upper bound for $P(F_n)$:

$$\exp\left[-\theta\left(M\alpha - \frac{M\beta}{2} + N - n \right) \right] \{ E(e^{\theta r_t}) \}^{(M\alpha + t\alpha - n - \beta M/2)/t\alpha}$$

for any $t > 0$. Fix $\varepsilon > 0$ so that $\varepsilon(1 + \beta/\alpha) < \beta/3$. Choose t so large that $E r_t < t(\alpha + \varepsilon)$, which is possible by Theorem 2.19(a). Then choose θ small and positive so that

$$E(e^{\theta r_t}) < e^{\theta t(\alpha + \varepsilon)},$$

which is possible by the choice of t and by (2.20). Using this, we get the following upper bound for $P(F_n)$:

$$\exp\left[\theta \alpha t - \theta N + \theta \varepsilon M + \theta \varepsilon t - \frac{\theta \varepsilon n}{\alpha} - \frac{\theta \varepsilon \beta M/2}{\alpha}\right].$$

Summing the geometric series yields the following upper bound for $\sum_{n=-3M\beta/2}^{\infty} P(F_n)$:

$$\frac{\exp[\theta \alpha t - \theta N + \theta \varepsilon M + \theta \varepsilon t + \theta \varepsilon M \beta/\alpha]}{1 - \exp(-\theta \varepsilon/\alpha)}.$$

By the choice of ε and the fact that $N \sim M\beta/3$, the limit of this expression is zero as $M \to \infty$. This proves that the probability of the third event in (3.18) tends to one also. Therefore

$$\lim_{M \to \infty} P(\mathscr{E}) = 1.$$

The statement of the lemma follows from this and the corresponding statement in which $\mathscr{E}$ is defined in terms of $L_{0,0}$ instead of $R_{0,0}$. $\square$

Next we will see that if p is sufficiently close to one, then there is percolation with positive probability.

Theorem 3.19. *For an oriented one-dependent percolation model with parameter* $p \geq 1 - 3^{-54}$, *the probability of percolation from* $(0, 0)$ *is at least* $17/18$.

Proof. Let

$$W = \{(j, k) \in I: \text{ there is an active path from } (0, 0) \text{ to } (j, k)\}.$$

For $(j, k) \in I$, let

$$D(j, k) = \{y \in R^2: \|(j, k) - y\| \leq \tfrac{1}{2}\},$$

which is a diamond centered at (j, k). Put

$$V = \bigcup_{(j,k) \in W} D(j, k),$$

so that V is obtained by putting diamonds around those points which can be reached from $(0, 0)$ via an active path. If W is a finite set, so that percolation does not occur from $(0, 0)$, let Γ be the boundary of the infinite component of V^c. Now let γ be any possible value of Γ of length $2n$ (using the metric $\|\cdot\|$, so that the edge of a $D(j, k)$ has length 1). Each segment of length one in γ is a boundary segment for exactly one $D(j, k)$ with $(j, k) \in W$. Say that it is an upper segment if that $D(j, k)$ lies below it, and a lower segment otherwise. Clearly γ consists of n upper and n lower segments. An upper segment must have values of $U_{(j,k)}$ on either side of it as given in Figure 3. The site above it at which there is a zero can be shared by at most two of these upper segments of γ. Therefore there are at least $n/2$ such sites. Since each point in I has at most eight neighbors, at least $n/18$ of the sites identified above at which there is a zero correspond to independent random variables. Therefore

$$P(\Gamma = \gamma) \le (1-p)^{n/18}.$$

The number of γ's of length $2n$ is at most 3^{2n-1}, since the right lower boundary of $D(0, 0)$ is always in γ, and then γ can be continued at each stage in at most three ways. Therefore

$$P(W \text{ is finite}) \le \sum_{\gamma} P(\Gamma = \gamma)$$

$$\le \sum_{n=2}^{\infty} 3^{2n-1}(1-p)^{n/18}$$

$$= \frac{27(1-p)^{1/9}}{1-9(1-p)^{1/18}},$$

which is at most $\frac{1}{18}$ if $p \ge 1-3^{-54}$. $\square$

Corollary 3.20. *For the contact process,* $\alpha(\lambda_c) = 0$.

Proof. By Theorem 2.27, $\alpha(\lambda_c) \ge 0$. Therefore it suffices to show that $\alpha(\lambda) > 0$ implies $\lambda > \lambda_c$. So, suppose $\alpha(\lambda_0) > 0$, and choose $\alpha = \alpha(\lambda_0)$ in the definition of $\mathscr{E}_{j,k}$ which precedes Lemma 3.15. By Lemma 3.17, M can be chosen so large that $P(\mathscr{E}_{j,k}) > 1-3^{-54}$, where P refers to the contact process with parameter λ_0. For that fixed α and M, $P(\mathscr{E}_{j,k})$ is a continuous function of

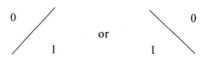

Figure 3

λ, since $\mathscr{E}_{j,k}$ involves only finitely many Poisson processes run for a finite amount of time. Therefore there is a $\lambda_1 < \lambda_0$ so that $P(\mathscr{E}_{j,k}) > 1 - 3^{-54}$ for the contact process with parameter λ_1. By Theorem 3.19 and Lemma 3.16, the contact process with parameter λ_1 and initial configuration $[-\alpha M, \alpha M]$ has positive probability of not dying out. Therefore $\lambda_1 \geq \lambda_c$, so that $\lambda_0 > \lambda_c$. $\square$

The next result gives an exponential estimate for the analogue of the right edge process in the percolation context. It will then be used to obtain similar estimates for the right edge of the supercritical contact process. For $n \geq 1$, let

$$\rho_n = \max\{j: \text{there is an active path in the percolation}$$
$$\text{model from } (m, 0) \text{ to } (j, n) \text{ for some } m \leq 0\}.$$

Theorem 3.21. *Suppose $q < 1$. For an oriented one-dependent percolation model with parameter $p > 1 - 3^{-72/(1-q)}$,*

$$P(\rho_n < nq) \leq 3^{-n+1}$$

for all $n \geq 1$.

Proof. The proof is similar to that of Theorem 3.19. This time let

$$W = \{(j, k) \in I: \text{there is an active path from } (m, 0) \text{ to}$$
$$(j, k) \text{ for some } m \leq 0\}, \quad \text{and}$$

$$V = \bigcup_{(j,k) \in W} D(j, k).$$

Let Γ be the boundary of the infinite component of $V^c \cap ((-\infty, \infty) \times [0, n])$. (Note that there is only one infinite component by Theorem 3.19.) Now let γ be any possible value of Γ of length $n + 2m$. We will refine the classification of the boundary segments of γ by saying that one is upper right, upper left, lower right or lower left according to how it is placed relative to the $D(j, k)$ with $(j, k) \in W$ of which it is a boundary. This is shown in Figure 4. Let

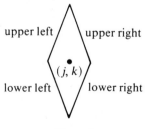

Figure 4

N_{ul}, N_{ur}, N_{ll}, and N_{lr} be the number of segments of each type in γ. Then

$$N_{ul} + N_{ur} + N_{ll} + N_{lr} = n + 2m, \quad \text{and}$$

$$N_{lr} + N_{ll} - N_{ul} - N_{ur} = \rho_n.$$

Taking the difference of these two expressions gives

$$N_{ul} + N_{ur} = \frac{n + 2m - \rho_n}{2}.$$

Therefore $\rho_n < nq$ implies that

$$N_{ul} + N_{ur} > \frac{n(1-q) + 2m}{2}.$$

As in the proof of Theorem 3.19, it follows that

$$P(\Gamma = \gamma) \leq (1 - p)^{\langle [n(1-q) + 2m]/36 \rangle}.$$

Since there are at most 3^{n+2m} such γ's, it follows that

$$P(\rho_n < nq) \leq \sum_{m=0}^{\infty} 3^{n+2m} [(1-p)^{1/36}]^{n(1-q) + 2m}$$

$$\leq 3^{-n+1}$$

provided that $p > 1 - 3^{-72/(1-q)}$. $\square$

Corollary 3.22. *For $\lambda > \lambda_c$ and $a < \alpha(\lambda)$,*

$$\lim_{t \to \infty} \frac{1}{t} \log P(r_t < at)$$

exists and is strictly negative.

Proof. Define $r_{s,t}$ for $0 < s < t$ as in the proof of Theorem 2.19. As observed there, $r_{s,t}$ is independent of r_s and has the same distribution as r_{t-s}. Furthermore, $r_t \leq r_s + r_{s,t}$ so that

$$P(r_t < at) \geq P(r_s < as \text{ and } r_{s,t} < a(t-s))$$

$$= P(r_s < as) P(r_{t-s} < a(t-s)).$$

So, by the argument used in Step 1 of the proof of Theorem 2.6,

$$\lim_{t \to \infty} \frac{1}{t} \log P(r_t < at) = \sup_{t > 0} \frac{1}{t} \log P(r_t < at).$$

It remains to show that the limit is strictly negative. Consider the percolation model with $U_{j,k}$ = indicator of $\mathscr{E}_{j,k}$, where $\alpha = \alpha(\lambda)$, $\beta \leq (\alpha - a)/2$, and M is so large that

$$p = P(\mathscr{E}_{j,k}) > 1 - 3^{-72/(1-q)},$$

where $1 > q > 2a/(\alpha + a)$. This is possible by Lemma 3.17. By Lemma 3.16,

$$r_{nM} \geq \rho_n(\alpha - \beta)M - 2\beta M.$$

Therefore for sufficiently large n,

$$P(r_{nM} < anM) < P(\rho_n < nq).$$

Hence by Theorem 3.21,

$$\liminf_{t \to \infty} \frac{1}{t} \log P(r_t < at) \leq -\frac{\log 3}{M}.$$

This completes the proof. $\square$

Corollary 3.22 has a number of immediate consequences. The first gives exponential convergence in (3.3), and hence provides a supercritical analogue of Theorem 3.4. One consequence of this exponential convergence is that ν_λ has exponentially decaying correlations. (See Theorem 4.20 of Chapter I.) As mentioned in Section 4 of Chapter I, this implies that under ν_λ, the random variables $\{\eta(x), x \in Z^1\}$ satisfy classical central limit behavior (see Malysev (1975)). Again let τ be the hitting time of $\varnothing$.

Theorem 3.23. *Suppose $\lambda > \lambda_c$. Then there are positive constants K and ε depending only on λ so that for all $t \geq 0$*

$$(3.24) \qquad\qquad P^A(t < \tau < \infty) \leq K|A| e^{-\varepsilon t}$$

for all $A \in Y$, and

$$(3.25) \qquad\qquad \left| S(t)f(1) - \int f \, d\nu_\lambda \right| \leq K e^{-\varepsilon t} \|f\|$$

for all $f \in D(X)$. Here 1 denotes the configuration $\eta \equiv 1$, and $\|\cdot\|$ is defined in Section 3 of Chapter I.

Proof. By (1.8), (3.24) and (3.25) are identical if

$$f(\eta) = \begin{cases} 1 & \text{if } \eta(x) = 1 \text{ for some } x \in A, \\ 0 & \text{otherwise.} \end{cases}$$

Furthermore, since $\delta_1 S(t) \geq \nu_\lambda$ for all t, (3.25) for general $f \in D(X)$ follows from it for $f(\eta) = \eta(0)$. (See the proof of Corollary 2.8 of Chapter II for an argument of this type.) Therefore it suffices to prove (3.24) for $A = \{0\}$. By Theorem 2.2,

$$P^{\{0\}}(t < \tau < \infty) = P(l_s \leq r_s \text{ for all } s \leq t \text{ and } l_s > r_s \text{ for some } s > t)$$

(3.26)
$$\leq P(l_s > r_s \text{ for some } s > t)$$

$$\leq 2P(r_s < 0 \text{ for some } s > t).$$

In what follows, K and ε are positive constants whose values will change from line to line. Now, by Corollary 3.22,

$$P\left(r_n < \frac{\alpha(\lambda)}{2} n\right) \leq K e^{-\varepsilon n}.$$

Therefore

(3.27)
$$P\left(r_m < \frac{\alpha(\lambda)}{2} m \text{ for some } m \geq n\right) \leq K e^{-\varepsilon n}.$$

By (2.20),

(3.28) $\quad P\left(r_t < 0 \text{ for some } t \in [m, m+1] \text{ and } r_{m+1} > \frac{\alpha(\lambda)}{2}(m+1)\right) \leq K e^{-\varepsilon m}.$

Combing (3.26), (3.27), and (3.28) gives (3.24) for $A = \{0\}$ as required. $\quad\square$

Theorem 3.29. *Suppose $\lambda > \lambda_c$. Then there are positive constants K and ε depending only on λ so that*

(3.30)
$$P^A(\tau < \infty) \leq K e^{-\varepsilon |A|}.$$

Here $|A|$ is the cardinality of A.

Proof. By Theorem 2.2 and (2.20).

(3.31)
$$P^{\{0\}}\left(\sup_{t>0} \frac{|A_t|}{t} < \infty\right) = 1.$$

Fix $M > 0$. By the Markov Property and part (a) of Theorem 1.9,

$$P^{\{0\}}(t < \tau < \infty) = E^{\{0\}}[P^{A_t}(\tau < \infty), A_t \neq \varnothing]$$

$$\geq P^{[-Mt, Mt]}(\tau < \infty) P^{\{0\}}(\varnothing \neq A_t \subset [-Mt, Mt]).$$

Therefore by (3.31) and Theorem 3.23,

$$P^{[-Mt, Mt]}(\tau < \infty) \le K e^{-\varepsilon t}$$

for some positive M, K, and ε. Therefore

(3.32) $$P^{\{1,2,\ldots,n\}}(\tau < \infty) \le K e^{-\varepsilon n}$$

for some (different) positive K and ε. This is enough to give (3.30) by part (c) of Theorem 1.9. $\square$

It is interesting to note that while the full percolation construction was needed to prove (3.30) for all $\lambda > \lambda_c$, it is easy to get it for $\lambda > 2$ as a consequence of Theorem 1.33. To see this, note that by part (b) of that theorem,

$$P^{\{1,2,\ldots,n\}}(\tau < \infty) \le \mu\{\eta: \eta(x) = 0 \text{ for all } 1 \le x \le n\}$$

$$= \sum_{k=n}^{\infty} \mu\{\eta: \eta(0) = 1, \eta(x) = 0 \text{ for all } 1 \le x \le k\}$$

$$= \mu\{\eta: \eta(0) = 1\} \sum_{k=n}^{\infty} F(k+1)$$

$$= \frac{1}{\lambda - \sqrt{\lambda^2 - 2\lambda}} \sum_{k=n}^{\infty} \frac{(2k)!}{k!(k+1)!} \frac{1}{(2\lambda)^k}$$

$$\le \left(\frac{2}{\lambda}\right)^n \frac{1}{\lambda - \sqrt{\lambda^2 - 2\lambda}} \sum_{k=0}^{\infty} \frac{(2k)!}{k!(k+1)!} \frac{1}{4^k}$$

$$= \frac{2}{\lambda - \sqrt{\lambda^2 - 2\lambda}} \left(\frac{2}{\lambda}\right)^n,$$

where μ and F are given in (1.16) and (1.18) respectively. This approach does not appear to give other exponential bounds such as (3.24).

Another application of our exponential estimates is to give a Strong Law for the cardinality of the finite contact process.

Theorem 3.33. *Suppose $\lambda > \lambda_c$ and $A_0 = \{0\}$. Then*

$$\lim_{t \to \infty} \frac{|A_t|}{t} = 2\alpha(\lambda)\rho(\lambda) \quad a.s. \text{ on } \{\tau = \infty\},$$

where $\rho(\lambda)$ is defined in (1.2) and $\alpha(\lambda)$ is as in the statement of Theorem 2.19.

Proof. By Theorem 2.2,

$$|A_t| = \sum_{x=l_t}^{r_t} \eta_t(x) \quad \text{on } \{\tau = \infty\},$$

where $\eta_t(x)$ is the contact process with $\eta_0 \equiv 1$. Therefore by Theorem 2.19, it suffices to prove that

$$\frac{1}{t} \sum_{|x| \le \alpha t} \eta_t(x) \to 2\alpha\rho(\lambda) \quad \text{a.s.}$$

for $\alpha = \alpha(\lambda)$. Let

$$Z_t = \sum_{|x| \le \alpha t} [\eta_t(x) - E\eta_t(x)].$$

Since $E\eta_t(x) \to \rho(\lambda)$ as $t \to \infty$, it then suffices to prove that $Z_t/t \to 0$ a.s. This will be proved by a fourth moment argument. By Theorem 4.20 of Chapter I and Theorem 3.23 of this chapter, there are positive constants K and δ so that whenever $x_1 \le x_2 \le x_3 \le x_4$,

$$\left| E \prod_{i=1}^{4} [\eta_t(x_i) - E\eta_t(x_i)] \right| \le K \exp[-\delta \max(x_2 - x_1, x_4 - x_3)].$$

The number of pairs (x_1, x_4) such that

$$-\alpha t \le x_1 \le x_4 \le \alpha t$$

is at most $(2\alpha t + 1)^2$. For fixed x_1, x_4, the number of pairs (x_2, x_3) such that $x_1 \le x_2 \le x_3 \le x_4$ and $\max(x_2 - x_1, x_4 - x_3) = m$ is at most $2m$. Therefore

(3.34) $$EZ_t^4 \le 24(2\alpha t + 1)^2 \sum_{m=1}^{\infty} (2m)K e^{-\delta m} \le ct^2$$

for some constant c. By Chebyshev's inequality and (3.34),

$$P\left(\frac{|Z_t|}{t} > \varepsilon \right) \le \frac{c}{t^2 \varepsilon^4},$$

so it follows that

(3.35) $$\lim_{n \to \infty} \frac{Z_{n^{3/4}}}{n^{3/4}} = 0 \quad \text{a.s.}$$

To fill in between successive values of $n^{3/4}$, note that

$P(Z_t > Z_{n^{3/4}} + \varepsilon n^{3/4}$ for some $t \in [n^{3/4}, (n+1)^{3/4}])$

$$\leq P\left(\sum_{|x| \leq \alpha n^{3/4}} [n_t(x) - \eta_{n^{3/4}}(x)] > \varepsilon n^{3/4} - 2[\alpha t] + 2[\alpha n^{3/4}] \right.$$

$$\left. \text{for some } t \in [n^{3/4}, (n+1)^{3/4}] \right)$$

$$\leq P(U_n > u_n),$$

where U_n has a binomial distribution with parameters $N_n = 2[\alpha n^{3/4}] + 1$ and $p_n = 1 - \exp[-2\lambda\{(n+1)^{3/4} - n^{3/4}\}]$, and $u_n = \varepsilon n^{3/4} - 2[\alpha(n+1)^{3/4}] + 2[\alpha n^{3/4}]$. To estimate $P(U_n > u_n)$, write

$$P(U_n > u_n) \leq 2^{-u_n} E 2^{U_n}$$

$$= 2^{-u_n} (1 + p_n)^{N_n}$$

$$\leq 2^{-u_n} e^{p_n N_n},$$

which is summable in n. Using a similar argument on the down side, we see that

$$\sum_{n=1}^{\infty} P\left(\frac{|Z_t - Z_{n^{3/4}}|}{n^{3/4}} > \varepsilon \text{ for some } t \in [n^{3/4}, (n+1)^{3/4}] \right) < \infty.$$

Combining this with (3.35) gives

$$\frac{Z_t}{t} \to 0 \quad \text{a.s.}$$

as required.   □

The ideas used in the proofs of Corollaries 3.20 and 3.22 can be combined to show that $\alpha(\lambda)$ is continuous on $[\lambda_c, \infty)$.

Theorem 3.36. $\alpha(\lambda)$ *is continuous on* $[\lambda_c, \infty)$.

Proof. By Theorem 2.24, $\alpha(\lambda)$ is strictly increasing on $[\lambda_c, \infty)$. As was observed in the proof of Theorem 2.27, $\alpha(\lambda)$ is upper semicontinuous. Therefore $\alpha(\lambda)$ is right continuous on $[\lambda_c, \infty)$. To show left continuity at $\lambda_0 > \lambda_c$, we proceed as in the proof of Corollary 3.22. Let $\alpha = \alpha(\lambda_0)$. For $a < \alpha$, choose $0 < \beta \leq (\alpha - a)/2$ and then M so large that for the contact process with parameter λ_0,

(3.37) $p = P(\mathscr{E}_{j,k}) > 1 - 3^{-72/(1-q)},$

where $1 > q > 2a/(\alpha + a)$, which is possible by Lemma 3.17. Since $P(\mathscr{E}_{j,k})$ is continuous in λ for fixed M and β, we can choose $\lambda_1 < \lambda_0$ so that (3.37) holds for the contact process with parameter λ_1 also. For that process, Lemma 3.16 gives

$$r_{nM} \geq \rho_n(\alpha - \beta)M - 2\beta M,$$

so that

$$\alpha(\lambda_1) \geq (\alpha - \beta)q$$

by Theorem 3.21. Since β can be made arbitrarily small and q can be made arbitrarily close to one, it follows that

$$\lim_{\lambda \uparrow \lambda_0} \alpha(\lambda) = \alpha(\lambda_0)$$

as required. $\square$

4. Higher Dimensions

The theory of the contact process on Z^d for $d \geq 2$ is much less complete than it is in one dimension. This section summarizes the current status of that theory. We begin by discussing the critical value λ_d of the d-dimensional contact process. It is defined in the statement of Theorem 2.14 of Chapter III.

Theorem 4.1. (a) *Let A_t be the (finite) d-dimensional contact process with parameter λ, and B_t be the (finite) one-dimensional contact process with parameter λd. Then*

(4.2) $$P^{\{0\}}(A_t \neq \varnothing) \geq P^{\{0\}}(B_t \neq \varnothing)$$

for all $t \geq 0$.
   (b) $d\lambda_d \leq \lambda_1$ *for all $d \geq 1$.*

Proof. Part (b) follows immediately from part (a), together with the fact that the contact process is self-dual in any number of dimensions. (Recall Example 4.18 of Chapter III.) To prove part (a), define $\pi_d : Z^d \to Z^1$ by

$$\pi_d(x_1, \ldots, x_d) = x_1 + \cdots + x_d.$$

For finite $A \subset Z^d$, let

$$\pi_d(A) = \{\pi_d(x): x \in A\} \subset Z^1.$$

The idea of the proof is to couple together A_t and B_t in such a way that

$$(4.3) \qquad\qquad\qquad B_t \subset \pi_d(A_t)$$

for all $t \geq 0$. Inequality (4.2) follows immediately from the existence of such a coupling. Suppose A and B are finite subsets of Z^d and Z^1 respectively, and that $B \subset \pi_d(A)$. Then for each $y \in B$, there is an $x = (x_1, \ldots, x_d) \in A$ so that $y = \pi_d(x)$. Associate a death at x with a death at y. Also associate births at $y-1$ in the B_t process (if $y-1 \notin B$) due to the presence of a particle at y with births in the A_t process at the d points $(x_1 - 1, x_2, \ldots, x_d), \ldots, (x_1, x_2, \ldots, x_d - 1)$. This is consistent with the birth rates of the processes A_t and B_t, since the parameter of the latter is d times the parameter of the former. Note also that π_d maps each of those d points onto $y-1$, so that these births preserve the relation (4.3). With analogous rules for births at $y+1$, this coupling has all the desired properties. (Of course, points x in A_t which are not needed in the above correspondence generate births independently of the B_t process.) $\square$

Corollary 4.4.

$$\frac{1}{2d-1} \leq \lambda_d \leq \frac{2}{d}.$$

Proof. The lower bound comes from Example 5.11 of Chapter III. The upper bound is the result of combining Theorems 1.33 and 4.1. $\square$

This corollary gives the best known upper bound in dimensions one and two. Better upper bounds are available for $d \geq 3$. They are obtained by comparing the contact process with other processes which take on positive real values at each site. These comparisons will be carried out in Section 6 of Chapter IX. To state these bounds, let ρ_d be the probability that the simple symmetric random walk on Z^d starting at 0 never returns to 0. Comparison with the smoothing process yields

$$(4.5) \qquad\qquad\qquad \lambda_d \leq \rho_d^{-1} - 1,$$

while comparison with the binary contact path process yields

$$(4.6) \qquad\qquad\qquad \lambda_d \leq [2d(2\rho_d - 1)]^{-1}$$

for $d \geq 3$. Since ρ_d is asymptotically equal to $1 - 1/2d$ as $d \to \infty$, either (4.5) or (4.6) combines with the lower bound in Corollary 4.4 to give

$$(4.7) \qquad\qquad\qquad \lim_{d \to \infty} d\lambda_d = \tfrac{1}{2}.$$

This result should be quite intuitive. The idea is that as d tends to ∞, the (finite) contact process should behave more and more like a branching process in which individuals die at rate one and produce offspring at rate $2d\lambda$ (i.e., at rate λ for each of the $2d$ neighbors). Such a branching process survives with positive probability if and only if $2d\lambda > 1$.

There is another method for obtaining an upper bound for λ_2 based on the known critical value of the two-dimensional stochastic Ising model. The idea is to compare the contact process with the stochastic Ising model using Theorem 1.9 of Chapter III. If one uses the second version of the stochastic Ising model in Example 4.3(a) of Chapter I for example, one finds that if it is nonergodic for a given β, then the contact process with

$$\lambda = \frac{(e^{4\beta} - 1)(e^{8\beta^*} + 1)}{8(e^{4\beta} + 1)}$$

is also nonergodic. Using the (deep) fact that $\beta_c = \frac{1}{2}\log(1 + \sqrt{2})$ in two dimensions, the resulting upper bound for λ_2 is approximately 2.8. This is of course not as good as the bound in Corollary 4.4.

Next we turn to a discussion of the class of invariant measures and convergence results. As in one dimension, let

$$\nu_\lambda = \lim_{t \to \infty} \delta_1 S(t)$$

be the upper invariant measure for the d-dimensional contact process. Ideally, one would like to have an analogue of Theorem 2.28 in higher dimensions, but such a result is not yet available. We do have as an immediate consequence of Theorem 5.18 of Chapter III, the following result for translation invariant initial distributions.

Theorem 4.8. *Suppose $\mu \in \mathscr{S}$. Then*

$$\lim_{t \to \infty} \mu S(t) = \gamma \delta_0 + (1 - \gamma)\nu_\lambda,$$

where $\gamma = \mu\{\eta: \eta \equiv 0\}$. In particular,

$$(\mathscr{I} \cap \mathscr{S})_e = \{\delta_0, \nu_\lambda\}.$$

Griffeath (1978c) observed that Harris' (1976) proof of Theorem 4.8 could be modified to show that

$$\lim_{t \to \infty} \mu S(t) = \nu_\lambda$$

for any initial distribution μ which concentrates on dense configurations.

For this purpose, a configuration $\eta \in X$ is said to be dense if there is an N so that every ball in Z^d of radius N contains an x so that $\eta(x) = 1$.

This is about all that is known for λ down to the critical value in dimensions greater than one. More is known for large λ. Harris (1978) proved that if λ is sufficiently large, then

$$P^A\left(\inf_{t>0}\frac{|A_t|}{t} > 0 \,\middle|\, A_t \neq \varnothing \text{ for all } t\right) = 1$$

for all $A \in Y$. More recently, Durrett and Griffeath (1982) proved a shape result for a class of growth models which generalizes Richardson's (1973) Theorem. This is a multidimensional analogue of Theorem 2.19. They then used the exponential estimates for the one-dimensional contact process which were proved in Section 3 to verify that the d-dimensional contact process satisfies the hypotheses of their shape result whenever $\lambda > \lambda_1$. From this they deduced versions of Theorems 2.28, 2.33 and 3.33 in d dimensions for $\lambda > \lambda_1$. These results remain unproved for $\lambda_d < \lambda \leq \lambda_1$. (See Open Problems 9 and 10.)

5. Notes and References

Section 1. Contact processes were first introduced and studied by Harris (1974). In that paper he obtained the lower bound $\lambda_c \geq 1.18$ in one dimension, and showed that λ_c is finite. He did not give an explicit upper bound for λ_c, and in fact, any upper bound obtained using his technique of proof would be rather large. Theorem 1.33 was proved by Holley and Liggett (1978). The proof given here of Lemma 1.24 is different from the one given there. While the technique used in proving this theorem gives a rather good upper bound for λ_c, its usefulness is limited to a very small class of systems including the contact process. A more generally applicable approach was developed by Gray and Griffeath (1982) in proving Theorem 3.32 of Chapter III. When applied to the contact process, it yields the result that $\lambda_c < 7$. Duality for the contact process was discovered independently by Holley and Liggett (1975) and Harris (1976). Theorem 1.5 was proved by Liggett (1978a). The weaker statement that $(\mathscr{I} \cap \mathscr{S})_e = \{\delta_0, \nu_\lambda\}$ was proved earlier by Harris (1976). Part (c) of Corollary 1.6 is Theorem 11 in Griffeath (1981). Its proof is attributed there to Gray. That paper by Griffeath is an excellent survey of contact processes, which was very useful in writing this chapter.

In the high-energy physics context, the contact process was introduced by Grassberger and de la Torre (1979). It was shown there to be equivalent to the reggeon spin model, which was introduced by Brower, Furman, and Subbarao (1977) and by Amati, Le Bellac, Marchesini, and Ciafaloni (1976). The reggeon spin model is a discretization of reggeon field theory. For a review of that subject, see Moshe (1978).

Section 2. Theorem 2.6 is an extension of Kingman's Subadditive Ergodic
Theorem. Good expositions of Kingman's theory can be found in Kingman
(1973) and Kingman (1976). The conclusions of his theorem are the same
as those of Theorem 2.6. His hypotheses are stronger however, in that (a)
is replaced by

$$X_{m,n} \le X_{m,k} + X_{k,n} \quad \text{for } m \le k \le n$$

and (b) and (c) are both replaced by

$$\{X_{m,n}, 0 \le m \le n\} = \{X_{m+1,n+1}, 0 \le m \le n\} \quad \text{in distribution.}$$

The extension is necessary since the process r_t satisfies neither of these
stronger hypotheses. Theorems 2.19 and 2.24, and their application to
Theorems 2.27, 2.28, and 2.33 are due to Durrett (1980). In fact, Theorem
2.6 and its proof, which appeared in Liggett (1985), are based on Durrett's
proof of Theorem 2.19. Other extensions of Kingman's Theorem have been
given by Derriennic (1983).

Prior to Durrett's paper, a number of weaker results of the type proved
in this section were obtained by others. Harris (1978) proved that for
sufficiently large λ (and in any number of dimensions),

$$P^A\left[\inf_{t>0} \frac{|A_t|}{t} > 0 \middle| \tau = \infty\right] = 1.$$

Furthermore, he proved a version of Theorem 2.33 for a large class of initial
η and for sufficiently large λ. Griffeath (1978c) proved Theorem 2.28 for λ
larger than the critical value of the one-sided contact process. Then in
(1979b), he proved Theorem 2.33 for all initial η, but with the same
restriction on λ. Part (d) of Theorem 2.27 is Theorem 4 from Griffeath (1981).

Section 3. Theorem 3.4 and its corollaries, and Theorem 3.10 come from
Griffeath (1981). (In the latter result, Griffeath proved only that the limit
in (3.11) is positive. The improved statement is due to Durrett.) All of the
results which follow them in this section are due to Durrett and Griffeath
(1983). Their paper is based to some extent on ideas developed by Russo
(1978) in his work on percolation. The construction given here which relates
the contact process to oriented one-dependent percolation incorporates a
number of improvements which are due to Gray. The use of percolation
techniques to study the contact process was pioneered by Harris (1974,
1978). See also Toom (1968). Oriented two-dimensional percolation can be
viewed as a discrete time version of the one-dimensional contact process.
Results in that context which are analogous to those presented in Sections
2 and 3 are given in Durrett (1984).

Section 4. Theorem 4.1 and its application to Corollary 4.4 are due to Holley and Liggett (1978). The bound in (4.5) is due to Holley and Liggett (1981), and the one in (4.6) is due to Griffeath (1983).

6. Open Problems

1. Find more effective ways to compute upper and lower bounds for λ_c in one dimension. Lower bounds are the easiest, and can be obtained by combining ingenuity and large computations. See the bounds due to Ziezold and Grillenberger which are quoted at the end of Section 2, for example. Excellent lower bounds have been obtained in the discrete time context, which are discussed in Durrett (1984). The real challenge is to improve the upper bound $\lambda_c \leq 2$. So far, no method is available which would accomplish this.

2. Determine whether or not the critical contact process in one dimension dies out. By Theorem 1.9, the answer is the same whether dying out is viewed as the statement that $\delta_1 S(t)$ converges weakly to δ_0 as $t \to \infty$ or the statement that $P^A(\tau < \infty) = 1$ for all $A \in Y$. This is probably the most important open problem for the one-dimensional contact process. Most workers in the field believe that the critical contact process does die out. This belief is based to some extent on analogies with other systems such as critical branching processes (see, for example, Theorem 1 in Chapter I of Athreya and Ney (1972)) and finite reversible nearest particle systems (see Theorem 1.10 and Problem 9 of Chapter VII). Another argument which suggests that the critical contact process dies out is based on the fact that $\alpha(\lambda_c) = 0$. (See Corollary 3.20). The idea is that when $\lambda = \lambda_c$, r_t and l_t should behave like random walks with no drift. If one could prove that

$$P[r_t < l_t \text{ for some } t] = 1,$$

which would be true if they were such random walks (and were independent), then the critical contact process would die out by Theorem 2.2.

3. Is $\rho(\lambda)$ as defined in (1.2) concave on $[\lambda_c, \infty)$? This is not true for some attractive reversible nearest-particle systems. See the latter part of Section 4 of Chapter VII for examples. Of course a major difficulty is proving the desired concavity is that ρ is clearly not concave on all of $[0, \infty)$.

4. Is the right derivative of $\rho(\lambda)$ at λ_c (which would exist if the answer to Problem 3 is affirmative) finite or infinite? If it is infinite (as is likely), is it the case that $\rho(\lambda) - \rho(\lambda_c)$ behaves like a constant multiple of $(\lambda - \lambda_c)^\beta$ for some $\beta \in (0, 1)$ as $\lambda \downarrow \lambda_c$? If so, this would be an example of a critical exponent. A brief definition of critical exponents is given at the end of

Section 1. See Durrett (1985) for a discussion of critical exponents in the percolation context. Brower, Furman, and Moshe (1978) have given (nonrigorous) computations of a number of critical exponents for the contact process.

5. As a consequence of Theorem 3.4,

$$\int_0^\infty P^{\{0\}}(A_t \neq \varnothing)\, dt < \infty$$

for $\lambda < \lambda_c$. Using also (2.20), it is not hard to show that for $\lambda < \lambda_c$,

$$\int_0^\infty E^{\{0\}}|A_t|\, dt < \infty.$$

How do these quantities behave as $\lambda \uparrow \lambda_c$? Answers to this would give two critical exponents for the finite process. Partial answers may be obtainable using the techniques discussed in Durrett (1985). In the case of reversible nearest-particle systems, the first of these quantities equals $\lambda_c(\lambda_c - \lambda)^{-1}$ and the second equals $\lambda_c^2(\lambda_c - \lambda)^{-2}$, as we will see in Chapter VII.

6. Theorem 2.19 gives a Strong Law for the edge process for $\lambda \geq \lambda_c$. Is there an associated Central Limit Theorem, at least for $\lambda > \lambda_c$?

7. Is $\alpha(\lambda)$ concave on $[\lambda_c, \infty)$? To explain the reason for conjecturing this, suppose that

$$D(\lambda) = \lim_{t \to \infty} E[r_t - \max\{x < r_t : \eta_t^-(x) = 1\}]$$

exists for $\lambda > \lambda_c$. Then one would expect that

$$\alpha(\lambda) = \lambda - D(\lambda)$$

would hold. It would follow from Corollary 3.20 that $D(\lambda_c) = \lambda_c$ and from Theorem 2.24(b) that $D(\lambda)$ is nonincreasing. Furthermore, $\lim_{\lambda \to \infty} D(\lambda) = 1$, so one would expect that $D(\lambda)$ is convex on $[\lambda_c, \infty)$. This would imply the concavity of $\alpha(\lambda)$.

8. Theorem 3.10 gives a lower bound for $P^1[\eta_t(x) = 1]$ when $\lambda = \lambda_c$. What is the asymptotic behavior of this quantity as $t \to \infty$?

9. Prove analogoues of Theorems 1.5, 2.28, 2.33, and 3.33 for all $\lambda > \lambda_d$ in d dimensions. In particular, show that the d-dimensional contact process has no invariant measures which are not translation invariant. One approach to doing this is to verify the assumptions of the Durrett–Griffeath (1982) shape theorem for the contact process with arbitrary $\lambda > \lambda_d$. (See the discussion at the end of Section 4.)

10. Consider a natural version of the contact process on $S = \{-k, -k+1, \ldots, k-1, k\}^{d-1} \times Z^1$. Does the critical value for this process converge as $k \to \infty$ to λ_d? A result of this type would probably be useful in carrying out the program described in Problem 9.

Chapter VII

Nearest-Particle Systems

Nearest-particle systems are one-dimensional spin systems in which the flip rates depend on η in a certain way which we will now describe. Configurations $\eta \in X = \{0, 1\}^{Z^1}$ will be given an occupancy interpretation: $\eta(x) = 1$ means that there is a particle at x, and $\eta(x) = 0$ means that x is vacant. For $x \in Z^1$ and $\eta \in X$, let $l_x(\eta)$ and $r_x(\eta)$ be the distances from x to the nearest particle to the left and right respectively:

$$l_x(\eta) = x - \max\{y < x: \eta(y) = 1\}, \quad \text{and}$$

$$r_x(\eta) = \min\{y > x: \eta(y) = 1\} - x.$$

If $\eta(y) = 0$ for all $y < x$, set $l_x(\eta) = \infty$, and if $\eta(y) = 0$ for all $y > x$, set $r_x(\eta) = \infty$.

Suppose $\{\beta(l, r), 1 \le l, r \le \infty\}$ is a collection of nonnegative numbers which satisfies

$$\sup_{l,r} \beta(l, r) < \infty, \qquad \beta(l, r) = \beta(r, l),$$

$$\beta(1, \infty) = \beta(\infty, 1) > 0, \quad \text{and} \quad \beta(\infty, \infty) = 0.$$

The flip rates for the nearest-particle system are then given by

$$c(x, \eta) = \begin{cases} \beta(l_x(\eta), r_x(\eta)) & \text{if } \eta(x) = 0, \\ 1 & \text{if } \eta(x) = 1. \end{cases}$$

The death rates are taken to be one for convenience only. Much of the analysis given in this chapter would apply equally well to systems in which the death rates are also allowed to depend on η through $l_x(\eta)$ and $r_x(\eta)$, provided that these rates are bounded above and below by positive constants.

The one-dimensional contact process is the special case in which

(0.1) $$\beta(l, r) = \begin{cases} 2\lambda & \text{if } l = r = 1, \\ \lambda & \text{if } l = 1 \text{ or } r = 1 \text{ and } l + r > 2, \\ 0 & \text{if } l \ge 2 \text{ and } r \ge 2. \end{cases}$$

This connection with the contact process is one motivation for the study of nearest-particle systems. While the contact process is not reversible, many other nearest-particle systems are. One class of examples which turn out to be reversible is given by

(0.2)
$$\beta(l, r) = c\left(\frac{1}{l} + \frac{1}{r}\right)^p$$

for $l, r < \infty$, and $\beta(n, \infty) = \beta(\infty, n) = cn^{-p}$, where $c > 0$ and $p > 0$ are constants. Additional tools become available in the reversible case, so we will be able to solve some problems in this context which are open in general. These solutions provide insight into what might be expected to be true for nonreversible systems such as the contact process. On the other hand, since nearest-particle systems do not in general satisfy nearest neighbor or finite range assumptions, we will not be able to do as well here on certain problems as we did for the contact process.

Just as in the cases of the contact process and the voter model, we will study both finite and infinite versions of nearest-particle systems. The first two sections deal with the finite version, in which $\sum_x \eta_t(x) < \infty$. In this case we will regard η_t as being a Markov chain on the collection Y of all finite subsets of Z^1 through the identification

$$A_t = \{x \in Z^1 : \eta_t(x) = 1\}.$$

In order to guarantee that A_t remains in Y at all times, we will assume in these two sections that

(0.3)
$$\sum_{n=1}^{\infty} \beta(n, \infty) < \infty.$$

This of course forces $p > 1$ in example (0.2). The assumption $\beta(\infty, \infty) = 0$ implies that $\varnothing$ is absorbing for A_t. The main problem of interest is to determine whether A_t survives in the sense that

(0.4)
$$P^A(A_t \neq \varnothing \text{ for all } t \geq 0) > 0$$

for all $A \neq \varnothing$. This problem will be completely solved for reversible systems in Section 1. The solution in the case of example (0.2) turns out to be that A_t survives if and only if

$$c \sum_{n=1}^{\infty} n^{-p} > 1.$$

The analysis in this section will also enable us to compute some critical exponents for A_t. The theory in the nonreversible case is much less complete, and is presented in Section 2.

In Section 3, we treat the construction problem for infinite nearest-particle systems. It is important to do this because the approach to the existence and uniqueness question which was presented in Chapter I would force us to impose unnatural and unnecesary assumptions on the birth rates $\beta(l, r)$. For example, (0.3) of Chapter III is not satisfied by the examples in (0.2) above for any choice of $c > 0$ and $p > 0$. Another example which would be ruled out is

$$(0.5) \qquad\qquad \beta(l, r) = \frac{b}{l+r-1}.$$

This example is interesting because it has the interpretation of spreading out uniformly a total birth rate of b over the sites in each maximal connected subset of $\{x: \eta(x) = 0\}$.

In Section 4, we will determine exactly which infinite nearest-particle systems have nontrivial reversible invariant measures. It turns out that these reversible measures are always stationary renewal processes. This connection with renewal theory provided the initial motivation for the study of nearest-particle systems. As corollaries of these results, we will be able to determine in example (0.2) exactly which choices of c and p lead to processes which die out in the sense that

$$(0.6) \qquad\qquad \lim_{t\to\infty} \delta_1 S(t) = \delta_0.$$

Section 5 is devoted to some partial solutions to the problem of determining which infinite nonreversible nearest-particle systems die out.

The first two and the next three sections are almost completely independent of each other. Within each group, the sections should be read in the order in which they appear.

1. Reversible Finite Systems

Let $q(\cdot, \cdot)$ be the Q matrix for the finite nearest-particle system A_t with birth rates determined by $\beta(\cdot, \cdot)$. Then

$$q(A, A\backslash\{x\}) = 1 \quad \text{for } x \in A, \quad \text{and}$$

$$q(A, A \cup \{x\}) = \beta(l_x(A), r_x(A)) \quad \text{for } x \notin A,$$

where $l_x(A)$ and $r_x(A)$ are the distances from x to the nearest points in A to the left and right respectively. Recall that we are assuming that (0.3) holds, so that A_t is a well-defined continuous time Markov chain on Y.

Our first objective is to determine what choices of $\beta(\cdot, \cdot)$ lead to reversible chains A_t. Since A_t is a continuous time chain, and since $\varnothing$ is a trap for it, it is necessary to modify Definition 5.4(b) if Chapter II somewhat.

Definition 1.1. A strictly positive function π on Y is said to be a *reversible measure* for the chain A_t provided that

$$\pi(A)q(A, B) = \pi(B)q(B, A)$$

for all $A, B \neq \varnothing$.

Theorem 1.2. A_t *has a reversible measure* π *if and only if there is a strictly positive function* $\beta(l)$ *on* $\{1, 2, \ldots\}$ *such that* $\sum_{l=1}^{\infty} \beta(l) < \infty$,

$$(1.3) \qquad \beta(l, r) = \frac{\beta(l)\beta(r)}{\beta(l+r)}$$

for all $1 \leq l, r < \infty$, *and*

$$(1.4) \qquad \beta(l, \infty) = \beta(l)$$

for all $1 \leq l < \infty$. *In this case* π *is given (up to constant multiples) by* $\pi(\{x\}) = 1$ *and*

$$(1.5) \qquad \pi(A) = \prod_{i=1}^{n-1} \beta(x_{i+1} - x_i),$$

where $A = \{x_1, \ldots, x_n\}$ *and* $x_1 < x_2 < \cdots < x_n$ *if* $n \geq 2$.

Proof. If $\beta(l, r)$ is given by (1.3) and (1.4) for some $\beta(l)$, then clearly the π given in (1.5) is a reversible measure for the chain. For the converse, suppose that there exists a reversible measure π for A_t. Define $\beta(l)$ by

$$\beta(l) = \beta(l, \infty).$$

Take $A = \{x_1, \ldots, x_n\}$ with $x_1 < x_2 < \cdots < x_n$ and $n \geq 2$. By reversibility,

$$(1.6) \qquad \pi(A) = \pi(A \setminus \{x_n\})\beta(x_n - x_{n-1}).$$

Since π is strictly positive, it follows that $\beta(l) > 0$ for all $1 \leq l < \infty$. Iterating (1.6) yields

$$(1.7) \qquad \pi(A) = \pi(\{x_1\}) \prod_{i=1}^{n-1} \beta(x_{i+1} - x_i).$$

If this procedure is done from left to right instead of right to left, we obtain

$$(1.8) \qquad \pi(A) = \pi(\{x_n\}) \prod_{i=1}^{n-1} \beta(x_{i+1} - x_i).$$

From (1.7) and (1.8) it follows that $\pi(\{x\})$ is independent of x, so that π is a constant multiple of the π given in (1.5). To prove (1.3), use reversibility applied to the sets $\{-l, 0, r\}$ and $\{-l, r\}$ to obtain

$$\pi\{-l, 0, r\} = \pi\{-l, r\}\beta(l, r),$$

and then use (1.5). $\quad\Box$

For the remainder of this section, we will assume that $\beta(l, r)$ is given by (1.3) and (1.4), where $\beta(l)$ is a strictly positive summable sequence. Note that the examples (0.2) satisfy this with $\beta(l) = cl^{-p}$, provided that $p > 1$.

In applying the results of Section 6 of Chapter II to A_t, it is convenient to modify the chain slightly. If $A, B \in Y$, say that $A \sim B$ if A is a translate of B. Let $\tilde{Y}$ be the collection of all equivalence classes determined by this relation. Since the mechanism of A_t on Y is translation invariant, it determines a well-defined transition mechanism on $\tilde{Y}$. We will denote this chain by $\tilde{A}_t$, and will call it the "shape" chain. It keeps track of the shape, but not the location of A_t. The element of $\tilde{Y}$ which corresponds to all the singletons in Y will be denoted by $*$. Finally, to make the chain irreducible on $\tilde{Y}$, we will add a transition from $\emptyset$ to $*$ at rate 1. Let $\tilde{q}(A, B)$ be the Q matrix for $\tilde{A}_t$, which is then given by

$$\tilde{q}(\emptyset, *) = \tilde{q}(*, \emptyset) = 1,$$

$$\tilde{q}(A, A\backslash\{x_i\}) = \begin{cases} 2 & \text{if } i = 1 \text{ or } n \text{ and } A\backslash\{x_1\} \sim A\backslash\{x_n\}, \\ 1 & \text{otherwise, and} \end{cases}$$

$$\tilde{q}(A\backslash\{x_i\}, A) = \begin{cases} 2\beta(x_i - x_{i-1}, x_{i+1} - x_i) & \text{if } i = 1 \text{ or } n \text{ and } A\backslash\{x_1\} \sim A\backslash\{x_n\}, \\ \beta(x_i - x_{i-1}, x_{i+1} - x_i) & \text{otherwise,} \end{cases}$$

if $A = \{x_1, \ldots, x_n\}$ with $x_1 < \cdots < x_n$ and $n \geq 2$. $(x_1 - x_0 = x_{n+1} - x_n = \infty$ by convention.) Note that in spite of the extra factors of 2 which appear above, $\tilde{A}_t$ is reversible with respect to the π given in (1.5) with $\pi(\emptyset) = 1$, which is now interpreted in the natural way as a function on $\tilde{Y}$.

We will say that the chain A_t survives if (0.4) holds, and that it dies out otherwise. Note that the survival of A_t is equivalent to the transience of $\tilde{A}_t$. Of course it would be relatively straightforward to determine when $\tilde{A}_t$ is transient and when it is recurrent if there were a natural integer-valued function g on $\tilde{Y}$, such as the cardinality or diameter of the set, with the property that $g(\tilde{A}_t)$ is Markovian. These functions of $\tilde{A}_t$ are not Markovian.

The idea to be exploited in the next result is therefore that $\tilde{A}_t$ can be modified in such a way that one or the other of these functions of the modified chain does have the Markov property. The results of Section 6 of Chapter II are then used to turn statements about the modification into analogous statements about $\tilde{A}_t$ itself. The criterion for survival is given in terms of

$$(1.9) \qquad\qquad \lambda = \sum_{l=1}^{\infty} \beta(l),$$

which plays roughly the same role that the parameter in the contact process did. Note that the next result gives much more precise information than we were able to obtain in the case of the contact process. This shows that reversibility can be an extremely useful property.

Theorem 1.10. (a) A_t *survives if $\lambda > 1$ and dies out if $\lambda \leq 1$.*
  (b)   *Let τ be the hitting time of $\varnothing$ (for either A_t or $\tilde{A}_t$). Then*

$$(1.11) \qquad \frac{\lambda - 1}{\lambda} \leq P^*(\tau = \infty) \leq \left| \lambda \, \log \frac{\lambda - 1}{\lambda} \right|^{-1} \quad for \, \lambda > 1.$$

Proof. We will apply Corollary 6.7 and Theorem 6.10 of Chapter II to the embedded discrete time chain corresponding to $\tilde{A}_t$. It has transition probabilities

$$p_2(A, B) = \frac{\tilde{q}(A, B)}{\tilde{q}(A)},$$

where

$$\tilde{q}(A) = \sum_{B \neq A} \tilde{q}(A, B),$$

and is reversible with respect to

$$\pi_2(A) = \pi(A) \tilde{q}(A).$$

In applying Corollary 6.7 of Chapter II, take

$$p_1(A, B) = \frac{p_2(A, B)}{p_2(A)}$$

if A and B have different diameters, and $p_1(A, B) = 0$ otherwise. The factor $p_2(A)$ is simply a normalization to make $p_1(A, B)$ stochastic. Thus $p_1(\cdot, \cdot)$ is obtained from $p_2(\cdot, \cdot)$ by suppressing all "interior" births and deaths.

This new chain is reversible with respect to

$$\pi_1(A) = \pi_2(A)p_2(A).$$

Note that

$$\pi_1(A)p_1(A, B) \le \pi_2(A)p_2(A, B)$$

for $A \ne B$, and

$$\pi_1(\varnothing) = \pi_2(\varnothing) = \pi(\varnothing) = 1.$$

The important observation is that the cardinality of the chain with transition probabilities $p_1(\cdot, \cdot)$ is a Markov chain on $\{0, 1, 2, \ldots\}$ with transition probabilities

$$n \to n+1 \text{ with probability} \begin{cases} \dfrac{\lambda}{\lambda+1} & \text{if } n \ge 2, \\ \dfrac{2\lambda}{2\lambda+1} & \text{if } n = 1, \end{cases}$$

$$n \to n-1 \text{ with probability} \begin{cases} \dfrac{1}{\lambda+1} & \text{if } n \ge 2, \\ \dfrac{1}{2\lambda+1} & \text{if } n = 1. \end{cases}$$

This chain is of course transient if $\lambda > 1$ and its probability of never returning to 0 when started at 1 is $\ge 1 - \lambda^{-1}$ (see page 32 of Hoel, Port, and Stone (1972), for example). Thus Corollary 6.7 of Chapter II gives the first part of (a) and the left inequality in (1.11). To prove the other half of (a) and (b), we will apply Theorem 6.10 of Chapter II to

$$S_k = \{A \in \tilde{Y}: |A| = k\},$$

where $\varnothing$ is the distinguished point called 0 in that theorem. Using the easily verified fact that

(1.12)
$$\sum_{\substack{A \in \tilde{Y} \\ |A|=k}} \pi(A) = \lambda^{k-1},$$

we have, in the notation of that theorem, that

$$\tilde{\pi}(k) = k\lambda^{k-1} + (k+1)\lambda^k,$$

$$\tilde{p}(k, k+1) = \frac{(k+1)\lambda^k}{\tilde{\pi}(k)}, \quad \text{and}$$

$$\tilde{p}(k, k-1) = \frac{k\lambda^{k-1}}{\tilde{\pi}(k)}.$$

This chain is recurrent if $\lambda \le 1$, and its probability of never returning to 0 when started at 1 is

$$\left[\sum_{k=0}^{\infty} \frac{1}{(k+1)\lambda^k} \right]^{-1} = \left| \lambda \log\left(1 - \frac{1}{\lambda}\right) \right|^{-1}$$

for $\lambda > 1$ (see page 32 of Hoel, Port, and Stone (1972), for example). Therefore Theorem 6.10 of Chapter II gives the second half of (a) and the right-hand side inequality in (1.11). $\square$

It is interesting to note that if $\beta(l, r)$ and $\tilde{\beta}(l, r)$ are given by (1.3) and (1.4) in terms of $\beta(l)$ and $\tilde{\beta}(l)$ respectively, where $\tilde{\beta}(l) = \beta(l)\theta^l$, then $\tilde{\beta}(l, r) = \beta(l, r)$ for $1 \le l, r < \infty$, but $\tilde{\lambda} = \sum_l \tilde{\beta}(l) \ne \lambda = \sum_l \beta(l)$ in general. Thus the survival of the chain depends quite strongly on the values of $\beta(l, \infty)$. When applied to the example (0.2), Theorem 1.10 says that the chain survives if and only if

$$c > \left(\sum_{n=1}^{\infty} \frac{1}{n^p} \right)^{-1}.$$

Theorem 1.10 settles completely the question of survival for reversible finite nearest-particle systems. However, the bounds on the survival probability given in (1.11) are not sufficiently good to determine how rapidly $P^*(\tau = \infty)$ tends to zero as $\lambda \downarrow 1$. One way of improving the upper bound is to apply Theorem 6.10 of Chapter II with

$$S_k = \{A \in \tilde{Y}: \text{diameter}(A) = k\}.$$

This leads to a comparison chain on $\{0, 1, 2, \ldots\}$ with the following transition probabilities:

$$\tilde{p}(0, 1) = 1, \qquad \tilde{p}(1, 0) = \frac{1}{\lambda + 1},$$

$$\tilde{p}(k, l) = \frac{\beta(l-k)}{\lambda + 1} \qquad \text{if } l > k \ge 1, \quad \text{and}$$

$$\tilde{p}(k, l) = \frac{\beta(k-l)\rho(l)}{(\lambda + 1)\rho(k)} \quad \text{if } 1 \le l < k,$$

where

$$\rho(k) = \sum_{\text{diameter}(A)=k} \pi(A)$$

can be computed recursively via $\rho(1) = 1$ and

$$\rho(k) = \sum_{l=1}^{k-1} \rho(l)\beta(k-l).$$

Liggett (1983b) applied this comparison to the one-parameter family

(1.13) $$\beta_\lambda(l) = \lambda f(l),$$

where $f(\cdot)$ is strictly positive and satisfies

(1.14) $$\sum_{n=1}^{\infty} f(n) = 1,$$

(1.15) $$\frac{f(n)}{f(n+1)} \downarrow 1 \quad \text{as } n \uparrow \infty, \quad \text{and}$$

(1.16) $$\sum_{n=1}^{\infty} n^2 f(n) < \infty.$$

Under these assumptions, he showed that there is a constant c depending on $\{f(l), l \geq 1\}$ so that

(1.17) $$P^*(\tau = \infty) \leq c(\lambda - 1) \quad \text{for } \lambda \geq 1.$$

In this case, (1.11) and (1.17) together imply that the critical exponent for $P^*(\tau = \infty)$ is one, in the sense that

$$\lim_{\lambda \downarrow 1} \frac{\log P^*(\tau = \infty)}{\log(\lambda - 1)} = 1.$$

Part (b) of Theorem 1.10 and (1.17) give some information about super-critical (i.e. $\lambda > 1$) systems near the critical value (i.e. near $\lambda = 1$). The same techniques give analogous results for the critical case. Using Corollary 6.6 and Theorem 6.10 of Chapter II as in the proof of Theorem 1.10, we obtain the following:

Theorem 1.18 *Suppose* $\lambda = 1$. *Then*

$$\frac{1}{n} \leq P^*(|\tilde{A}_t| = n \text{ for some } t < \tau) \leq \frac{1}{\log n} \quad \text{for } n \geq 1.$$

The critical analogue of (1.17) was also proved in Liggett (1983b). It states that if $f(\cdot)$ satisfies (1.14), (1.15), and (1.16) and $\beta(l) = f(l)$ for all l, then

(1.19) $P^*(|\tilde{A}_t| = n \text{ for some } t < \tau) \leq \dfrac{c}{n}$ for some constant c.

We next consider critical exponents in the subcritical ($\lambda < 1$) case. The idea is to find some quantity which is finite for $\lambda < 1$ but diverges as $\lambda \uparrow 1$, and to determine the rate of this divergence. We will do this for the expected time to die out and for the expected total space time occupation measure. The next result states that the critical exponents for these two quantities are 1 and 2 respectively. An interesting feature here is that these quantities can be computed exactly for all $\lambda < 1$, so the conclusions are much more precise than those in (1.11) and (1.17).

Theorem 1.20. *Assume $\lambda < 1$. Then*

   (a) $E^*(\tau) = (1 - \lambda)^{-1}$, *and*
   (b) $E^* \int_0^\tau |\tilde{A}_t| \, dt = (1 - \lambda)^{-2}$.

Proof. When the initial state of the shape chain $\tilde{A}_t$ is $\varnothing$, τ is to be interpreted in what follows as the first time to hit $\varnothing$ after leaving it. By Theorem 1.10, $\tilde{A}_t$ is recurrent. By (1.12),

(1.21) $\displaystyle\sum_{A \in \tilde{Y}} \pi(A) = \frac{2 - \lambda}{1 - \lambda} < \infty$,

so that $\tilde{A}_t$ is positive recurrent. Therefore

(1.22) $\displaystyle E^\varnothing \int_0^\tau h(\tilde{A}_t) \, dt = \sum_{A \in \tilde{Y}} h(A) \pi(A)$

for any nonnegative function on $\tilde{Y}$. (Recall here that $\pi(\varnothing) = 1$.) Applying (1.22) to $h \equiv 1$ gives

$$E^\varnothing(\tau) = \sum_{A \in \tilde{Y}} \pi(A) = \frac{2 - \lambda}{1 - \lambda}$$

by (1.21). Since $\tilde{q}(\varnothing, *) = 1$ and $\tilde{q}(\varnothing, A) = 0$ for $A \neq *$,

$$E^\varnothing(\tau) = 1 + E^*(\tau),$$

and (a) follows. For (b), apply (1.22) to $h(A) = |A|$, to obtain

$$E^* \int_0^\tau |\tilde{A}_t| \, dt = E^\varnothing \int_0^\tau |\tilde{A}_t| \, dt = \sum_{A \in \tilde{Y}} |A| \pi(A)$$

$$= \sum_{k=1}^\infty k \sum_{\substack{A \in \tilde{Y} \\ |A| = k}} \pi(A)$$

$$= \sum_{k=1}^\infty k\lambda^{k-1} = (1-\lambda)^{-2},$$

where (1.12) was used in the next to last step. $\square$

The lack of dependence of the answers in Theorem 1.20 on the rates is a striking illustration of the universality principle, which was discussed briefly at the end of Section 1 of Chapter VI. One could describe the results of that theorem as asserting that all finite reversible nearest-particle systems are in the same universality class.

2. General Finite Systems

In this section, we consider the survival problem for general finite nearest-particle systems. The reversible systems treated in Section 1 provide an important class of examples for which a complete solution is available. A nonreversible example which we have discussed in some detail is the finite contact process.

The conditions for surviving or dying out will be formulated in terms of the total birth rate in maximal connected subsets of the complement of A_t. This is natural, since the complement of a general $A \in Y$ of cardinality n has approximately n such maximal connected subsets. Since each element of A_t has death rate one, one would expect that A_t would survive if these total birth rates are large, and die out if they are small. Thus we will define $b(n)$ for $1 \le n \le \infty$ by

$$b(n) = \sum_{l+r=n+1} \beta(l, r) \quad \text{if } 1 \le n < \infty, \quad \text{and}$$

$$b(\infty) = \sum_{l=1}^\infty \beta(l, \infty) + \sum_{r=1}^\infty \beta(\infty, r).$$

Note that $b(n) = 2\lambda$ for all n in the case of the contact process, while $b(\infty) = 2\lambda$ in the case described by (1.3), (1.4), and (1.9).

Theorem 2.1. *If* $b(n) \leq 1$ *for all* $1 \leq n \leq \infty$, *then* A_t *dies out.*

Proof. A_t has total death rate $|A_t|$, and a total birth rate which is at most $|A_t|$ if $b(n) \leq 1$ for all n. Therefore $|A_t|$ is a (nonnegative) supermartingale, so

$$(2.2) \qquad\qquad \lim_{t\to\infty} |A_t|$$

exists a.s. Since $\beta(1, \infty) = \beta(\infty, 1) > 0$, the only possible limit in (2.2) is zero. Therefore $|A_t| = 0$ for all sufficiently large t a.s., so the process dies out. $\square$

Theorem 2.3. *For every number* $b > 2$, *there is a reversible nearest-particle system which survives and satisfies* $b(n) = b$ *for all* $1 \leq n \leq \infty$.

Proof. Choose $\beta(l, r)$ of the form given by (1.3) and (1.4), where

$$\beta(n) = b \frac{(2n-2)!}{(n-1)!\, n!} \frac{1}{4^n}.$$

Then $\beta(n) = (b/4)F(n)$, where F is given in (1.18) of Chapter VI (with $\lambda = 2$). By (1.20) and (1.23) of Chapter VI, $b(n) = b$ for all $1 \leq n \leq \infty$ as required. By Theorem 1.10, A_t survives if and only if $b > 2$. $\square$

The next main result guarantees the survival of A_t whenever $b(n) \geq 4$ for all $1 \leq n \leq \infty$. The proof will be based on some of the work we did in Section 1 of Chapter VI. To prepare for the proof, let $f(n)$ be the probability density on $\{1, 2, \ldots\}$ which is defined by $f(n) = F(n) - F(n+1)$, where

$$F(n) = \frac{(2n-2)!}{(n-1)!\, n!} 4^{-n+1}, \qquad n \geq 1.$$

Then

$$(2.4) \qquad\qquad \sum_{n=1}^{\infty} F(n) = \sum_{n=1}^{\infty} nf(n) = 2$$

by (1.18) and (1.23) of Chapter VI. Let μ be the stationary renewal measure on $\{0, 1\}^{Z^1}$ corresponding to f, which is defined in (1.16) of Chapter VI.

Lemma 2.5.

$$\mu\{\eta: \eta(0) = 1, \eta(n) = 0\} = \tfrac{1}{8} \sum_{k=1}^{n} F(k) \quad \text{for } n \geq 1.$$

Proof. The proof is by induction on n. For $n = 1$, write

$$\mu\{\eta: \eta(0) = 1, \eta(1) = 0\} = \mu\{\eta: \eta(0) = 1\} - \mu\{\eta: \eta(0) = \eta(1) = 1\}$$
$$= \tfrac{1}{2} - \tfrac{1}{2}f(1) = \tfrac{1}{8}.$$

For $n \geq 2$, use translation invariance and the induction hypothesis to write

$$\mu\{\eta: \eta(0) = 1, \eta(n) = 0\} = \mu\{\eta: \eta(0) = 1, \eta(k) = 0 \text{ for all } 1 \leq k \leq n\}$$

$$+ \sum_{k=1}^{n-1} \mu\{\eta: \eta(0) = 1, \eta(j) = 0 \text{ for all } 1 \leq j < k, \eta(k) = 1, \eta(n) = 0\}$$

$$= \tfrac{1}{2}F(n+1) + \sum_{k=1}^{n-1} f(k)\mu\{\eta(k) = 1, \eta(n) = 0\}$$

$$= \tfrac{1}{2}F(n+1) + \tfrac{1}{8} \sum_{k=1}^{n-1} f(k) \sum_{j=1}^{n-k} F(j)$$

$$= \tfrac{1}{2}F(n+1) + \tfrac{1}{8} \sum_{j=1}^{n-1} F(j)[1 - F(n-j+1)]$$

$$= \tfrac{1}{8} \sum_{k=1}^{n} F(k).$$

The last equality follows from (1.20) of Chapter VI.   □

Now let h be defined as in (1.15) of Chapter VI in terms of this μ, and for $A \in Y$ and $x \notin A$ define

$$g_A(x) = h(A \cup \{x\}) - h(A)$$
$$= \mu\{\eta: \eta(x) = 1 \text{ and } \eta(y) = 0 \text{ for all } y \in A\}.$$

Lemma 2.6. *Suppose that* $y \geq x + 2$ *for all* $y \in A$. *Then*

(2.7) $g_A(x) \geq g_A(x+1)$ *and* $2g_A(x) \geq g_A(x+1) + g_A(x-1)$.

Proof. The first statement is a consequence of Lemma 1.24 of Chapter VI. However it does not seem possible to prove the second statement by the technique used there, so we will use a different approach which proves both statements simultaneously. If $A = \{y\}$, use Lemma 2.5 to write

$$g_A(x) - g_A(x+1) = \tfrac{1}{8}F(y-x) > 0, \quad \text{and}$$

$$2g_A(x) - g_A(x+1) - g_A(x-1) = \tfrac{1}{8}[F(y-x) - F(y-x+1)]$$
$$= \tfrac{1}{8}f(y-x) > 0,$$

so (2.7) holds for singleton A's. The proof in general is by induction on the cardinality of A. Given A with $|A| \geq 2$, let y be the leftmost element of A. If $B = A \backslash \{y\}$, then $|B| = |A| - 1$. Using elementary properties of renewal measures, write

$$
\begin{aligned}
g_A(x) &= g_B(x) - \mu\{\eta: \eta(x) = \eta(y) = 1, \eta(z) = 0 \text{ for all } z \in B\} \\
&= g_B(x) - 2g_B(y)\mu\{\eta: \eta(x) = \eta(y) = 1\} \\
&= g_B(x) - g_B(y)[1 - 2g_{\{y\}}(x)] \\
&= g_B(x) + 2g_B(y)g_{\{y\}}(x) - g_B(y).
\end{aligned}
$$

Thus (2.7) for g_A follows from (2.7) for g_B and $g_{\{y\}}$. $\square$

Lemma 2.8. *Suppose that* $x - 1, x, x + 1 \in A^c$. *Then*

$$(2.9) \qquad\qquad 2g_A(x) \geq g_A(x+1) + g_A(x-1).$$

Proof. Let $B = A \cap (-\infty, x-1)$ and $C = A \cap (x+1, \infty)$. Then since μ is a renewal measure and $\mu\{\eta(x) = 1\} = \frac{1}{2}$ by (2.4),

$$(2.10) \qquad\qquad g_A(x) = 2g_B(x)g_C(x).$$

Therefore

$$
\begin{aligned}
2g_A(x) &- g_A(x+1) - g_A(x-1) \\
&= 4g_B(x)g_C(x) - 2g_B(x+1)g_C(x+1) - 2g_B(x-1)g_C(x-1) \\
&= [g_B(x) + \tfrac{1}{2}g_B(x+1) + \tfrac{1}{2}g_B(x-1)][2g_C(x) - g_C(x+1) - g_C(x-1)] \\
&\quad + [g_C(x) + \tfrac{1}{2}g_C(x+1) + \tfrac{1}{2}g_C(x-1)][2g_B(x) - g_B(x+1) - g_B(x-1)] \\
&\quad + [g_B(x-1) - g_B(x+1)][g_C(x+1) - g_C(x-1)],
\end{aligned}
$$

which is nonnegative by Lemma 2.6 and the fact that μ is invariant under reflection about 0 in Z^1. $\square$

Theorem 2.11. *If* $b(n) \geq 4$ *for all* $1 \leq n \leq \infty$, *then* A_t *survives.*

Proof. Let B_t be the contact process with $\lambda = 2$, and denote its birth rates by $\tilde{\beta}(l, r)$. These are given in (0.1). By Lemma 1.25 of Chapter VI,

$$(2.12) \qquad\qquad \frac{d}{dt} E^A h(B_t)\Big|_{t=0} \geq 0$$

for all $A \in Y$. Since A_t and B_t have the same death rates,

(2.13)

$$\frac{d}{dt} E^A h(A_t)\bigg|_{t=0} - \frac{d}{dt} E^A h(B_t)\bigg|_{t=0}$$

$$= \sum_{x \notin A} [\beta(l_x(A), r_x(A)) - \tilde{\beta}(l_x(A), r_x(A))] g_A(x).$$

We want to show that the sum on the right side of (2.13) is nonnegative. This will be done by showing that for each maximal connected subset of A^c, the sum over those x in that subset is nonnegative. By Lemmas 2.6 and 2.8, it suffices to show that if g is any concave function on $\{1, \ldots, n\}$ with $n \geq 3$, then

(2.14)
$$\sum_{l+r=n+1} g(l)\beta(l, r) \geq 2[g(1) + g(n)],$$

while if g is any increasing function on $\{1, 2, \ldots\}$, then

(2.15)
$$\sum_{l=1}^{\infty} g(l)\beta(l, \infty) \geq 2g(1).$$

Inequality (2.15) follows immediately from the assumption that $b(\infty) = 2\sum_{l=1}^{\infty} \beta(l, \infty) \geq 4$. To prove (2.14), use the concavity of g to write

$$g(l) \geq \frac{n-l}{n-1} g(1) + \frac{l-1}{n-1} g(n).$$

Therefore

$$\sum_{l+r=n+1} g(l)\beta(l, r) \geq g(1) \sum_{l=1}^{n} \beta(l, n+1-l) \frac{n-l}{n-1}$$

$$+ g(n) \sum_{l=1}^{n} \beta(l, n+1-l) \frac{l-1}{n-1}.$$

Since $\beta(l, r) = \beta(r, l)$ for all r, l, and $b(n) \geq 4$,

$$\sum_{l=1}^{n} \beta(l, n+1-l) \frac{n-l}{n-1} = \sum_{l=1}^{n} \beta(l, n+1-l) \frac{l-1}{n-1} = \tfrac{1}{2} b(n) \geq 2$$

as required. We have now proved that the sum in (2.13) is nonnegative. This together with (2.12) implies that

(2.16)
$$\frac{d}{dt} E^A h(A_t)\bigg|_{t=0} \geq 0.$$

for all $A \in Y$, and hence that

(2.17) $E^A h(A_t) \geq h(A) > 0$

for all $t \geq 0$ and all $A \neq \emptyset$. Therefore A_t survives.   □

3. Construction of Infinite Systems

This section is devoted to the construction of infinite nearest-particle systems. As was pointed out in the introduction to this chapter, the approach to the construction problem which was used in Chapter I is not effective for many of these systems. In fact, we will see here that in order to carry out a useful construction it is generally necessary to restrict the process to the subset

$$X' = \left\{ \eta \in X : \sum_{x \geq 0} \eta(x) = \sum_{x \leq 0} \eta(x) = \infty \right\}$$

of the space X. This point will be discussed further following the proof of Theorem 3.6.

For $m \leq n$, let

$$Z_{m,n} = \{m, m+1, \ldots, n\},$$

and $X_{m,n} = \{0, 1\}^{Z_{m,n}}$. Let $C_{m,n}$ be the set of all functions on $X_{m,n}$, and let $C(X')$ be the collection of all bounded continuous functions on X'. If g_n, $g \in C(X')$, we will say that $g_n \to g$ provided that $\sup_n \|g_n\| < \infty$ and g_n converges to g uniformly on compact subsets of X'. Here $\|\cdot\|$ denotes the supremum norm on X'. Consider the continuous time Markov chain $\eta_t^{m,n}$ on $X_{m,n}$ which evolves according to the mechanism of the nearest-particle system on $Z_{m,n}$. In defining it, use the convention that there are fixed ones at $m-1$ and $n+1$, which are used in determining the nearest particle to an $x \in Z_{m,n}$ when there are no ones in $Z_{m,n}$ to the left or right of x. This Markov chain is of course well defined for $m \leq n$. In will be interpreted as a Markov process on X' by not permitting any flips off $Z_{m,n}$. The basic coupling which was described in Section 1 of Chapter III will be used to construct simultaneously copies of $\eta_t^{m,n}$ for various choices of m and n and various initial configurations. Denote the semigroup and generator corresponding to $\eta_t^{m,n}$ by $S_{m,n}(t)$ and $\Omega_{m,n}$ respectively.

Since the death rates are identically equal to one, the coupled processes have the following important property: for $m \leq j \leq k \leq n$,

(3.1)
$$P^\eta [\eta_s^{m,n}(x) \neq \eta_s^{m',n'}(x) \text{ for some } s \leq t, \text{ some } m' \leq m \text{ and } n' \geq n,$$
$$\text{and some } x \in Z_{j,k}]$$
$$\leq (1 - e^{-t})^L + (1 - e^{-t})^R,$$

where

$$L = \sum_{m<x<j} \eta(x), \quad \text{and} \quad R = \sum_{k<x<n} \eta(x).$$

For $\eta \in X'$, L and R both tend to ∞ as $m \to -\infty$ and $n \to +\infty$. This permits the definition of η_t via

(3.2)
$$\eta_t(x) = \lim_{\substack{m \to -\infty \\ n \to \infty}} \eta_t^{m,n}(x).$$

Of course, $\eta_t \in X'$ with probability one for $\eta \in X'$. Note that while we are assuming for simplicity in this chapter that the death rates are identically one, the same construction works provided that the death rates are uniformly bounded. In that case, (3.1) would be valid with e^{-t} replaced by $e^{-\delta t}$, where δ is a bound on the death rates.

For $g \in C(X')$, let $S(t)g(\eta) = E^\eta g(\eta_t)$. For $g \in \mathcal{D} = \bigcup_{m \leq n} C_{m,n}$, let

$$\Omega g(\eta) = \sum_{\eta(x)=1} [g(\eta_x) - g(\eta)]$$

$$+ \sum_{\eta(x)=0} \beta(l_x(\eta), r_x(\eta))[g(\eta_x) - g(\eta)],$$

where as usual, $\eta_x(y) = \eta(y)$ for $y \neq x$ and $\eta_x(x) = 1 - \eta(x)$. Note that $\Omega g \in C(X')$, since $\beta(l, r)$ is bounded, and

(3.3)
$$\Omega g = \lim_{\substack{m \to -\infty \\ n \to +\infty}} \Omega_{m,n} g \quad \text{in } C(X')$$

for $g \in \mathcal{D}$. Furthermore, $S(t): C(X') \to C(X')$ and

(3.4)
$$S_{m,n}(t)g \to S(t)g \quad \text{as } m \to -\infty \text{ and } n \to \infty$$

as elements of $C(X')$ for each $g \in C(X')$. The first result states that $S(t)$ satisfies the semigroup property, and also establishes the connection between $S(t)$ and Ω.

Theorem 3.5. (a) $S(t+s) = S(t)S(s)$.
    (b) $S(t)g(\eta) - g(\eta) = \int_0^t S(s)\Omega g(\eta) \, ds$ for $t \geq 0$, $\eta \in X'$ and $g \in \mathcal{D}$.
    (c) $S(t)g \to g$ as $t \downarrow 0$ for all $g \in C(X')$.
    (d) $(S(t)g - g)/t \to \Omega g$ as $t \downarrow 0$ as elements of $C(X')$ for any $g \in \mathcal{D}$.

Proof: To prove (a), start with the semigroup property for $S_{m,n}(t)$:

$$S_{m,n}(t+s)g = S_{m,n}(t)S_{m,n}(s)g$$

for $g \in C(X')$. The left side converges to $S(t+s)g$ as $m \to -\infty$ and $n \to +\infty$ by (3.4). To pass to the limit on the right side, write

$$S_{m,n}(t)S_{m,n}(s)g(\eta) = E^{\eta}(S_{m,n}(s)g)(\eta_t^{m,n}).$$

By (3.2), (3.4) and the Bounded Convergence Theorem, this converges to

$$E^{\eta}(S(s)g)(\eta_t) = S(t)S(s)g(\eta).$$

For part (b), begin with

$$S_{m,n}(t)g(\eta) - g(\eta) = \int_0^t S_{m,n}(s)\Omega_{m,n}g(\eta)\,ds,$$

and pass to the limit as $m \to -\infty$ and $n \to +\infty$. For the left side, use (3.4). For the right side, write

$$S_{m,n}(s)\Omega_{m,n}g(\eta) = E^{\eta}(\Omega_{m,n}g)(\eta_s^{m,n}),$$

and use (3.2), (3.3), and the Bounded Convergence Theorem. Part (c) follows from the right continuity of η_t, which is a consequence of (3.1) and (3.2). Finally, part (d) follows from (b) and (c). $\square$

It is clear from (3.2) and (3.3) that the η_t which we have constructed is a solution to the martingale problem for Ω on $\mathscr{D}$ for any initial $\eta \in X'$ (see Definition 5.1 of Chapter I). Gray (1978) proved that the solution to this martingale problem is unique. Analogous results for nearest-particle systems on R^1 were proved by Holley and Stroock (1978c) and by Cocozza and Kipnis (1980). In the second of these papers, it was assumed as well that $\beta(l, r)$ is bounded below, which rules out many interesting examples. We will not prove this uniqueness here, but will instead take the process constructed above to be the nearest-particle system which is the object of study.

The nearest-particle system with birth rates $\beta(\cdot, \cdot)$ will be called attractive if $\beta(l, r)$ is a nonincreasing function of l and r. (Compare this with Definition 2.1 of Chapter III.) It is interesting that in the attractive case, the nearest-particle system extends by continuity from X' to X, and thus defines a Feller process on X. This extension can be quite useful, since it is then possible to start the process at $\eta \equiv 0$ and let it increase in time. The next result gives a formal statement of this observation. Note that $C(X)$ is naturally embedded in $C(X')$.

Theorem 3.6. *Consider the attractive case. Then Ωg extends continuously to X for $g \in \mathscr{D}$, and $S(t)g$ extends continuously to X for $g \in C(X)$. Furthermore, $S(t)g \to g$ uniformly on X as $t \downarrow 0$ for any $g \in C(X)$.*

Proof. The statement about Ωg is immediate from the definition of Ω and the fact that by the attractiveness assumption, $\beta(l, r)$ extends continuously to a function on $\{1, 2, \ldots, \infty\} \times \{1, 2, \ldots, \infty\}$. For the next statement, it suffices to prove that $S(t)g$ extends continuously to X for $g \in \mathcal{M}$ (the class of increasing continuous functions on X). To do so, note that by the attractiveness assumption, if $\eta, \zeta \in X'$ and $\eta \leq \zeta$, then

$$S(t)g(\eta) \leq S(t)g(\zeta)$$

for any $g \in \mathcal{M}$. Therefore we can define

$$(3.7) \qquad\qquad S(t)g(\eta) = \lim_{\substack{\zeta \in X' \\ \zeta \downarrow \eta}} S(t)g(\zeta)$$

for $\eta \in X$. Clearly this agrees with $S(t)g(\eta)$ for $\eta \in X'$, and is an increasing function on X. To prove the continuity of $S(t)g$ on X, it suffices by monotonicity to prove that

$$(3.8) \qquad\qquad \lim_{\zeta \uparrow \eta} S(t)g(\zeta) = S(t)g(\eta) = \lim_{\zeta \downarrow \eta} S(t)g(\zeta).$$

The second equality is immediate from (3.7). To prove the first equality in (3.8), use (3.7) and the argument which led to (3.1) to obtain the following estimate for $g \in C_{j,k}$ and $\zeta, \eta \in X$ such that $\eta(x) = \zeta(x)$ for all $x \in Z_{j,k}$:

$$(3.9) \qquad |S(t)g(\eta) - S(t)g(\zeta)| \leq [(1 - e^{-t})^L + (1 - e^{-t})^R] \|g\|,$$

where

$$L = \text{number of } x < j \text{ such that } \eta(x) = \zeta(x) = 1 \text{ and}$$

$$\eta(y) = \zeta(y) \text{ for all } x \leq y \leq j, \quad \text{and}$$

$$R = \text{number of } x > k \text{ such that } \eta(x) = \zeta(x) = 1 \text{ and}$$

$$\eta(y) = \zeta(y) \text{ for all } k \leq y \leq x.$$

Now the first equality in (3.8) for $g \in \mathcal{M} \cap C_{j,k}$ follows quickly by considering separately the four cases:

  (i) $\sum_x \eta(x) < \infty$,
  (ii) $\sum_{x<0} \eta(x) < \infty$ and $\sum_{x>0} \eta(x) = \infty$,
  (iii) $\sum_{x<0} \eta(x) = \infty$ and $\sum_{x>0} \eta(x) < \infty$, and
  (iv) $\eta \in X'$.

In case (i), the result is immediate since there are only finitely many $\zeta \leq \eta$. In case (iv) the right side of (3.9) tends to zero as $\zeta \uparrow \eta$. The other two cases follow from an easy modification of (3.9), since only one of the two terms on the right side of (3.9) is needed, and it tends to zero as $\zeta \uparrow \eta$. We have now proved (3.8) for $g \in \mathcal{M} \cap \mathcal{D}$. It follows for $g \in \mathcal{M}$ by the fact that $\mathcal{M} \cap \mathcal{D}$ is dense in $\mathcal{M}$ and $S(t)$ is a contraction. For the final statement of the theorem, simply use part (b) of Theorem 3.5 for $g \in \mathcal{D}$, and then approximate $g \in C(X)$ by elements of $\mathcal{D}$. $\square$

Theorem 3.6 allows us to define η_t as a Feller process on all of X. It is the only Feller extension since X' is dense in X. Its generator is an extension of Ω by Theorem 3.5. In general, $\mathcal{D}$ is not a core for the generator. (See Definition 2.11 of Chapter I.) If it were, then the solution to the martingale problem for Ω on $\mathcal{D}$ would be unique by Theorem 5.2 of Chapter I. To find examples of nonuniqueness, it would be enough to find attractive choices of $\beta(l, r)$ which have the following two properties:

$$(3.10) \qquad \qquad \lim_{l,r \to \infty} \beta(l, r) = 0, \quad \text{and}$$

$$(3.11) \qquad \qquad \delta_0 S(t) \neq \delta_0,$$

where δ_0 is the pointmass on $\eta \equiv 0$. By (3.10), $\Omega g(\eta) = 0$ for $\eta \equiv 0$, so that $\eta_t \equiv 0$ would be a solution to the martingale problem for the initial configuration $\eta \equiv 0$. On the other hand, (3.11) would imply that the η_t which we have constructed in this section is a different solution.

Liggett (1983a) gave a criterion for (3.11) in the context of attractive nearest-particle systems. To state that result, let

$$\alpha(n) = \sum_{l+r=n+1} (l \wedge r) \beta(l, r), \quad \text{and}$$

$$\tilde{\alpha}(n) = \max_{k \leq n} \alpha(k).$$

His result is that $\delta_0 S(t) = \delta_0$ if

$$(3.12) \qquad \qquad \sum_n \frac{1}{\tilde{\alpha}(n)} = \infty, \quad \text{and}$$

$\delta_0 S(t) \neq \delta_0$ if

$$(3.13) \qquad \qquad \sum_n \frac{1}{\alpha(n)} < \infty.$$

In example (0.2), $\alpha(n)$ grows asymptotically like a constant multiple of n^{2-p} for $0 < p < 2$. Therefore both (3.10) and (3.11) hold for $0 < p < 1$.

4. Reversible Infinite Systems

In this section, we consider problems for infinite nearest-particle systems which are analogous to those solved in Section 1 for the finite systems. The main problem is to determine which systems have reversible invariant measures which concentrate on X'. As corollaries to the solution of this problem, we will be able to compute exactly the critical value for a large number of infinite nearest-particle systems, including those with rates given by (0.2), much as we did in Theorem 1.10 for the finite systems. The techniques of proof will be entirely different.

Modifying Definition 5.1 of Chapter II slightly to account for the fact that our process is defined only on X', we will say that a probability measure μ on X' is reversible for the process if

$$(4.1) \qquad \int fS(t)g \, d\mu = \int gS(t)f \, d\mu$$

for all $f, g \in C(X')$. Here $S(t)$ is the semigroup which was defined just before Theorem 3.5. It will turn out that any probability measure on X' which is reversible for a nearest-particle system is automatically a stationary renewal measure. Therefore, if β is a probability density function on $\{1, 2, \ldots\}$ with finite mean, we will let μ_β be the corresponding stationary renewal measure, which was defined in (1.16) of Chapter VI.

Theorem 4.2. *Suppose that the birth rates have the form*

$$\beta(l, r) = \frac{\beta(l)\beta(r)}{\beta(l+r)} \quad \text{for } 1 \le l, r < \infty,$$

where $\beta(n)$ is a strictly positive probability density function on $\{1, 2, \ldots\}$ with finite mean. Then μ_β is reversible for the process.

Proof. Consider the approximating Markov chains $\eta_t^{m,n}$ which were defined in Section 3. Let $\mu_{m,n}$ be the probability measure on $X_{m,n}$ which is defined by

$$\mu_{m,n}\{\zeta\} = \mu_\beta\{\eta : \eta = \zeta \text{ on } Z_{m,n} | \eta(m-1) = \eta(n+1) = 1\}.$$

Then by (1.16) of Chapter VI,

$$(4.3) \qquad \mu_{m,n}\{\zeta\} = c \prod_{i=1}^{k} \beta(x_i - x_{i-1})$$

for some constant c, where $x_0 = m-1$, $x_k = n+1$, and $x_1 < \cdots < x_{k-1}$ are those points x in $Z_{m,n}$ with $\zeta(x) = 1$. Note the similarity with (1.5). From this it is clear that if $\zeta(x) = 0$ for some $x \in Z_{m,n}$, then

$$(4.4) \qquad \mu_{m,n}(\zeta)\beta(l_x(\zeta), r_x(\zeta)) = \mu_{m,n}(\zeta_x).$$

Of course, in defining $l_x(\zeta)$ and $r_x(\zeta)$, we use the convention $\zeta(m-1) = \zeta(n+1) = 1$. Equation (4.4) implies that $\mu_{m,n}$ is reversible for $\eta_t^{m,n}$, so that

$$(4.5) \qquad \int f S_{m,n}(t) g \, d\mu_{m,n} = \int g S_{m,n}(t) f \, d\mu_{m,n}$$

whenever $f, g \in C_{m,n}$. By the Renewal Theorem (see Chapter XI of Feller (1971), for example),

$$(4.6) \qquad \lim_{\substack{m \to -\infty \\ n \to +\infty}} \mu_{m,n} = \mu_\beta$$

in the topology of weak convergence on X. Since $\mu_\beta(X') = 1$, this together with (3.4) means that we can take the limit in (4.5) as $m \to -\infty$ and $n \to +\infty$ to conclude that

$$\int f S(t) g \, d\mu_\beta = \int g S(t) f \, d\mu_\beta$$

for all $f, g \in \mathcal{D}$. This implies that μ_β is reversible, since $\mathcal{D}$ is dense in $C(X')$. $\square$

Next, we state and prove the more important converse to Theorem 4.2.

Theorem 4.7. *Suppose that μ is a probability measure on X' which is reversible for the nearest-particle system with strictly positive birth rates $\beta(l, r)$. Then there exists a strictly positive probability density function $\beta(n)$ on $\{1, 2, \ldots\}$ with finite mean so that $\mu = \mu_\beta$. Furthermore,*

$$\beta(l, r) = \frac{\beta(l)\beta(r)}{\beta(l+r)} \quad \textit{for } 1 \le l, r < \infty.$$

Proof. The first part of the proof follows the proof of Proposition 2.7 of Chapter IV. For a finite subset A of Z^1 and an $x \in A$, define $f, g \in \mathcal{D}$ by

$$f(\eta) = \prod_{y \in A} \eta(y) \quad \text{and} \quad g(\eta) = f(\eta_x).$$

Then by the definition of Ω which precedes Theorem 3.5,

$$(4.8) \quad g(\eta)\Omega f(\eta) = f(\eta_x) \sum_{y \in A} c(y, \eta)[f(\eta_y) - f(\eta)] = c(x, \eta)f(\eta_x),$$

and

$$(4.9) \quad f(\eta)\Omega g(\eta) = f(\eta) \sum_{y \in A} c(y, \eta)[g(\eta_y) - g(\eta)] = c(x, \eta)f(\eta).$$

Since μ is reversible, (4.1) and part (d) of Theorem 3.5 imply that

$$\int f\Omega g \, d\mu = \int g\Omega f \, d\mu,$$

so that by (4.8) and (4.9),

$$(4.10) \qquad \int c(x, \eta)[f(\eta_x) - f(\eta)] \, d\mu = 0.$$

By linearity, (4.10) then holds for all $f \in \mathscr{D}$ and all $x \in Z^1$. Since $\mathscr{D}$ is dense in $C(X')$, it therefore holds for $f \in C(X')$ as well. The next step is to turn (4.10) into the following statement about the conditional probabilities of μ:

$$(4.11) \qquad \mu\{\eta(x) = 1 | \eta(y), y \neq x\} = \frac{\beta(l_x(\eta), r_x(\eta))}{\beta(l_x(\eta), r_x(\eta)) + 1}.$$

Note that $l_x(\eta) < \infty$ and $r_x(\eta) < \infty$ a.s. with respect to μ since μ concentrates on X'. In order to prove (4.11), it suffices to show that

$$(4.12) \qquad \int \eta(x)f(\eta) \, d\mu = \int \frac{\beta(l_x(\eta), r_x(\eta))}{\beta(l_x(\eta), r_x(\eta)) + 1} f(\eta) \, d\mu$$

for all $f \in \mathscr{D}$ which do not depend on the coordinate $\eta(x)$. Writing

$$g(\eta) = \frac{f(\eta)}{\beta(l_x(\eta), r_x(\eta)) + 1},$$

we see that it then suffices to show that

$$\int \eta(x)g(\eta)[\beta(l_x(\eta), r_x(\eta)) + 1] \, d\mu = \int \beta(l_x(\eta), r_x(\eta))g(\eta) \, d\mu$$

for all $g \in C(X')$ which do not depend on $\eta(x)$, or equivalently that

$$\int \eta(x)g(\eta) \, d\mu = \int [1 - \eta(x)]\beta(l_x(\eta), r_x(\eta))g(\eta) \, d\mu.$$

But this follows from (4.10) by setting the f there equal to $\eta(x)g(\eta)$.

Therefore we have now proved (4.11). From (4.11) and the strict positivity of $\beta(l, r)$, it follows easily that μ assigns positive probability to each subset of X which depends on finitely many coordinates. For $x \in Z^1$ and $k_1, \ldots, k_n \geq 1$, define subsets of X by

$$A_x(k_1, \ldots, k_n) = \{\eta: \eta(x) = \eta(x + k_1) = \cdots = \eta(x + k_1 + \cdots + k_n) = 1$$

$$\text{and } \eta(y) = 0 \text{ for all other } x < y < x + k_1 + \cdots + k_n\}.$$

By (4.11),

(4.13) $$\frac{\mu(A_x(j, k))}{\mu(A_x(j+k))} = \beta(j, k)$$

for all $x \in Z^1$ and all $j, k \geq 1$. Note that we are using here the translation invariance of the conditional probabilities in (4.11). We are not using translation invariance of μ itself, which is not being assumed. Using (4.11) again, write

$$\frac{\mu(A_x(j, k, l))}{\mu(A_x(j+k+l))} = \frac{\mu(A_x(j, k, l))}{\mu(A_x(j+k, l))} \frac{\mu(A_x(j+k, l))}{\mu(A_x(j+k+l))} = \beta(j, k)\beta(j+k, l).$$

By a similar argument, this quantity also equals

$$\beta(k, l)\beta(j, k+l).$$

Therefore we conclude that $\beta(\cdot, \cdot)$ satisfies

(4.14) $$\beta(j, k)\beta(j+k, l) = \beta(k, l)\beta(j, k+l)$$

for all $j, k, l \geq 1$. To solve this relation, set

$$\gamma(n) = \frac{\beta(n, 1)}{\prod_{i=1}^{n} \beta(i, 1)}$$

and observe by induction using (4.14) that

(4.15) $$\beta(l, r) = \frac{\gamma(l)\gamma(r)}{\gamma(l+r)}.$$

Now let $c_x(0) = \mu\{\eta: \eta(x) = 1\}$ and

$$c_x(n) = \frac{\mu(A_x(n))}{\gamma(n)} \quad \text{for } n \geq 1.$$

By (4.11),

$$\mu(A_x(k_1, \ldots, k_n)) = \beta(k_1, k_2)\mu(A_x(k_1 + k_2, k_3, \ldots, k_n)),$$

so that by (4.15),

$$\frac{\mu(A_x(k_1, \ldots, k_n))}{\gamma(k_1)\gamma(k_2)} = \frac{\mu(A_x(k_1 + k_2, k_3, \ldots, k_n))}{\gamma(k_1 + k_2)}.$$

Using induction, it then follows that

(4.16) $\mu(A_x(k_1, \ldots, k_n)) = c_x(k_1 + k_2 + \cdots + k_n)\gamma(k_1)\gamma(k_2)\ldots \gamma(k_n).$

Comparing this with (1.16) of Chapter VI, and recalling (4.15), it should be clear that in order to complete the proof of the theorem, it suffices to show that there is a $\theta > 0$ so that

(4.17) $$\sum_{n=1}^{\infty} \gamma(n)\theta^n = 1,$$

(4.18) $$K = \sum_{n=1}^{\infty} n\gamma(n)\theta^n < \infty, \quad \text{and}$$

(4.19) $$c_x(n) = \frac{\theta^n}{K} \quad \text{for all } x \in Z^1, n \geq 0.$$

Once this is done, the statement of the theorem follows by setting $\beta(n) = \gamma(n)\theta^n$. To begin, note since $\mu(X') = 1$, that

$$\mu(A_x(n)) = \sum_{k=1}^{\infty} \mu(A_x(n, k)).$$

Using (4.16) and the definition of $c_x(n)$, this becomes

(4.20) $$c_x(n) = \sum_{k=1}^{\infty} c_x(k + n)\gamma(k).$$

Let H be the collection of all nonnegative functions $c(n)$ on $\{0, 1, 2, \ldots\}$ such that $c(0) = 1$ and

(4.21) $$\sum_{k=1}^{\infty} c(n + k)\gamma(k) \leq c(n) \quad \text{for all } n \geq 0.$$

Then H is convex and compact in the topology of pointwise convergence, since $c(n)\gamma(n)\le 1$. H is nonempty since by (4.20), $c(n) = c_x(n)/c_x(0)$ is in H for any $x \in Z^1$. The topology on H is metrizable, so that by Choquet's Representation Theorem (see Section 3 of Phelps (1966)), every element of H has an integral representation in terms of the extreme points of H. Let H_0 be the collection of all $c(\cdot)$ in H such that equality holds in (4.21) for all $n \ge 0$. Then the integral representation of any $c(\cdot)$ in H_0 can only involve extreme points of H which are in H_0. Thus H_0 is the closed convex hull of its extreme points. Let $c(\cdot)$ be an extreme point of H_0. Then (note that $c(n) > 0$ for all n)

(4.22)
$$c(n) = \sum_{k=1}^{\infty} \gamma(k)c(k+n)$$

$$= \sum_{k=1}^{\infty} \gamma(k)c(k)\frac{c(k+n)}{c(k)}.$$

Since $\sum_{k=1}^{\infty} \gamma(k)c(k) = c(0) = 1$, and

$$\frac{c(k+\cdot)}{c(k)} \in H_0$$

for each $k \ge 1$, (4.22) exhibits $c(\cdot)$ as a convex combination of elements of H_0. So, since $c(\cdot)$ was assumed to be extremal, it follows that

$$c(n) = \frac{c(k+n)}{c(k)}$$

for all $k, n \ge 1$. Therefore $c(n) = [c(1)]^n$ for all $n \ge 0$. Since $c(\cdot)$ is in H_0, $c(1)$ must be a solution θ of

$$1 = \sum_{k=1}^{\infty} \gamma(k)\theta^k.$$

This solution is clearly unique, and defines θ for us so that (4.17) is satisfied. Therefore H_0 consists only of the function $c(n) = \theta^n$ for that θ. Therefore by (4.20),

(4.23) $c_x(n) = c_x(0)\theta^n$

for all $x \in Z^1$ and $n \ge 0$. By the definition of $c_x(n)$, it follows from this that

$$\mu(A_x(n)) = \gamma(n)\theta^n c_x(0).$$

By interchanging the roles of right and left, we would obtain similarly that

$$\mu(A_x(n)) = \gamma(n)\theta^n c_{x+n}(0).$$

Thus $c_x(0) = c$ is independent of x. To complete the proof of (4.18) and (4.19) it remains to show that

$$\sum_{n=1}^{\infty} n\gamma(n)\theta^n = c^{-1}.$$

To do this write

$$1 = \mu(X') = \sum_{x \leq 0 < x+n} \mu(A_x(n))$$

$$= \sum_{n=1}^{\infty} \gamma(n) \sum_{x=-n+1}^{0} c_x(n)$$

$$= \sum_{n=1}^{\infty} \gamma(n) \sum_{x=-n+1}^{0} c\theta^n$$

$$= c \sum_{n=1}^{\infty} n\gamma(n)\theta^n. \quad \square$$

Of course, if the birth rates have the form

$$(4.24) \qquad \beta(l, r) = \frac{\beta(l)\beta(r)}{\beta(l+r)}$$

for some positive sequence $\beta(\cdot)$, this sequence is not uniquely determined. It can always be replaced by $\beta(n)\theta^n$ for some $\theta > 0$. An interesting question which is suggested by Theorems 4.2 and 4.7 is the following. What happens if $\beta(l, r)$ has the form (4.24) for some positive sequence $\beta(n)$, but there is no choice of $\theta > 0$ so that $\beta(n)\theta^n$ is a probability density with finite mean? The process has no reversible measure by Theorem 4.7, but might it have a nontrivial invariant measure? The answer is no, at least in the attractive case, as can be seen from the next result. Recall that by Theorem 3.6, attractive nearest-particle systems are defined as Feller processes on all of X. They are said to die out if (0.6) holds. Just as in the context of Section 2 of Chapter III, this implies that

$$\lim_{t \to \infty} \mu S(t) = \delta_0$$

for any probability measure on X, so that the process has no invariant measure other than δ_0. If the system does not die out, it is said to survive.

Theorem 4.25. *Consider an attractive nearest-particle system with rates given by (4.24) for some positive sequence $\beta(\cdot)$. Suppose that there is no $\theta > 0$ such that $\beta_\theta(n) = \beta(n)\theta^n$ is a probability density with finite mean. Then the system dies out.*

Proof. Define $\mu_{m,n}$ on $X_{m,n}$ by (4.3). Just as in the proof of Theorem 4.2, $\mu_{m,n}$ is reversible with respect to the approximating process $\eta_t^{m,n}$, so that (4.5) holds. Since the rates are attractive,

$$(4.26) \qquad\qquad \lim_{t\to\infty} \delta_1 S(t) = \lim_{\substack{m\to-\infty \\ n\to\infty}} \mu_{m,n}$$

just as in the context of Theorem 2.7 of Chapter III. Let μ be the common limit in (4.26), which we wish to show is equal to δ_0. We can then pass to the limit in (4.5) to conclude that

$$(4.27) \qquad\qquad \int fS(t)g\, d\mu = \int gS(t)f\, d\mu$$

for all $t \geq 0$ and all $f, g \in C(X)$. Any measure in $\mathscr{S}$ is a convex combination of δ_0 and a measure concentrating on X'. Since $P^\eta(\eta_t \in X') = 1$ for $\eta \in X'$, it follows from this that

$$\delta_0 S(t)(X') = 1 - e^{-ct}$$

for some $0 \leq c \leq \infty$. Therefore either δ_0 is invariant or

$$(4.28) \qquad\qquad \lim_{t\to\infty} \delta_0 S(t)(X') = 1.$$

If $\mu \neq \delta_0$, write

$$\mu = (1-\varepsilon)\delta_0 + \varepsilon\tilde{\mu},$$

where $\varepsilon > 0$ and $\tilde{\mu}(X') = 1$. If δ_0 is invariant, then δ_0 and hence $\tilde{\mu}$ satisfies (4.27). If on the other hand (4.28) holds, then $\tilde{\mu} = \mu$. Hence in either case $\tilde{\mu}$ is reversible and satisfies $\tilde{\mu}(X') = 1$. This is impossible by the assumption of the theorem and Theorem 4.7. Therefore $\mu = \delta_0$ as required. $\square$

If the birth rates are given by (4.24) for some positive sequence $\beta(\cdot)$, then the rates are attractive if and only if $\beta(n)/\beta(n+1)$ is nonincreasing. If

$$\lim_{n\to\infty} \frac{\beta(n)}{\beta(n+1)} = 0,$$

then Theorem 4.25 implies that the process dies out. On the other hand, if

$$(4.29) \qquad\qquad \lim_{n\to\infty} \frac{\beta(n)}{\beta(n+1)} > 0,$$

we can replace $\beta(n)$ by $\beta(n)\theta^n$ for some θ so that the limit in (4.29) is one without changing the birth rates. With these comments, we can summarize the results obtained so far in this section in the following way.

Corollary 4.30. *Suppose*

$$\beta(l, r) = \frac{\beta(l)\beta(r)}{\beta(l+r)},$$

where $\beta(n) > 0$ and

(4.31)
$$\frac{\beta(n)}{\beta(n+1)} \downarrow 1 \quad as \ n \uparrow \infty.$$

Let $\lambda = \sum_{n=1}^{\infty} \beta(n)(\leq \infty)$. Then the process survives if either (a) $\lambda > 1$ or (b) $\lambda = 1$ and $\sum_{n=1}^{\infty} n\beta(n) < \infty$. Otherwise the process dies out. In particular, if

(4.32)
$$\beta(l, r) = c\left(\frac{1}{l} + \frac{1}{r}\right)^{p}$$

for some $c > 0$ and $p > 0$, then we have the following conclusions: If $p \leq 1$, the process survives for all c. If $1 < p \leq 2$, the process survives if and only if $c\sum_{n=1}^{\infty} 1/n^{p} > 1$. If $p > 2$, the process survives if and only if $c\sum_{n=1}^{\infty} 1/n^{p} \geq 1$.

Proof. Let

(4.33)
$$\phi(\theta) = \sum_{n=1}^{\infty} \beta(n)\theta^{n},$$

which has radius of convergence 1 by (4.31). By Theorems 4.2, 4.7, and 4.25, the process survives if and only if there is a $\theta > 0$ so that

$$\phi(\theta) = 1 \quad and \quad \phi'(\theta) < \infty.$$

Since $\phi(0) = 0$, $\phi(1) = \lambda$, and $\phi(\theta) = \infty$ for $\theta > 1$, there is such a θ if $\lambda > 1$ and there is no such θ if $\lambda < 1$. On the other hand, if $\lambda = 1$, we must take $\theta = 1$. Of course $\phi'(1) < \infty$ if and only if $\sum_{n=1}^{\infty} n\beta(n) < \infty$. □

Comparing Corollary 4.30 with Theorem 1.10 we see that if the birth rates are given by

$$\beta(l, r) = \lambda \frac{\beta(l)\beta(r)}{\beta(l+r)} \quad for \ 1 \leq l, r < \infty \quad and$$

$$\beta(l, \infty) = \lambda\beta(l) \quad for \ 1 \leq l < \infty,$$

where $\beta(l)$ is a strictly positive probability density such that

$$\frac{\beta(n)}{\beta(n+1)} \downarrow 1 \quad \text{as } n \uparrow \infty,$$

then we can draw the following conclusions:

(a)   The critical values for the finite and infinite systems agree.
(b)   At the critical value, the finite system always dies out, but the infinite system dies out if and only if $\sum_{n=1}^{\infty} n\beta(n) = \infty$.

Thus there are examples in which the critical finite system dies out while the critical infinite system survives.

Returning to the examples in (4.32), one can raise the question of whether, in the case of survival, the process might be ergodic (with then the unique invariant measure being the appropriate renewal measure). By criterion (3.12), δ_0 is invariant if $p \geq 1$. Therefore if the process survives, it cannot be ergodic. On the other hand, if $p < 1$, criterion (3.13) implies that δ_0 is not invariant, so $\lim_{t \to \infty} \delta_0 S(t)$ (which exists by attractiveness) is an invariant measure which is translation invariant and concentrates on X'. In the context of Corollary 4.30, Liggett (1983a) has shown that if either (a) $\lambda > 1$ or if (b) $\lambda = 1$,

$$\sum_{n=1}^{\infty} n\beta(n) < \infty \quad \text{and} \quad \sum_{n=1}^{\infty} \frac{\beta^2(n)}{\beta(2n)} < \infty,$$

then any such measure must be the appropriate reversible renewal measure. Thus in example (4.32), if $p < 1$ then the process is ergodic. The proof of this theorem is based on the relative entropy technique, which is used much as in Section 5 of Chapter IV.

The fact that the upper invariant measure for an attractive reversible nearest-particle system is a known renewal measure allows one to answer a number of questions much more completely than is possible in nonreversible situations. For the remainder of this section, we will adopt the context and assumptions of Corollary 4.30. We will assume in addition that $\lambda < \infty$, since we are interested in how the upper invariant measure depends on λ for $1 < \lambda < \infty$.

Fix a strictly positive $\beta(\cdot)$ which satisfies $\beta(n)/\beta(n+1) \downarrow 1$ as $n \uparrow \infty$ and $\sum_{n=1}^{\infty} \beta(n) = 1$, and consider the one-parameter family

$$\beta_\lambda(n) = \lambda \beta(n).$$

By Corollary 4.30, the critical value for this family is 1, and the critical system dies out if and only if $\sum_{n=1}^{\infty} n\beta(n) = \infty$. The upper invariant measure μ_λ for $\lambda > 1$ is the renewal measure with density $\lambda\beta(n)\theta^n$, where $\theta = \theta(\lambda)$

is the solution of

(4.34)
$$\phi(\theta) = \sum_{n=1}^{\infty} \beta(n)\theta^n = \lambda^{-1}.$$

Define

$$\rho(\lambda) = \mu_\lambda\{\eta: \eta(x) = 1\}$$

as in (1.2) of Chapter VI. Then

(4.35)
$$\rho(\lambda) = \frac{1}{\sum_{n=1}^{\infty} n\lambda\beta(n)\theta^n} = \frac{\phi(\theta)}{\theta\phi'(\theta)}.$$

Suppose that

(4.36)
$$\beta(n) \sim cn^{-p} \quad \text{as } n \to \infty$$

for some $p > 1$ and $c > 0$. Let

$$M = \sum_{n=1}^{\infty} n\beta(n) \qquad \text{if } p > 2, \quad \text{and}$$

$$V = \sum_{n=1}^{\infty} n^2\beta(n) - M^2 \quad \text{if } p > 3$$

be the mean and variance respectively of $\beta(\cdot)$. Then as $\theta \uparrow 1$,

$$1 - \phi(\theta) \sim \frac{c\Gamma(2-p)}{p-1}(1-\theta)^{p-1} \qquad \text{if } 1 < p < 2,$$

$$1 - \phi(\theta) \sim c(1-\theta)|\log(1-\theta)| \qquad \text{if } p = 2,$$

$$1 - \phi(\theta) \sim M(1-\theta) \qquad \text{if } p > 2,$$

$$\phi'(\theta) \sim c\Gamma(2-p)(1-\theta)^{p-2} \qquad \text{if } 1 < p < 2,$$

$$\phi'(\theta) \sim c|\log(1-\theta)| \qquad \text{if } p = 2,$$

$$\phi'(1) - \phi'(\theta) \sim \frac{c\Gamma(3-p)}{p-2}(1-\theta)^{p-2} \qquad \text{if } 2 < p < 3,$$

$$\phi'(1) - \phi'(\theta) \sim c(1-\theta)|\log(1-\theta)| \qquad \text{if } p = 3, \quad \text{and}$$

$$\phi'(1) - \phi'(\theta) \sim (V + M(M-1))(1-\theta) \quad \text{if } p > 3,$$

where Γ denotes the usual gamma function. Therefore by (4.34) and (4.35), we have the following asymptotic expressions for $\rho(\lambda)$ as $\lambda \downarrow 1$:

$$\rho(\lambda) \sim \left[\frac{(p-1)^{2-p}}{c\Gamma(2-p)} \right]^{1/(p-1)} (\lambda - 1)^{(2-p)/(p-1)} \quad \text{if } 1 < p < 2,$$

$$\rho(\lambda) \sim \frac{1}{c} |\log(\lambda - 1)|^{-1} \qquad\qquad\qquad\qquad\quad \text{if } p = 2,$$

$$\rho(\lambda) - \rho(1) \sim \frac{c\Gamma(3-p)}{(p-2)M^p} (\lambda - 1)^{p-2} \qquad\quad \text{if } 2 < p < 3,$$

$$\rho(\lambda) - \rho(1) \sim \frac{c}{M^3} (\lambda - 1)|\log(\lambda - 1)| \qquad\quad \text{if } p = 3, \quad \text{and}$$

$$\rho(\lambda) - \rho(1) \sim \frac{V}{M^3} (\lambda - 1) \qquad\qquad\qquad\qquad \text{if } p > 3.$$

This gives very explicit critical exponents for $\rho(\lambda)$ in this context.

Much more can of course be said about μ_λ by referring to the theory of renewal processes. For example, if $p > 3$ then

$$\frac{1}{\sqrt{n}} \sum_{x=-n}^{n} [\eta(x) - \rho(1)]$$

has a limiting normal distribution under μ_1, while if $2 < p < 3$, then

$$\frac{1}{n^{(p-1)^{-1}}} \sum_{x=-n}^{n} [\eta(x) - \rho(1)]$$

has a limiting stable distribution of index $p - 1$ under μ_1. For these and related facts, see Chapter XI of Feller (1971).

5. General Infinite Systems

This section is devoted to the infinite analogues of the problems and results which were discussed for finite systems in Section 2. We will begin with the analogues of Theorems 2.1 and 2.3, which are again quite easy. As of now, there is no known analogue of Theorem 2.11, so the survival results for nonreversible systems will take a different form here. For $1 \le n < \infty$, define

$$b(n) = \sum_{l+r=n+1} \beta(l, r)$$

as in Section 2. Also take $b(0) = 0$. For $\mu \in \mathcal{S}$, let

$$m_\mu = \mu\{\eta: \eta(0) = 1\},$$

$$m_\mu(n) = \mu\{\eta: \eta(0) = \eta(n) = 1, \eta(x) = 0 \text{ for } 0 < x < n\}, \quad \text{and}$$

$$m_\mu(k, l) = \mu\{\eta: \eta(0) = \eta(k) = \eta(k+l) = 1, \eta(x) = 0 \text{ for } 0 < x < k$$
$$\text{and for } k < x < k+l\}$$

for $k, l, n \geq 1$. If in addition $\mu(X') = 1$, then m_t, $m_t(n)$, and $m_t(k, l)$ will denote the corresponding quantities evaluated for the measure $\mu S(t)$. At various times in this section, we will need to compute the time derivatives of these quantities, so they will be recorded here for future reference. In each case, they follow from Theorem 3.5(d) applied to an appropriate $g \in \mathcal{D}$:

$$(5.1) \qquad \frac{d}{dt} m_t = -m_t + \sum_{n=1}^{\infty} b(n) m_t(n+1).$$

$$(5.2) \qquad \frac{d}{dt} m_t(n) = -m_t(n)[2 + b(n-1)] + \sum_{k=1}^{n-1} m_t(k, n-k)$$
$$+ 2 \sum_{k=1}^{\infty} \beta(k, n) m_t(k+n).$$

It will also be useful to keep in mind the following relations, which hold for any $\mu \in \mathcal{S}$:

$$(5.3) \qquad \sum_{n=1}^{\infty} m_\mu(n) \leq m_\mu, \qquad \sum_{n=1}^{\infty} n m_\mu(n) \leq 1.$$

$$(5.4) \qquad \sum_{k=1}^{\infty} m_\mu(k, l) \leq m_\mu(l), \qquad \sum_{l=1}^{\infty} m_\mu(k, l) \leq m_\mu(k).$$

If in addition $\mu(X') = 1$, then equality holds in both (5.3) and (5.4). In particular, it follows from (5.1), (5.2), (5.3), and (5.4) that m_t has uniformly bounded first and second derivatives, provided that $\sup_n b(n) < \infty$. The same is true for $m_t(n)$ and $m_t(k, l)$, as can be seen from analogous identities for higher-order probabilities.

Theorem 5.5. *If $b(n) \leq 1$ for all $1 \leq n < \infty$, then*

$$(5.6) \qquad \lim_{t \to \infty} \mu S(t) = \delta_0$$

for every $\mu \in \mathcal{S}$ such that $\mu(X') = 1$. If in addition the process is attractive, then (5.6) holds for all $\mu \in \mathcal{P}$.

Proof. The second statement follows immediately from the first, since in the attractive case,

$$\mu S(t) \leq \delta_1 S(t)$$

for all $t \geq 0$. To prove the first statement, use (5.1) and (5.3) to write

$$\frac{d}{dt} m_t = -m_t + \sum_{n=1}^{\infty} b(n) m_t(n+1)$$

(5.7)
$$\leq -m_t + \sum_{n=1}^{\infty} m_t(n+1)$$

$$= -m_t(1) \leq 0.$$

Therefore m_t is decreasing. Since its second derivative is uniformly bounded, it follows that

$$\lim_{t \to \infty} \frac{d}{dt} m_t = 0.$$

So, by (5.7),

$$\lim_{t \to \infty} m_t(1) = 0.$$

This is the first step in proving that

(5.8)
$$\lim_{t \to \infty} m_t(n) = 0$$

for all $n \geq 1$, which implies (5.6). To prove (5.8) for larger n, suppose it is true for all $n \leq N$ but false for $n = N+1$. By (5.7),

$$\lim_{t \to \infty} [1 - b(N)] m_t(N+1) = 0,$$

so it follows that $b(N) = 1$. Choose then k, l so that $\beta(k, l) > 0$ and $k + l = N + 1$. Use Theorem 3.5(d) again to write

(5.9) $$\frac{d}{dt} m_t(k, l) \geq -[3 + b(k-1) + b(l-1)] m_t(k, l) + \beta(k, l) m_t(N+1).$$

Since $m_t(k, l) \leq m_t(k)$, the inductive hypothesis implies that $\lim_{t \to \infty} m_t(k, l) = 0$. Since $m_t(k, l)$ has a bounded second derivative, it then follows that

$$\lim_{t \to \infty} \frac{d}{dt} m_t(k, l) = 0.$$

Therefore by (5.9), (5.8) holds for $n = N+1$ as required.   □

Note that the above proof would be greatly simplified if we strengthened the hypothesis to $b(n) \le 1 - \varepsilon$ for all n and some $\varepsilon > 0$. In this case, (5.7) would become

$$\frac{d}{dt} m_t \le -\varepsilon m_t,$$

from which $\lim_{t \to \infty} m_t = 0$ follows immediately.

Theorem 5.10. *For every number $b > 2$, there is an infinite reversible nearest-particle system which survives and satisfies $b(n) = b$ for all $1 \le n < \infty$.*

Proof. As in the proof of Theorem 2.3, choose $\beta(l, r)$ of the form (4.24) with $\beta(n)$ given by

$$\beta(n) = b \frac{(2n-2)!}{(n-1)! \, n!} \frac{1}{4^n}.$$

It was shown there that $b(n) = b$ for all $1 \le n < \infty$. The infinite system with these rates survives for $b > 2$ by Corollary 4.30, since the λ there is $b/2$. Note that $\beta(n)$ is asymptotic to a constant multiple of $n^{-3/2}$, so that this system dies out if $b \le 2$. $\square$

In connection with the previous theorem, it should be noted that if $b \le 2$, there is no infinite reversible nearest-particle system with strictly positive rates satisfying $b(n) = b$ for all n which survives. In order to show this, suppose that $\beta(n)$ is a strictly positive probability density on $\{1, 2, \ldots\}$ with finite mean and

$$\beta(l, r) = \frac{\beta(l)\beta(r)}{\beta(l+r)}$$

satisfies $b(n) = b$ for all $1 \le n < \infty$. (Recall from Theorem 4.7 that any proposed counterexample to the above statement must be of this form.) Then

$$b\beta(n) = \sum_{l+r=n} \beta(l)\beta(r)$$

for $2 \le n < \infty$. Multiplying by n and summing for $n \ge 2$ yields

$$b \sum_{n=2}^{\infty} n\beta(n) = \sum_{l,r=1}^{\infty} (l+r)\beta(l)\beta(r)$$

$$= 2 \sum_{n=1}^{\infty} n\beta(n).$$

Therefore

$$b = 2 \frac{\sum\limits_{n=1}^{\infty} n\beta(n)}{\sum\limits_{n=2}^{\infty} n\beta(n)} > 2$$

as required.

As mentioned earlier, no exact analogue of Theorem 2.11 is known for infinite nearest-particle systems. At this point, the only nonreversible infinite nearest-particle systems which we know survive are those which dominate the contact process with parameter 2 in the sense that $\beta(1, 1) \geq 4$ and $\beta(1, n) \geq 2$ for $n \geq 2$. This of course does not cover examples such as (0.5) for any value of b. In the remainder of this section, we will develop a technique for demonstrating the survival of such processes.

It is easy to see from (5.1) and (5.2) why the analysis of m_t, $m_t(n)$, etc. presents difficulties. The problem is that derivatives of quantities involving j arguments are expressed in terms of similar quantities involving $j+1$ arguments. We would like to "close up" the system of differential equations by expressing quantities involving $j+1$ arguments back in terms of those involving j arguments. The following result provides a means for doing so.

Lemma 5.11. *If $\mu \in \mathcal{S}$ and $\mu(X') = 1$, then*

$$\sum_{1 \leq k, l \leq N} m_\mu(k, l)(k+l) \log(k+l) \leq 2 \sum_{1 \leq k \leq N} m_\mu(k)k \log k + 2 \log 2.$$

for any $N \geq 1$.

Proof. Let $\phi(t) = t \log t$, which is convex. Then by (5.3) and (5.4),

$$\sum_{1 \leq k, l \leq N} m_\mu(k, l)(k+l) \log(k+l) = \sum_{1 \leq k, l \leq N} m_\mu(k, l)\phi(k+l)$$

$$\leq \tfrac{1}{2} \sum_{1 \leq k, l \leq N} m_\mu(k, l)[\phi(2k) + \phi(2l)]$$

$$\leq \sum_{k=1}^{N} \phi(2k)m_\mu(k)$$

$$= 2 \sum_{k=1}^{N} (k \log 2k)m_\mu(k)$$

$$\leq 2 \sum_{k=1}^{N} (k \log k)m_\mu(k) + 2 \log 2. \quad \square$$

For $\mu \in \mathcal{S}$ with $\mu(X') = 1$, set

$$h(\mu) = \sum_{n=1}^{\infty} m_\mu(n)n \log n.$$

Lemma 5.12. *If* $\mu \le \nu$, *then* $h(\mu) \ge h(\nu)$.

Proof. By writing

$$m_\mu(n) = M(n-1) - 2M(n) + M(n+1),$$

where

$$M(n) = \mu\{\eta(0) = \eta(1) = \cdots = \eta(n-1) = 0\},$$

we can rewrite

$$h(\mu) = \sum_{n=1}^{\infty} M(n)[(n+1)\log(n+1) - 2n\log n + (n-1)\log(n-1)].$$

The result then follows from the convexity of the function $n\log n$ and the fact that the indicator function of the set $\{\eta : \eta(0) = \eta(1) = \cdots = \eta(n-1) = 0\}$ is a decreasing function on X. □

Lemma 5.13. *If* $h(\mu) < \infty$, *then for all* $t \ge 0$,

$$h(\mu S(t)) \le h(\mu) + 2t\log 2.$$

Proof. By Lemma 5.12, it suffices to prove the result in case $\beta(l, r) = 0$ for all l, r. In this case, (5.2) and Lemma 5.11 imply that

$$\frac{d}{dt} \sum_{n=1}^{N} n(\log n) m_t(n) \le 2\log 2$$

for all $t \ge 0$ and $N \ge 1$. Therefore

$$\sum_{n=1}^{N} n(\log n) m_t(n) \le h(\mu) + 2t\log 2. \quad \square$$

Theorem 5.14. *Suppose that*

$$\liminf_{n\to\infty} \frac{1}{n} \sum_{l+r=n} \beta(l, r)[n\log n - l\log l - r\log r] > 2\log 2.$$

Then the process survives in the sense that

$$\inf_{t>0} \frac{1}{t} \int_0^t m_s \, ds > 0$$

whenever $\mu \in \mathcal{S}$ *satisfies* $\mu(X') = 1$ *and* $h(\mu) < \infty$.

Proof. Take $\mu \in \mathscr{S}$ such that $\mu(X')=1$ and $h(\mu)<\infty$, and set $h(t)=h(\mu S(t))$, which is finite by Lemma 5.13. Integrate (5.2) from 0 to t, then multiply by $n \log n$ and sum on n to obtain

$$h(t)-h(0)=-2\int_0^t h(s)\,ds-\int_0^t \sum_{n=1}^{\infty}(n\log n)m_s(n)b(n-1)\,ds$$

(5.15)
$$+\int_0^t \sum_{k,l=1}^{\infty}m_s(k,l)(k+l)\log(k+l)\,ds$$

$$+2\int_0^t \sum_{l,r=1}^{\infty}\beta(l,r)m_s(l+r)r\log r\,ds.$$

By the assumption of the theorem, there exists $c>2\log 2$ and a finite $K>c$ so that

$$\sum_{l+r=n}\beta(l,r)[n\log n-l\log l-r\log r]\geq cn-K$$

for all $n\geq 2$. Rewrite this using the symmetry of $\beta(l,r)$ and the definition of $b(n-1)$ to get

$$2\sum_{l+r=n}\beta(l,r)r\log r\leq b(n-1)n\log n-cn+K.$$

Using this bound for the last term in (5.15) and Lemma 5.11 for the next to last term, we have

$$h(t)-h(0)\leq\int_0^t\left\{2\log 2-c\sum_{n=2}^{\infty}nm_s(n)+K\sum_{n=2}^{\infty}m_s(n)\right\}ds.$$

By (5.3) (with equality since $\mu S(t)(X')=1$),

$$h(t)-h(0)\leq(2\log 2-c)t+K\int_0^t m_s\,ds.$$

Since $h(t)\geq 0$, it follows that

$$K\frac{1}{t}\int_0^t m_s\,ds\geq c-2\log 2-\frac{h(0)}{t}$$

as required. $\square$

To better understand the statement of Theorem 5.14, we should consider some examples. The centered birth model is the system in which

$$\beta(l, l) = b \qquad\qquad \text{for } 1 \le l < \infty,$$

(5.16) $$\beta(l, l+1) = \beta(l+1, l) = \frac{b}{2} \quad \text{for } 1 \le l < \infty, \quad \text{and}$$

$$\beta(l, r) = 0 \qquad\qquad \text{otherwise.}$$

In this case, Theorem 5.14 implies survival for $b > 2$. Thus this gives a nonreversible example with the properties asserted in Theorem 5.10. The uniform birth process is the one in which

(5.17) $$\beta(l, r) = \frac{b}{l+r-1} \quad \text{for all } l, r \ge 1.$$

Since

$$\sum_{l=1}^{n-1} l \log l \le \int_1^n x \log x \, dx = \frac{n^2}{2} \log n - \tfrac{1}{4}n^2 + \tfrac{1}{4},$$

Theorem 5.14 guarantees survival for $b > 4 \log 2$, which is approximately 2.77. Theorem 5.14 gives no information at all in the case of the contact process. Finally, consider the examples in (4.32). Theorem 5.14 implies that the process survives for all c if $p < 1$, and that it survives for $c > 6 \log 2 / \pi^2$ if $p = 1$. It gives no information at all for $p > 1$. Comparing with the exact results which Corollary 4.30 provides in this case, we see that Theorem 5.14 is far from being best possible. However, it is the only technique presently available for demonstrating survival for infinite nearest-particle systems which satisfy

$$\lim_{n \to \infty} \beta(1, n) = 0.$$

6. Notes and References

Section 1. Theorems 1.10, 1.18, and 1.20 are due to Griffeath and Liggett (1982).

Section 2. This section is based on Liggett (1984).

Section 3. Infinite nearest-particle systems were introduced by Spitzer (1977). In fact, this preceded and motivated the study of the finite systems which we considered in the first two sections. Infinite nearest-particle systems were first constructed by Gray (1978). Analogous systems on the real line were constructed and studied by Holley and Stroock (1978c) and by Cocozza and Kipnis (1980). The treatment in this section is based on Liggett (1983a).

Section 4. Theorems 4.2 and 4.7 are due to Spitzer (1977). Theorem 4.25 was proved by Holley for birth rates of the form (0.2) using a somewhat different technique. (His result is stated in Spitzer's paper.) The general form of Theorem 4.25 was proved by Liggett (1983a). The asymptotics for $\rho(\lambda)$ at the end of the section are more precise versions of results in Griffeath and Liggett (1982). Renormalization results for infinite reversible nearest-particle systems were proved by Holley and Stroock (1978b, 1979a).

Section 5. Theorem 5.14 is a refinement of the results in Bramson and Gray (1981), which dealt with the special cases (5.16) and (5.17). They used

$$h_\beta(\mu) = \sum_{n=1}^{\infty} m_\mu(n) n^{\beta+1}$$

for small $\beta > 0$ in place of the $h(\mu)$ used here.

7. Open Problems

1. Does

(7.1)
$$\lim_{\lambda \downarrow 1} \frac{P^*(\tau = \infty)}{\lambda - 1}$$

exist for the one-parameter family of reversible finite nearest-particle systems given by (1.13) where $f(\cdot)$ satisfies (1.14), (1.15), and (1.16)? By (1.11) and (1.17), such a limit would be finite and positive.

2. How does $P^*(\tau = \infty)$ behave as $\lambda \downarrow 1$ for the family given by (1.13) if (1.14) and (1.15) hold but (1.16) fails? For example, if $f(n)$ is a constant multiple of n^{-p} for $1 < p \le 3$, does $P^*(\tau = \infty)$ tend to zero like a constant multiple of some power of $\lambda - 1$, where that power depends on p in a nontrivial way? It would already be interesting to show that the limit in (7.1) is infinite if (1.16) fails. Based on the asymptotics for $\rho(\lambda)$ given at the end of Section 4, one might guess that $P^*(\tau = \infty)$ behaves like a constant multiple of $(\lambda - 1)|\log(\lambda - 1)|$ if $p = 3$ and of $(\lambda - 1)^{p-2}$ if $2 < p < 3$. Of course, there are analogous questions in the critical case based on Theorem 1.18 and (1.19).

3. What can be said about the distribution of τ for a finite reversible system in the critical case $\lambda = 1$? By Theorem 1.10, $\tau < \infty$ a.s., while by Theorem

1.20 (or more precisely, by (1.12)), $E^*(\tau) = \infty$. What moments does τ have? Is it in the domain of attraction of a stable law?

4. In the context of Theorem 1.20,

$$(1 - \lambda)tP^*(\tau \in dt)$$

is a probability distribution on $[0, \infty)$. As $\lambda \uparrow 1$, this distribution tends to the pointmass at ∞. When properly normalized, is there a limiting distribution? In order to answer this question via the method of moments, for example, one would have to find the asymptotic behavior of the higher moments of τ as $\lambda \uparrow 1$. A first step is then to determine how $E^*\tau^2$ diverges as $\lambda \uparrow 1$.

5. Define $g_A(x)$ by

$$g_A(x) = \nu_\lambda\{\eta\colon \eta(x) = 1 \text{ and } \eta(y) = 0 \text{ for all } y \in A\},$$

where ν_λ is the upper invariant measure for the one-dimensional contact process (see (1.1) of Chapter VI). This g_A satisfies the first conclusion of Lemma 2.6 by Theorems 1.9(c) and 1.10(a) of Chapter VI. Prove that this g_A satisfies the conclusion of Lemma 2.8 as well. A new proof would be required, since (2.10) fails for this g_A. If this could be proved, then the proof of Theorem 2.11 could be carried out with this choice of g_A. The advantage would be that the hypothesis of Theorem 2.11 could be weakened to $\inf_n b(n) > 2\lambda_c$, where λ_c is the critical value for the one-dimensional contact process.

6. For each number $b \geq 0$, determine which of the following three statements is true:

(a) Every finite nearest-particle system satisfying $b(n) = b$ for all $1 \leq n \leq \infty$ dies out.
(b) Some finite nearest-particle systems satisfying $b(n) = b$ for all $1 \leq n \leq \infty$ die out, and some survive.
(c) All finite nearest-particle systems satisfying $b(n) = b$ for all $1 \leq n \leq \infty$ survive.

By Theorem 2.1, (a) is true for $b \in [0, 1]$. By Theorem 2.3, (b) is true for $b \in (2, 2\lambda_c)$, where λ_c is the critical value for the one-dimensional contact process. By Theorem 2.11, (c) is true for $b \geq 4$. A solution to Problem 5 above would imply that (c) is true for $b > 2\lambda_c$. The answer for $b = 2\lambda_c$ is probably (b). (See Problem 2 of Chapter VI.) The real mystery is what happens if $b \in (1, 2]$. No reversible finite nearest-particle system can survive for b in this range by Theorem 1.10. Can a nonreversible one survive?

7. Prove an analogue of Theorem 2.28 of Chapter VI for nearest-particle systems. This would be of interest either in the reversible case or in general.

8. Consider a one-parameter family of finite nearest-particle systems, with birth rates given by

$$\beta_\lambda(l, r) = \lambda\beta(l, r)$$

for some fixed choice $\beta(l, r)$. Assume that $\beta(l, r)$ is nonincreasing in l and r. Then there is a critical value $\lambda_c \in [0, \infty]$ so that the system survives for $\lambda > \lambda_c$ and dies out for $\lambda < \lambda_c$. Find necessary and sufficient conditions on $\beta(l, r)$ so that $0 < \lambda_c < \infty$. By Theorems 2.1 and 2.11, a sufficient condition is that $\{b(n), 1 \le n \le \infty\}$ be bounded away from zero and infinity. If $\beta(l, r)$ is given by (1.3) and (1.4) for a positive summable sequence $\beta(n)$, then $0 < \lambda_c < \infty$ is always true by Theorem 1.10.

9. Suppose that $0 < \lambda_c < \infty$ in the context of Problem 8. Is it the case that the critical system with $\lambda = \lambda_c$ always dies out? This is true in the reversible case by Theorem 1.10. In the special case of the one-dimensional contact process, this conjecture appeared before as Problem 2 of Chapter VI.

10. Suppose that $\beta(l, r)$ is attractive and that

$$\beta(l, \infty) = \lim_{r \to \infty} \beta(l, r)$$

and is summable in l. Prove that if the finite system survives, then so does the infinite system. This is true for the contact process by duality, and is true whenever $\beta(l, r)$ is of the form

(7.2) $$\beta(l, r) = \frac{\beta(l)\beta(r)}{\beta(l+r)}$$

for a positive $\beta(\cdot)$ by Theorem 1.10 and Corollary 4.30.

11. In the contexts of Problems 8 and 10, show that the finite and infinite systems have the same critical value. This is true for the contact process by duality, and is the case when the rates are of the form (7.2) by Theorem 1.10 and Corollary 4.30. Note that in the latter case, the result is proved by computing the critical values for the finite and infinite systems separately, rather than by comparing the two systems directly.

12. Prove the analogue of Theorem 2.11 for infinite systems. A solution to Problem 10 would yield a solution to this problem in the attractive case.

13. Is there an infinite nearest-particle system which survives and satisfies $b(n) = b < 2$ for all $1 \le n < \infty$? If so, what is the smallest b for which there is one? As a result of the comments following Theorem 5.10, such a system would necessarily be nonreversible. Reasonable examples to consider are those given in (5.16) and (5.17). Problems 15 and 16 discuss a possible approach to this problem.

14. Prove a comparison theorem which would say that if one set of birth rates is obtained from another by appropriately increasing $\beta(l, r)$ when l and r are approximately equal and decreasing $\beta(l, r)$ when they are not, then the second system is more likely to survive than the first. As a more concrete version of this problem, consider the following four one-parameter families of infinite nearest-particle systems:

(1) the contact process with $\lambda = b/2$;
(2) the reversible process constructed in the proof of Theorem 5.10;
(3) the uniform birth process in (5.17); and
(4) the centered birth process in (5.16).

Note that each of these systems satisfies $b(n) = b$ for all $1 \leq n < \infty$. Let b_1, b_2, b_3, b_4 be the critical values for each of these examples. (The critical values exist by attractiveness in the first three cases. Since the centered birth process is not attractive, b_4 is not known to exist. For purposes of this problem, we will assume that it does.) Show that $b_4 \leq b_3 \leq b_2 \leq b_1$. These inequalities are consistent with the information we have about these critical values: $1 \leq b_4 \leq 2$, $1 \leq b_3 \leq 2.77$, $b_2 = 2$, and $3 \leq b_1 \leq 4$, and in fact the last one follows from this information.

15. For $n \geq 1$ and $t \geq 0$, set

$$f_n(t) = \delta_1 S(t)\{\eta: \eta(1) = \eta(2) = \cdots = \eta(n) = 0\}, \quad \text{and}$$

$$f_{k,l,n}(t) = \delta_1 S(t)\{\eta: \eta(1) = \cdots = \eta(k) = \eta(k+l+1)$$

$$= \cdots = \eta(k+l+n) = 0\}.$$

Show under reasonable conditions on the birth rates that

(7.3) $f_{k,l,n}(t) \leq f_{k+n}(t)$

for all $k, l, n \geq 1$. These conditions should be satisfied by the uniform birth process defined in (5.17), for example. Perhaps attractiveness is the right assumption. A potential application of (7.3) will be described in Problem 16. There are several reasons for expecting (7.3) to be true in some generality. First, it is true for the contact process. Theorems 1.9(c) and 1.10(a) of Chapter VI together give (7.3) in the limit as $t \to \infty$. However, it was shown in the proofs of those theorems that (7.3) holds for finite t as well. A second reason for believing (7.3) is that it is true for $k = l = n = 1$ whenever the birth rates are attractive and satisfy

(7.4) $\beta(2, n) - \beta(2, n+1) \geq \beta(1, n+1) - \beta(1, n+2)$ for $n \geq 1$.

(For an example which satisfies this, see (5.17).) To see this, compute $f'_{1,1,1}(t)$ and $f'_2(t)$ and use (7.4) to obtain

$$\frac{d}{dt}[f_2(t) - f_{1,1,1}(t)] \geq -2[f_2(t) - f_{1,1,1}(t)][1 + \beta(1, 1)],$$

from which it follows that

$$f_2(t) - f_{1,1,1}(t) \geq 0$$

for all t. A final reason for believing (7.3) is the following. Suppose that the process is attractive. Then

$$f_{k,l,n}(t) \geq f_k(t)f_n(t)$$

by Theorem 2.14 of Chapter II. Also

(7.5) $$\lim_{l \to \infty} f_{k,l,n}(t) = f_k(t)f_n(t)$$

for fixed t. Therefore, if the convergence in (7.5) were monotone, it would follow that $f_{k,l,n}(t)$ is decreasing in l. Since

$$f_{k,0,n}(t) = f_{k+n}(t),$$

(7.3) would be a consequence of this monotonicity.

16. Give better bounds on the critical values b_3 and b_4 which were defined in Problem 14 than those provided by Theorem 5.14. In the proof of that theorem we used Lemma 5.11 to "close up" the infinite system of differential equations in (5.2). There is another way to close it up based on (7.3), which would give better results (provided one could prove (7.3)). The idea is to rewrite (5.2) in terms of $f_n(t)$ and $f_{k,1,n}(t)$ using the relations

$$m_t(n) = f_{n-1}(t) - 2f_n(t) + f_{n+1}(t), \quad \text{and}$$

$$m_t(k, l) = f_{k-1,1,l-1}(t) - f_{k,1,l-1}(t) - f_{k-1,1,l}(t)$$
$$- f_{k+l-1}(t) + 2f_{k+l}(t) + f_{k,1,l}(t) - f_{k+l+1}(t).$$

If this is done, then using (7.3) we obtain the following system of differential inequalities:

(7.6)
$$f'_n(t) \leq n f_{n-1}(t) - [n + b(n)] f_n(t)$$
$$+ \sum_{j=1}^{\infty} f_{n+j}(t)[2\gamma_n(j) - \gamma_n(j-1) - \gamma_n(j+1)],$$

where

$$\gamma_n(j) = \sum_{l+r=n+j} \beta(l, r)[l \wedge r \wedge j \wedge n],$$

and $\wedge$ denotes the minimum. Suppose now that $\gamma_n(j)$ is concave in j for each n. Consider the Markov chain X_t on $\{0, 1, 2, \ldots\}$ with an absorbing state at 0 which has the following transition rates for $n, j \geq 1$:

$$n \to n-1 \quad \text{at rate } n, \quad \text{and}$$

$$n \to n+j \quad \text{at rate } 2\gamma_n(j) - \gamma_n(j-1) - \gamma_n(j+1).$$

Since

$$\sum_{j=1}^{\infty} [2\gamma_n(j) - \gamma_n(j-1) - \gamma_n(j+1)] = \gamma_n(1) = b(n),$$

the right side of (7.6) is the result of applying the generator of this chain to $f_n(t)$ (as a function of n). Noting that $f_n(0) = 0$ for $n \geq 1$ and $f_0(0) = 1$, it then follows that

$$f_n(t) \leq P^n(X_t = 0),$$

so that the transience of X_t implies the survival of the nearest-particle system. For the uniform birth process in (5.17),

$$\gamma_n(j) = n \frac{jb}{n+j-1},$$

so X_t goes from n to $n+j$ at rate

$$\frac{2n(n-1)b}{(n+j)(n+j-1)(n+j-2)}.$$

In this case, the chain is transient if and only if $b > 2$. For the centered birth process in (5.16),

$$\gamma_n(j) = (n \wedge j)b,$$

where $[\cdot]$ denotes the integer part function. Therefore X_t goes from n to $2n$ at rate b. This chain is transient if and only if $b > 1/\log 2 \approx 1.44$. If this program could be carried out, we would then obtain the improved upper bounds $b_3 \leq 2$ and $b_4 \leq 1/\log 2$. Of course, since the centered birth process is not attractive, there is some doubt about whether it satisfies (7.3) as would be required. Another approach to the proof of $b_3 \leq 2$ is described in Problem 14.

17. What limit theorems can be proved in the critical attractive nonergodic reversible case which is the subject of part (b) of Corollary 4.30? If μ is stochastically larger than the invariant renewal measure μ_β, then we know that $\lim_{t\to\infty} \mu S(t) = \mu_\beta$. On the other hand, if μ concentrates on finite configurations, then $\lim_{t\to\infty} \mu S(t) = \delta_0$ by Theorem 1.10. What happens for other μ? In particular, what happens if μ is a product measure with a constant density?

18. How rapidly does $\delta_1 S(t)$ tend to δ_0 in the subcritical reversible case (i.e., in Corollary 4.30 with $\lambda < 1$)? Presumably the rate is exponential. If so, how rapidly does

$$\int_0^\infty P^1[\eta_t(0) = 1]\, dt$$

tend to ∞ as $\lambda \uparrow 1$?

Chapter VIII

The Exclusion Process

The exclusion process differs from the spin systems which are the subject of the previous five chapters in that two (rather than one) coordinates of η_t change at a time. In order to describe this process, let $p(x, y)$ be the transition probabilities for a discrete time Markov chain on the countable set S:

$$p(x, y) \geq 0, \quad \text{and} \quad \sum_y p(x, y) = 1.$$

Particles move on S according to the following rules:

(a) there is always at most one particle per site;
(b) a particle at x waits an exponential time with parameter one, and then chooses a y with probability $p(x, y)$;
(c) if y is vacant at that time, it moves to y, while if y is occupied, it remains at x.

Let $\eta_t(x) = 1$ if x is occupied at time t, and $\eta_t(x) = 0$ if x is vacant at time t. Then η_t is the exclusion process corresponding to $p(x, y)$. Rule (c) above describes the exclusion interaction which gives this process its name.

The exclusion process is described (and constructed) more formally in the following way. For $\eta \in X = \{0, 1\}^S$ and $x, y \in S$, let η_{xy} be the element of X which is defined by

$$\eta_{xy}(u) = \begin{cases} \eta(y) & \text{if } u = x, \\ \eta(x) & \text{if } u = y, \\ \eta(u) & \text{if } u \neq x, y. \end{cases}$$

For $f \in \mathcal{D}$, let

(0.1) $$\Omega f(\eta) = \sum_{\substack{\eta(x)=1 \\ \eta(y)=0}} p(x, y)[f(\eta_{xy}) - f(\eta)].$$

By Theorem 3.9 of Chapter I, the closure of Ω is the generator of the semigroup $S(t)$ of a Feller process on X provided that

(0.2) $$\sup_{y} \sum_{x} p(x, y) < \infty.$$

(See Example 3.15 of Chapter I.) In most of this chapter we will assume either that $p(x, y)$ is symmetric in x and y, or that $S = Z^d$ for some d and $p(x, y) = p(0, y - x)$. In either case, (0.2) is automatically satisfied. Thus we can take η_t to be the process constructed in Chapter I.

The first problems to be solved for the exclusion process are similar to those we have studied in the context of spin systems:

  (a) characterize $\mathscr{I}$ as explicitly as possible; and
  (b) for each $\nu \in \mathscr{I}$, find all $\mu \in \mathscr{P}$ such that

$$\lim_{t \to \infty} \mu S(t) = \nu.$$

Unlike most of the spin systems which have appeared earlier (the exception is the voter model), the exclusion process has the property that the "density" of particles in η_t is preserved in time. Thus we would expect to have for each $\alpha \in [0, 1]$ an invariant measure which concentrates on configurations of density α. This turns out to be the case whenever $p(x, y)$ is doubly stochastic, and hence in particular in the symmetric and translation invariant cases which will be considered in this chapter. In fact, the invariant measure with density $\alpha \in [0, 1]$ is simply the product measure with that density. The closed convex hull of the set of product measures with constant density is the set of exchangeable measures. Therefore problem (a) can be rephrased by asking when $\mathscr{I}$ is equal to the set of exchangeable measures on X.

Problems (a) and (b) are essentially completely resolved for symmetric systems in Section 1. The most important tool which is used there is duality. It permits the reduction of these problems to others involving harmonic functions and convergence for systems of finitely many particles. It turns out that $\mathscr{I}$ consists exactly of the exchangeable probability measures on X if and only if $p(x, y)$ has no nonconstant bounded harmonic functions.

Duality fails in the asymmetric case, so other techniques are required for the analysis of the general translation invariant systems which are the subject of Section 3. The main tool used there is coupling. The results are much less complete than in the symmetric case. However, it will be shown there that $\mathscr{I} \cap \mathscr{S}$ consists exactly of the exchangeable measures. All of $\mathscr{I}$ will be described in certain special cases, and some convergence theorems will be proved. Several ideas which are needed in this section are described in Section 2.

Section 4 treats the problem of describing the motion of a tagged particle in the exclusion system. Rather precise information will be obtained in one dimension in the cases $p(x, x+1) = 1$ for all x and $p(x, x+1) = p(x, x-1) = \frac{1}{2}$ for all x. Results in greater generality will be described, but not proved.

The ergodic theory in Sections 1 and 3 describes the behavior of the exclusion process at fixed sites as $t \to \infty$. A natural problem is to determine the limiting behavior of the system at sites near $x(t)$ at time t, where $x(t) \to \infty$. Section 5 presents a solution to this problem.

1. Ergodic Theorems for Symmetric Systems

Throughout this section, we will assume that $p(x, y) = p(y, x)$, and that the Markov chain on S with these transition probabilities is irreducible. The main tool which will be exploited here is the self-duality of the symmetric exclusion process, which is described in the first theorem. The symmetry assumption is essential in this result. As usual, let Y denote the collection of all finite subsets of S, and identify $\eta \in X$ such that $\sum_x \eta(x) < \infty$ with $A \in Y$ by

$$A = \{x \in S : \eta(x) = 1\}.$$

Then A_t will denote the finite exclusion process with the same transition probabilities $p(x, y)$.

Theorem 1.1. *If $A \in Y$ and $\eta \in X$, then*

$$P^{\eta}[\eta_t = 1 \text{ on } A] = P^{A}[\eta = 1 \text{ on } A_t]$$

for all $t \geq 0$.

Proof. The proof is essentially the same as that of Theorem 4.13 of Chapter III. Let

$$u_{\eta}(t, A) = P^{\eta}[\eta_t = 1 \text{ on } A] = S(t)H(\cdot, A)(\eta),$$

where

$$H(\eta, A) = \begin{cases} 1 & \text{if } \eta(x) = 1 \text{ for all } x \in A, \\ 0 & \text{otherwise.} \end{cases}$$

Then $H(\cdot, A) \in \mathcal{D}$ for each $A \in Y$, so by (0.1),

$$
\begin{aligned}
\Omega H(\cdot, A)(\eta) &= \sum_{\substack{\eta(x)=1 \\ \eta(y)=0}} p(x, y)[H(\eta_{xy}, A) - H(\eta, A)] \\
&= \tfrac{1}{2} \sum_{x,y} p(x, y)[H(\eta_{xy}, A) - H(\eta, A)] \\
&= \tfrac{1}{2} \sum_{x,y} p(x, y)[H(\eta, A_{xy}) - H(\eta, A)] \\
&= \sum_{\substack{x \in A \\ y \notin A}} p(x, y)[H(\eta, A_{xy}) - H(\eta, A)].
\end{aligned}
$$

(1.2)

Here A_{xy} is obtained from A in the same way that η_{xy} is obtained from η. The symmetry of $p(x, y)$ was used in the second and fourth steps above. By Theorem 2.9(c) of Chapter I and (1.2),

$$\frac{d}{dt} u_\eta(t, A) = S(t)\Omega H(\,\cdot\,, A)(\eta)$$

$$= \sum_{\substack{x \in A \\ y \notin A}} p(x, y)[S(t)H(\,\cdot\,, A_{xy})(\eta) - S(t)H(\,\cdot\,, A)(\eta)]$$

$$= \sum_{\substack{x \in A \\ y \notin A}} p(x, y)[u_\eta(t, A_{xy}) - u_\eta(t, A)].$$

The unique solution to these differential equations with initial condition $H(\eta, A)$ is

$$E^A H(\eta, A_t) = P^A[\eta = 1 \text{ on } A_t],$$

so the result follows. $\square$

For any probability measure μ on X, let

$$\hat{\mu}(A) = \int H(\eta, A)\mu(d\eta) = \mu\{\eta \colon \eta = 1 \text{ on } A\}$$

for all $A \in Y$. Integrating the identity in the statement of Theorem 1.1 with respect to μ gives the following useful reformulation of the self-duality relation.

Corollary 1.3. *For any $\mu \in \mathcal{P}$, set $\mu_t = \mu S(t)$. Then*

$$(1.4) \qquad\qquad\qquad \hat{\mu}_t(A) = E^A \hat{\mu}(A_t)$$

for all $A \in Y$. In particular, $\mu \in \mathcal{I}$ if and only if $\hat{\mu}$ is harmonic for the chain A_t on Y.

Note that the cardinality $|A_t|$ is independent of time. Therefore one way of viewing (1.4) is to say that the n-dimensional marginals of μ_t depend only on the n-dimensional marginals of μ for the same n. This property is satisfied only infrequently in the theory of interacting particle systems, but it certainly provides a powerful tool for analyzing μ_t when it holds. The voter model satisfies the similar, and equally useful, property that the n-dimensional marginals of μ_t depend only on the k-dimensional marginals of μ for $k \leq n$. In fact, some of the analysis of the symmetric exclusion process draws on results from Section 1 of Chapter V.

A probability measure μ on X is said to be exchangeable if $\hat{\mu}(A)$ depends on A only through its cardinality. The class of all exchangeable measures is a compact convex set in $\mathcal{P}$. Its extreme points are exactly the product measures on X with constant density by de Finetti's Theorem (see Section VII.4 of Feller (1971)). In fact, any exchangeable measure can be written (uniquely) in the form

$$\int_0^1 \nu_\alpha \gamma(d\alpha),$$

where ν_α is the product measure with density α, and γ is a probability measure on $[0, 1]$. Since $|A_t|$ is independent of time, one immediate consequence of Corollary 1.3 is that every exchangeable measure is invariant for the symmetric exclusion process. It is easy to check that the exchangeable measures are even reversible for the process, but we will not need this fact.

Let $X_1(t), X_2(t), \ldots$ be independent Markov chains on S with transition probabilities

$$p_t(x, y) = e^{-t} \sum_{n=0}^{\infty} \frac{t^n}{n!} p^{(n)}(x, y).$$

where $p^{(n)}(x, y)$ are the n-step transition probabilities corresponding to $p(x, y)$. Then $\vec{X}(t) = (X_1(t), \ldots, X_n(t))$ is a Markov chain on S^n. Let

$$T_n = \{\vec{x} \in S^n : x_i \neq x_j \text{ for all } 1 \leq i \neq j \leq n\}.$$

The following function on $\bigcup_{n=2}^{\infty} S^n$ will play much the same role that the one defined in (1.14) of Chapter V did for the voter model: if $\vec{x} \in S^n$, then

(1.5) $g(\vec{x}) = P^{\vec{x}}[\vec{X}(t) \notin T_n \text{ for some } t \geq 0].$

Two cases will arise naturally in this section: $g \equiv 1$ and $g \not\equiv 1$. In the translation invariant case, $g \equiv 1$ is equivalent to the recurrence of $X_i(t)$, since $X_1(t) - X_2(t)$ is the same chain run at twice the speed. In general, however, $g \equiv 1$ implies the recurrence of $X_i(t)$, but not conversely. To see that $g \equiv 1$ implies the recurrence of $X_i(t)$, use the Chapman–Kolmogorov equation and the assumed symmetry to write

$$p_{2t}(x, x) = \sum_y p_t(x, y) p_t(y, x)$$

$$= \sum_y [p_t(x, y)]^2$$

$$= P^{(x,x)}[X_1(t) = X_2(t)].$$

Thus if $X_i(t)$ is transient,

$$(1.6) \qquad \int_0^\infty P^{(x,x)}[X_1(t) = X_2(t)] \, dt < \infty,$$

so that $g \neq 1$. An example of a symmetric irreducible recurrent Markov chain which satisfies $g \neq 1$ is given in Liggett (1974a).

In both cases, but particularly in case $g \neq 1$, we will need to compare the finite exclusion process A_t with the finite independent system $\vec{X}(t)$. The latter has semigroup

$$U_n(t)f(\vec{x}) = \sum_{\vec{y} \in S^n} \prod_{i=1}^n p_t(x_i, y_i)f(\vec{y}).$$

The former will be regarded as the chain on T_n which is obtained from $\vec{X}(t)$ by suppressing transitions to points in $S^n \backslash T_n$. Let $V_n(t)$ be its semigroup. The comparison we need is based on the following result, which says that in some sense, the particles in the exclusion process are more spread out than those in the independent system. The formal statement involves positive definite functions on S^n. A bounded symmetric function f on S^2 is said to be positive definite provided that

$$\sum_{x,y \in S} f(x, y)\beta(x)\beta(y) \geq 0$$

whenever $\sum_x |\beta(x)| < \infty$ and $\sum_x \beta(x) = 0$. A bounded symmetric function f on S^n is said to be positive definite if it is a positive definite function of each pair of variables.

Proposition 1.7. *Suppose that f is a bounded symmetric positive definite function on S^n for some $n \geq 2$. Then*

$$V_n(t)f(\vec{x}) \leq U_n(t)f(\vec{x})$$

for all $\vec{x} \in T_n$.

Proof. Let U and V be the generators of $U_n(t)$ and $V_n(t)$ respectively. Then

$$Uf(\vec{x}) - Vf(\vec{x})$$

$$= \sum_{i,j=1}^n p(x_i, x_j)[f(x_1, \ldots, x_{i-1}, x_j, x_{i+1}, \ldots, x_n) - f(\vec{x})]$$

$$= \tfrac{1}{2} \sum_{i,j=1}^n p(x_i, x_j)[f(x_1, \ldots, x_{i-1}, x_j, x_{i+1}, \ldots, x_n)$$

$$+ f(x_1, \ldots, x_{j-1}, x_i, x_{j+1}, \ldots, x_n) - 2f(\vec{x})]$$

for $\vec{x} \in T_n$ by the symmetry of $p(x, y)$. Therefore

$$(1.8) \qquad\qquad Vf(\vec{x}) \le Uf(\vec{x}) \quad \text{for } \vec{x} \in T_n$$

whenever f is bounded, symmetric, and positive definite. This class of functions is mapped into itself by $U_n(t)$, since

$$\sum_{x_1, x_2} \beta(x_1)\beta(x_2) U_n(t) f(\vec{x}) = \sum_{\vec{y} \in S^n} \gamma(y_1)\gamma(y_2) \prod_{j=3}^{n} p_t(x_j, y_j) f(\vec{y}),$$

where $\gamma(y) = \sum_x \beta(x) p_t(x, y)$. Combining this observation with (1.8), we see that

$$VU_n(t) f(\vec{x}) \le UU_n(t) f(\vec{x}) \quad \text{for } \vec{x} \in T_n$$

for any bounded, symmetric, positive definite function. The result follows now from the integration by parts formula

$$U_n(t) - V_n(t) = \int_0^t V_n(t-s)[U-V] U_n(s) \, ds$$

and the fact that $V_n(t)$ maps functions which are nonnegative on T_n to functions which are nonnegative on T_n. $\quad\square$

Corollary 1.9. *Suppose that μ is a probability measure on X, and set $\mu_t = \mu S(t)$. Then*

$$\hat{\mu}_t(A) \le E^{\vec{x}} \hat{\mu}(\{X_1(t), \dots, X_n(t)\}),$$

where $|A| = n$ and $A = \{x_1, \dots, x_n\}$.

Proof. Let $f(\vec{x}) = f(x_1, \dots, x_n) = \mu\{\eta: \eta(x_1) = \cdots = \eta(x_n) = 1\}$ for $\vec{x} \in S^n$. Then f is positive definite, so that by Corollary 1.3 and Proposition 1.7,

$$\begin{aligned}
\hat{\mu}_t(A) &= E^A \hat{\mu}(A_t) \\
&= V_n(t) f(x_1, \dots, x_n) \\
&\le U_n(t) f(x_1, \dots, x_n) \\
&= E^{\vec{x}} \hat{\mu}(\{X_1(t), \dots, X_n(t)\}). \quad\square
\end{aligned}$$

We are now prepared to prove the main results in case $g \equiv 1$, where g is the function defined in (1.5).

Theorem 1.10. *Suppose that $g \equiv 1$. If f is a bounded symmetric function on T_n such that $V_n(t) f = f$, then f is constant on T_n.*

Proof. The basic idea of the proof is to construct a successful coupling of two copies of A_t which have the same cardinality. This idea was discussed at some length in Section 1 of Chapter II in a general discrete time context. In order to show that f is constant on T_n, it is enough to prove that $f(\vec{x}) = f(\vec{y})$ whenever $\vec{x}$ and $\vec{y}$ differ at one coordinate. To do so, define the Markov chain $\vec{Z}(t)$ on

$$T = \{\vec{z} \in S^{n+1} \colon z_i \neq z_j \text{ whenever } i \neq j \text{ and } \{i, j\} \neq \{n, n+1\}\}$$

in the following way:

(a)   the set $\{\vec{z} \in T \colon z_n = z_{n+1}\}$ is closed for the chain, and on this set the process $(Z_1(t), \ldots, Z_n(t))$ has semigroup $V_n(t)$; and

(b)   on the set $\{\vec{z} \in T \colon z_n \neq z_{n+1}\}$, the chain has the transition rates below:

Transition	Rate
$\vec{z} \to (z_1, \ldots, z_{i-1}, u, z_{i+1}, \ldots, z_n, z_{n+1})$	$p(z_i, u)$ for $u \neq z_j$ for all j
$\vec{z} \to (z_1, \ldots, z_{n-1}, z_n, z_n)$	$p(z_{n+1}, z_n)$
$\vec{z} \to (z_1, \ldots, z_{n-1}, z_{n+1}, z_{n+1})$	$p(z_n, z_{n+1})$
$\vec{z} \to (z_1, \ldots, z_{i-1}, z_n, z_{i+1}, \ldots, z_{n-1}, z_i, z_{n+1})$	$p(z_i, z_n)$ for $i < n$
$\vec{z} \to (z_1, \ldots, z_{i-1}, z_{n+1}, z_{i+1}, \ldots, z_{n-1}, z_n, z_i)$	$p(z_i, z_{n+1})$ for $i < n$.

The easily verified properties of this coupling are:

(a) $(Z_1(t), \ldots, Z_n(t))$ and $(Z_1(t), \ldots, Z_{n-1}(t), Z_{n+1}(t))$ are Markovian and have semigroup $V_n(t)$; and

(b) $(Z_n(t), Z_{n+1}(t))$ is Markovian and has the same law as the process with semigroup $V_2(t)$ until the first time that $Z_n(t) = Z_{n+1}(t)$; after that time $Z_n(t) = Z_{n+1}(t)$.

Note that the symmetry of $p(x, y)$ is used in verifying property (b). Now suppose that $\vec{x}, \vec{y} \in T_n$ satisfy $x_i = y_i$ for $i < n$ and $x_n \neq y_n$. Take $\vec{z} = (x_1, x_2, \ldots, x_n, y_n)$ as the initial state of $\vec{Z}(t)$. By property (a) above and the fact that $V_n(t)f = f$ on T_n,

(1.11)
$$f(\vec{x}) = V_n(t)f(\vec{x}) = Ef(Z_1(t), \ldots, Z_n(t)) \quad \text{and}$$
$$f(\vec{y}) = V_n(t)f(\vec{y}) = Ef(Z_1(t), \ldots, Z_{n-1}(t), Z_{n+1}(t)).$$

By property (b) above and the fact that $g \equiv 1$,

$$P[Z_n(t) = Z_{n+1}(t) \text{ for all large } t] = 1.$$

Since f is bounded, this and (1.11) imply that $f(\vec{x}) = f(\vec{y})$. $\square$

If $\alpha(x)$ is a function from S to $[0, 1]$, let ν_α be the product measure on X with marginals

$$\nu_\alpha\{\eta: \eta(x) = 1\} = \alpha(x).$$

In the next result, only constant α's will be relevant. Nonconstant α's will arise in the treatment of the case $g \not\equiv 1$.

Theorem 1.12. *Suppose that* $g \equiv 1$. *Then*

$$\mathcal{I}_e = \{\nu_\alpha: \alpha \in [0, 1]\}.$$

Proof. If $\alpha \in [0, 1]$, $\hat{\nu}_\alpha(A) = \alpha^{|A|}$. Since this depends on A only through its cardinality and $|A_t|$ is independent of t, it follows from Corollary 1.3 that $\nu_\alpha \in \mathcal{I}$. Conversely, suppose $\nu \in \mathcal{I}$. By Corollary 1.3 and Theorem 1.10, there are constants $c(n)$ for $n \geq 1$ so that

$$\hat{\nu}(A) = c(|A|).$$

Therefore ν is exchangeable, and hence is a mixture of $\{\nu_\alpha: \alpha \in [0, 1]\}$ by de Finetti's Theorem. (See Section VII.4 of Feller (1971).) □

We now know $\mathcal{I}$ completely in case $g \equiv 1$. Next we will prove the main convergence theorem in this case.

Theorem 1.13. *Assume that* $g \equiv 1$, *and take* $\mu \in \mathcal{P}$ *and* $\alpha \in [0, 1]$. *Then*

$$(1.14) \qquad \qquad \lim_{t \to \infty} \mu S(t) = \nu_\alpha$$

if and only if

$$(1.15) \qquad \qquad \lim_{t \to \infty} E^x \hat{\mu}(\{X_1(t)\}) = \alpha \quad and$$

$$(1.16) \qquad \qquad \lim_{t \to \infty} E^{\{x,y\}} \hat{\mu}(A_t) = \alpha^2$$

for all $x \neq y$. *A sufficient condition for* (1.14) *is that* (1.15) *and*

$$(1.17) \qquad \qquad \lim_{t \to \infty} E^{\{x,y\}} \hat{\mu}(\{X_1(t), X_2(t)\}) = \alpha^2$$

hold.

Proof. That (1.14) implies (1.15) and (1.16) is an immediate consequence of Corollary 1.3. For the converse, assume that (1.15) and (1.16) are satisfied, and take a sequence $t_n \to \infty$ such that $\nu = \lim_{n \to \infty} \mu S(t_n)$ exists. Then by Lemma 1.26 of Chapter V and Corollary 1.3, $\hat{\nu}$ is harmonic for A_r. Hence by Theorem 1.10, there are constants $c(n)$ for $n \geq 1$, so that

$$\hat{\nu}(A) = c(|A|).$$

It follows that ν is exchangeable, so that by de Finetti's Theorem, there is a probability measure γ on $[0, 1]$ for which

$$\nu = \int_0^1 \nu_\rho \gamma(d\rho).$$

By (1.15), $\int_0^1 \rho \, d\gamma = \alpha$, while by (1.16), $\int_0^1 \rho^2 \, d\gamma = \alpha^2$. Therefore γ is the pointmass at α, so that $\nu = \nu_\alpha$ as required. If we assume (1.17) instead of (1.16), we can use Corollary 1.9 to conclude that $\int_0^1 \rho^2 \, d\gamma \leq \alpha^2$, which is good enough. $\square$

Condition (1.17) is more convenient to work with than (1.16), since the latter involves two interacting particles, while the former involves two independent ones. For example, (1.15) and (1.17) together can be restated in the following form:

$$\sum_y p_t(x, y) \eta(y) \to \alpha$$

in probability relative to μ. We will find later in this section that in the result which corresponds to Theorem 1.13 in the case $g \neq 1$, (1.17) can be used in place of (1.16) in the necessary and sufficient condition for convergence.

We now turn to the case $g \neq 1$. Here it will be even more important than in the previous case that we be able to compare the finite interacting system with the finite independent system. The first few results carry out this comparison. Let g_n be the restriction of g to S^n.

Lemma 1.18. *Suppose that $g \neq 1$. Then*

    (a) $g_2(\vec{x}) < 1$ *for all* $x \in T_2$,

    (b) $\lim_{t \to \infty} U_2(t) g_2(\vec{x}) = 0$ *for all* $x \in S^2$, *and*

    (c) $\lim_{t \to \infty} U_n(t) g_n(\vec{x}) = 0$ *for all* $\vec{x} \in S^n$ *for any* $n \geq 2$.

Proof. For part (a), note first that

$$g_2(x_1, x_2) \leq g_n(x_1, x_2, \ldots, x_n),$$

so that $g_2 \equiv 1$ implies $g \equiv 1$. Thus we need to show that if $g_2 < 1$ somewhere on T_2, it follows that $g_2 < 1$ everywhere on T_2. Suppose then that $g_2(x, y) < 1$

and let $z \neq x, y$ be a point in S. By irreducibility, either a particle can go from z to y without passing through x, or it can go from x to y without passing through z and from z to x without passing through y. In either case, $\{X_1(t), X_2(t)\}$ can go from $\{x, z\}$ to $\{x, y\}$ without passing through $S^2 \backslash T_2$, so that it follows that $g_2(x, z) < 1$. For part (b), write

$$
\begin{aligned}
& \lim_{t \to \infty} U_2(t) g_2(x, y) \\
& = P^{(x,y)}[\text{there are arbitrarily large } t \text{ so that } X_1(t) = X_2(t)],
\end{aligned}
\tag{1.19}
$$

as in Section 1 of Chapter V. If $p(x, y)$ is transient, then the right side of (1.19) is zero by (1.6). If $p(x, y)$ is recurrent, then it has no nonconstant bounded harmonic functions by Proposition 6.3 of Kemeny, Snell, and Knapp (1976). Therefore by Corollary 7.3 of Chapter II, $(X_1(t), X_2(t))$ has no nonconstant bounded harmonic functions. So, by Proposition 5.19 of Kemeny, Snell, and Knapp (1976), the right side of (1.19) is identically zero or identically one. Hence it is identically zero by part (a). For part (c), start with the simple inequality

$$
g_n(\vec{x}) \leq \tilde{g}_n(\vec{x}),
\tag{1.20}
$$

where

$$
\tilde{g}_n(\vec{x}) = \sum_{1 \leq i < j \leq n} g_2(x_i, x_j).
\tag{1.21}
$$

Applying $U_n(t)$ to both sides of (1.20) yields

$$
U_n(t) g_n(\vec{x}) \leq U_n(t) \tilde{g}_n(x) = \sum_{1 \leq i < j \leq n} U_2(t) g_2(x_i, x_j).
\tag{1.22}
$$

Therefore (c) follows from (b). $\square$

Lemma 1.23. *Suppose that $g \not\equiv 1$. Then*

$$
\lim_{t \to \infty} V_n(t) g_n(\vec{x}) = 0
$$

for all $\vec{x} \in T_n$ and $n \geq 2$.

Proof. By (1.20), it suffices to prove this with g_n replaced by the $\tilde{g}_n$ defined in (1.21). This will follow from Proposition 1.7 and Lemma 1.18 once we show that $\tilde{g}_n$ is positive definite. To show this, take $\beta(x)$ such that $\sum_x |\beta(x)| < \infty$ and $\sum_x \beta(x) = 0$, and note that

$$
\sum_{x_1, x_2} \beta(x_1)\beta(x_2)\tilde{g}_n(x_1, \ldots, x_n) = \sum_{x,y} \beta(x)\beta(y)g_2(x, y).
$$

(It is for this application that we included the somewhat unnatural condition $\sum_x \beta(x) = 0$ in the definition of positive definiteness.) Therefore it is enough to show that $g_2(x, y)$ is positive definite. To do this, take $0 < t_1 < \cdots < t_n$, and set

$$h(x, y) = P^{(x,y)}[X_1(t_i) = X_2(t_i) \text{ for some } 1 \le i \le n].$$

Decomposing this event according to the time and place of the last occurrence of $X_1(t_i) = X_2(t_i)$, we see that

$$h(x, y) = \sum_{i=1}^{n} P^{(x,y)}[X_1(t_i) = X_2(t_i), X_1(t_j) \ne X_2(t_j) \text{ for all } j > i]$$

$$= \sum_{i=1}^{n} \sum_{u \in S} p_{t_i}(x, u) p_{t_i}(y, u) P^{(u,u)}[X_1(t_j - t_i) \ne X_2(t_j - t_i) \text{ for all } j > i],$$

so that

$$\sum_{x,y} \beta(x)\beta(y)h(x, y)$$

$$= \sum_{i=1}^{n} \sum_{u \in S} \left[\sum_x \beta(x) p_{t_i}(x, u) \right]^2 P^{(u,u)}[X_1(t_j - t_i) \ne X_2(t_j - t_i) \text{ for all } j > i].$$

This expression is clearly nonnegative, so $h(x, y)$ is positive definite. Since $g_2(x, y)$ is the limit of $h(x, y)$ as $\{t_1, \ldots, t_n\}$ becomes dense in $[0, \infty)$, it follows that $g_2(x, y)$ is positive definite as required. $\square$

As in Section 1 of Chapter V, let

$$\mathcal{H} = \{\alpha(\cdot) \text{ on } S: 0 \le \alpha(x) \le 1 \text{ and } \sum_y p(x, y)\alpha(y) = \alpha(x) \text{ for all } x\}.$$

The following is the main convergence theorem in case $g \ne 1$.

Theorem 1.24. *Suppose* $g \ne 1$.

(a)   *For each* $\alpha \in \mathcal{H}$,

$$\mu_\alpha = \lim_{t \to \infty} \nu_\alpha S(t)$$

   exists and is in $\mathcal{I}$.

(b)   *If* $\alpha \in \mathcal{H}$ *and* $\mu \in \mathcal{P}$, *then*

$$\lim_{t \to \infty} \mu S(t) = \mu_\alpha$$

if and only if

(1.25) $\quad \lim\limits_{t\to\infty} \sum\limits_{y} p_t(x, y)\hat{\mu}(\{y\}) = \alpha(x) \quad$ *for all* $x \in S$, *and*

(1.26) $\quad \lim\limits_{t\to\infty} \sum\limits_{u,v} p_t(x, u)p_t(y, v)\hat{\mu}(\{u, v\}) = \alpha(x)\alpha(y) \quad$ *for all* $x, y \in S$.

Proof. For part (a), apply Proposition 1.7 to

$$h_n(x_1, \ldots, x_n) = \prod_{i=1}^{n} \alpha(x_i),$$

which is clearly positive definite. Since $\alpha \in \mathcal{H}$,

(1.27) $\qquad\qquad\qquad U_n(t)h_n = h_n.$

Therefore Proposition 1.7 implies that

$$V_n(t)h_n \le h_n$$

on T_n. From the semigroup property of $V_n(t)$ it then follows that $V_n(t)h_n$ is decreasing in t on T_n, so that

(1.28) $\qquad\qquad\qquad \tilde{h}_n(\vec{x}) = \lim\limits_{t\to\infty} V_n(t)h_n(\vec{x})$

exists on T_n. By Corollary 1.3, we then have

$$\mu_\alpha = \lim\limits_{t\to\infty} \nu_\alpha S(t)$$

exists, and

(1.29) $\qquad\qquad\qquad \hat{\mu}_\alpha(\{x_1, \ldots, x_n\}) = \tilde{h}_n(\vec{x})$

for $\vec{x} = (x_1, \ldots, x_n) \in T_n$. The limit μ_α is invariant either by Corollary 1.3 or by Proposition 1.8(d) of Chapter I. Turning to part (b), couple copies of the chains with semigroups $U_n(t)$ and $V_n(t)$ together by having them move together until just before they hit $S^n \setminus T_n$, and then having them move independently. From this it is clear that if $0 \le f \le 1$, then

(1.30) $\qquad\qquad\qquad |V_n(t)f - U_n(t)f| \le g_n$

on T_n. Applying this to $f = h_n$ we see from (1.27) and (1.28) that

(1.31) $\qquad\qquad\qquad |\tilde{h}_n(\vec{x}) - h_n(\vec{x})| \le g_n(\vec{x})$

for all $\bar{x} \in T_n$. Now take $\mu \in \mathcal{P}$ and define

(1.32)
$$f_n(x_1, \ldots, x_n) = \mu\{\eta: \eta(x_1) = \cdots = \eta(x_n) = 1\}$$
$$= \hat{\mu}(\{x_1, \ldots, x_n\})$$

on S^n. By Corollary 1.3, $\lim_{t\to\infty} \mu S(t) = \mu_\alpha$ is equivalent to

(1.33)
$$\lim_{t\to\infty} V_n(t)f_n = \tilde{h}_n$$

on T_n for each $n \geq 1$. On the other hand, (1.25) and (1.26) together are equivalent to the assertion that

$$\lim_{t\to\infty} \sum_y p_t(x, y)\eta(y) = \alpha(x)$$

in probability relative to μ for each $x \in S$, so it follows that they are in turn equivalent to

(1.34)
$$\lim_{t\to\infty} U_n(t)f_n = h_n$$

on S^n for each $n \geq 1$. Thus to prove part (b) we need to show that (1.33) and (1.34) are equivalent. The proofs of the two implications are almost the same, so we will prove one only. Suppose that (1.33) holds. Then

(1.35)
$$|U_n(t)f_n - h_n| \leq |U_n(t)f_n - V_n(t)f_n|$$
$$+ |V_n(t)f_n - \tilde{h}_n| + |\tilde{h}_n - h_n|$$
$$\leq 2g_n + |V_n(t)f_n - \tilde{h}_n|$$

on T_n by (1.30) and (1.31). Take $t_k \uparrow \infty$ such that

$$F_n = \lim_{k\to\infty} U_n(t_k)f_n$$

exists on S^n. Then $U_n(t)F_n = F_n$ by Lemma 1.26 of Chapter V. Passing to the limit in (1.35), we see that

$$|F_n - h_n| \leq 2g_n$$

on T_n, and hence on S^n, since $g_n = 1$ on $S^n \setminus T_n$. Since $U_n(t)F_n = F_n$ and $U_n(t)h_n = h_n$ on S^n, $F_n = h_n$ by this and Lemma 1.18(c). Therefore (1.34) holds as required. In proof of the opposite implication, Lemma 1.23 is used in place of Lemma 1.18(c) in the last step of the argument. $\quad\square$

The next result says that relative to $\mu \in \mathscr{I}_e$, the coordinate random variables $\eta(x)$ are negatively correlated. This property plays an important role in the proof of the complete characterization of $\mathscr{I}_e$.

Lemma 1.36. *If* $\mu \in \mathscr{I}_e$, *then*

$$\mu\{\eta: \eta(x) = \eta(y) = 1\} \le \mu\{\eta: \eta(x) = 1\}\mu\{\eta: \eta(y) = 1\}$$

for $x \neq y$.

Proof. This is immediate if $g \equiv 1$ by Theorem 1.12, so we may assume that $g \not\equiv 1$. Define $\alpha \in \mathscr{H}$ by

$$\alpha(x) = \mu\{\eta: \eta(x) = 1\}.$$

By the irreducibility of $p(x, y)$, either $\alpha \equiv 0$ or $\alpha \equiv 1$ or $0 < \alpha(x) < 1$ for all x. We can assume the latter case, since the result is clear in the first two cases. Fix $x \in S$ and write

$$\mu = \alpha(x)\mu_1 + [1 - \alpha(x)]\mu_0,$$

where μ_1 and μ_0 are obtained by conditioning μ on $\{\eta(x) = 1\}$ and $\{\eta(x) = 0\}$ respectively. Then

(1.37) $$\mu S(t) = \alpha(x)\mu_1 S(t) + [1 - \alpha(x)]\mu_0 S(t).$$

By Lemma 1.26 of Chapter V and Corollary 1.3, if $t_n \uparrow \infty$ and $\lim_{n \to \infty} \mu_1 S(t_n)$ exists, then this limit is in $\mathscr{I}$. Since $\mu \in \mathscr{I}_e$, we see by passing to the limit in (1.37) that

$$\lim_{t \to \infty} \mu_1 S(t) = \mu.$$

In particular,

$$\lim_{t \to \infty} \mu_1 S(t)\{\eta: \eta(y) = 1\} = \alpha(y).$$

By Corollary 1.3,

$$\mu_1 S(t)\{\eta: \eta(y) = 1\} = \sum_z p_t(y, z)\mu_1\{\eta: \eta(z) = 1\},$$

so that we conclude that

(1.38) $$\lim_{t \to \infty} \sum_z p_t(y, z)\hat{\mu}(\{x, z\}) = \alpha(x)\alpha(y).$$

Now let $f(x, y) = \mu\{\eta: \eta(x) = 1, \eta(y) = 1\}$ for $x, y \in S$. Since $\mu \in \mathcal{I}$, Corollary 1.9 implies that

$$(1.39) \qquad f(x, y) \leq U_2(t)f(x, y) \quad \text{for } x \neq y.$$

We wish to use (1.38) to show that the limit of the right side of (1.39) is $\alpha(x)\alpha(y)$. To do so, note that when we showed that $g_2(x, y)$ is positive definite in the proof of Lemma 1.23, we actually showed that $\sum_{x,y} \beta(x)\beta(y)g_2(x, y) \geq 0$ whenever $\sum_x |\beta(x)| < \infty$. The requirement $\sum_x \beta(x) = 0$ was not used there. Applying this to

$$\beta(x) = p_t(u, x) - \varepsilon p_0(v, x),$$

we see that

$$0 \leq \sum_{x,y} [p_t(u, x) - \varepsilon p_0(v, x)][p_t(u, y) - \varepsilon p_0(v, y)]g_2(x, y)$$
$$= U_2(t)g_2(u, u) - 2\varepsilon \sum_y p_t(u, y)g_2(y, v) + \varepsilon^2 g_2(v, v).$$

Passing to the limit first on $t \uparrow \infty$ using Lemma 1.18 and then dividing by ε and letting $\varepsilon \downarrow 0$, we see that

$$(1.40) \qquad \lim_{t \to \infty} \sum_y p_t(u, y)g_2(y, v) = 0.$$

By (1.30),

$$|V_2(t)f - U_2(t)f| \leq g_2$$

on T_2. Since $\mu \in \mathcal{I}$, $V_2(t)f = f$ on T_2, so that this gives

$$(1.41) \qquad |f - U_2(t)f| \leq g_2$$

on T_2. Let h be the limit of $U_2(t)f$ along any sequence of times which tend to ∞. Then

$$(1.42) \qquad |f - h| \leq g_2$$

by (1.41), and $U_2(t)h = h$ for all t by Lemma 1.26 of Chapter V. Therefore

$$(1.43) \qquad \sum_y p_t(x, y)h(y, z) = h(x, z)$$

by Corollary 7.3 of Chapter II. Combining (1.38), (1.40), (1.42), and (1.43), we see that $h(x, y) = \alpha(x)\alpha(y)$. This shows that

$$\lim_{t\to\infty} U_2(t)f(x, y) = \alpha(x)\alpha(y),$$

which, when combined with (1.39), implies that $f(x, y) \le \alpha(x)\alpha(y)$ as required. $\square$

Finally we obtain a complete description of $\mathscr{I}_e$ in the following form.

Theorem 1.44. $\mathscr{I}_e = \{\mu_\alpha : \alpha \in \mathscr{H}\}$.

Proof. If $g \equiv 1$, then $\mathscr{H}$ consists of the constants in $[0, 1]$ by the continuous time version of Theorem 1.5 of Chapter II. For constant α, $\mu_\alpha = \nu_\alpha$ by Corollary 1.3. Therefore in this case, this result is identical with Theorem 1.12. So, we may assume that $g \not\equiv 1$. In this case, we showed that $\mu_\alpha \in \mathscr{I}$ in Theorem 1.24(a). To show that μ_α is extremal, write

(1.45) $$\mu_\alpha = \lambda\mu_1 + (1 - \lambda)\mu_2,$$

where $0 < \lambda < 1$ and $\mu_1, \mu_2 \in \mathscr{I}$. Then

(1.46) $$\sum_y p_t(x, y)\eta(y) \to \alpha(x)$$

in probability relative to μ_α by Theorem 1.24(b). By (1.45), the same is true relative to μ_1 and μ_2. Therefore using Theorem 1.24(b) again, we see that

$$\lim_{t\to\infty} \mu_1 S(t) = \lim_{t\to\infty} \mu_2 S(t) = \mu_\alpha.$$

Since $\mu_1, \mu_2 \in \mathscr{I}$, it follows that $\mu_1 = \mu_2 = \mu_\alpha$. For the converse, take $\mu \in \mathscr{I}_e$ and define $\alpha \in \mathscr{H}$ by

$$\alpha(x) = \mu\{\eta : \eta(x) = 1\}.$$

By Lemma 1.36,

$$\int \left[\sum_y p_t(x, y)\eta(y) - \alpha(x)\right]^2 d\mu$$

$$= \sum_{y,z} p_t(x, y)p_t(x, z)\hat{\mu}(\{y, z\}) - [\alpha(x)]^2$$

$$\le \sum_{y\ne z} p_t(x, y)p_t(x, z)\alpha(y)\alpha(z) + \sum_y p_t^2(x, y)\alpha(y) - [\alpha(x)]^2$$

$$= \sum_y p_t^2(x, y)[\alpha(y) - \alpha^2(y)] \le p_{2t}(x, x),$$

which tends to zero as $t\uparrow\infty$. Therefore

$$\lim_{t\to\infty} \mu S(t) = \mu_\alpha$$

by Theorem 1.24(b). Since $\mu \in \mathcal{I}$, it follows that $\mu = \mu_\alpha$. □

For the next result, take $S = Z^d$ and $p(x, y) = p(0, y - x) = p(y, x)$. As usual, $\mathcal{S}$ denotes the translation invariant probability measures on X, and $\mathcal{S}_e$ is the set of elements of $\mathcal{S}$ which are extremal (or equivalently, ergodic).

Theorem 1.47. *Suppose* $\mu \in \mathcal{S}_e$, *and* $\alpha \in [0, 1]$ *is defined by* $\alpha = \mu\{\eta\colon \eta(x) = 1\}$. *Then*

$$\lim_{t\to\infty} \mu S(t) = \mu_\alpha.$$

Proof. By Theorems 1.13 and 1.24 and translation invariance, it suffices to check that

$$\lim_{t\to\infty} \sum_y p_t(x, y)\mu\{\eta\colon \eta(y) = \eta(0) = 1\} = \alpha^2.$$

This follows from Corollary 8.20 of Chapter II and the remark following its proof. In that result, simply take $p(x, y) = q(x, y)$ and $h \equiv 1$. □

We will conclude this section by giving an application of Theorem 1.12 to a certain spin system.

Example 1.48. Consider the spin system on $S = Z^1$ with flip rates given by

$$c(x, \eta) = \begin{cases} \frac{1}{2} & \text{if } \eta(x - 1) \neq \eta(x + 1), \\ 0 & \text{if } \eta(x - 1) = \eta(x + 1), \end{cases}$$

and let $\mathcal{I}$ be its invariant measures. Then

$$(1.49) \qquad \mathcal{I}_e = \{\delta_0, \delta_1, \delta_2, \delta_3\} \cup \{\nu^\alpha, 0 < \alpha < 1\},$$

where δ_0, δ_1, δ_2, and δ_3 are the pointmasses on $\eta_0 \equiv 0$, $\eta_1 \equiv 1$,

$$\eta_2(x) = \begin{cases} 1 & \text{if } x \text{ is even}, \\ 0 & \text{if } x \text{ is odd}, \end{cases}$$

and $\eta_3(x) = 1 - \eta_2(x)$, and ν^α is the distribution on $\{0, 1\}^S$ of the stationary two-state Markov chain with transition matrix

$$\begin{pmatrix} 1 - \alpha & \alpha \\ \alpha & 1 - \alpha \end{pmatrix}.$$

One reason for being interested in this spin system is that it provides an example for which $\mathscr{I}_e$ is not closed and for which

$$(\mathscr{I} \cap \mathscr{S})_e \not\subset \mathscr{I}_e.$$

We will only outline the argument which leads to (1.49), leaving the details to the interested reader. The first observation is that if η_t is this spin system and $T: \{0, 1\}^S \to \{0, 1\}^S$ is defined by

$$T\eta(x) = \begin{cases} 1 & \text{if } \eta(x) \neq \eta(x+1), \\ 0 & \text{if } \eta(x) = \eta(x+1), \end{cases}$$

then $T\eta_t$ is the exclusion process corresponding to $p(x, x+1) = p(x, x-1) = \frac{1}{2}$. It is not difficult to conclude then from Theorem 1.12 that if $\mu \in \mathscr{I}_e \setminus \{\delta_0, \delta_1, \delta_2, \delta_3\}$, then $\mu T^{-1} = \nu_\alpha$ for some $0 < \alpha < 1$. We need to show that it follows from this that $\mu = \nu^\alpha$. This is not completely obvious because T is two-to-one, rather than one-to-one. What is clear is that

$$\frac{\mu + \tilde{\mu}}{2} = \nu^\alpha,$$

where $\tilde{\mu}$ is obtained from μ by interchanging the roles of 0 and 1. Therefore μ is absolutely continuous with respect to ν^α, and the Radon–Nikodym derivative $d\mu/d\nu^\alpha$ is bounded above by 2. Note next that except for the requirement of strict positivity of the rates, $c(x, \eta)$ satisfies the definition of a stochastic Ising model with respect to any translation invariant nearest-neighbor two-body potential. Using the arguments from Section 4 of Chapter IV (see in particular Proposition 4.1 and Lemma 4.3 of that chapter), it then follows that

$$\frac{d\mu}{d\nu^\alpha}(\eta_x) = \frac{d\mu}{d\nu^\alpha}(\eta) \quad \text{a.s. } (\nu^\alpha)$$

on the set $\{\eta(x-1) \neq \eta(x+1)\}$. As a consequence of this,

(1.50) $$\frac{\mu\{\eta: \eta(x) = \zeta(x) \text{ for } x = 1, \dots, n\}}{\nu^\alpha\{\eta: \eta(x) = \zeta(x) \text{ for } x = 1, \dots, n\}}$$

depends on ζ only through $\zeta(1)$ and

$$d(\zeta) = \sum_{i=1}^{n-1} |\zeta(i) - \zeta(i+1)|.$$

Let $F(k, n)$ be the value of the ratio in (1.50) for those ζ such that $\zeta(1) = 1$

and $d(\zeta) = k$. Then $F(k, n)$ is bounded and satisfies

$$F(k, n) = \alpha F(k+1, n+1) + (1-\alpha)F(k, n+1).$$

Iterating this yields

$$F(k, n) = \sum_{j=0}^{l} \binom{l}{j} \alpha^j (1-\alpha)^{l-j} F(k+j, n+l),$$

from which it follows that F must be constant. Returning to the definition of F, it is easy to see that $F = 1$, so that $\mu = \nu^\alpha$.

2. Coupling and Invariant Measures for General Systems

In the previous section, we used duality to check that the exchangeable measures on X are invariant for symmetric exclusion processes. This duality is not valid in the absence of symmetry, so we will begin here by checking that certain product measures are invariant for certain exclusion systems by direct computation, using the generator of the process. Then we will describe the basic coupling for exclusion processes, which is the main tool used in the ergodic theory of asymmetric systems. In this section, we will assume only that (0.2) holds, so that the corresponding exclusion process is well defined. Recall that ν_α is the product measure on X with marginals

$$\nu_\alpha\{\eta: \eta(x) = 1\} = \alpha(x).$$

Theorem 2.1. (a) *Suppose that*

(2.2) $\sum_{x} p(x, y) = 1$ *for each $y \in S$.*

Then $\nu_\alpha \in \mathscr{I}$ for every constant $\alpha \in [0, 1]$.
   (b)   *Suppose that $\pi(\cdot)$ satisfies*

(2.3) $\pi(x)p(x, y) = \pi(y)p(y, x)$ *for all $x, y \in S$.*

Then $\nu_\alpha \in \mathscr{I}$, where $\alpha(x) = \pi(x)/[1 + \pi(x)]$.

Proof. By Proposition 2.13 of Chapter I, to show that $\nu_\alpha \in \mathscr{I}$, it suffices to check that

(2.4) $\int \Omega f \, d\nu_\alpha = 0$

for all $f \in \mathscr{D}$ (the class of functions on X which depend on finitely many coordinates). By linearity, it then is enough to check (2.4) for f of the form

(2.5)
$$f(\eta) = \begin{cases} 1 & \text{if } \eta(x) = 1 \text{ for all } x \in A, \\ 0 & \text{otherwise,} \end{cases}$$

where $A \in Y$. Using the expression for Ω in (0.1), (2.4) becomes

(2.6)
$$\sum_{x,y} p(x,y) \int [f(\eta_{xy}) - f(\eta)] \eta(x)[1 - \eta(y)] \, d\nu_\alpha = 0.$$

For the f given in (2.5) and $x \neq y$,

$$\int f(\eta)\eta(x)[1-\eta(y)] \, d\nu_\alpha = \begin{cases} 0 & \text{if } y \in A, \\ [1-\alpha(y)] \prod_{u \in A \cup \{x\}} \alpha(u) & \text{if } y \notin A, \end{cases}$$

and

$$\int f(\eta_{xy})\eta(x)[1-\eta(y)] \, d\nu_\alpha = \begin{cases} 0 & \text{if } x \in A, \\ [1-\alpha(y)] \prod_{u \in A \cup \{x\}\setminus\{y\}} \alpha(u) & \text{if } x \notin A. \end{cases}$$

Therefore the left side of (2.6) becomes

$$\sum_{\substack{x \notin A \\ y \in A}} p(x,y)\alpha(x)[1-\alpha(y)] \prod_{u \in A\setminus\{y\}} \alpha(u)$$

(2.7)
$$- \sum_{\substack{x \in A \\ y \notin A}} p(x,y)\alpha(x)[1-\alpha(y)] \prod_{u \in A\setminus\{x\}} \alpha(u)$$

$$= \sum_{\substack{x \in A \\ y \notin A}} \left\{ \prod_{u \in A\setminus\{x\}} \alpha(u) \right\} \{\alpha(y)[1-\alpha(x)]p(y,x) - \alpha(x)[1-\alpha(y)]p(x,y)\}.$$

Part (b) of the theorem is immediate from this, since the assumptions there imply that each term in the above sum is zero. For part (a), take $\alpha(x) \equiv \alpha$. Then the sum on the right of (2.7) is a constant multiple of

$$\sum_{\substack{x \in A \\ y \notin A}} [p(y,x) - p(x,y)],$$

which is zero for any $A \in Y$ in the doubly stochastic case (i.e, when (2.2) holds). $\square$

Example 2.8. Suppose $S = Z^1$, $p(x, x+1) = p$, and $p(x, x-1) = q$, where $0 < p < 1$ and $p + q = 1$. Then (2.2) holds, so $\nu_\alpha \in \mathcal{I}$ for constant $\alpha \in [0, 1]$.

In addition, (2.3) holds with

(2.9)
$$\pi(x) = c\left(\frac{p}{q}\right)^x$$

for any constant $c > 0$. Therefore $\nu_\alpha \in \mathscr{I}$ for

(2.10)
$$\alpha(x) = \frac{cp^x}{q^x + cp^x},$$

where $c > 0$ is arbitrary. Thus this translation invariant system has invariant measures which are not translation invariant, provided that $p \neq \frac{1}{2}$. Note however that these latter invariant measures have a countable support. If $p > \frac{1}{2}$ for example, the support is

> $\{\eta \in X: \eta(x) = 1$ for all sufficiently large positive x, and $\eta(x) = 0$ for all sufficiently large negative $x\}$.

A further discussion of this example will be given at the end of Section 3.

The basic coupling for the exclusion process is given by the Feller process (η_t, ζ_t) on $X \times X$ whose generator is the closure of the operator defined on $\tilde{\mathscr{D}}$ by

$$\tilde{\Omega}f(\eta, \zeta) = \sum_{\substack{\eta(x)=\zeta(x)=1 \\ \eta(y)=\zeta(y)=0}} p(x, y)[f(\eta_{xy}, \zeta_{xy}) - f(\eta, \zeta)]$$

(2.11)
$$+ \sum_{\substack{\eta(x)=1,\eta(y)=0 \text{ and} \\ \zeta(y)=1 \text{ or } \zeta(x)=0}} p(x, y)[f(\eta_{xy}, \zeta) - f(\eta, \zeta)]$$

$$+ \sum_{\substack{\zeta(x)=1,\zeta(y)=0 \text{ and} \\ \eta(y)=1 \text{ or } \eta(x)=0}} p(x, y)[f(\eta, \zeta_{xy}) - f(\eta, \zeta)].$$

That this closure is a Markov generator under (0.2) is again a consequence of Theorem 3.9 of Chapter I. To put ourselves in the context of that theorem, it suffices to identify $X \times X$ with $\{(0, 0), (0, 1), (1, 0), (1, 1)\}^S$ in the natural way. The interpretation of (2.11) is simply that particles in η_t and ζ_t move together whenever they can. Clearly the marginal processes η_t and ζ_t are copies of the exclusion process corresponding to $p(x, y)$. Note that with this coupling,

$$P^{(\eta, \eta)}(\eta_t = \zeta_t) = 1, \qquad P^{(\eta, \zeta)}(\eta_t \leq \zeta_t) = 1 \quad \text{if } \eta \leq \zeta, \quad \text{and}$$

$$P^{(\eta, \zeta)}(\eta_t \geq \zeta_t) = 1 \quad \text{if } \eta \geq \zeta.$$

This observation leads immediately to the following three propositions. The proof of the third one is the same as that of Theorem 2.15 of Chapter III. We will denote quantities involving the coupled process by tildes. Thus $\tilde{S}(t)$ is the semigroup corresponding to (η_t, ζ_t), $\tilde{\mathscr{I}}$ is its invariant measures, and $\tilde{\mathscr{S}}$ is the set of translation invariant probability measures on $X \times X$ in case $S = Z^d$.

Proposition 2.12. η_t *is a monotone process.* (*See Definition 2.3 of Chapter* II.)

Proposition 2.13. *If* $\nu \in \tilde{\mathscr{I}}_e$, *then* $\nu\{(\eta, \zeta): \eta = \zeta\}$, $\nu\{(\eta, \zeta): \eta \leq \zeta\}$, *and* $\nu\{(\eta, \zeta): \eta \geq \zeta\}$ *are each either zero or one. The same is true for* $\nu \in (\tilde{\mathscr{I}} \cap \tilde{\mathscr{S}})_e$ *if* $S = Z^d$ *and* $p(x, y) = p(0, y - x)$.

Proposition 2.14. (a) *If* $\nu_1, \nu_2 \in \mathscr{I}$, *there is a* $\nu \in \tilde{\mathscr{I}}$ *with marginals* ν_1 *and* ν_2.
(b) *If* $\nu_1, \nu_2 \in \mathscr{I}_e$, *then* ν *can be taken in* $\tilde{\mathscr{I}}_e$.
(c) *In each of the preceding two statements, if in addition* $\nu_1 \leq \nu_2$, *then* ν *can be taken so that*

$$\nu\{(\eta, \zeta): \eta \leq \zeta\} = 1.$$

(d) *If* $S = Z^d$ *and* $p(x, y) = p(0, y - x)$, *then each of the preceding three statements holds if* $\mathscr{I}$ *and* $\tilde{\mathscr{I}}$ *are replaced by* $\mathscr{I} \cap \mathscr{S}$ *and* $\tilde{\mathscr{I}} \cap \tilde{\mathscr{S}}$ *respectively.*

The basic coupling can be implemented via a graphical representation which is similar to that used for spin systems in Section 6 of Chapter III. To do so, let $N_{x,y}(t)$ be independent Poisson processes, defined for each $(x, y) \in S^2$ such that $p(x, y) > 0$. The rate of $N_{x,y}(t)$ is to be $p(x, y)$. At each event time of $N_{x,y}(t)$, a particle is moved from x to y, provided that x is occupied and y is vacant just before that time. The advantage of this construction is that versions of the exclusion process for all initial configurations in X are defined simultaneously on the same probability space. Note that this is not an additive coupling in the sense of Definition 6.1 of Chapter III. A different graphical representation for symmetric exclusion processes which does yield an additive coupling is described and used in Section 4.

The coupled process will be used extensively in the next section. To illustrate its use in a simple context, suppose that $p(x, y)$ is positive recurrent and reversible with stationary measure π:

(2.15) $\qquad \pi(x)p(x, y) = \pi(y)p(y, x), \qquad \sum_x \pi(x) = 1.$

Assume also that $p(x, y)$ is irreducible. By Theorem 2.1(b), $\nu_\alpha \in \mathscr{I}$ whenever $\alpha(x) = c\pi(x)/[1 + c\pi(x)]$ for some $c > 0$. Since $\sum_x \pi(x) = 1$, these measures concentrate on $\{\eta: \sum_x \eta(x) < \infty\}$. The exclusion process restricted to $\{\eta: \sum_x \eta(x) = n\}$ is therefore an irreducible positive recurrent Markov chain.

Let ν_n be its stationary distribution. Then ν_n is obtained from the invariant ν_α's by conditioning on $\{\eta: \sum_x \eta(x) = n\}$. It follows that if $\alpha(x) = c\pi(x)/[1 + c\pi(x)]$,

$$\frac{c\pi(x)}{1 + c\pi(x)} = \nu_\alpha\{\eta: \eta(x) = 1\} = \sum_{n=0}^{\infty} \nu_n\{\eta: \eta(x) = 1\} \nu_\alpha\left\{\eta: \sum_x \eta(x) = n\right\}.$$

Since the left side of this identity tends to one as $c\uparrow\infty$ and since $\nu_n \le \nu_{n+1}$ by Proposition 2.12, it follows that

(2.16) $$\lim_{n\to\infty} \nu_n = \nu_\infty,$$

where ν_∞ is the pointmass on $\eta \equiv 1$. These observations lead to the following result.

Theorem 2.17. *Suppose that $p(x, y)$ is positive recurrent, reversible, and irreducible.*
    (a) $\mathcal{I}_e = \{\nu_n, 0 \le n \le \infty\}$.
    (b) *If $\mu\{\eta: \sum_x \eta(x) = \infty\} = 1$, then*

$$\lim_{t\to\infty} \mu S(t) = \nu_\infty.$$

Proof. Part (a) follows immediately from part (b). To prove part (b), take μ so that

$$\mu\left\{\eta: \sum_x \eta(x) = \infty\right\} = 1.$$

For every $1 \le n < \infty$, there is a probability measure μ_n on X so that $\mu_n\{\eta: \sum_x \eta(x) = n\} = 1$ and $\mu_n \le \mu$. By Proposition 2.12,

(2.18) $$\mu_n S(t) \le \mu S(t)$$

for any $t \ge 0$. By the convergence theorem for positive recurrent Markov chains (the continuous time analogue of Theorem 1.2 of Chapter II),

(2.19) $$\lim_{t\to\infty} \mu_n S(t) = \nu_n$$

for each n. Part (b) follows from (2.16), (2.18), and (2.19). $\square$

3. Ergodic Theorems for Translation Invariant Systems

In this section, we consider the case in which $S = Z^d$, and $p(x, y) = p(0, y - x)$. Rather than full irreducibility, we will assume that for every $x, y \in S$,

(3.1) $$p_t(x, y) + p_t(y, x) > 0.$$

This will allow us to include the case $d = 1$, $p(x, x + 1) = 1$ for example.

Our first objective is to say as much as possible about the limiting behavior of $\mu S(t)$ for $\mu \in \mathscr{S}$. For $x, y \in S$, let

$$f_x(\eta, \zeta) = \begin{cases} 1 & \text{if } \eta(x) = 0, \zeta(x) = 1, \\ 0 & \text{otherwise}, \end{cases}$$

and

$$f_{xy}(\eta, \zeta) = \begin{cases} 1 & \text{if } \eta(x) = \zeta(y) = 1 \text{ and } \eta(y) = \zeta(x) = 0, \\ 0 & \text{otherwise}. \end{cases}$$

Note that if ν is a probability measure on $X \times X$ with marginals μ_1 and μ_2, then $\int f_x \, d\nu = 0$ for all x implies $\mu_2 \leq \mu_1$, while $\int f_{xy} \, d\nu = 0$ for all x, y implies that ν concentrates on $\{(\eta, \zeta): \eta \leq \zeta \text{ or } \eta \geq \zeta\}$.

Say that there is a discrepancy at x if $\eta(x) \neq \zeta(x)$. The idea which will be exploited in this section is the following. Transitions in the coupled process can move discrepancies or eliminate them, but cannot create new discrepancies. Therefore in some sense, the "number" of discrepancies should decrease in time. Of course this "number" is usually infinite, but still the idea should be clear. In equilibrium, this number cannot strictly decrease. Therefore invariant measures for the coupled process should concentrate on configurations from which the elimination of discrepancies is not possible—i.e., configurations (η, ζ) so that $\eta \leq \zeta$ or $\eta \geq \zeta$. The first lemma implements this idea for translation invariant measures.

Lemma 3.2. *Suppose* $\nu \in \tilde{\mathscr{S}}$ *and let* $\nu_t = \nu \tilde{S}(t)$. *Then*
(a) $\int f_x \, d\nu_t$ *is nonincreasing in* t, *and*
(b) $\lim_{t \to \infty} \int f_{xy} \, d\nu_t = 0$ *for each* x, y.

Proof. First use the expression for $\tilde{\Omega}$ in (2.11) to compute

$$\tilde{\Omega} f_x(\eta, \zeta) = \eta(x)\zeta(x) \sum_y p(x, y) f_y(\eta, \zeta)$$

$$+ [1 - \eta(x)][1 - \zeta(x)] \sum_y p(y, x) f_y(\eta, \zeta)$$

$$- f_x(\eta, \zeta) \sum_y p(x, y)[1 - \zeta(y)]$$

$$- f_x(\eta, \zeta) \sum_y p(y, x) \eta(y).$$

Writing

$$\eta(y) f_x(\eta, \zeta) = \eta(y)\zeta(y) f_x(\eta, \zeta) + f_{yx}(\eta, \zeta), \quad \text{and}$$

$$[1 - \zeta(y)] f_x(\eta, \zeta) = [1 - \eta(y)][1 - \zeta(y)] f_x(\eta, \zeta) + f_{yx}(\eta, \zeta),$$

we see that

$$\tilde{\Omega}f_x(\eta, \zeta) = \sum_y \{p(x, y)\eta(x)\zeta(x)f_y(\eta, \zeta) - p(y, x)\eta(y)\zeta(y)f_x(\eta, \zeta)\}$$

$$+ \sum_y \{p(y, x)[1 - \eta(x)][1 - \zeta(x)]f_y(\eta, \zeta)$$

(3.3)

$$- p(x, y)[1 - \eta(y)][1 - \zeta(y)]f_x(\eta, \zeta)\}$$

$$- \sum_y [p(x, y) + p(y, x)]f_{yx}(\eta, \zeta).$$

If $\nu \in \tilde{\mathscr{S}}$, then

$$\int p(x, y)\eta(x)\zeta(x)f_y(\eta, \zeta)\, d\nu, \quad \text{and}$$

$$\int p(y, x)[1 - \eta(x)][1 - \zeta(x)]f_y(\eta, \zeta)\, d\nu$$

are functions of $y - x$, so that the integrals of the first two terms on the right of (3.3) with respect to ν vanish. Hence

(3.4)
$$\int \tilde{\Omega}f_x\, d\nu = -\sum_y [p(x, y) + p(y, x)] \int f_{yx}\, d\nu \le 0.$$

Therefore by Theorem 2.9(c) of Chapter I,

$$\frac{d}{dt} \int f_x\, d\nu_t = \frac{d}{dt} \int \tilde{S}(t)f_x\, d\nu$$

$$= \int \tilde{S}(t)\tilde{\Omega}f_x\, d\nu$$

$$= \int \tilde{\Omega}f_x\, d\nu_t \le 0,$$

which proves (a). By applying $\tilde{\Omega}$ again, it is not hard to see that

$$\sup_{t \ge 0} \left| \frac{d^2}{dt^2} \int f_x\, d\nu_t \right| < \infty.$$

Therefore it follows from part (a) that

$$\lim_{t \to \infty} \frac{d}{dt} \int f_x\, d\nu_t = 0.$$

Recalling (3.4), this implies that

(3.5)
$$\lim_{t\to\infty} \int f_{xy}\, d\nu_t = 0$$

for all x, y for which $p(x, y) + p(y, x) > 0$. Using (3.1) and an induction argument, (3.5) follows for all $x, y \in S$. The details of this argument are left to the reader. The following is the basic idea used in the induction step. Suppose that $p^{(2)}(x, y) > 0$, but $p(x, y) = p(y, x) = 0$. Then there is a $z \in S$ so that $p(x, z) > 0$ and $p(z, y) > 0$. For this choice,

$$\lim_{t\to\infty} \int f_{xz}\, d\nu_t = 0, \qquad \lim_{t\to\infty} \int f_{zy}\, d\nu_t = 0, \quad \text{and}$$

$$\frac{d}{dt} \int \eta(y)\zeta(y)f_{xz}\, d\nu_t \geq p(z, y) \int \eta(z)\zeta(z)f_{xy}\, d\nu_t - 6 \int \eta(y)\zeta(y)f_{xz}\, d\nu_t.$$

Therefore

$$\lim_{t\to\infty} \int \eta(z)\zeta(z)f_{xy}\, d\nu_t = 0.$$

A similar argument shows that

$$\lim_{t\to\infty} \int [1 - \eta(z)][1 - \zeta(z)]f_{xy}\, d\nu_t = 0.$$

Now use the inequality

$$f_{xy} \leq \eta(z)\zeta(z)f_{xy} + [1 - \eta(z)][1 - \zeta(z)]f_{xy} + f_{xz} + f_{zy}$$

to conclude the proof of (3.5) for this x, y. $\square$

We will show that measures are exchangeable by using the following criterion.

Lemma 3.6. *Suppose that $\mu \in \mathcal{S}$, and that for every $\alpha \in [0, 1]$, μ can be written as a convex combination*

$$\mu = \lambda_\alpha \mu_1^\alpha + (1 - \lambda_\alpha)\mu_2^\alpha$$

where $\mu_1^\alpha \leq \nu_\alpha$ and $\mu_2^\alpha \geq \nu_\alpha$. Then μ is exchangeable. Furthermore

$$\mu = \int_0^1 \nu_\alpha \gamma(d\alpha)$$

where γ is a probability measure on $[0, 1]$ which satisfies

(3.7) $\gamma[0, \alpha] = \lambda_\alpha$

whenever $\gamma(\{\alpha\}) = 0$.

Proof. Since $\mu \in \mathcal{S}$, the Ergodic Theorem (Theorem 4.9 of Chapter I) implies that

$$L = \lim_{n \to \infty} \frac{1}{(2n+1)^d} \sum_{|x| \leq n} \eta(x)$$

exists a.s. with respect to μ, where $|(x_1, \ldots, x_d)| = \max_{1 \leq i \leq d} |x_i|$. By the assumption, $0 < \lambda_\alpha < 1$ implies that L exists a.s. with respect to μ_1^α and μ_2^α, and that $L \leq \alpha$ a.s. with respect to μ_1^α and $L \geq \alpha$ a.s. with respect to μ_2^α. Therefore the assumption can be restated by saying that the conditional measures $\mu(\cdot | L \leq \alpha)$ and $\mu(\cdot | L \geq \alpha)$ satisfy

                                        $\mu(\cdot | L \leq \alpha) \leq \nu_\alpha$ and

(3.8)

                                        $\mu(\cdot | L \geq \alpha) \geq \nu_\alpha$

whenever the conditioning events have positive probability and $\mu\{L = \alpha\} = 0$. Furthermore $\lambda_\alpha = \mu\{L \leq \alpha\}$, so that (3.7) will follow once we show that μ is exchangeable. To show that μ is exchangeable, it is enough to extend (3.8) to

$$\nu_\alpha \leq \mu(\cdot | \alpha \leq L \leq \beta) \leq \nu_\beta$$

whenever $\mu\{\alpha \leq L \leq \beta\} > 0$ and $\mu\{L = \alpha\} = \mu\{L = \beta\} = 0$. But this is easy to show using the fact that $\nu_\alpha \in \mathcal{S}_e$ (by Corollary 4.14 of Chapter I) and Theorem 2.4 of Chapter II. $\square$

Theorem 3.9. (a) $(\mathcal{I} \cap \mathcal{S})_e = \{\nu_\alpha : \alpha \in [0, 1]\}$.
    (b) *If $\mu \in \mathcal{S}$, then $\mu_\infty = \lim_{t \to \infty} \mu S(t)$ exists, is exchangeable, and satisfies*
      $\mu_\infty\{\eta : \eta(x) = 1\} = \mu\{\eta : \eta(x) = 1\}$.

Proof. While part (a) follows from part (b) and Theorem 2.1(a), we will prove (a) first since its proof is much simpler than the proof of part (b). Take $\mu \in (\mathcal{I} \cap \mathcal{S})_e$ and $\alpha \in [0, 1]$. By Theorem 2.1(a), $\nu_\alpha \in \mathcal{I}$. Since $\nu_\alpha \in \mathcal{S}_e$, it follows that $\nu_\alpha \in (\mathcal{I} \cap \mathcal{S})_e$ as well. By Proposition 2.14, there is a $\nu \in (\tilde{\mathcal{I}} \cap \tilde{\mathcal{S}})_e$ with marginals μ and ν_α. By Lemma 3.2(b), $\int f_{xy} \, d\nu = 0$ for all $x, y \in S$. Therefore by Proposition 2.13, either $\mu \leq \nu_\alpha$ or $\mu \geq \nu_\alpha$. Let

$$\beta = \sup\{\alpha \in [0, 1] : \mu \geq \nu_\alpha\}.$$

Since ν_α is weakly continuous in α, it follows that $\mu = \nu_\beta$ as required. To prove part (b), take $\mu \in \mathcal{S}$, $\alpha \in [0, 1]$, and $t_n \uparrow \infty$ so that

$$\nu = \lim_{n \to \infty} (\mu \times \nu_\alpha) \tilde{S}(t_n)$$

exists. By Lemma 3.2(b), $\int f_{xy}\, d\nu = 0$ for all $x, y \in S$. Therefore

$$\nu\{(\eta, \zeta): \eta \leq \zeta \text{ or } \eta \geq \zeta\} = 1,$$

so that we can write

$$\nu = \lambda_1 \nu_1 + \lambda_2 \nu_2 + \lambda_3 \nu_3,$$

where

$$\nu_1, \nu_2, \nu_3 \in \tilde{\mathcal{S}}, \qquad \lambda_1, \lambda_2, \lambda_3 \geq 0, \qquad \lambda_1 + \lambda_2 + \lambda_3 = 1,$$

$$\nu_1\{(\eta, \zeta): \eta \leq \zeta, \eta \neq \zeta\} = 1, \qquad \nu_2\{(\eta, \zeta): \eta = \zeta\} = 1,$$

$$\text{and} \quad \nu_3\{(\eta, \zeta): \eta \geq \zeta, \eta \neq \zeta\} = 1.$$

Since $\nu_\alpha \in \mathcal{S}_e$ and is the second marginal of ν, the second marginal of each ν_i for which $\lambda_i > 0$ must also be ν_α. Therefore by Lemma 3.6, any weak limit of $\mu S(t)$ as $t \uparrow \infty$ is exchangeable. So, let

(3.10)
$$\mu_\infty = \int_0^1 \nu_\beta \gamma(d\beta)$$

be the limit of $\mu S(t_n)$ as $n \uparrow \infty$, where γ is a probability measure on $[0, 1]$. We need to show that γ, and hence μ_∞, is independent of the sequence $\{t_n\}$. By Lemma 3.2(a),

$$\lim_{t \to \infty} \int f_x \, d((\mu \times \nu_\alpha) \tilde{S}(t))$$

exists, so it will be enough to show that

(3.11)
$$\lim_{t \to \infty} \int f_x \, d((\mu \times \nu_\alpha) \tilde{S}(t)) = \int_0^\alpha (\alpha - \beta) \gamma(d\beta)$$

for any α such that $\gamma(\{\alpha\}) = 0$. To prove (3.11), pass to a subsequence if necessary so that

$$\nu = \lim_{n \to \infty} (\mu \times \nu_\alpha) \tilde{S}(t_n)$$

exists. We then need to show that

(3.12) $$\int f_x \, d\nu = \int_0^\alpha (\alpha - \beta)\gamma(d\beta).$$

Recalling the first part of the proof, and noting that $\gamma(\{\alpha\}) = 0$ implies that $\nu\{(\eta, \zeta): \eta = \zeta\} = 0$, we see that there are $\nu_1, \nu_2 \in \mathscr{S}$ and $0 \leq \lambda \leq 1$ so that

(3.13) $$\nu = \lambda \nu_1 + (1 - \lambda)\nu_2,$$

where ν_1 concentrates on $\{(\eta, \zeta): \eta \leq \zeta\}$, and ν_2 concentrates on $\{(\eta, \zeta): \eta \geq \zeta\}$. Suppose that $0 < \lambda < 1$. The second marginal of ν is ν_α. Therefore since $\nu_\alpha \in \mathscr{S}_e$, the second marginals of ν_1 and ν_2 must be ν_α also. The first marginal of ν is μ_∞, so that by (3.10) and the fact that the first marginal of ν_1 lies below ν_α and the first marginal of ν_2 lies above ν_α,

$$\lambda = \gamma[0, \alpha]$$

and ν_1 and ν_2 have first marginals

$$\frac{1}{\lambda} \int_0^\alpha \nu_\beta \gamma(d\beta) \quad \text{and} \quad \frac{1}{1-\lambda} \int_\alpha^1 \nu_\beta \gamma(d\beta)$$

respectively. Now,

$$\begin{aligned}
\int f_x \, d\nu &= \nu\{(\eta, \zeta): \eta(x) = 0, \zeta(x) = 1\} \\
&= \lambda \nu_1\{(\eta, \zeta): \eta(x) = 0, \zeta(x) = 1\} \\
&= \lambda \nu_1\{(\eta, \zeta): \zeta(x) = 1\} - \lambda \nu_1\{(\eta, \zeta): \eta(x) = 1\} \\
&= \lambda \nu_\alpha\{\eta: \eta(x) = 1\} - \int_0^\alpha \nu_\beta\{\eta: \eta(x) = 1\}\gamma(d\beta) \\
&= \lambda\alpha - \int_0^\alpha \beta\gamma(d\beta) = \int_0^\alpha (\alpha - \beta)\gamma(d\beta).
\end{aligned}$$

This proves (3.12) for all α so that $\gamma(\{\alpha\}) = 0$ and the λ in (3.13) is not zero or one. If $\lambda = 1$, the proof proceeds just as before, while in case $\lambda = 0$, both sides of (3.12) are zero. The last statement follows from the fact that $\mu S(t)\{\eta: \eta(x) = 1\} = \mu\{\eta: \eta(x) = 1\}$ for all t. $\square$

Theorem 3.9(a) gives a complete description of the invariant measures which are translation invariant. In the symmetric case, there are no nontranslation invariant invariant measures by Theorem 1.44 and the fact that $\mathscr{H}$ consists only of constants (by Corollary 7.2 of Chapter II). We saw in

Example 2.8 that there are asymmetric cases in which $\mathscr{I} \not\subset \mathscr{S}$. It would be nice to determine exactly when $\mathscr{I} \not\subset \mathscr{S}$. A reasonable guess, based on that example and the next result, is that if the underlying random walk has a finite first moment, then $\mathscr{I} \subset \mathscr{S}$ if and only if its mean is zero.

Theorem 3.14. *Suppose* $d = 1$, $\sum_x |x| p(0, x) < \infty$, *and* $\sum_x x p(0, x) = 0$. *Then* $\mathscr{I}_e = \{\nu_\alpha, 0 \le \alpha \le 1\}$, *and hence* $\mathscr{I} \subset \mathscr{S}$.

Proof. Take $\nu \in \tilde{\mathscr{I}}$. Then

(3.15)
$$\int \tilde{\Omega} f_x \, d\nu = 0$$

for all $x \in S$ by Proposition 2.13 of Chapter I. Define

$$g_{xy}(\eta, \zeta) = \eta(x)\zeta(x) f_y(\eta, \zeta) \quad \text{and} \quad h_{xy}(\eta, \zeta) = [1 - \eta(x)][1 - \zeta(x)] f_y(\eta, \zeta).$$

By (3.3) and (3.15),

(3.16)
$$\sum_y [p(x, y) + p(y, x)] \int f_{yx} \, d\nu$$
$$= \sum_y \left\{ p(x, y) \int g_{xy} \, d\nu - p(y, x) \int g_{yx} \, d\nu \right\}$$
$$+ \sum_y \left\{ p(y, x) \int h_{xy} \, d\nu - p(x, y) \int h_{yx} \, d\nu \right\}.$$

Take $\sigma_n(x)$ so that $\sigma_n(x) \ge 0$ and $\sum_x \sigma_n(x) < \infty$ for each $n \ge 1$. Multiplying (3.16) by $\sigma_n(x)$ and summing on x gives

(3.17)
$$\sum_{x,y} [p(x, y) + p(y, x)] \sigma_n(x) \int f_{yx} \, d\nu$$
$$= \sum_{x,y} p(x, y) [\sigma_n(x) - \sigma_n(y)] \left\{ \int g_{xy} d\nu - \int h_{yx} \, d\nu \right\}$$
$$= \sum_z p(0, z) \sum_x [\sigma_n(x) - \sigma_n(x + z)] \int (g_{0z} - h_{z0}) \, d\nu_x,$$

where ν_x is obtained by shifting ν in the following way:

$$\nu_x\{(\eta, \zeta): \eta(y_i) = 1, \zeta(z_j) = 1\} = \nu\{(\eta, \zeta): \eta(y_i + x) = 1, \zeta(z_j + x) = 1\}.$$

Now take

$$\sigma_n(x) = \begin{cases} 1 - |x|/n & \text{if } |x| \le n \\ 0 & \text{if } |x| > n. \end{cases}$$

Let n_k be a sequence tending to ∞ so that the following limits exist for each $z \neq 0$

$$\nu^+ = \lim_{k \to \infty} \frac{1}{z} \sum_{x \geq 1} [\sigma_{n_k}(x) - \sigma_{n_k}(x+z)]\nu_x, \quad \text{and}$$

$$\nu^- = -\lim_{k \to \infty} \frac{1}{z} \sum_{x \leq -1} [\sigma_{n_k}(x) - \sigma_{n_k}(x+z)]\nu_x.$$

Then $\nu^+, \nu^- \in \tilde{\mathcal{I}} \cap \tilde{\mathcal{G}}$, and passing to the limit in (3.17) gives

$$(3.18) \quad \sum_{x,y} [p(x, y) + p(y, x)] \int f_{yx} \, d\nu = \sum_z zp(0, z) \int (g_{0z} - h_{z0}) \, d(\nu^+ - \nu^-).$$

Since $\nu^+ \in \tilde{\mathcal{I}} \cap \tilde{\mathcal{G}}$, Lemma 3.2(b) implies that $\int f_{xy} \, d\nu^+ = 0$ for all x, y. Therefore ν^+ can be written in the form

$$\nu^+ = \lambda \nu_1 + (1 - \lambda)\nu_2,$$

where ν_1 concentrates on $\{(\eta, \zeta): \eta \leq \zeta\}$ and ν_2 concentrates on $\{(\eta, \zeta): \eta \geq \zeta, \eta \neq \zeta\}$. Then

$$\int (g_{0z} - h_{z0}) \, d\nu^+ = \lambda \int (g_{0z} - h_{z0}) \, d\nu_1$$

$$(3.19) \qquad\qquad = \lambda \nu_1\{(\eta, \zeta): \eta(0) = 1, \eta(z) = 0\}$$

$$\qquad\qquad - \lambda \nu_1\{(\eta, \zeta): \zeta(0) = 1, \zeta(z) = 0\}.$$

Of course, $\nu_1 \in \tilde{\mathcal{I}} \cap \tilde{\mathcal{G}}$ as well, so that its first and second marginals are in $\mathcal{I} \cap \mathcal{G}$ and hence are exchangeable by Theorem 3.9(a). Therefore the right side of (3.19) is a constant c^+ which is independent of z. Similarly,

$$\int (g_{0z} - h_{z0}) \, d\nu^- = c^-,$$

where c^- is independent of z. By (3.18) and the mean zero assumption,

$$\sum_{x,y} [p(x, y) + p(y, x)] \int f_{yx} \, d\nu = 0,$$

so that $\int f_{xy} \, d\nu = 0$ whenever $p(x, y) + p(y, x) > 0$. Using this and (3.1), it then follows that $\int f_{xy} \, d\nu = 0$ for all $x, y \in S$. Thus we have proved that any $\nu \in \tilde{\mathcal{I}}$ satisfies

$$(3.20) \qquad\qquad \nu\{(\eta, \zeta): \eta \leq \zeta \text{ or } \eta \geq \zeta\} = 1.$$

Now take $\mu \in \mathscr{I}_e$ and $\alpha \in [0, 1]$. By Proposition 2.14, there is a $\nu \in \tilde{\mathscr{I}}$ with marginals μ and ν_α respectively. By the first part of this proof, ν satisfies (3.20). Since $\mu \in \mathscr{I}_e$, it follows that there are $\mu_1, \mu_2 \in \mathscr{I}$ such that $\mu_1 \le \mu \le \mu_2$ and

$$(3.21) \qquad \nu_\alpha = \lambda\mu_1 + (1-\lambda)\mu_2$$

for some $0 \le \lambda \le 1$. Suppose that $0 < \lambda < 1$. Then μ_1 and μ_2 are absolutely continuous with respect to ν_α, so that by the Strong Law of Large Numbers,

$$\mu_i\left\{\eta: \lim_{n\to\infty} \frac{1}{n} \sum_{x=1}^{n} \eta(x) = \alpha\right\} = 1$$

for $i = 1, 2$. Since $\mu_1 \le \mu \le \mu_2$, it follows that

$$\mu\left\{\eta: \lim_{n\to\infty} \frac{1}{n} \sum_{x=1}^{n} \eta(x) = \alpha\right\} = 1.$$

Therefore, given $\mu \in \mathscr{I}_e$, there is at most one $\alpha \in [0, 1]$ for which the λ in (3.21) is neither 0 nor 1. Hence there is an $\alpha_0 \in [0, 1]$ so that $\mu \le \nu_\alpha$ if $\alpha > \alpha_0$ and $\mu \ge \nu_\alpha$ if $\alpha < \alpha_0$. Since the family ν_α is continuous in α, it follows that $\mu = \nu_{\alpha_0}$ as required. $\square$

Besides the cases covered by Theorems 1.44 and 3.14, the only translation invariant exclusion process for which $\mathscr{I}$ has been described completely is the one corresponding to the asymmetric simple random walk on Z^1. To state the result, consider Example 2.8 with $\frac{1}{2} < p < 1$. Let

$$(3.22) \qquad W_n = \left\{\eta \in X: \sum_{x>n} [1 - \eta(x)] = \sum_{x\le n} \eta(x) < \infty\right\}$$

for $-\infty < n < \infty$. The exclusion process restricted to W_n is a Markov chain, which is irreducible and positive recurrent by the observations made in that example. Let ν_n be the unique stationary measure for this Markov chain. Using coupling techniques similar to those used in the proofs of Theorems 3.9 and 3.14, Liggett (1976) showed that

$$(3.23) \qquad \mathscr{I}_e = \{\nu_\alpha: \alpha \in [0, 1]\} \cup \{\nu_n: n \in Z^1\}.$$

If $p = 1$, the same result holds with ν_n being the pointmass on the configuration in which $\eta(x) = 1$ if and only if $x > n$.

4. The Tagged Particle Process

In this section we will focus attention on one of the particles in an exclusion process, and will try to determine how its motion is affected by its interaction with the other particles in the system. The only thing that is clear at the outset is that the interaction tends to slow down the motion of the tagged particle, since many of the transitions it would otherwise make are not permitted. The main problem is to determine how much slowing there is, and then to determine the limiting distribution of the position of the tagged particle as $t \uparrow \infty$. The analysis is complicated substantially by the fact that the motion of the tagged particle is not Markovian. We will assume throughout this section that $S = Z^d$ and $p(x, y) = p(0, y - x)$.

In order to follow the motion of the tagged particle it is convenient to regard the exclusion process as a Markov process (X_t, η_t) on the locally compact space

$$V = \{(x, \eta) \in S \times X : \eta(x) = 1\},$$

so that x is the position of the tagged particle and η is the entire configuration. The generator of the process is the closure of the operator which is given by

$$\Omega f(x, \eta) = \sum_{\substack{\eta(u)=1 \\ \eta(v)=0 \\ u \neq x}} p(u, v)[f(x, \eta_{uv}) - f(x, \eta)]$$

(4.1)

$$+ \sum_{\substack{y \\ \eta(y)=0}} p(x, y)[f(y, \eta_{xy}) - f(x, \eta)]$$

for functions which depend on finitely many coordinates and have compact support. The existence of the process does not follow directly from the results of Chapter I, since the processes considered there had compact state spaces. However, the same approach, with estimates analogous to those used in that chapter, does imply that this process is well defined. Alternatively, one could use the graphical representation described in Section 2 to construct the tagged particle process.

Suppose that initially, the tagged particle is placed at some $x \in S$, and particles are placed at each of the sites $y \in S \setminus \{x\}$ independently with the same probability $\alpha \in [0, 1]$. The next result asserts that this system is stationary when viewed from the position of the tagged particle. In other words,

$$E \prod_{y \in A} \eta_t(X_t + y) = \alpha^{|A|}$$

for all finite $A \subset S \setminus \{0\}$. For finite $A \subset S \setminus \{0\}$ and $(x, \eta) \in V$, let

$$H_A(x, \eta) = \begin{cases} 1 & \text{if } \eta(x + y) = 1 \text{ for all } y \in A, \\ 0 & \text{otherwise.} \end{cases}$$

Theorem 4.2. *Suppose that* $x \in S$ *and* $\alpha \in [0, 1]$. *Let* ν *be the probability measure on* V *whose first marginal is the pointmass at* x *and which satisfies*

$$\int_V H_A(x, \eta)\, d\nu = \alpha^{|A|}$$

for all finite $A \in S\backslash\{0\}$. *If* (X_t, η_t) *is the tagged particle process with initial distribution* ν, *then*

$$EH_A(X_t, \eta_t) = \alpha^{|A|}$$

for all $t \geq 0$ *and all finite* $A \in S\backslash\{0\}$.

Proof. For any $(x, \eta) \in V$, and finite $A \in S\backslash\{0\}$, compute

$$\frac{d}{dt} E^{(x, \eta)} H_A(X_t, \eta_t)\Big|_{t=0}$$

$$= \sum_{u, v \neq x} p(u, v)\eta(u)[1 - \eta(v)][H_A(x, \eta_{uv}) - H_A(x, \eta)]$$

$$+ \sum_{y \neq x} p(x, y)[1 - \eta(y)][H_A(y, \eta_{xy}) - H_A(x, \eta)]$$

$$= - \sum_{\substack{u-x \in A \\ 0 \neq v-x \notin A}} p(u, v)[H_A(x, \eta) - H_{A \cup \{v-x\}}(x, \eta)]$$

$$+ \sum_{\substack{0 \neq u-x \notin A \\ v-x \in A}} p(u, v)[H_{A\backslash\{v-x\} \cup \{u-x\}}(x, \eta) - H_{A \cup \{u-x\}}(x, \eta)]$$

(4.3) $$\qquad + \sum_{0 \neq x-y \notin A} p(x, y)[H_{A-x+y}(x, \eta) - H_{(A-x+y) \cup (y-x)}(x, \eta)]$$

$$- \sum_{0 \neq y-x \notin A} p(x, y)[H_A(x, \eta) - H_{A \cup (y-x)}(x, \eta)].$$

$$= - \sum_{\substack{u \in A \\ 0 \neq v \notin A}} p(u, v)[H_A(x, \eta) - H_{A \cup \{v\}}(x, \eta)]$$

$$+ \sum_{\substack{0 \neq u \notin A \\ v \in A}} p(u, v)[H_{A\backslash\{v\} \cup \{u\}}(x, \eta) - H_{A \cup \{u\}}(x, \eta)]$$

$$+ \sum_{0 \neq u \notin A} p(u, 0)[H_{A-u}(x, \eta) - H_{(A-u) \cup \{-u\}}(x, \eta)]$$

$$- \sum_{0 \neq u \notin A} p(0, u)[H_A(x, \eta) - H_{A \cup \{u\}}(x, \eta)].$$

The right-hand side of (4.3) is a linear combination of $H_B(x, \eta)$, with coefficients which do not depend on (x, η). The sum of the absolute values of these coefficients is at most $4(|A| + 1)$. Furthermore, the B's which appear in this linear combination have cardinalities at most $|A| + 1$. Therefore there exist $\gamma_t(A, B)$ which are summable in B for each A and t so that

(4.4) $$E^{(x,\eta)} H_A(X_t, \eta_t) = \sum_B \gamma_t(A, B) H_B(x, \eta).$$

Suppose now that ν is any probability measure on V such that

(4.5) $$\int H_A(x, \eta) \, d\nu$$

depends on A only through its cardinality. Then the integral of the right side of (4.3) with respect to ν is easily seen to be zero. This, together with (4.4), implies that

$$\frac{d}{dt} \int E^{(x,\eta)} H_A(X_t, \eta_t) \, d\nu = 0.$$

The proof of the theorem is complete, since the measure ν which appears in its statement satisfies

$$\int H_A \, d\nu = \alpha^{|A|}. \quad \square$$

Corollary 4.6. *Suppose that* $\sum_y \|y\| p(0, y) < \infty$. *If the tagged particle process has the initial distribution* ν *described in the statement of Theorem 4.2, then*

$$EX_t = x + (1 - \alpha) t \sum_y y p(0, y), \quad and$$

$$\lim_{t \to \infty} \frac{X_t}{t} \quad exists \ a.s. \ and \ in \ L_1.$$

Proof. By Theorem 4.2, X_t has stationary increments. Therefore EX_t is a linear function of t. The first statement follows from this and

$$\lim_{t \downarrow 0} \frac{EX_t - x}{t} = (1 - \alpha) \sum_y y p(0, y).$$

The convergence of X_n/n is a consequence of the Ergodic Theorem (Theorem 4.9 of Chapter I). To extend to convergence along all t, simply note that

$$E \sup_{0 \le t \le 1} \|X_t\| < \infty,$$

and use a Borel–Cantelli argument. $\quad \square$

According to Theorem 4.2, $\{\eta_t(X_t + x), x \in Z^d \setminus \{0\}\}$ are independent and identically distributed Bernoulli random variables with parameter α at time t, provided that this is the case at time 0. In general, these random variables are not independent of X_t. If they were, the analysis of X_t would be quite simple. There is one case in which this type of independence does occur. It is the subject of the next result.

Theorem 4.7. *Suppose that $d = 1$ and $p(x, x+1) = 1$. Assume that for some $\alpha \in [0, 1]$, the initial distribution of the tagged particle system has the following property: X_0 and $H_A(X_0, \eta_0)$ are independent, and*

$$(4.8) \qquad EH_A(X_0, \eta_0) = \alpha^{|A|}$$

for every finite $A \subset \{1, 2, \ldots\}$. Then the distribution at time t has the same property.

Proof. We need to show that

$$Ef(X_t)[H_A(X_t, \eta_t) - \alpha^{|A|}] = 0$$

for all bounded f on S and all finite $A \subset \{1, 2, \ldots\}$. Proceed as in the proof of Theorem 4.2 to compute

$$\frac{d}{dt} E^{(x,\eta)} f(X_t)[H_A(X_t, \eta_t) - \alpha^{|A|}]\Big|_{t=0}$$

$$= -f(x) \sum_{\substack{u \in A \\ u+1 \notin A}} [H_A(x, \eta) - H_{A \cup \{u+1\}}(x, \eta)]$$

$$+ f(x) \sum_{\substack{0 \neq u \notin A \\ u+1 \in A}} [H_{A \setminus \{u+1\} \cup \{u\}}(x, \eta) - H_{A \cup \{u\}}(x, \eta)]$$

$$+ f(x+1)[H_{A+1}(x, \eta) - H_{(A+1) \cup \{1\}}(x, \eta)]$$

$$- f(x)[H_A(x, \eta) - H_{A \cup \{1\}}(x, \eta)]$$

$$- \alpha^{|A|}[1 - H_{\{1\}}(x, \eta)][f(x+1) - f(x)]$$

$$= -f(x) \sum_{\substack{u \in A \\ u+1 \notin A}} [H_A(x, \eta) - \alpha^{|A|} - H_{A \cup \{u+1\}}(x, \eta) + \alpha^{|A|+1}]$$

$$+ f(x) \sum_{\substack{0 \neq u \notin A \\ u+1 \in A}} [H_{A \setminus \{u+1\} \cup \{u\}}(x, \eta) - \alpha^{|A|} - H_{A \cup \{u\}}(x, \eta) + \alpha^{|A|+1}]$$

$$+ f(x+1)[H_{A+1}(x, \eta) - \alpha^{|A|} - H_{(A+1) \cup \{1\}}(x, \eta) + \alpha^{|A|+1}]$$

$$- f(x)[H_A(x, \eta) - \alpha^{|A|} - H_{A \cup \{1\}}(x, \eta) + \alpha^{|A \cup \{1\}|}]$$

$$+ \alpha^{|A|}[f(x+1) - f(x)][H_{\{1\}}(x, \eta) - \alpha].$$

The right side of this identity is a linear combination of functions on V of the form

$$g(x)[H_A(x, \eta) - \alpha^{|A|}].$$

Furthermore, the integral of any function of this form with respect to any probability measure on V which has property (4.8) is zero. Thus property (4.8) is preserved in time. □

Corollary 4.9. *Under the assumptions of Theorem 4.7, $X_t - X_0$ is a Poisson process with rate $(1 - \alpha)$.*

Proof. We may as well assume that $X_0 = 0$. Then

$$\frac{d}{dt} P(X_t = x) = P(X_t = x - 1, \eta_t(x) = 0) - P(X_t = x, \eta_t(x + 1) = 0).$$

By Theorem 4.7,

$$\frac{d}{dt} P(X_t = x) = (1 - \alpha)[P(X_t = x - 1) - P(X_t = x)],$$

so that

$$P(X_t = x) = e^{-(1-\alpha)t} \frac{[(1 - \alpha)t]^x}{x!}$$

for $x = 0, 1, \ldots$. Since the process

$$(X_t; \eta_t(X_t + x), x \geq 1)$$

is Markovian, this together with another application of Theorem 4.7 yields the Markov property for X_t. □

Corollary 4.9 tells us that in the special case $d = 1$, $p(x, x + 1) = 1$, the effect of the interaction on X_t is to slow it down by a factor of $1 - \alpha$. In case $d = 1$, $p(x, x + 1) = p(x, x - 1) = \frac{1}{2}$, the interaction slows X_t down much more. In fact, the variance of X_t is asymptotic to a constant multiple of $\sqrt{t}$ as $t \uparrow \infty$, rather than to a multiple of t, which would be the variance in the absence of the interaction. The rest of this section is devoted to the proof of this result, together with the associated Central Limit Theorem for X_t.

The first step in this proof is to express the exclusion process in terms of a stirring system. While this is possible whenever $p(x, y)$ is symmetric, it is only helpful in the analysis of the tagged particle in the special case $d = 1$, $p(x, x + 1) = p(x + 1, x) = \frac{1}{2}$. Therefore we will consider only this case in the remainder of the section.

The representation of the exclusion process in terms of the stirring system is an analogue of the graphical representation for spin systems which was described in Section 6 of Chapter III. It is somewhat different from the representation discussed in Section 2, and exhibits the exclusion process as an additive process in the sense of Definition 6.1 from Chapter III. The idea is to associate independent rate $\frac{1}{2}$ Poisson processes with the pairs $\{(x, x+1), x \in Z^1\}$. Initially, there is a particle at each point of Z^1. At the event times of the Poisson process associated with the pair $(x, x+1)$, the particles at x and $x+1$ are interchanged. The stirring system $\{\xi_t^x, x \in Z^1\}$ is then defined by

$$\xi_t^x = \text{position at time } t \text{ of the particle initially at } x.$$

Note that for each x, ξ_t^x is a simple random walk on Z^1. Of course ξ_t^x and ξ_t^y are not independent. However, these random walks have negative correlations, as we shall see in Lemma 4.12.

To see the connection with the exclusion process, take $\eta \in X$. Define $\eta_t \in X$ by

$$\eta_t(x) = 1 \quad \text{if and only if there is a } y \in Z^1 \text{ so that } \xi_t^y = x \text{ and } \eta(y) = 1.$$

Then η_t is a version of the exclusion process with initial configuration η corresponding to $p(x, x+1) = p(x+1, x) = \frac{1}{2}$ for all x. This provides a simultaneous coupling for the exclusion process for all initial configurations. Note however that this coupling does not agree with the basic coupling which was described in Section 2. This construction of the exclusion processes will be used in the rest of this section without further comment. The next result identifies the position of the tagged particle for the exclusion process in terms of the stirring system. In what follows, 1_A will denote the indicator function of A.

Lemma 4.10. (a) *For any $\eta \in X$ such that $\eta(0) = 1$, let η_t be the corresponding exclusion process and X_t be the position at time t of the particle which was originally at 0. Then for any $z \in Z^1$, the events $\{X_t \geq z\}$ and*

$$\left\{ \sum_{y \leq 0} \eta(y) 1_{\{\xi_t^y \geq z\}} > \sum_{y > 0} \eta(y) 1_{\{\xi_t^y < z\}} \right\}$$

are identical.

(b)   *For every realization of the stirring system and every $z \in Z^1$,*

$$\sum_{y \leq 0} 1_{\{\xi_t^y \geq z\}} - \sum_{y > 0} 1_{\{\xi_t^y < z\}} = -(z - 1).$$

Proof. Take $\eta \in X$ so that $\eta(0) = 1$ and $\sum_x \eta(x) < \infty$. Then

$$\sum_{y \leq 0} \eta(y) 1_{\{\xi_t^y \geq z\}} - \sum_{y > 0} \eta(y) 1_{\{\xi_t^y < z\}}$$

(4.11)
$$= \sum_y \eta(y) 1_{\{\xi_t^y \geq z\}} - \sum_{y > 0} \eta(y)$$

$$= \sum_{y \geq z} \eta_t(y) - \sum_{y > 0} \eta(y).$$

Since particles cannot jump over each other in the exclusion process, the right side of (4.11) is positive if and only if $X_t \geq z$. This proves (a) for finite configurations. The general case is obtained by a simple limiting argument. For part (b), apply (4.11) to the configuration η given by

$$\eta(x) = \begin{cases} 1 & \text{if } |x| \leq n, \\ 0 & \text{if } |x| > n \end{cases}$$

and then let n tend to ∞. $\quad \square$

Lemma 4.12. *For any $A \subset Z^1$ and finite $T \subset Z^1$,*

$$P(\xi_t^x \in A \text{ for all } x \in T) \leq \prod_{x \in T} P(\xi_t^x \in A).$$

Proof. Define f on Z^n, where $n = |T|$, by

$$f(x_1, \ldots, x_n) = \begin{cases} 1 & \text{if } x_i \in A \text{ for all } 1 \leq i \leq n, \\ 0 & \text{otherwise.} \end{cases}$$

This function is positive definite, so that by Proposition 1.7,

$$Ef(\xi_t^{x_1}, \ldots, \xi_t^{x_n}) \leq \prod_{i=1}^n P(\xi_t^{x_i} \in A)$$

whenever the x_i are distinct. This is the desired result. $\quad \square$

Theorem 4.13. *Suppose that initially, $\eta(0) = 1$ and $\{\eta(x), x \neq 0\}$ are independent and satisfy*

$$P[\eta(x) = 1] = \alpha > 0 \quad \text{for all } x \neq 0.$$

Let X_t be the position at time t of the particle which was originally at 0. Then $t^{-1/4} X_t$ converges in distribution to the normal law with mean zero and variance $\sqrt{2/\pi}[(1 - \alpha)/\alpha]$. Furthermore,

(4.14)
$$\lim_{t \to \infty} \frac{\text{var}(X_t)}{\sqrt{t}} = \sqrt{\frac{2}{\pi}} \frac{1 - \alpha}{\alpha}.$$

Proof. Let

$$N_t(z) = \sum_{y \leq 0} 1_{\{\xi_t^y \geq z\}},$$

and let $\{\gamma_n, -\infty < n < \infty\}$ be independent random variables which are independent of the stirring process and have distribution

$$P(\gamma_n = 1) = \alpha, \ P(\gamma_n = 0) = 1 - \alpha.$$

By Lemma 4.10, if z is a positive integer,

(4.15)
$$P(X_t \geq z) = P\left(\sum_{n=1}^{N_t(z)} \gamma_n + 1_{\{\xi_t^0 \geq z, \gamma_1 = 0\}} > \sum_{n=-N_t(z)-z+1}^{-1} \gamma_n \right)$$

$$= P\left(\sum_{n=1}^{N_t(z)} (\gamma_n - \gamma_{-n}) + 1_{\{\xi_t^0 \geq z, \gamma_1 = 0\}} > \sum_{n=-N_t(z)-z+1}^{-N_t(z)-1} \gamma_n \right).$$

By Lemma 4.12, $N_t(z)$ is a sum of negatively correlated indicator random variables. Therefore

(4.16)
$$\text{var } N_t(z) \leq \sum_{y \leq 0} \text{var}(1_{\{\xi_t^y \geq z\}})$$

$$\leq \sum_{y \leq 0} E 1_{\{\xi_t^y \geq z\}} = EN_t(z).$$

On the other hand,

(4.17)
$$EN_t(z) = \sum_{y \leq 0} \sum_{x \geq z} p_t(y, x)$$

$$= \sum_{u \geq z} p_t(0, u)(u - z + 1)$$

$$= E(\xi_t^0 - z + 1)^+,$$

which is asymptotic to $\sqrt{t/2\pi}$ by the Central Limit Theorem, provided that $z/\sqrt{t} \to 0$. Now let $z = z(t)$ depend on t in such a way that $z(t) \sim at^{1/4}$ for some $a > 0$. Then (4.16), (4.17), and Chebyshev's inequality imply that

$$\lim_{t \to \infty} P\left(\left| \frac{N_t(z)}{\sqrt{t}} - \frac{1}{\sqrt{2\pi}} \right| > t^{1/4} \right) = 0.$$

Therefore by the Weak Law of Large Numbers,

$$\lim_{t \to \infty} \frac{1}{t^{1/4}} \sum_{n=-N_t(z)-z+1}^{-N_t(z)-1} \gamma_n = a\alpha.$$

in probability, and by the Central Limit Theorem,

$$\frac{1}{t^{1/4}} \sum_{n=1}^{N_t(z)} (\gamma_n - \gamma_{-n})$$

is asymptotically normal with mean zero and variance $\sqrt{2/\pi}[\alpha(1-\alpha)]$. The first statement follows from this and (4.15), and then by using the fact that the distribution of X_t is symmetric. For (4.14), it suffices to prove that $\{t^{-1/2}X_t^2, t \geq 1\}$ is uniformly integrable. This is left to the reader. $\square$

Corollary 4.9 and Theorem 4.13 give a rather precise description of the motion of the tagged particle in two very special cases. In recent work, the asymptotic normality (with nonzero asymptotic variance) of

(4.18) $$\frac{X_t - EX_t}{\sqrt{t}}$$

has been proved under two sets of assumptions. Kipnis and Varadhan (1985) proved it in any number of dimensions assuming $p(x, y) = p(y, x) = p(0, y-x)$, the irreducibility of the random walk, and

$$\sum_x \|x\|^2 p(0, x) < \infty,$$

but excluding of course the case $d = 1$, $p(x, x+1) = p(x+1, x) = \frac{1}{2}$. In this case X_t is symmetric, so that $EX_t = 0$. Their proof is based on a general Central Limit Theorem for functionals of reversible Markov processes. The reversibility requirement forces them to assume that $p(x, y)$ is symmetric. Kipnis (1985) proved the asymptotic normality of (4.18) in case $d = 1$, $p(x, x+1) = p$, $p(x, x-1) = q$, with $p + q = 1$ and $p \neq \frac{1}{2}$. The case $p = 1$ is trivial in view of Corollary 4.9. It appears likely from these results that the only case in which $\text{var}(X_t)$ is of smaller order than t is that covered in Theorem 4.13.

5. Nonequilibrium Behavior

Throughout this section, we will take $S = Z^1$, and $p(x, y) = p(0, y-x)$ such that $\sum_x |x| p(0, x) < \infty$, and will study the behavior of the exclusion process with initial configuration η given by

(5.1) $$\eta^0(x) = \begin{cases} 1 & \text{if } x \leq 0, \\ 0 & \text{if } x > 0. \end{cases}$$

The objective is to describe the distribution of the process at sites near x_t

for large times, where $x_t \to \pm\infty$ at an appropriate rate. For $x \in Z^1$ and $t \geq 0$, let μ_t^x be the probability measure given by

$$\mu_t^x\{\zeta: \zeta(x_1) = \cdots = \zeta(x_n) = 1\} = P^{\eta^0}[\eta_t(x + x_1) = \cdots = \eta_t(x + x_n) = 1].$$

When $x = 0$, we will write $\mu_t = \mu_t^0$.

As we saw earlier, duality is a powerful tool in the symmetric case. Thus rather precise results can be obtained in this case. The proof of the following theorem is essentially the same as that of Theorem 1.13, and will therefore be omitted.

Theorem 5.2. *Suppose that $p(0, x) = p(x, 0)$ and that*

$$0 < \sigma^2 = \sum_x x^2 p(0, x) < \infty.$$

Then

$$\lim_{t \to \infty} \mu_t^{[c\sqrt{t}]} = \nu_{\alpha(c)},$$

and

$$\lim_{t \to \infty} \frac{1}{\sqrt{t}} \sum_{y \geq c\sqrt{t}} \eta_t(y) = \frac{\sigma}{\sqrt{2\pi}} e^{-c^2/2\sigma^2} - c\alpha(c) = \int_c^\infty \alpha(u) \, du$$

in probability relative to P^{η^0}, where

$$\alpha(c) = \frac{1}{\sigma\sqrt{2\pi}} \int_c^\infty e^{-u^2/2\sigma^2} \, du,$$

and $[\cdot]$ is the greatest integer function.

According to Theorem 5.2, the disturbance at time t from the initial configuration spreads out at rate $\sqrt{t}$ if $p(\cdot, \cdot)$ is symmetric and has a finite variance. Our main interest in this section is the case in which $p(\cdot, \cdot)$ has a positive drift. In this case, one would expect the disturbance to spread out at rate t. Precise versions of this statement will be proved later in this section. To begin, let

$$\sigma_t(x) = \sum_{y > x} \eta_t(y),$$

which is clearly finite for all $t \geq 0$ and $x \in Z^1$ since $\sum_x |x| p(0, x) < \infty$.

Theorem 5.3. *Consider the process with initial configuration η^0, which is defined in (5.1). For any $c \in (-\infty, \infty)$,*

$$\sigma(c) = \lim_{t \to \infty} \frac{1}{t} \sigma_t([ct])$$

exists in L^1 and with probability one, and is nonrandom.

Proof. The proof is based on Theorem 2.6 of Chapter VI. Using the graphical representation described in Section 2, versions of the exclusion process starting from an arbitrary configuration at an arbitrary time can be constructed simultaneously on the same probability space. For simplicity, take $c > 0$. The cases $c = 0$ and $c < 0$ are handled similarly. For integers $0 \le m \le n$, let

$$X_{m,n} = \sum_{y > n} \eta_{n/c}(y)$$

for the process which starts at time m/c in the configuration

(5.4)
$$\eta^m(x) = \begin{cases} 1 & \text{if } x \le m, \\ 0 & \text{otherwise.} \end{cases}$$

We wish to verify that the collection $\{X_{m,n}, m \le n\}$ satisfies the hypotheses of Theorem 2.6 of Chapter VI. To check (a), let η_t be the process which satisfies $\eta_0 = \eta^0$ (given in (5.1)), and ζ_t for $t \ge m/c$ be the process which starts at time m/c in the configuration $\eta^m \vee \eta_{m/c}$. Then

(5.5)
$$X_{0,n} = \sum_{y > n} \eta_{n/c}(y) \le \sum_{y > n} \zeta_{n/c}(y)$$
$$\le X_{0,m} + X_{m,n}.$$

Hypotheses (b) and (c) are clearly satisfied. In fact, the stationary processes in (b) consist of independent and identically distributed random variables, and are hence ergodic. For hypothesis (d), compare the exclusion process with independent particles which are only allowed to move to the right. The result of this comparison is

(5.6)
$$EX_{0,n}^+ \le \sum_{\substack{x \le 0 \\ y > n}} q_{n/c}(x, y) = \sum_{y > n} (y - n) q_{n/c}(0, y),$$

where $q_t(x, y)$ are the transition probabilities for a random walk with jump rates

$$q(x, y) = \begin{cases} p(x, y) & \text{if } y > x, \\ 0 & \text{otherwise.} \end{cases}$$

Since $X_{0,n} = \sigma_{n/c}(n)$, Theorem 2.6 of Chapter VI gives the desired convergence along the sequence of times $t = n/c$. The extension to convergence along all t is routine, and is left to the reader. $\square$

The next objective is to say as much as possible about $\sigma(c)$. First we need to record some simple properties of the distribution of the process with initial configuration η^0.

Theorem 5.7. (a) $\sum_{y>x} \eta_t(y)$ and $\sum_{y>-x} \eta_t(y) - x$ have the same distribution for all $x \in Z^1$ and $t \geq 0$.
   (b) $\mu_t^x \geq \mu_t^{x+1}$ for all $x \in Z^1$ and $t \geq 0$.
   (c) Let $M_s = \max\{y: \eta_s(y) = 1\}$. Then

$$\mu_{t+s}^x \leq \sum_y P(M_s = y)\mu_t^{x-y}, \quad and$$

$$\mu_{t+s}^x \geq \sum_y P(M_s = y)\mu_t^{x+y}.$$

Proof. Let ζ_t be defined by $\zeta_t(x) = 1 - \eta_t(x)$. Then ζ_t is an exclusion process corresponding to the transpose $p(y, x)$ of $p(x, y)$. Therefore

$$\sum_{y>x} \eta_t(y) \quad and \quad \sum_{y\leq -x} \zeta_t(y)$$

have the same distribution. Since

$$\sum_{y\leq -x} \zeta_t(y) = \sum_{y\leq -x} [1 - \eta_t(y)]$$

$$= -x + \sum_{y>-x} \eta_t(y),$$

part (a) follows. Part (b) is an immediate consequence of Proposition 2.12, the translation invariance of the mechanism, and the fact that $\eta^{-1} \leq \eta^0$, where η^{-1} is defined as in (5.4). For the first inequality in part (c), simply compare the process with the one which is obtained by filling in with ones at time s all the sites to the left of M_s, and then using Proposition 2.12. The proof of the second inequality is similar. In this case, compare the process with the one which is obtained by filling in with zeros at time s all the sites to the right of the leftmost zero. The leftmost zero in η_s is the leftmost one in ζ_s, and therefore its position has the same distribution as $-M_s + 1$. $\square$

Theorem 5.8. (a) $\sigma(c)$ is decreasing, convex and satisfies

(5.9) $$\sigma(c) = -c + \sigma(-c).$$

(b)   *Suppose that σ is differentiable at c and that $x_t/t \to c$. Then any weak limit μ^* of the measures $\mu_t^{x_t}$ as $t \to \infty$ is exchangeable and satisfies*

$$\mu^*\{\eta: \eta(x) = 1\} = -\sigma'(c).$$

Proof. Take $c_1 < c_2$. By the definition of σ,

(5.10)
$$\sigma(c_1) - \sigma(c_2) = \lim_{t \to \infty} \frac{E\sigma_t([c_1 t]) - E\sigma_t([c_2 t])}{t}$$

$$= \lim_{t \to \infty} \frac{1}{t} \sum_{c_1 t < x < c_2 t} \mu_t\{\eta: \eta(x) = 1\}.$$

This gives the monotonicity of σ, and when combined with Theorem 5.7(b), it gives the convexity of σ also. For (5.9), it suffices by symmetry to take $c > 0$. Let $c_1 = -c$ and $c_2 = c$ in (5.10), and use Theorem 5.7(a) to obtain (5.9). For part (b), use Theorem 5.7(b) and (5.10) to write

$$\sigma(c_1) - \sigma(c_2) \leq (c_2 - c_1) \liminf_{t \to \infty} \mu_t\{\eta: \eta([c_1 t]) = 1\}, \quad \text{and}$$

$$\sigma(c_1) - \sigma(c_2) \geq (c_2 - c_1) \limsup_{t \to \infty} \mu_t\{\eta: \eta([c_2 t]) = 1\}.$$

Therefore if σ is differentiable at c and $x_t/t \to c$,

(5.11)
$$\lim_{t \to \infty} \mu_t\{\eta: \eta(x_t) = 1\} = -\sigma'(c).$$

Let μ^* be the weak limit of the measures $\mu_{t_n}^{x_{t_n}}$ as $t \to \infty$. By Theorem 5.7(b), μ^* lies above its image under the shift. Both μ^* and this image have one point probabilities $-\sigma'(c)$ by (5.11), so the two measures must coincide by Corollary 2.8 of Chapter II. Therefore $\mu^* \in \mathcal{S}$. Using this fact and Theorem 5.7(c), we conclude that

$$\lim_{n \to \infty} \mu_{t_n + s}^{x_{t_n}} = \mu^*$$

for all $s \geq 0$. Therefore

$$\mu^* S(s) = \mu^*,$$

so that $\mu^* \in \mathcal{I}$. Hence μ^* is exchangeable by Theorem 3.9(a). $\square$

The next step is to evaluate $\sigma(c)$ explicitly. Let

$$\gamma = \sum_x x p(0, x).$$

Theorem 5.12. (a) *If* $\gamma \leq 0$, *then*

$$\sigma(c) = 0 \quad for\ c \geq 0 \quad and \quad \sigma(c) = -c \quad for\ c \leq 0.$$

(b) *Suppose* $\gamma > 0$. *Then*

$$\sigma(c) = 0 \quad for\ c \geq \gamma, \qquad \sigma(c) = -c \quad for\ c \leq -\gamma, \quad and$$

(5.13) $$\sigma(c) = \frac{1}{4\gamma}(\gamma - c)^2 \quad for\ |c| \leq \gamma.$$

Furthermore, if $\sum_x x^2 p(0, x) < \infty$, *then*

$$\lim_{t \to \infty} \mu_t^{[ct]} = \nu_{\alpha(c)},$$

where

$$\alpha(c) = -\sigma'(c) = \frac{1}{2\gamma}(\gamma - c) \quad for\ |c| \leq \gamma.$$

Proof. Apply (0.1) to the function

$$f(\eta) = \sum_{y=x}^{z} \eta(y)$$

and then let z tend to ∞ to obtain

$$\frac{d}{dt} \sum_{y \geq x} \mu_t\{\eta: \eta(y) = 1\}$$

(5.14) $$= \sum_{u < x \leq v} p(u, v)\mu_t\{\eta: \eta(u) = 1, \eta(v) = 0\}$$

$$- \sum_{u < x \leq v} p(v, u)\mu_t\{\eta: \eta(v) = 1, \eta(u) = 0\}.$$

Fix $c > 0$ and integrate (5.14) from x/c to $(x+1)/c$. This gives

$$\sum_{y \geq x} \mu_{(x+1)/c}\{\eta: \eta(y) = 1\} - \sum_{y \geq x} \mu_{x/c}\{\eta: \eta(y) = 1\}$$

(5.15) $$= \sum_{u < x \leq v} p(u, v) \int_{x/c}^{(x+1)/c} \mu_s\{\eta: \eta(u) = 1, \eta(v) = 0\}\, ds$$

$$- \sum_{u < x \leq v} p(v, u) \int_{x/c}^{(x+1)/c} \mu_s\{\eta: \eta(v) = 1, \eta(u) = 0\}\, ds.$$

Replacing u and v by $u + x$ and $v + x$ respectively, and summing on x for $0 \le x < z$ yields

$$\sum_{x=1}^{z} \mu_{x/c}\{\eta: \eta(x-1) = 1\} + \sum_{y \ge z} \mu_{z/c}\{\eta: \eta(y) = 1\} - 1$$

$$= \sum_{u<0\le v} p(u, v) \sum_{x=0}^{z-1} \int_{x/c}^{(x+1)/c} \mu_s^x\{\eta: \eta(u) = 1, \eta(v) = 0\}\, ds$$

$$- \sum_{u<0\le v} p(v, u) \sum_{x=0}^{z-1} \int_{x/c}^{(x+1)/c} \mu_s^x\{\eta: \eta(v) = 1, \eta(u) = 0\}\, ds.$$

Since $x = [cs]$ for $x/c \le s < (x+1)/c$, this can be rewritten as

$$\sum_{x=1}^{z} \mu_{x/c}\{\eta: \eta(x-1) = 1\} + \sum_{y \ge z} \mu_{z/c}\{\eta: \eta(y) = 1\} - 1$$

$$= \sum_{u<0\le v} p(u, v) \int_0^{z/c} \mu_s^{[cs]}\{\eta: \eta(u) = 1, \eta(v) = 0\}\, ds$$

$$- \sum_{u<0\le v} p(v, u) \int_0^{z/c} \mu_s^{[cs]}\{\eta: \eta(v) = 1, \eta(u) = 0\}\, ds.$$

Now take $z = [ct]$, divide by t, and let $t \to \infty$. If σ is differentiable at c, then Theorems 5.3 and 5.8(b) imply that

$$(5.16) \quad -c\sigma'(c) + \sigma(c) = \gamma \lim_{t \to \infty} \frac{1}{t} \int_0^t \mu_s^{[cs]}\{\eta: \eta(0) = 1, \eta(1) = 0\}\, ds.$$

Since σ is convex, it is differentiable at all but at most countably many points. If $\gamma \le 0$, the right side of (5.16) is nonpositive, while each term on the left is nonnegative. Therefore $\sigma(c) = 0$ for a dense set of $c > 0$. Part (a) follows from this, the continuity of σ, and (5.9). Assume then that $\gamma > 0$, and continue to take $c > 0$ so that σ is differentiable at c. By Theorem 5.8(b), any weak limit of $\mu_s^{[cs]}$ as $s \to \infty$ is exchangeable, and hence is a mixture

$$(5.17) \qquad\qquad \int_0^1 \nu_\alpha \tau(d\alpha)$$

for some probability measure τ on $[0, 1]$ which satisfies

$$(5.18) \qquad\qquad \int_0^1 \alpha \tau(d\alpha) = -\sigma'(c).$$

By Hölder's inequality,

(5.19)
$$\int_0^1 \alpha^2 \tau(d\alpha) \geq [\sigma'(c)]^2,$$

so that

(5.20) $\limsup\limits_{s\to\infty} \mu_s^{[cs]}\{\eta: \eta(0)=1,\ \eta(1)=0\} \leq -\sigma'(c) - [\sigma'(c)]^2.$

Using this in (5.16) yields

$$\sigma(c) \leq c\sigma'(c) - \gamma\sigma'(c) - \gamma[\sigma'(c)]^2.$$

Therefore since $-1 \leq \sigma'(c) \leq 0$,

$$\sigma(c) \leq \sup_{0\leq b\leq 1} (-bc + \gamma b - \gamma b^2)$$

$$= \begin{cases} (1/4\gamma)(\gamma-c)^2 & \text{if } 0<c<\gamma, \\ 0 & \text{if } c\geq\gamma. \end{cases}$$

To prove the opposite inequality, let ν_t be the distribution at time t for the exclusion process whose initial distribution is the product measure with marginals given by

$$\nu_0\{\eta(x)=1\} = \begin{cases} \lambda & \text{if } x\leq 0, \\ 0 & \text{if } x>0, \end{cases}$$

where $0\leq\lambda\leq 1$. Let $B_x(t)$ be the expression on the right of (5.14) when μ_t is replaced by ν_t. By (5.14),

$$\sum_{y\geq x} \nu_t\{\eta: \eta(y)=1\} - \lambda(|x|+1) = \int_0^t B_x(s)\,ds$$

for $x\leq 0$. Therefore for such x,

$$\int_0^t B_0(s)\,ds - \int_0^t B_x(s)\,ds = \lambda|x| - \sum_{x\leq y<0} \nu_t\{\eta: \eta(y)=1\}.$$

This expression is nonnegative since $\nu_t\{\eta: \eta(y)=1\} \leq \lambda$ for all y by Theorem 2.1 and Proposition 2.12. But

$$\lim_{x\to-\infty} B_x(s) = \gamma\lambda(1-\lambda)$$

for any s, so it follows that

$$\int_0^t B_0(s) \, ds \geq \gamma\lambda(1-\lambda)t$$

for all $t \geq 0$. Using (5.14) with μ_t replaced by ν_t again, we see that

$$\sum_{y \geq 0} \nu_t\{\eta: \eta(y) = 1\} = \int_0^t B_0(s) \, ds + \lambda \geq \gamma\lambda(1-\lambda)t.$$

On the other hand, we know that

$$\sum_{0 \leq y \leq x} \nu_t\{\eta: \eta(y) = 1\} \leq (x+1)\lambda$$

for $x \geq 0$, so that

$$\sum_{y > x} \nu_t\{\eta: \eta(y) = 1\} \geq \gamma\lambda(1-\lambda)t - (x+1)\lambda.$$

Since $\nu_t \leq \mu_t$, it then follows that

$$\sigma(c) = \lim_{t \to \infty} E\left[\frac{1}{t} \sigma_t([ct])\right] \geq \gamma\lambda(1-\lambda) - \lambda c.$$

Maximizing over $\lambda \in [0, 1]$ as before, we obtain the other inequality in (5.13). Thus we have proved (5.13) for all $c > 0$ at which σ is differentiable. Using the convexity of σ and (5.9), this gives the first statement in part (b) of the theorem. For the second statement, note that by (5.16),

$$(5.21) \qquad \lim_{t \to \infty} \frac{1}{t} \int_0^t \mu_s^{[cs]}\{\eta: \eta(0) = 1, \eta(1) = 0\} ds = \frac{\gamma^2 - c^2}{4\gamma^2},$$

while by (5.20),

$$(5.22) \qquad \limsup_{s \to \infty} \mu_s^{[cs]}\{\eta: \eta(0) = 1, \eta(1) = 0\} \leq \frac{\gamma^2 - c^2}{4\gamma^2}.$$

In order to complete the proof of part (b) it suffices to show that

$$(5.23) \qquad \lim_{s \to \infty} \mu_s^{[cs]}\{\eta: \eta(0) = 1, \eta(1) = 0\} = \frac{\gamma^2 - c^2}{4\gamma^2},$$

since then equality would hold in (5.19). This implies that the τ in (5.17) is the pointmass on $-\sigma'(c)$, and hence any weak limit of $\mu_s^{[cs]}$ would be ν_α

with $\alpha = -\sigma'(c)$. To deduce (5.23) from (5.21) and (5.22), begin with the observation that if μ_1 and μ_2 are two measures which satisfy $\mu_1 \leq \mu_2$, then

$$0 \leq \mu_2\{\eta: \eta(x) = 1, \eta(y) = 1\} - \mu_1\{\eta: \eta(x) = 1, \eta(y) = 1\}$$
$$\leq |\mu_t^x\{\eta: \eta(0) = 1\} - \mu_t^y\{\eta: \eta(0) = 1\}|$$
$$+ |\mu_t^x\{\eta: \eta(1) = 1\} - \mu_t^y\{\eta: \eta(1) = 1\}|.$$

This follows from Theorem 2.4 of Chapter II, as did Corollary 2.8 of that chapter. Applying this inequality to the measures μ_t^x and μ_t^y, we have by Theorem 5.7(b) that

$$|\mu_t^x\{\eta: \eta(0) = \eta(1) = 1\} - \mu_t^y\{\eta: \eta(0) = \eta(1) = 1\}|$$
$$\leq |\mu_t^x\{\eta: \eta(0) = 1\} - \mu_t^y\{\eta: \eta(0) = 1\}|$$
$$+ |\mu_t^x\{\eta: \eta(1) = 1\} - \mu_t^y\{\eta: \eta(1) = 1\}|.$$

This and Theorem 5.7(c) give

$$\mu_{t+s}^{[ct+cs]}\{\eta: \eta(0) = \eta(1) = 1\}$$
$$\geq \sum_y P(M_s = y)\mu_t^{[ct+cs]+y}\{\eta: \eta(0) = \eta(1) = 1\}$$
$$\geq \mu_t^{[ct]}\{\eta: \eta(0) = \eta(1) = 1\}$$
$$- \sum_y P(M_s = y)|\mu_t^{[ct+cs]+y}\{\eta: \eta(0) = 1\} - \mu_t^{[ct]}\{\eta: \eta(0) = 1\}|$$
$$- \sum_y P(M_s = y)|\mu_t^{[ct+cs]+y}\{\eta: \eta(1) = 1\} - \mu_t^{[ct]}\{\eta: \eta(1) = 1\}|.$$

Since the distribution of M_t/t is tight, it then follows from Theorem 5.8(b) that

$$\liminf_{\substack{s,t \to \infty \\ s/t \to 0}} [\mu_{t+s}^{[ct+cs]}\{\eta: \eta(0) = \eta(1) = 1\} - \mu_t^{[ct]}\{\eta: \eta(0) = \eta(1) = 1\}] \geq 0.$$

Using Theorem 5.8(b) again, this translates into

$$\limsup_{\substack{s,t \to \infty \\ s/t \to 0}} [\mu_{t+s}^{[ct+cs]}\{\eta: \eta(0) = 1, \eta(1) = 0\} - \mu_t^{[ct]}\{\eta: \eta(0) = 1, \eta(1) = 0\}] \leq 0$$

This is enough to deduce (5.23) from (5.21) and (5.22) as required. □

The results in this section have an interesting geometric interpretation, which leads to the evaluation of the asymptotic shape for a certain class of growth models. Start with the exclusion process corresponding to

$p(x, x+1) = p$, $p(x, x-1) = 1-p$ for $0 \le p \le 1$. Let η_t be the configuration at time t when the initial configuration is the η^0 defined in (5.1). Partition the plane into unit squares, and write $C_{k,l}$ for the square whose upper right corner is at (k, l). Say that $C_{k,l}$ is black at time t if

$$(5.24) \qquad\qquad \sigma_t(k-l) = \sum_{y>k-l} \eta_t(y) \ge l,$$

and let it be white otherwise. Equivalently, $C_{k,l}$ is black if and only if the displacement of the lth particle (the one originally at $-(l-1)$) from its initial position is at least k. Initially, the first quadrant is white. The complement of the first quadrant is black at all times. At all times, if $C_{k,l}$ is black, then $C_{m,n}$ is also black for all (m, n) such that $m \le k$ and $n \le l$. The dynamics of the growth model are that white squares become black at rate p and black squares become white at rate $1-p$, provided that the transition does not violate the monotonicity statement in the previous sentence. In other words, a white square is only allowed to turn black if its two neighbors below and to the left are black, while a black square is only allowed to turn white if its two neighbors above and to the right are white.

Let B_t be the union of the black squares in the first quadrant at time t, and consider the three cases $p < \frac{1}{2}$, $p = \frac{1}{2}$, and $p > \frac{1}{2}$. The result described in Example 2.8 implies that B_t has a limiting distribution if $p < \frac{1}{2}$. To treat the other two cases, note that by (5.24), the boundary of B_t is given by

$$\{(k, l): \sigma_t(k-l) = l, \sigma_t(k+1-l) = l-1\}.$$

Therefore if $p = \frac{1}{2}$, $B_t/\sqrt{t}$ converges in probability to the unbounded convex set which lies between the coordinate axes and the curve parametrized by

$$x - y = c, \qquad y = \int_c^\infty \alpha(u)\, du,$$

where

$$\alpha(c) = \frac{1}{\sqrt{2\pi}} \int_c^\infty e^{-u^2/2}\, du.$$

On the other hand, if $p > \frac{1}{2}$, Theorems 5.3 and 5.12 imply that B_t/t converges almost surely to the convex set bounded by the coordinate axes and the curve parametrized by

$$x - y = c, \qquad y = \frac{1}{4\gamma}(\gamma - c)^2,$$

where $\gamma = 2p - 1$. This curve is the parabola

$$\sqrt{x} + \sqrt{y} = \sqrt{\gamma}.$$

6. Notes and References

Section 1. The exclusion process was introduced by Spitzer (1970). With a somewhat different description, the symmetric exclusion process appeared also as the stirring process in Harris (1972) and Lee (1974), and as the swapping process in Clifford and Sudbury (1973). Barner (1983) has studied the stirring process on W^S for rather general W.

A generalization of the exclusion process which has been studied is known as the exclusion process with speed change. In this model, the rate at which a particle at x attempts to move to y depends on the configuration at other sites. If this rate is chosen appropriately, canonical Gibbs states (see Georgii (1979)) arise as reversible invariant measures for the process, much as Gibbs states did for the stochastic Ising model in Chapter IV. The exclusion process with speed change was introduced by Spitzer (1970) as a model of a lattice gas. A slightly different version of it was used as a model of a binary alloy by Bortz, Kalos, Lebowitz, and Zendejas (1974) and then by Bortz, Kalos, Lebowitz, and Marro (1975). Other papers on the exclusion process with speed change are Holley (1970), Harris (1972), Holley (1972c), Logan (1974), Liggett (1977b), and De Masi, Presutti, Spohn, and Wick (1985).

Spitzer (1970) introduced two other types of exclusion processes. In one of them, a particle continues searching for a vacant site until it finds one. This version was studied by Liggett (1980b). In the other, which is necessarily one dimensional, jumps which would change the order of the particles are excluded. This process was discussed further in Spitzer (1977).

Theorem 1.1 is due to Spitzer (1970). He suggested there that this duality relation should be the key to proving ergodic theorems for symmetric exclusion systems. Theorems 1.12 and 1.13 are due to Spitzer (1974b). Proposition 1.7 and Theorems 1.24, 1.44, and 1.47 are due to Liggett (1973a, 1974a).

Arratia (1985) has used an extension of Proposition 1.7 to prove a large deviation theorem for the symmetric exclusion process. His result is the following. Let $\rho_x(t)$ be the expected range of $X_1(\cdot)$ up to time t when the initial state is x:

$$\rho_x(t) = \sum_y P^x(X_1(s) = y \text{ for some } s \le t).$$

Take η_t to be the symmetric exclusion process with initial distribution ν_α for some constant $\alpha \in (0, 1)$. Then

$$(1-\alpha)^{\rho_x(t)} \le P[\eta_s(x) = 0 \text{ for all } s \le t] \le e^{-\alpha \rho_x(t)}.$$

Section 2. Propositions 2.13 and 2.14 are from Liggett (1976). Theorem 2.17 is due to Liggett (1974b).

Section 3. Part (a) of Theorem 3.9 is due to Liggett (1976), and part (b) is due to Andjel (1981). Theorem 3.14 is due to Liggett (1976). Coupling techniques similar to those used in this section were applied by Barner (1983) to obtain analogous results for certain discrete time processes on W^S. A version of Theorem 3.9 for one-dimensional reversible nearest-neighbor exclusion processes with speed change was proved by Holley (1972c) using the relative entropy technique, much as it was used in Section 5 of Chapter IV.

Section 4. Theorem 4.2 and Corollary 4.6 are due to Spitzer (1970). Corollary 4.9 is stated without proof in the same paper, and is credited there to H. Kesten. Theorem 4.13 is due to Arratia (1983a). It answers a question which was raised by Spitzer (1970). Both Spitzer's conjecture and the part of Arratia's proof which is contained in Lemma 4.10 were based on a similar result due to Harris (1965) concerning the behavior of a tagged particle in a system of reflecting one-dimensional Brownian motions. In his paper, Arratia credits Andjel for the proof of Lemma 4.12. The original version of this result asserted that for $x \neq y$,

$$P(\xi_t^x \geq u, \xi_t^y \geq v) \leq P(\xi_t^x \geq u) P(\xi_t^y \geq v),$$

and was proved by using Theorem 2.14 of Chapter II, much as it was used in the proof of Proposition 4.2 of Chapter V.

In the same paper, Arratia used similar techniques to prove a limit theorem for M_t, the position of the rightmost particle in the exclusion process on Z^1 with $p(x, x+1) = p(x+1, x) = \frac{1}{2}$ and initial configuration η given by

$$\eta(x) = \begin{cases} 1 & \text{if } x < 0, \\ 0 & \text{if } x \geq 0. \end{cases}$$

His result is that

$$\frac{M_t}{\sqrt{t}} - \sqrt{\log t} \to 0$$

with probability one.

Section 5. This section is based on Rost (1981), who treated the case $p(x, x+1) = 1$. The proofs given here are essentially the same as his. The main difference is in the proof of the second inequality in (5.13), which is due to Andjel.

Several authors have discussed the asymptotic shape for a variety of growth models. Among these are Richardson (1973), Biggins (1978), Schürger (1979), Bramson and Griffeath (1980c, d, 1981), Durrett and Liggett (1981), and Durrett and Griffeath (1982). The exact evaluation of the asymptotic shape which was carried out for a very special growth model at the end of this section is not possible in most of the other models which have been treated. In most cases, all that is known is that the asymptotic shape has some symmetries.

Results of the type proved in this section, which give the asymptotic profile of the disturbance in a system, are sometimes referred to as being hydrodynamical in nature. Similar hydrodynamical results have been obtained for a number of particle systems. Examples are given in De Masi, Ianiro, and Presutti (1982) and De Masi, Ferrari, Ianiro, and Presutti (1982) for the symmetric exclusion process; De Masi, Presutti, Spohn, and Wick (1985) for the exclusion process with speed change; Galves, Kipnis, Marchioro, and Presutti (1981) for the stochastic Ising model; Presutti and Spohn (1983) for the voter model; Andjel and Kipnis (1984) and Brox and Rost (1984) for the zero range process; and Greven (1984) for the coupled branching process. A survey of this aspect of the theory of interacting particle systems is given in De Masi, Ianiro, Pellegrinotti, and Presutti (1984).

7. Open Problems

1. In the context of Theorem 1.13, show that (1.14) implies (1.17). There are two related reasons for wanting to prove this. First, (1.17) is a more useful condition than (1.16) since it involves two independent chains rather than the interacting pair. Secondly, this would make it possible to unify Theorem 1.13 and Theorem 1.24, giving a single necessary and sufficient condition for convergence in both cases. In order to solve this problem, it is enough to show that if f is a bounded symmetric positive definite function on S^2 for which

$$(7.1) \qquad \lim_{t \to \infty} V_2(t)f = \alpha^2,$$

then

$$(7.2) \qquad \lim_{t \to \infty} U_2(t)f = \alpha^2$$

as well. Under some assumptions, the chains with semigroups $U_2(t)$ and $V_2(t)$ have been compared by Bertein and Galves (1977a) and De Masi and Presutti (1983). See also Bougerol and Kipnis (1980). These results do not yield the conclusion that (7.1) implies (7.2) except in rather special cases. (Note that (7.1) does not imply (7.2) if S is finite. However, S is assumed to be infinite throughout this chapter.)

2. Consider the exclusion process in which the transition probabilities $p(x, y)$ satisfy (0.2) and are irreducible and positive recurrent. Prove under weak assumptions that

(7.3) $$\lim_{t\to\infty} P^\eta[\eta_t(x)=1]=1$$

for all $x\in S$ and all $\eta\in X$ which satisfy $\sum_x \eta(x)=\infty$.

Under the additional assumption that $p(x, y)$ is reversible, this is proved in Theorem 2.17(b). An example which shows that (7.3) is not always true in the positive recurrent case is given by $S=Z^1$ with $p(n, n-1)=p(n,|n|)=\frac{1}{2}$ for all n. Note that for this example, there is an invariant measure which concentrates on the set of configurations containing a single zero.

3. Consider two irreducible continuous time Markov chains on the countable set S whose Q matrices are related by

$$q_2(x, y)=q_1(y, x)$$

for $x\neq y$. Suppose that

$$\sup_x \sum_{y:y\neq x} q_i(x, y)<\infty$$

for $i=1, 2$. Show under weak assumptions that not both can be positive recurrent. This is easy to show in the reversible case, since if $\pi_i(x)q_i(x, y)=\pi_i(y)q_i(y, x)$ for $i=1, 2$, then $\pi_1(x)\pi_2(x)$ is independent of x, so that not both π_1 and π_2 can be summable. This problem is the first step in solving Problem 2. To see the connection, consider the motion of the position of the single zero in the exclusion process if initially, and hence for all times, $\sum_x[1-\eta_t(x)]=1$. The example given in Problem 2 above shows that some assumptions are needed in this problem as well.

4. Theorem 3.9 is unsatisfactory as it stands. It seems fairly clear that if $\mu\in\mathcal{S}_e$, then the limit of $\mu S(t)$ as $t\uparrow\infty$ should be ν_α, where $\alpha=\mu\{\eta:\eta(x)=1\}$. Prove this. All Theorem 3.9 gives is that this limit is of the form

$$\int_0^1 \nu_\beta\gamma(d\beta)$$

for some probability measure γ on $[0, 1]$ such that $\int_0^1 \beta\gamma(d\beta)=\alpha$. In the symmetric case, the limit is ν_α by Theorem 1.47. The only asymmetric case in which this has been proved is $S=Z^1$, $p(x, x+1)=p$, $p(x, x-1)=q$, with $p+q=1$ (see Andjel (1981)). A solution to this problem would immediately lead to the identification of the limit of $\mu S(t)$ for any $\mu\in\mathcal{S}$. The reason for this is that any $\mu\in\mathcal{S}$ can be written as an integral average of elements of $\mathcal{S}_e$. (See Section 10 of Phelps (1966).)

5. Prove Theorem 3.14 for $d>1$.

6. Prove an analogue of (3.23) for a general one-dimensional random walk satisfying $\sum_x |x| p(0, x) < \infty$ and $\sum_x x p(0, x) > 0$. The first step is to prove that the restriction of the exclusion process to the W_n in (3.22) is positive recurrent. It is not clear how to do this when (2.3) is satisfied for no choice of $\pi(\cdot)$. Together with Theorem 3.14, a solution to this problem would provide a fairly complete ergodic theory for one-dimensional exclusion systems.

7. Prove convergence theorems for translation invariant exclusion systems when the initial distribution is not translation invariant. Outside of the symmetric case, the only result of this type which has been proved is the following, which is due to Liggett (1977a). (The special case in which the initial configuration is the η^0 in (5.1) is proved in Section 5.) Suppose $d = 1$, $\sum_x |x| p(0, x) < \infty$, and $\sum_x x p(0, x) > 0$. Let μ be a product measure on X for which the following limits exist:

(7.4) $\lambda = \lim_{x \to -\infty} \mu\{\eta: \eta(x) = 1\}$ and $\rho = \lim_{x \to +\infty} \mu\{\eta: \eta(x) = 1\}$.

Then

(a) $\lim_{t \to \infty} \mu S(t) = \nu_{1/2}$ if $\lambda \geq \frac{1}{2}$ and $\rho \leq \frac{1}{2}$,
(b) $\lim_{t \to \infty} \mu S(t) = \nu_\rho$ if $\rho \geq \frac{1}{2}$ and $\lambda + \rho > 1$, and
(c) $\lim_{t \to \infty} \mu S(t) = \nu_\lambda$ if $\lambda \leq \frac{1}{2}$ and $\lambda + \rho < 1$.

Note that the case $\lambda < \frac{1}{2}$ and $\lambda + \rho = 1$ is not covered by this result. In fact it was shown in Liggett (1975) that $\mu S(t)$ can have both ν_λ and ν_ρ as limit points as $t \to \infty$ in that case. The above result shows that the ergodic theory for asymmetric exclusion systems must be substantially different from the theory in the symmetric case. For comparison purposes, note that in the symmetric case, Theorem 1.13 implies that $\lim_{t \to \infty} \mu S(t) = \nu_{(\lambda + \rho)/2}$ if μ is a product measure satisfying (7.4).

8. Prove the asymptotic normality of the position of the tagged particle in cases not covered by Theorem 4.13 or by the results of Kipnis (1985) and Kipnis and Varadhan (1985) which are quoted at the end of Section 4.

9. Referring to Corollary 4.6, show that the limit of X_t/t is

$$(1 - \alpha) \sum_y y p(0, y).$$

This result is stated in Spitzer (1970), but no proof is given there. In special cases, this follows from Corollary 4.9 and Theorem 4.13.

10. Prove a version of Theorem 5.2 in which the symmetry assumption is replaced by $\sum_x x p(0, x) = 0$. A very small first step is given by part (a) of Theorem 5.12.

Chapter IX

Linear Systems with Values in $[0, \infty)^s$

All the processes considered in previous chapters have the property that each coordinate $\eta(x)$ can take on only two values. When the set of possible values per site is allowed to be noncompact, new problems and different phenomena occur. The literature contains many types of models in which the set of possible values per site is either the nonnegative integers or the nonnegative real numbers. The oldest and simplest of these is a system of particles which move independently on S. This process has been modified by adding a speed change interaction and/or by allowing branching. In these cases, $\eta(x)$ is interpreted as the number of particles at x. In other models, one can view $\eta(x)$ as being a nonnegative real-valued characteristic of the particle at x, which is updated in some way which involves interactions among the various sites.

In this chapter, we will analyze one broad class of systems within which one can see many of the main problems, techniques, and results which have occurred in this general area. In these systems, $\eta(x)$ will take on nonnegative real values, and η_t will depend linearly on the initial configuration η. While at first glance there may appear to be little connection between these linear systems and the spin systems treated earlier, there are in fact several points of contact between them. The voter model will appear as a special example of a linear system, while the contact process can be obtained as a function of another linear system.

The linear systems on the countable set S will be defined in terms of a deterministic collection of numbers $a(x, y)$ indexed by $x, y \in S$ which satisfy

$$(0.1) \qquad\qquad a(x, y) \geq 0 \quad \text{for } x \neq y,$$

and for each $x \in S$, a collection of nonnegative random variables $A_x(u, v)$ indexed by $u, v \in S$. Thus A_x can be thought of as an infinite random matrix. The evolution is described as follows. Let $\{N_x(t), x \in S\}$ be independent rate one Poisson processes, and let $\{A_x^i(u, v), u, v \in S\}$ have the same joint distributions as $\{A_x(u, v), u, v \in S\}$ for each $x \in S$ and $i = 1, 2, \ldots$. For different x and i these are to be independent of one another and independent of the Poisson processes. At the ith event time t of $N_x(\cdot)$, η_{t-} is replaced by the η_t given by

$$\eta_t(u) = \sum_v A_x^i(u, v)\eta_{t-}(v).$$

Between event times of the Poisson processes, η_t evolves according to the system of linear differential equations

$$\frac{d}{dt}\eta_t(u) = \sum_v a(u, v)\eta_t(v).$$

Thus the formal generator of the process is

$$(0.2) \qquad \Omega f(\eta) = \sum_x \{Ef(A_x\eta) - f(\eta)\} + \sum_{u,v} f_u(\eta)a(u, v)\eta(v),$$

where

$$(A_x\eta)(u) = \sum_v A_x(u, v)\eta(v)$$

and $f_u(\eta)$ is the partial derivative of $f(\eta)$ with respect to the coordinate $\eta(u)$. In order to obtain a well-defined Markov process η_t, it will be necessary to impose additional restrictions on $a(x, y)$ and $A_x(u, v)$, and to restrict the class of configurations which are permitted. This will be done in Section 1 (see (1.3), (1.4), and (1.5)).

The voter model corresponding to $p(x, y)$ is the special case in which $a(x, y) \equiv 0$ and for each x, with probability $p(x, y)$,

$$(0.3) \qquad A_x(u, v) = \begin{cases} 1 & \text{if } u = x \text{ and } v = y, \text{ or if } u = v \neq x, \\ 0 & \text{otherwise.} \end{cases}$$

With this choice, at rate $p(x, y)$, the value of η at x is replaced by its value at y. If initially η takes only the values 0 and 1, then the same will be true at later times, and η_t is the voter model which was the subject of Chapter V. If η is allowed to take more values, this process can be regarded as a multitype voter model.

To see the connection between linear systems and the contact process, suppose $S = Z^d$, and let η_t be the linear system in which $a(x, y) = 0$ for all $x, y \in S$, and for each $x \in S$,

$$(0.4) \qquad A_x(u, v) = \begin{cases} 1 & \text{if } u = v \neq x, \\ 0 & \text{otherwise,} \end{cases}$$

with probability $(2\lambda d + 1)^{-1}$, and for each of the $2d$ neighbors y of x,

$$(0.5) \qquad A_x(u, v) = \begin{cases} 1 & \text{if } u = v, \text{ or if } u = y \text{ and } v = x, \\ 0 & \text{otherwise,} \end{cases}$$

with probability $\lambda(2\lambda d+1)^{-1}$. This system is known as the binary contact path process. At rate $(2\lambda d+1)^{-1}$ for each x, $\eta(x)$ is replaced by 0. For each nearest-neighbor pair x and y, $\eta(y)$ is replaced by $\eta(y)+\eta(x)$ at rate $\lambda(1+2\lambda d)^{-1}$. Define

$$\zeta_t(x) = \begin{cases} 1 & \text{if } \eta_t(x) > 0, \\ 0 & \text{if } \eta_t(x) = 0. \end{cases}$$

Except for a deterministic time change, ζ_t is simply the contact process with parameter λ. This connection will enable us to obtain information about the contact process from some of our results concerning linear systems.

The normalized binary contact path process is the one which is modified by setting

$$a(x, x) = \frac{1-2\lambda d}{1+2\lambda d}$$

for each x. It has the advantage that $E^{\eta}\eta_t(x) = 1$ for all x and t if $\eta \equiv 1$, and is obtained from the binary contact process by multiplying each coordinate by $\exp([(1-2\lambda d)/(1+2\lambda d)]t)$. Thus we see in this example that one role of the part of the mechanism which involves $a(u, v)$ is to normalize the process so that it is stable, at least insofar as the mean is concerned.

Other special cases which are of interest are the smoothing and potlatch processes. In each of these cases, $a(x, y) = 0$ for all x and y. Take W to be a nonnegative random variable with mean one and $p(x, y)$ to be the transition probabilities for a discrete time irreducible Markov chain on S. In the smoothing process, let

(0.6) $$A_x(u, v) = \begin{cases} 1 & \text{if } u = v \neq x, \\ Wp(u, v) & \text{if } u = x, \\ 0 & \text{otherwise,} \end{cases}$$

while in the potlatch process, let

(0.7) $$A_x(u, v) = \begin{cases} 1 & \text{if } u = v \neq x, \\ Wp(v, u) & \text{if } v = x, \\ 0 & \text{otherwise.} \end{cases}$$

In the first case, the coordinate $\eta(x)$ is replaced at exponential times by

$$W \sum_y p(x, y)\eta(y).$$

The averaging with respect to $p(x, y)$ tends to reduce the variability of the coordinates, while the multiplication by W tends to increase this variability.

In the second case, the coordinate $\eta(x)$ is multiplied by W at exponential times, and then the quantity $W\eta(x)$ is distributed to the coordinates $\eta(y)$ according to $p(x, y)$. In both cases, independent copies of W are used each time they are needed.

Linear systems will be constructed on an appropriate subset of $[0, \infty)^S$ in the first section. The approach to the construction problem which is used there is applicable to many other processes with values in $[0, \infty)^S$ or $\{0, 1, 2, \ldots\}^S$. As a result of the linear nature of the transition mechanism in this case, there is a natural coupling for versions of the system with different initial configurations. This is very much in the spirit of the graphical representation for spin systems, which was described in Section 6 of Chapter III.

Linear systems also have a simple and useful duality theory. The dual system is obtained from the original one by replacing the matrices $a(x, y)$ and $A_x(u, v)$ by their transposes. The duality relation asserts that $\sum_x \eta_t(x)\zeta(x)$ and $\sum_x \eta(x)\zeta_t(x)$ have the same distribution if η_t and ζ_t have initial configurations η and ζ respectively. Recalling the examples mentioned above, it is easy to check that the binary contact path process is self-dual and that the smoothing and potlatch process are duals of one other. The dual of the multitype voter model (0.3) is the process in which at the event times of $N_x(\cdot)$, a y is chosen with probability $p(x, y)$ and η is replaced by the $\tilde{\eta}$ given for $x \neq y$ by

$$(0.8) \qquad \tilde{\eta}(u) = \begin{cases} 0 & \text{if } u = x, \\ \eta(x) + \eta(y) & \text{if } u = y, \\ \eta(u) & \text{if } u \neq x, y. \end{cases}$$

This can be thought of as coalescing system, so that this duality is consistent with the coalescing duality for the voter model which was used in Chapter V.

Beginning with Section 2, we will restrict our attention to translation invariant systems. For these, there is a constant c so that

$$(0.9) \qquad E^\eta \sum_x \eta_t(x) = e^{ct} \sum_x \eta(x),$$

and

$$(0.10) \qquad E^1 \eta_t(x) = e^{ct},$$

where 1 is the configuration which is identically one. By adjusting the number $a(x, x)$, we can take $c = 0$. Once this is done, (0.9) implies that if the initial configuration satisfies $\sum_x \eta(x) < \infty$ then $\sum_x \eta_t(x)$ is a (nonnegative) martingale. Therefore

$$\lim_{t \to \infty} \sum_x \eta_t(x)$$

exists with probability one. We will see in Section 2 that there are only two possibilities: either the limit is identically zero, or the convergence occurs in L_1. These two possibilities will be referred to respectively as the extinction or survival of the finite system. Similarly, by (0.10),

$$E^1 \eta_t(x) = 1$$

for all x and t, and it turns out that either $\{\eta_t(x), t \geq 0\}$ is uniformly integrable or $\eta_t(x) \to 0$ as $t \uparrow \infty$ in probability when initially $\eta_0 \equiv 1$. We will say that the infinite system survives in the former case, and dies out in the latter case. Note that if η_t and ζ_t are duals of one another, then the survival of the finite η_t system is equivalent to the survival of the infinite ζ_t system.

Sections 3, 4, and 5 develop criteria for the survival and extinction of linear systems. Without stating precise results here, it is perhaps useful to describe the general picture which emerges in those sections. In one and two dimensions, linear systems usually die out if there is any nontrivial randomness in the matrices $A_x(u, v)$. In higher dimensions, systems with a small amount of randomness survive, while those with a lot of randomness die out.

Section 6 has two objectives. The first is to specialize the results in the previous three sections to several classes of examples including the smoothing and potlatch systems. It is easier to see exactly what has been proved when attention is restricted to these examples than when hypotheses are stated for the general linear systems. The second objective is to carry out the comparisons between certain linear systems and the contact process which lead to improved upper bounds for the critical value of the contact process in dimensions $d \geq 3$.

1. The Construction; Coupling and Duality

The first step in the construction is to decide what the state space of the process will be. Fix a strictly positive function $\alpha(x)$ on S such that

(1.1) $$\sum_x \alpha(x) < \infty,$$

and let

$$X = \left\{ \eta \in [0, \infty)^S : \sum_x \eta(x)\alpha(x) < \infty \right\}.$$

X is given the smallest σ-algebra such that the map $\eta \to \eta(x)$ is measurable for each $x \in S$. Note that if $\{\eta(x), x \in S\}$ are nonnegative random variables with uniformly bounded means, then $\eta \in X$ with probability one. Thus X

is large enough to contain many distributions of interest. For $\eta, \zeta \in X$, let

$$\|\eta - \zeta\| = \sum_x |\eta(x) - \zeta(x)| \alpha(x).$$

This provides the topology on X.

Next we need to choose a class of functions on which the generator Ω will act. Let $\mathscr{L}$ be the class of all continuous functions f on X which have continuous first partial derivatives f_x with respect to $\eta(x)$, and such that

$$L(f) = \sup_x \frac{\|f_x\|}{\alpha(x)} < \infty,$$

where $\|f_x\|$ is the supremum norm of f_x. From this it follows that $f \in \mathscr{L}$ satisfies

(1.2) $$|f(\eta) - f(\zeta)| \le L(f)\|\eta - \zeta\|$$

for all $\eta, \zeta \in X$. In order to define Ω on $\mathscr{L}$ by (0.2), it is necessary to make the following assumption: There is a constant M so that

(1.3) $$\sum_u \alpha(u)|a(u, v)| \le M\alpha(v),$$

(1.4) $$E \sum_x |A_x(u, u) - 1| \le M, \quad \text{and}$$

(1.5) $$E \sum_x \sum_{u:u \ne v} \alpha(u) A_x(u, v) \le M\alpha(v).$$

Note that (1.4) and (1.5) together assert that A_x is rather close to the identity matrix.

The first result gives a sufficient condition for the existence of an $\alpha(\cdot)$ with the required properties. The examples (0.3), (0.4), (0.5), (0.6), and (0.7) all satisfy this sufficient condition, provided that the $p(x, y)$ which appears in some of them is, say, doubly stochastic. The proof of this result also provides a means of constructing such an $\alpha(\cdot)$.

Lemma 1.6. *Suppose that*

$$\sup_u \sum_v |a(u, v)| < \infty \quad \text{and} \quad \sup_u \sum_x \sum_{v:v \ne u} EA_x(u, v) < \infty.$$

Then there is a strictly positive $\alpha(\cdot)$ *on S which satisfies* (1.1), (1.3), *and* (1.5) *for some $M > 0$.*

Proof. Let

$$b(u, v) = \begin{cases} a(u, v) + \sum\limits_{x} EA_x(u, v) & \text{if } u \neq v, \\ |a(u, u)| & \text{if } u = v, \end{cases}$$

and define $b^{(n)}(u, v)$ by

$$b^{(0)}(u, v) = \begin{cases} 1 & \text{if } u = v, \\ 0 & \text{if } u \neq v, \quad \text{and} \end{cases}$$

$$b^{(n+1)}(u, v) = \sum\limits_{w} b(u, w) b^{(n)}(w, v).$$

Take

$$\alpha(v) = \sum\limits_{n=0}^{\infty} M^{-n} \sum\limits_{u} \beta(u) b^{(n)}(u, v)$$

where

$$M > \sup\limits_{u} \sum\limits_{v} b(u, v),$$

and $\beta(u)$ is chosen so that $\beta(u) \geq 0$, $\sum_u \beta(u) < \infty$, and sufficiently many $\beta(u) > 0$ so that $\alpha(v) > 0$ for all v. This choice has all the desired properties. $\square$

From now on, we will assume that α and M have been chosen so that (1.3), (1.4), and (1.5) hold.

Lemma 1.7. *For any $f \in \mathscr{L}$, the series defining Ωf in (0.2) converges absolutely and uniformly on compact subsets of X. Furthermore*

(1.8) $$|\Omega f(\eta)| \leq 3ML(f)\|\eta\|.$$

Proof. By (1.2) and the definition of $L(f)$,

$$|Ef(A_x\eta) - f(\eta)| \leq L(f)E\|A_x\eta - \eta\|$$

$$= L(f)E \sum\limits_{u} \left| \sum\limits_{v} A_x(u, v)\eta(v) - \eta(u) \right| \alpha(u)$$

$$\leq L(f)E \left\{ \sum\limits_{u} |A_x(u, u) - 1|\eta(u)\alpha(u) \right.$$

$$\left. + \sum\limits_{u \neq v} \alpha(u)A_x(u, v)\eta(v) \right\}$$

and

$$|f_u(\eta)| \le L(f)\alpha(u).$$

Therefore by (1.3), (1.4), and (1.5),

$$|\Omega f(\eta)| \le \sum_x |Ef(A_x\eta) - f(\eta)| + \sum_{u,v} |f_u(\eta)||a(u, v)|\eta(v)$$

$$\le 3ML(f)\|\eta\|.$$

The same estimates show the uniform convergence on compact subsets of X. $\square$

The strategy we will use in constructing the process is very similar to that used in Chapters I and VII. The process is easily defined in case S is finite. Estimates are then developed in that case which enable us to carry out a limiting procedure which defines the process when S is countably infinite.

To define the process in case S has cardinality n, let $\tau_1, \tau_2, \ldots$ be the event times for a Poisson process of rate n, and let $A^{(1)}, A^{(2)}, \ldots$ be independent and identically distributed random $n \times n$ matrices with distribution

$$P(A^{(i)} \in G) = \frac{1}{n} \sum_{x \in S} P(A_x \in G).$$

Let A be the matrix with entries $a(x, y)$, and as usual, define

$$e^{tA} = \sum_{k=0}^{\infty} \frac{t^k}{k!} A^k.$$

Note that e^{tA} has nonnegative entries, since $a(x, y) \ge 0$ for $x \ne y$. If η is the initial configuration of the process, define η_t on the event $\tau_k \le t < \tau_{k+1}$ by

(1.9) $\qquad e^{(t-\tau_k)A} A^{(k)} e^{(\tau_k - \tau_{k-1})A} A^{(k-1)} \cdots A^{(1)} e^{\tau_1 A} \eta.$

Note that this defines η_t simultaneously for all initial configurations η in terms of $\tau_1, \tau_2, \ldots$ and $A^{(1)}, A^{(2)}, \ldots$, and hence provides the basic coupling for these processes. Define

$$S(t)f(\eta) = E^\eta f(\eta_t).$$

Lemma 1.10. *Suppose S is finite. If $f \in \mathcal{L}$, then $S(t)f \in \mathcal{L}$ and*

$$L(S(t)f) \le L(f) e^{3Mt}.$$

Furthermore,

$$E^{\eta} \eta_t(u) \le e^{Bt} \eta(u),$$

where B is the matrix with entries

$$B(u, v) = a(u, v) + \begin{cases} \sum_x EA_x(u, v) & \text{if } u \ne v, \\ \sum_x E|A_x(u, u) - 1| & \text{if } u = v. \end{cases}$$

Proof. Take $f \in \mathcal{L}$ and $\eta, \zeta \in X$. By (1.2),

$$\begin{aligned} |S(t)f(\eta) - S(t)f(\zeta)| &= |E^{\eta}f(\eta_t) - E^{\zeta}f(\eta_t)| \\ &\le E|f(\eta_t) - f(\zeta_t)| \\ &\le L(f)E\|\eta_t - \zeta_t\|, \end{aligned}$$

(1.11)

where η_t and ζ_t are defined as in (1.9) in terms of η and ζ respectively. By (1.3),

$$\|e^{tA}\eta - e^{tA}\zeta\| \le e^{tM}\|\eta - \zeta\|,$$

while by (1.4) and (1.5),

$$E\|A^{(k)}\eta - A^{(k)}\zeta\| \le \|\eta - \zeta\|\left(1 + \frac{2M}{n}\right),$$

where n is the cardinality of S. Therefore by (1.9) and (1.11),

$$\begin{aligned} \|S(t)f(\eta) - S(t)f(\zeta)\| \\ &\le L(f) \sum_{k=0}^{\infty} E(\|\eta_t - \zeta_t\|, \tau_k \le t \le \tau_{k+1}) \\ &\le \|\eta - \zeta\|L(f) \sum_{k=0}^{\infty} e^{tM}\left(1 + \frac{2M}{n}\right)^k P(\tau_k \le t < \tau_{k+1}) \\ &= \|\eta - \zeta\|L(f) e^{3Mt}, \end{aligned}$$

which implies that

$$L(S(t)f) \le L(f) e^{3Mt}.$$

The proof of the second statement is left to the reader. □

Now let S be infinite, and take finite sets $S_n \uparrow S$. Let $S_n(t)$ be the semigroup on $\mathscr{L}$ constructed as indicated above corresponding to $A_x(u, v)$ and $a(x, y)$ which have been modified by setting $a(x, y) = 0$ if x or $y \notin S_n$ and

$$A_x(u, v) = \begin{cases} 1 & \text{if } u = v, \\ 0 & \text{if } u \neq v, \end{cases}$$

if $x \notin S_n$. Then

$$(1.12) \qquad S_n(t)f(\eta) = f(\eta) + \int_0^t \Omega_n S_n(s)f(\eta) \, ds$$

for $f \in \mathscr{L}$ and $\eta \in X$, where

$$(1.13) \qquad \Omega_n f(\gamma) = \sum_{x \in S_n} \{Ef(A_x\eta) - f(\eta)\} + \sum_{u,v \in S_n} f_u(\eta) a(u, v) \eta(v).$$

Note that the modified matrices satisfy (1.3), (1.4), and (1.5) with the same $\alpha(\cdot)$ and M for which they are satisfied by the original matrices.

Theorem 1.14. (a) $S(t)f(\eta) = \lim_{n \to \infty} S_n(t)f(\eta)$ *exists for $f \in \mathscr{L}$ and $\eta \in X$. The convergence is uniform on bounded t sets, compact η sets, and sets of f with bounded $L(f)$.*
  (b) $S(t)f \in \mathscr{L}$ *for $f \in \mathscr{L}$, and*

$$L(S(t)f) \le L(f) \, e^{3Mt}.$$

  (c) $S(t+s) = S(t)S(s)$.
  (d) $S(t)f(\eta) = f(\eta) + \int_0^t \Omega S(s)f(\eta) \, ds$ *for any $f \in \mathscr{L}$ and $\eta \in X$.*
  (e) $|S(t)f(\eta) - f(\eta)| \le L(f)\|\eta\|[e^{3Mt} - 1]$ *for any $f \in \mathscr{L}$ and $\eta \in X$.*
  (f) $\displaystyle \lim_{t \downarrow 0} \frac{S(t)f(\eta) - f(\eta)}{t} = \Omega f(\eta)$ *for any $f \in \mathscr{L}$ and $\eta \in X$.*
  (g) $\Omega S(t)f(\eta) = E^\eta \Omega f(\eta_t)$ *for any $f \in \mathscr{L}$ and $\eta \in X$, where η_t is the process with semigroup $S(t)$.*

Proof. For $m < n$, write

$$(1.15) \qquad S_n(t)f - S_m(t)f = \int_0^t S_n(s)[\Omega_n - \Omega_m]S_m(t-s)f \, ds.$$

By Lemma 1.10,

$$L(S_m(t-s)f) \le L(f) \, e^{3M(t-s)}.$$

Therefore by (1.2) and (1.13), using the estimates which were used in the proof of Lemma 1.7, we see that

$$|[\Omega_n - \Omega_m]S_m(t-s)f(\eta)|$$

$$\leq L(f)\, e^{3M(t-s)} \sum_{x \notin S_m} E\Big\{\sum_u |A_x(u, u) - 1|\eta(u)\alpha(u)$$

$$+ \sum_{u \neq v} \alpha(u)A_x(u, v)\eta(v)\Big\}$$

$$+ L(f)\, e^{3M(t-s)} \sum_{\substack{u \in S_m \\ v \notin S_m}} \{\alpha(u)|a(u, v)|\eta(v) + \alpha(v)|a(v, u)|\eta(u)\}$$

$$= L(f)\, e^{3M(t-s)} \sum_u \gamma_m(u)\eta(u),$$

where $\gamma_m(u)$ is defined by this equality. By (1.3), (1.4), and (1.5),

$$(1.16) \qquad\qquad\qquad \sup_{u,m} \frac{\gamma_m(u)}{\alpha(u)} < \infty, \quad \text{and}$$

$$(1.17) \qquad\qquad\qquad \lim_{m \to \infty} \gamma_m(u) = 0 \quad \text{for each } u \in S.$$

Using this in (1.15), and then applying the second estimate in Lemma 1.10, yields the following conclusion: $S_n(t)f(\eta)$ is a Cauchy sequence, uniformly in bounded t sets, compact η sets, and sets of f with bounded $L(f)$. This proves part (a). Part (b) is obtained by passing to the limit in the first part of Lemma 1.10. To prove part (c), start with the semigroup property for $S_n(t)$:

$$(1.18) \qquad\qquad\qquad S_n(t+s)f = S_n(t)S_n(s)f.$$

By Lemma 1.10 and the uniformity statement in part (a),

$$(1.19) \qquad\qquad\qquad \lim_{n \to \infty}[S(t) - S_n(t)]S_n(s)f(\eta) = 0$$

for $f \in \mathcal{L}$ and $\eta \in X$. By (1.2) and Lemma 1.10,

$$(1.20) \qquad\qquad\qquad |S_n(s)f(\eta) - f(0)| \leq L(f)\, e^{3Ms} \|\eta\|.$$

For fixed t and η, there is a probability measure μ on $[0, \infty)^S$ so that $\int \|\zeta\|\mu(d\zeta) < \infty$ and

$$S(t)g(\eta) = \int g(\zeta)\mu(d\zeta)$$

for all $g \in \mathcal{L}$. Therefore by (1.20) and the Dominated Convergence Theorem,

$$\lim_{n \to \infty} S(t)[S_n(s) - S(s)]f(\eta) = 0.$$

Part (c) follows from this, (1.18) and (1.19). For part (d), we need to justify the passage to the limit in (1.12). The left side is no problem. For the right side, we will use the Dominated Convergence Theorem. The domination comes from the estimate

(1.21) $$|\Omega_n S_n(s)f(\eta)| \le 3 M e^{3Ms} L(f)\|\eta\|,$$

which results from combining Lemmas 1.7 and 1.10. The convergence of $\Omega_n S_n(s)f(\eta)$ to $\Omega S(s)f(\eta)$ is not difficult to show from the explicit expressions for Ω and Ω_n, using Lemma 1.10. Part (e) follows from part (d) and (1.21). Again by Lemma 1.10,

$$\lim_{s \to 0} \Omega S(s)f = \Omega f.$$

Therefore part (f) follows from part (d). For the last part, write

$$\Omega S(t)f(\eta) = \lim_{s \downarrow 0} \frac{S(s)S(t)f(\eta) - S(t)f(\eta)}{s}$$

$$= \lim_{s \downarrow 0} S(t)\left[\frac{S(s)f - f}{s}\right](\eta)$$

$$= S(t)\Omega f(\eta).$$

The first equality above comes from part (f), the second from part (c), and the third from parts (f), (e), (b) and the Dominated Convergence Theorem. □

If $f(\eta) = \|\eta\|$, then $L(f) = 1$. This together with (1.2) and part (b) of Theorem 1.14 implies that for this f,

$$S(t)f(\eta) \le \|\eta\| \exp(3Mt).$$

Therefore we have now constructed a Markov process η_t on X which satisfies

(1.22) $$E^{\eta}\|\eta_t\| \le e^{3Mt}\|\eta\|$$

and

$$E^{\eta}f(\eta_t) = S(t)f(\eta)$$

for all $f \in \mathcal{L}$. Recalling the explicit way in which the process was defined for finite S, it is clear that the same approximation techniques lead to a joint construction of versions of the process with different initial configurations which are linearly related in the following sense. If $\eta_t^1, \ldots, \eta_t^n$ are the processes with initial configurations $\eta^1, \ldots, \eta^n$ which satisfy

$$\sum_{k=1}^n c_k \eta^k = 0$$

for some constants c_k, then

$$\sum_{k=1}^n c_k \eta_t^k = 0$$

with probability one for each t. This is the basic coupling for linear systems.

As usual, $\mathcal{I}$ will denote the set of all probability measures on X which are invariant for the process. It is easy to deduce from Theorem 1.14 the now familiar criterion for invariance.

Corollary 1.23. *Suppose* $\int \|\eta\| \, d\mu < \infty$. *Then* $\mu \in \mathcal{I}$ *if and only if* $\int \Omega f \, d\mu = 0$ *for all* $f \in \mathcal{L}$.

Proof. Suppose that $\int \Omega f \, d\mu = 0$ for all $f \in \mathcal{L}$. Integrate the identity in (d) of Theorem 1.14 with respect to μ. Using Fubini's Theorem on the rightmost term, which is justified by (1.21), it follows that

$$(1.24) \qquad \int S(t) f \, d\mu = \int f \, d\mu$$

for all $f \in \mathcal{L}$, so that $\mu \in \mathcal{I}$. Conversely, suppose that $\mu \in \mathcal{I}$. Then (1.24) holds for all $f \in \mathcal{L}$ and $t \geq 0$. Using this, parts (e) and (f) of Theorem 1.14, and the Dominated Convergence Theorem gives $\int \Omega f \, d\mu = 0$ as required. $\square$

The next topic for this section is duality, which takes a particularly simple form in the context of linear systems. Suppose $a(u, v)$, $b(u, v)$, $A_x(u, v)$, and $B_x(u, v)$ are transposes of one another:

$$a(u, v) = b(v, u), \qquad A_x(u, v) = B_x(v, u).$$

Assume that $\alpha(\cdot)$ and M are such that (1.3), (1.4), and (1.5) are satisfied for both $a(u, v)$, $A_x(u, v)$, and $b(u, v)$, $B_x(u, v)$. Let η_t be the process corresponding to $a(u, v)$ and $A_x(u, v)$, and ζ_t be the process corresponding to $b(u, v)$ and $B_x(u, v)$. Then η_t and ζ_t are said to be dual to one another. For example, the binary contact path process is self-dual, and the potlatch and smoothing processes are dual to each other. The following result expresses the probabilistic connection between the dual processes.

Theorem 1.25. *Suppose η_t has initial configuration η and ζ_t has initial configuration ζ. Then*

(1.26) $$\sum_x \eta(x)\zeta_t(x) \quad and \quad \sum_x \eta_t(x)\zeta(x)$$

have the same distribution for each t.

Proof. Note first that the series in (1.26) may diverge, but they are well defined in any case since the summands are nonnegative. It suffices to prove the result if S is finite, since one can then pass to the limit using Theorem 1.14. In the finite case, recall the explicit construction in (1.9). The equality in distribution of the series in (1.26) then follows from the fact that, conditioned on the event $\{\tau_k \leq t < \tau_{k+1}\}$, the joint distributions of

$$(\tau_1, \tau_2 - \tau_1, \ldots, \tau_k - \tau_{k-1}, t - \tau_k) \quad and \quad (t - \tau_k, \tau_k - \tau_{k-1}, \ldots, \tau_2 - \tau_1, \tau_1)$$

agree. $\square$

For use in the next section, we will record the equations which govern the evolution of the first moments of the system.

Theorem 1.27. *If $\eta \in X$, then $E^\eta \eta_t(x) < \infty$ for all $t \geq 0$ and $x \in S$, and satisfies*

$$\frac{d}{dt} E^\eta \eta_t(x) = \sum_y \gamma(x, y) E^\eta \eta_t(y),$$

where

(1.28) $$\gamma(x, y) = a(x, y) + \begin{cases} E\sum_u A_u(x, y) & \text{if } x \neq y, \\ \\ E\sum_u [A_u(x, x) - 1] & \text{if } x = y. \end{cases}$$

Note that the series in (1.28) converge by (1.4) and (1.5).

Proof. The first statement follows from (1.22). By Theorems 1.14(b) and 1.25, $S(t)$ maps the class of all functions of the form

$$f(\eta) = \sum_x \beta(x)\eta(x)$$

for $\beta(\cdot)$ satisfying $\sup_x(|\beta(x)|/\alpha(x)) < \infty$ into itself. Therefore by Theorem

1.14(d), it suffices to compute

$$\Omega f(\eta) = \sum_x \{Ef(A_x\eta) - f(\eta)\} + \sum_{u,v} f_u(\eta)a(u, v)\eta(v)$$

$$= E \sum_x \sum_{u \neq v} \beta(u)A_x(u, v)\eta(v) + E \sum_{x,u} \beta(u)[A_x(u, u) - 1]\eta(u)$$

$$+ \sum_{u,v} \beta(u)a(u, v)\eta(v)$$

$$= \sum_{x,y} \beta(x)\gamma(x, y)\eta(y)$$

for such an f.   □

2. Survival and Extinction

For the remainder of this chapter, we will assume that $S = Z^d$, and that the mechanism is translation invariant in the sense that $a(x, y) = a(0, y - x)$ and the joint distributions of

$$\{A_x(u + x, v + x), u, v \in S\}$$

are independent of x. Of course, we continue to assume that $\alpha(\cdot)$ and M have been chosen so that (1.1), (1.3), (1.4), and (1.5) are satisfied. Furthermore, we will assume that $\alpha(x) = \alpha(-x)$, so that the dual process satisfies (1.3), (1.4), and (1.5) as well. From Theorem 1.27,

$$E^1\eta_t(x) = e^{ct},$$

where 1 is the configuration $\eta \equiv 1$, and

$$c = \sum_y \gamma(0, y)$$

for the $\gamma(x, y)$ defined in (1.28). We would like $E^1\eta_t(x)$ to be independent of t, so the final assumption we will make from now on is that

$$(2.1) \qquad \sum_y a(0, y) + \sum_u \left\{ \sum_y EA_u(0, y) - 1 \right\} = 0.$$

Note that these series converge absolutely by (1.1), (1.3), (1.4), and (1.5). Under this assumption, $\gamma(x, y)$ is the Q matrix for a continuous time random walk on Z^d. Let $\gamma_t(x, y)$ be its transition probabilities.

A configuration $\eta \in X$ will be called finite if $\sum_x \eta(x) < \infty$. A finite system will be one in which $\sum_x \eta_t(x) < \infty$ for all t. $\mathcal{S}$ will denote the class of probability measures on X which are translation invariant.

Theorem 2.2. (a) *For any $\eta \in X$,*

(2.3)
$$E^{\eta}\eta_t(x) = \sum_y \gamma_t(x, y)\eta(y).$$

(b) *For any finite η,*

$$\sum_x \eta_t(x)$$

is a (nonnegative) martingale under P^{η}, and hence converges as $t \uparrow \infty$ with probability one.

Proof. Part (a) is an immediate consequence of Theorem 1.27. For part (b), sum (2.3) on x to obtain

$$E^{\eta} \sum_x \eta_t(x) = \sum_{x,y} \gamma_t(x, y)\eta(y) = \sum_y \eta(y).$$

The martingale property follows from this and the Markov property of η_t. □

Theorem 2.4. (a) *Either the martingale $\sum_x \eta_t(x)$ is uniformly integrable for all finite initial η, or*

$$\lim_{t \to \infty} \sum_x \eta_t(x) = 0 \quad a.s.$$

for all finite initial η. The finite system is said to survive in the first case, and die out in the second.
(b) *The limit*

$$\nu_{\beta} = \lim_{t \to \infty} \delta_{\beta}S(t) \in \mathcal{I}$$

exists for any $\beta \geq 0$, where δ_{β} is the pointmass on the constant configuration $\eta \equiv \beta$. Either

$$\int \eta(x) \, d\nu_{\beta} = \beta$$

for all β, or $\nu_{\beta} = \delta_0$ for all β. The infinite system is said to survive in the first case, and die out in the second.
(c) *Suppose η_t and ζ_t are dual to each other. Then the finite η_t system survives if and only if the infinite ζ_t system survives.*

Proof. Let

$$V = \lim_{t\to\infty} \sum_x \eta_t(x),$$

where initially $\eta(0) = 1$ and $\eta(x) = 0$ for $x \neq 0$. This limit exists by part (b) of Theorem 2.2. By the linearity of the system,

$$(2.5) \qquad\qquad V \overset{d}{=} \sum_x \eta_t(x) V_x,$$

where $\overset{d}{=}$ denotes equality in distribution and $\{V_x, x \in S\}$ is a collection of random variables which have the same distribution as V and are independent of η_t. By Jensen's inequality,

$$E\left[\exp\left(-\sum_x \eta_t(x) V_x\right)\,\middle|\, \eta_t\right] \geq \exp\left(-(EV)\sum_x \eta_t(x)\right).$$

Taking expected values of both sides and using (2.5) yields,

$$E\,e^{-V} \geq E\,\exp\left(-(EV)\sum_x \eta_t(x)\right).$$

Now let $t \to \infty$ to obtain

$$(2.6) \qquad\qquad E\,e^{-V} \geq E\,e^{-VEV}.$$

Since $0 \leq EV \leq 1$, equality must hold in (2.6), and then it follows that for the particular initial configuration η such that $\eta(0) = 1$ and $\eta(x) = 0$ for $x \neq 0$, either $\sum_x \eta_t(x)$ is uniformly integrable or $\lim_{t\to\infty} \sum_x \eta_t(x) = 0$ a.s. The general statement in (a) now follows by translation invariance and linearity. For parts (b) and (c), let η_t and ζ_t be dual to each other. Take $\eta_0(x) = \beta$ for all x and ζ_0 any finite configuration. By Theorem 1.25,

$$\sum_x \eta_t(x)\zeta(x) \quad \text{and} \quad \beta \sum_x \zeta_t(x)$$

have the same distribution. If the finite ζ_t system dies out, it follows that $\eta_t(x) \to 0$ in probability as $t \to \infty$ for each x. On the other hand, if the finite ζ_t system survives, then

$$\sum_x \eta_t(x)\zeta(x)$$

has a limiting distribution with mean

$$\beta \sum_x \zeta(x).$$

Therefore the limit ν_β of $\delta_\beta S(t)$ exists and has mean β. To check that ν_β is invariant, show directly that $\nu_\beta S(t) = \nu_\beta$, using part (b) of Theorem 1.14. This completes the proofs of parts (b) and (c). $\quad \square$

Theorem 2.7. *Suppose the infinite system dies out. If μ is a translation invariant measure on X such that $\int \eta(x)\, d\mu < \infty$, then*

$$\mu S(t) \to \delta_0$$

as $t \to \infty$. In particular, $\mathcal{I} \cap \mathcal{S}$ contains no elements of finite mean (other than δ_0).

Proof. Take η to have distribution μ, and write

$$\eta = \eta_1 + \eta_2,$$

where $\eta_1(x) = \min\{\eta(x), \beta\}$ for some fixed β. Let μ_1 and μ_2 be the distributions of η_1 and η_2 respectively. Then

$$\lim_{t\to\infty} \mu_1 S(t) = \delta_0$$

by assumption, since $\mu_1 \le \delta_\beta$. On the other hand for all t,

$$\int \eta(x)\, d\mu_2 S(t) = \int \eta(x)\, d\mu_2 = \int (\eta(x) - \beta)^+\, d\mu.$$

Therefore by linearity, any weak limit ν of $\mu S(t)$ satisfies

$$\int \eta(x)\, d\nu \le \int (\eta(x) - \beta)^+\, d\mu.$$

The result follows by letting $\beta \to \infty$. $\quad \square$

Next, we will prove some general facts about infinite systems which survive.

Lemma 2.8. *Suppose the infinite system survives. Take $\mu \in \mathcal{S}$ with $\int \eta(x)\, d\mu < \infty$, and let η_t be the linear system with initial distribution μ. Then $\{\eta_t(0), t \ge 0\}$ is uniformly integrable.*

Proof. Using the same truncation argument as in the proof of Theorem 2.7, it should be clear that it suffices to prove the result in case μ concentrates on $\{\eta: \eta(x) \le \beta \text{ for all } x \in S\}$ for some β. But then $\mu \le \delta_\beta$, so that

$$\mu S(t) \le \delta_\beta S(t)$$

for all $t \geq 0$. Since $\{\eta_t(0), t \geq 0\}$ is uniformly integrable under δ_β, the same is true under μ. $\square$

Lemma 2.9. *Suppose the infinite system survives. If $\nu_\beta = \lambda \mu_1 + (1 - \lambda) \mu_2$ for some $0 < \lambda < 1$ and $\mu_1, \mu_2 \in \mathcal{S}$, then*

$$\lim_{t \to \infty} \mu_i S(t) = \nu_\beta.$$

Proof. Let $\beta_i = \int \eta(x) \, d\mu_i$. Then

$$(2.10) \qquad \beta = \lambda \beta_1 + (1 - \lambda) \beta_2.$$

By Jensen's inequality,

$$(2.11) \qquad \exp\left[-\beta_i \sum_x \zeta(x) \right] \leq \int \exp\left[-\sum_x \eta(x) \zeta(x) \right] d\mu_i$$

for any finite configuration ζ. Let ζ_t be the dual process. Replacing ζ by ζ_t in (2.11) and taking expected values yields

$$E^\zeta \exp\left[-\beta_i \sum_x \zeta_t(x) \right] \leq \int E^\zeta \exp\left[-\sum_x \eta(x) \zeta_t(x) \right] d\mu_i.$$

By Theorem 1.25, this can be rewritten as

$$(2.12) \quad \int \exp\left[-\frac{\beta_i}{\beta} \sum_x \eta(x) \zeta(x) \right] d\delta_\beta S(t) \leq \int \exp\left[-\sum_x \eta(x) \zeta(x) \right] d\mu_i S(t).$$

Fix the finite configuration ζ, and let

$$\phi(\gamma) = \int \exp\left[-\gamma \sum_x \eta(x) \zeta(x) \right] d\nu_\beta.$$

Let $\bar{\mu}_i$ the limit of $\mu_i S(t)$ along any sequence of t's tending to ∞. Passing to the limit in (2.12), we obtain

$$(2.13) \qquad \phi\left(\frac{\beta_i}{\beta} \right) \leq \int \exp\left[-\sum_x \eta(x) \zeta(x) \right] d\bar{\mu}_i.$$

Since $\nu_\beta = \lambda \bar{\mu}_1 + (1 - \lambda) \bar{\mu}_2$, this implies that

$$(2.14) \qquad \lambda \phi\left(\frac{\beta_1}{\beta} \right) + (1 - \lambda) \phi\left(\frac{\beta_2}{\beta} \right) \leq \phi(1).$$

If $\zeta \neq 0$, then ϕ is strictly convex (since $\nu_\beta \neq \delta_0$). Therefore (2.10) and (2.14) imply that $\beta_1 = \beta_2 = \beta$ and that equality holds in (2.13). Therefore $\bar{\mu}_i = \nu_\beta$ as required. $\square$

Corollary 2.15. *Suppose the infinite system survives. Then ν_β is extremal in $\mathcal{I} \cap \mathcal{S}$.*

Under weak assumptions, we would expect that $\{\nu_\beta, \beta \geq 0\}$ gives all the extremal members of $\mathcal{I} \cap \mathcal{S}$. This is not always true, however. A simple counterexample is provided by the multitype voter model (0.3). In this case, $\nu_\beta = \delta_\beta$. If the random walk with transition probabilities $\frac{1}{2}[p_t(x, y) + p_t(y, x)]$ is transient, then Theorem 1.8 of Chapter V implies that there are many other extremal invariant measures. In fact, there is at least one corresponding to each probability measure on $[0, \infty)$ of finite mean, rather than to each element of $[0, \infty)$. The next results give sufficient conditions for $(\mathcal{I} \cap \mathcal{S})_e = \{\nu_\beta, \beta \geq 0\}$ to be correct. In the proofs, it will be convenient to write in a more explicit form the coupling which was discussed in Section 1. By the linearity of the mechanism, there is a family of random matrices $\{A_t(x, y), t \geq 0\}$ so that

$$(2.16) \qquad \eta_t(x) = \sum_y A_t(x, y) \eta_0(y)$$

for any initial configuration η_0. By Theorem 2.2(a),

$$(2.17) \qquad EA_t(x, y) = \gamma_t(x, y).$$

The basic assumption will be: for every $y, z \in S$,

$$(2.18) \qquad P(A_t(x, y) > 0, A_t(x, z) > 0) > 0 \quad \text{for some } x \in S.$$

Note that this assumption is not satisfied by the multitype voter model, since in that case,

$$P(A_t(x, y) = 1) = \gamma_t(x, y)$$

and $\sum_y A_t(x, y) = 1$ a.s. On the other hand, (2.18) is satisfied by the smoothing and potlatch processes.

Theorem 2.19. *Assume that the infinite system survives and that (2.18) is satisfied. Then*

$$(\mathcal{I} \cap \mathcal{S})_e = \{\nu_\beta, \beta \geq 0\}.$$

Proof. Let $\mu_1, \mu_2 \in (\mathcal{I} \cap \mathcal{S})_e$ satisfy

$$\int \eta(x) \, d\mu_1 < \infty \quad \text{and} \quad \int \eta(x) \, d\mu_2 < \infty.$$

By starting the coupled process with distribution $\mu_1 \times \mu_2$ and taking a limit of Cesaro averages of the distributions at time t, we see that there is a μ on $X \times X$ which is extremal among the measures which are both translation invariant and invariant for the coupled process and has marginals μ_1 and μ_2 respectively. (For details, see the proof of Theorem 2.15 of Chapter III. The required compactness is provided by the fact that

$$\int \eta(x)\, d\mu_i S(t)$$

is independent of t.) Let (η_t^1, η_t^2) be the coupled process with initial distribution μ. Then by (2.16),

$$(2.20) \qquad \eta_t^1(x) - \eta_t^2(x) = \sum_y A_t(x, y)[\eta_0^1(y) - \eta_0^2(y)],$$

where $\{A_t(x, y), t \geq 0, x, y \in S\}$ and $\{(\eta_0^1(y), \eta_0^2(y)), y \in S\}$ are independent. Therefore

$$(2.21) \qquad |\eta_t^1(x) - \eta_t^2(x)| \leq \sum_y A_t(x, y)|\eta_0^1(y) - \eta_0^2(y)|.$$

Taking expected values in (2.21) we see using (2.17) that

$$(2.22) \qquad E|\eta_t^1(x) - \eta_t^2(x)| \leq \sum_y \gamma_t(x, y) E|\eta_0^1(y) - \eta_0^2(y)|.$$

Since μ is translation invariant and invariant for the coupled process, equality holds in (2.22). Therefore equality must hold a.s. in (2.21). It follows that, with probability one, for each x and t, $\eta_0^1(y) - \eta_0^2(y)$ has the same sign for all y such that $A_t(x, y) > 0$. Therefore by assumption (2.18), $\eta_0^1(y) - \eta_0^2(y)$ has the same sign for all y, and hence

$$\mu\{(\eta^1, \eta^2) \in X \times X \colon \eta^1 \leq \eta^2 \text{ or } \eta^1 \geq \eta^2\} = 1.$$

(Recall in this argument that $\{A_t(x, y)\}$ and $\{\eta_0^1(y) - \eta_0^2(y)\}$ are independent.) Since μ is extremal invariant, either

$$\mu\{(\eta^1, \eta^2) \in X \times X \colon \eta^1 \leq \eta^2\} = 1, \quad \text{or}$$

$$\mu\{(\eta^1, \eta^2) \in X \times X \colon \eta^1 \geq \eta^2\} = 1.$$

Therefore either $\mu_1 \leq \mu_2$ or $\mu_1 \geq \mu_2$. In particular, if $\int \eta(x)\, d\mu_1 = \int \eta(x)\, d\mu_2$, then $\mu_1 = \mu_2$. (See Corollary 2.8 of Chapter II for a similar statement.) By Corollary 2.15, it now follows that ν_β is the only element of $(\mathscr{I} \cap \mathscr{S})_e$ which has mean β. It remains to show that there are no elements of $(\mathscr{I} \cap \mathscr{S})_e$ which have infinite mean. Suppose $\mu \in (\mathscr{I} \cap \mathscr{S})_e$ and $\int \eta(x) \, d\mu = \infty$. For $\gamma > 0$, let μ_γ be the measure induced by μ under the mapping

$$\eta \to \tilde\eta, \quad \text{where} \quad \tilde\eta(x) = \min\{\eta(x), \gamma\}.$$

Let $\bar\mu_\gamma$ be any weak limit of Cesaro averages of $\mu_\gamma S(t)$. By Lemma 2.8,

$$\int \eta(x) \, d\bar\mu_\gamma = \int \min\{\eta(x), \gamma\} \, d\mu.$$

On the other hand, since $\bar\mu_\gamma \in \mathscr{I} \cap \mathscr{S}$ and has finite mean, it is an average of $\{\nu_\beta, \beta \geq 0\}$. Since γ is arbitrary and $\int \eta(x) \, d\mu = \infty$, it follows that for every $\beta > 0$ there is an $\varepsilon > 0$ so that

$$\mu \geq \varepsilon \nu_\beta + (1 - \varepsilon)\delta_0.$$

Therefore there exist $\mu_1, \mu_2 \in \mathscr{S}$ such that

$$\mu = \varepsilon \mu_1 + (1 - \varepsilon)\mu_2$$

and $\mu_1 \geq \nu_\beta$. Since $\mu \in (\mathscr{I} \cap \mathscr{S})_e$, the Cesaro averages of $\mu_1 S(t)$ converge to μ. Therefore

$$\mu = \lim_{T \to \infty} \frac{1}{T} \int_0^T \mu_1 S(t) \, dt \geq \nu_\beta.$$

Since ν_β is obtained from ν_1 by multiplying the configurations by β, and $\nu_1 \neq \delta_0$, it follows that

$$\mu\{\eta: \eta(x) = \infty\} \geq \nu_1\{\eta: \eta(x) > 0\} > 0,$$

which is impossible. $\quad \square$

In the next three sections, we will develop sufficient conditions for survival and extinction. Generally speaking, systems with more randomness are more likely to die out than systems with less randomness. The following comparison theorem gives a formal statement along these lines. It can also be used to show that various natural one-parameter families of systems have critical values, as we will see following the proof. A final reason for the importance of the next theorem is technical. In Section 4, for example, it will be convenient to prove extinction under a second moment assumption first. This assumption is then removed by a comparison argument, using the next theorem.

Theorem 2.23. *Consider two linear systems with generators Ω_1 and Ω_2 of the form* (0.2). *For any $\zeta \in X$ such that $\zeta(x) > 0$ for only finitely many x, let*

$$(2.24) \qquad f_\zeta(\eta) = \exp\left(-\sum_x \zeta(x)\eta(x)\right).$$

Suppose that there are positive constants c_1 and c_2 so that

$$c_2 \Omega_2 f_\zeta(\eta) \geq c_1 \Omega_1 f_\zeta(\eta)$$

for all such ζ and all $\eta \in X$. If the system with generator Ω_2 survives, then so does the one with generator Ω_1. In this conclusion, survival can be interpreted either in terms of the finite system or in terms of the infinite system.

Proof. Let $S_1(t)$ and $S_2(t)$ be the corresponding semigroups. By Theorem 1.25, for either process,

$$S(t)f_\zeta(\eta) = E^\eta \exp\left(-\sum_x \zeta(x)\eta_t(x)\right)$$

$$= E^\zeta \exp\left(-\sum_x \zeta_t(x)\eta(x)\right)$$

$$= E^\zeta f_{\zeta_t}(\eta),$$

which is a limit of convex combinations of functions of the form (2.24). Therefore by assumption,

$$(2.25) \qquad (c_2 \Omega_2 - c_1 \Omega_1) S_1(s) f_\zeta \geq 0$$

for all $s \geq 0$ and all ζ. By integration by parts,

$$(2.26) \quad S_2(c_2 t)f - S_1(c_1 t)f = \int_0^t S_2(c_2 s)(c_2 \Omega_2 - c_1 \Omega_1)S_1(c_1(t-s))f \, ds.$$

Combining (2.25) and (2.26) gives

$$S_1(c_1 t)f_\zeta(\eta) \leq S_2(c_2 t)f_\zeta(\eta).$$

Evaluating this at $\eta \equiv \beta$ and letting $t \uparrow \infty$, we conclude that

$$\int \exp\left(-\sum_x \zeta(x)\eta(x)\right)\nu^1_\beta(d\eta) \leq \int \exp\left(-\sum_x \zeta(x)\eta(x)\right)\nu^2_\beta(d\eta),$$

where ν^1_β and ν^2_β are the invariant measures defined in Theorem 2.4(b).

This gives the result for the infinite systems. Since the dual processes satisfy the same hypotheses, the result for finite systems follows from the result for infinite systems, together with part (c) of Theorem 2.4. $\quad\square$

We will conclude this section by illustrating the use of Theorem 2.23. Consider first the smoothing process (0.6). In this case,

$$\Omega f_\zeta(\eta) = \sum_x \{Ef_\zeta(A_x\eta) - f_\zeta(\eta)\}$$

$$= \sum_x \left\{E \exp\left(-\sum_{u,v} A_x(u,v)\zeta(u)\eta(v)\right) - f_\zeta(\eta)\right\}$$

$$= \sum_x \exp\left(-\sum_{u \neq x} \zeta(u)\eta(u)\right)$$

$$\times \left\{E \exp\left(-\zeta(x)W\sum_v p(x,v)\eta(v)\right) - \exp(-\zeta(x)\eta(x))\right\}.$$

Thus, in comparing two smoothing processes corresponding to the same $p(x,y)$ but different W_1 and W_2, it is enough to assume that

$$E(e^{-\lambda W_1}) \leq E(e^{-\lambda W_2})$$

for all $\lambda \geq 0$, and then to take $c_1 = c_2 = 1$ in Theorem 2.23. A similar computation (or an appeal to duality) yields the same conclusion for potlatch processes.

To see why it is convenient to allow $c_1 \neq c_2$ in Theorem 2.23, consider the normalized binary contact path process. Then

$$\Omega f_\zeta(\eta) = \frac{f_\zeta(\eta)}{1+2\lambda d} \sum_x \left\{\exp(\eta(x)\zeta(x))\right.$$

$$+\lambda \sum_{|y-x|=1} \exp(-\zeta(y)\eta(x)) - (1+2\lambda d)\right\}$$

$$+\frac{2\lambda d - 1}{2\lambda d + 1}f_\zeta(\eta)\sum_x \zeta(x)\eta(x),$$

so that

$$\frac{1+2\lambda d}{2\lambda d}\Omega f_\zeta(\eta) = f_\zeta(\eta)\sum_x \left\{\frac{\exp(\eta(x)\zeta(x))}{2\lambda d}\right.$$

$$+\frac{1}{2d}\sum_{|y-x|=1} \exp(-\zeta(y)\eta(x)) - \frac{1+2\lambda d}{2\lambda d}\right\}$$

$$+\frac{2\lambda d - 1}{2\lambda d}f_\zeta(\eta)\sum_x \zeta(x)\eta(x).$$

The derivative with respect to λ of this expression is

$$-\frac{1}{2d\lambda^2}f_\zeta(\eta) \sum_x \{\exp(\eta(x)\zeta(x)) - 1 - \eta(x)\zeta(x)\},$$

which is ≤ 0. So, if $\lambda_2 \leq \lambda_1$, we can apply Theorem 2.23 with

$$c_i = \frac{1+2\lambda_i d}{2\lambda_i d}$$

to conclude that if the process with parameter λ_2 survives, then so does the one with parameter λ_1. This gives the existence of a well-defined critical value in $[0, \infty]$ for the normalized binary contact path process.

3. Survival via Second Moments

In order to prove that an infinite linear system survives, it is necessary to show that if initially $\eta_0 \equiv 1$, then $\{\eta_t(0), t \geq 0\}$ is uniformly integrable. A frequently used method for proving uniform integrability of a collection of random variables is to show that the second moments are uniformly bounded. In this section we will use this approach to obtain a sufficient condition for survival. When this condition is satisfied, we will be able to find the limit of $\mu S(t)$ for any $\mu \in \mathcal{S}$ which has finite first moments. By part (c) of Theorem 2.4, the results in this section can be used to give a sufficient condition for the survival of a finite linear system.

The usefulness of second moments is evident from the first result, which provides a way of computing the second moments at time t in terms of the second moments of the initial distribution. It is the linearity of the mechanism which makes this possible.

Theorem 3.1. *Suppose that*

$$(3.2) \qquad E \sum_x \left[|A_x(u, u) - 1| + \sum_{v:v \neq u} A_x(u, v) \right]^2 < \infty, \quad \text{and let}$$

$q((x, y), (u, v))$

$$= \begin{cases} E \sum_z A_z(x, u) A_z(y, v) & \text{if } u \neq x, v \neq y, \\[2mm] E \sum_z A_z(x, x) A_z(y, v) + a(y, v) & \text{if } u = x, v \neq y, \\[2mm] E \sum_z A_z(x, u) A_z(y, y) + a(x, u) & \text{if } u \neq x, v = y, \\ & \text{and} \\ E \sum_z [A_z(x, x) A_z(y, y) - 1] + a(x, x) + a(y, y) & \text{if } u = x, v = y. \end{cases}$$

Take $\mu \in \mathcal{S}$ satisfying $\int \eta^2(0)\, d\mu < \infty$, and let

(3.3)
$$f(t, x, y) = \int \eta(x)\eta(y)\, d\mu S(t).$$

Then $f(t, x, y)$ is the unique solution of

(3.4)
$$\frac{d}{dt} f(t, x, y) = \sum_{u,v \in S} q((x, y), (u, v)) f(t, u, v)$$

subject to $f(0, x, y) = \int \eta(x)\eta(y)\, d\mu$ and

(3.5)
$$\sup_{0 \le t \le T} \sup_{x,y \in S} f(t, x, y) < \infty$$

for each T.

Proof. First note that $q((x, y), (u, v)) \ge 0$ unless $(u, v) = (x, y)$, and that by (1.1), (1.3), (1.4), (1.5), and (3.2),

(3.6)
$$\sup_{x,y} \sum_{u,v} |q((x, y), (u, v))| < \infty.$$

Therefore (3.4) has a unique solution $\bar{f}(t, x, y)$ subject to the given constraints. The rest of the proof consists of two parts: showing that the $f(t, x, y)$ defined in (3.3) satisfies (3.5), and then using that fact, showing that $f(t, x, y)$ satisfies (3.4). For the first part, consider the approximating semigroups $S_n(t)$ defined just prior to Theorem 1.14, which have the generators Ω_n given in (1.13). Let $f_n(t, x, y)$ and $q_n((x, y), (u, v))$ be the quantities corresponding to the approximating system. Then f_n satisfies (3.5) by the explicit construction of the process given in (1.9), and assumption (3.2). Using the explicit form of Ω_n given in (1.13), one can then check that

$$\frac{d}{dt} f_n(t, x, y) = \sum_{u,v} q_n((x, y), (u, v)) f_n(t, u, v).$$

Since $q_n((x, y), (u, v)) \le q((x, y), (u, v))$ for $(u, v) \ne (x, y)$ and

$$q_n((x, y), (x, y)) \le q((x, y), (x, y)) + c$$

for some constant c, it follows that

$$f_n(t, x, y) \le e^{ct}\bar{f}(t, x, y).$$

By part (a) of Theorem 1.14 and Fatou's Lemma, it follows that

$$f(t, x, y) \le e^{ct}\bar{f}(t, x, y),$$

so that $f(t, x, y)$ satisfies (3.5). For the second part of the proof, we need a different type of approximation. Let $\phi: [0, \infty) \to [0, \infty)$ be a bounded continuously differentiable function which satisfies $\phi(0) = 0$ and $0 \le \phi'(u) \le 1$. Then

$$F(\eta) = \phi(\eta(x))\phi(\eta(y))$$

for fixed $x, y \in S$ is an element of $\mathcal{L}$. Therefore by parts (d) and (g) of Theorem 1.14,

(3.7) $$\int F d\mu S(t) = \int F d\mu + \int_0^t \int \Omega F d\mu \, S(s) \, ds.$$

Using the expression (0.2) for Ω, compute

$$\Omega F(\eta) = \sum_z \left\{ E\phi\left(\sum_u A_z(x, u)\eta(u)\right)\phi\left(\sum_v A_z(y, v)\eta(v)\right) - \phi(\eta(x))\phi(\eta(y)) \right\}$$

$$+ \phi'(\eta(x))\phi(\eta(y)) \sum_u a(x, u)\eta(u)$$

$$+ \phi'(\eta(y))\phi(\eta(x)) \sum_v a(y, v)\eta(v).$$

Therefore, since

$$|\phi(a)\phi(b) - \phi(c)\phi(d)| \le b|a - c| + c|b - d|,$$

$$|\Omega F(\eta)| \le \sum_z E \left| \sum_u A_z(x, u)\eta(u) - \eta(x) \right| \sum_v A_z(y, v)\eta(v)$$

$$+ \sum_z E \left| \sum_v A_z(y, v)\eta(v) - \eta(y) \right| \eta(x)$$

$$+ \eta(y) \sum_u |a(x, u)|\eta(u) + \eta(x) \sum_v |a(y, v)|\eta(v).$$

This estimate, together with the fact that $f(t, x, y)$ satisfies (3.5) will give the needed domination when we take the limit in (3.7) for a sequence ϕ_n which increase to $\phi(x) = x$. In passing to this limit, note that the limit of $\Omega F(\eta)$ is then

$$\sum_z \left\{ E \sum_{u,v} A_z(x, u)A_z(y, v)\eta(u)\eta(v) - \eta(x)\eta(y) \right\}$$

$$+ \eta(y) \sum_u a(x, u)\eta(u) + \eta(x) \sum_v a(y, v)\eta(v).$$

Thus, after passing to the limit, (3.7) becomes the integrated form of (3.4) as required. □

Since $\mu \in \mathcal{S}$ in Theorem 3.1, $f(t, x, y)$ is a function only of $y - x$. Also, by the translation invariance of the mechanism,

$$q((x + w, y + w), (u + w, v + w)) = q((x, y), (u, v))$$

for all $x, y, u, v, w \in S$. Therefore (3.4) can be simplified by setting

$$f(t, x) = f(t, x, 0), \quad \text{and}$$

$$q(x, y) = \sum_u q((0, x), (u, u + y)).$$

Then (3.4) becomes

(3.8) $$\frac{d}{dt} f(t, x) = \sum_{y \in S} q(x, y) f(t, y),$$

where

(3.9) $$q(x, y) = a(x, y) + a(y, x) + \sum_{u, z} EA_z(0, u) A_z(x, u + y)$$

for $x \neq y$, and

(3.10) $$q(x, x) = 2a(x, x) + \sum_z \left\{ \sum_u EA_z(0, u) A_z(x, u + x) - 1 \right\}.$$

Define $q_t(x, y)$ by giving the solution of (3.8) in the form

(3.11) $$f(t, x) = \sum_y q_t(x, y) f(0, y).$$

Note that $q_t(x, y) \geq 0$ for all t, x, y, since $q(x, y) \geq 0$ for all $x \neq y$. There is no reason to expect $\sum_y q_t(x, y) = 1$, however.

Theorem 3.12. *Suppose that (3.2) holds, and that there is a strictly positive function $h(x)$ on S such that*

(3.13) $$\lim_{x \to \infty} h(x) = 1, \quad and$$

(3.14) $$\sum_y q(x, y) h(y) = 0 \quad for \ all \ x \in S.$$

Then the infinite system survives. Furthermore, $\int [\eta(0)]^2 \, dv_\beta < \infty$.

Proof. Let η_t be the linear system with initial configuration $\eta_0 \equiv 1$. Then

(3.15) $$E\eta_t(x) \eta_t(0) = \sum_y q_t(x, y)$$

by (3.11). By (3.14),

(3.16) $$\sum_{y} q_t(x, y) h(y) = h(x).$$

Combining (3.15) and (3.16) yields the bound

$$E[\eta_t(0)]^2 = \sum_{y} q_t(0, y) \leq \frac{h(0)}{\inf_{x} h(x)} < \infty.$$

Therefore $\{\eta_t(0), t \geq 0\}$ is uniformly integrable, so that the infinite system survives by the criterion in part (b) of Theorem 2.4   □

Next, we come to the main convergence theorem for linear systems. For its statement and proof, it is convenient to let $\mathcal{S}_\beta$ be the set of all mixtures of the ergodic elements μ of $\mathcal{S}$ which satisfy $\int \eta(x) \, d\mu = \beta$. Note that any element μ of $\mathcal{S}$ with finite mean can be written as a mixture

$$\mu = \int_0^\infty \mu_\beta \lambda \, (d\beta)$$

of elements $\mu_\beta \in \mathcal{S}_\beta$. Here $\lambda(d\beta)$ is the distribution of

$$\lim_{x \to \infty} \frac{\sum_{0 \leq y \leq x} \eta(y)}{|\{y \in S : 0 \leq y \leq x\}|},$$

which exists (in L_1 relative to μ) by Theorem 4.9 of Chapter I. As a consequence of these remarks, it is sufficient in the following convergence theorem to consider only initial distributions which lie in $\mathcal{S}_\beta$ for some $\beta \geq 0$.

Theorem 3.17. *Assume* (2.18) *and the hypotheses of Theorem* 3.12. *Suppose further that*

(3.18) $$\lim_{t \to \infty} \sum_{y} q_t(x, y) \int \eta(y) \eta(0) \, d\mu = h(x) \beta^2$$

for any $\mu \in \mathcal{S}_\beta$ *satisfying* $\int [\eta(x)]^2 \, d\mu < \infty$, *where* $h(x)$ *is the function which appears in the statement of Theorem* 3.12. *Then*

(3.19) $$\lim_{t \to \infty} \mu S(t) = \nu_\beta$$

for any $\mu \in \mathcal{S}_\beta$. *Here* ν_β *is the invariant measure with density* β *which was defined in part* (b) *of Theorem* 2.4. *Furthermore*

(3.20)
$$\int \eta(x)\eta(0)\,d\nu_\beta = h(x)\beta^2.$$

Proof. The proof is similar to the proof of Theorem 2.19. In addition to guaranteeing the survival of the system, the second moment assumptions are used to rule out the possibility that a potential limit of $\mu S(t)$ for $\mu \in \mathcal{S}_\beta$ might be a nontrivial mixture of different ν_β's. Begin by applying Corollary 8.20 of Chapter II to show that

(3.21)
$$\lim_{t\to\infty} \int \left[\sum_y \gamma_t(x,y)\eta(y) - \beta \right]^2 d\mu = 0$$

for any $\mu \in \mathcal{S}_\beta$ with $\int [\eta(0)]^2\,d\mu < \infty$, where γ_t is defined at the beginning of Section 2. The irreducibility required to prove this comes from (2.18). Now take $\mu \in \mathcal{S}_\beta$ with $\int [\eta(0)]^2\,d\mu < \infty$. By (3.11) and (3.18),

(3.22)
$$\lim_{t\to\infty} \int \eta(x)\eta(0)\,d\mu S(t) = \beta^2 h(x).$$

Let ν be any weak limit of $\mu S(t)$ along a sequence of times tending to ∞. (Such limits exist by the boundedness of the moments.) By uniform integrability,

(3.23)
$$\int \eta(x)\,d\nu = \beta,$$

while by (3.22) and Fatou's lemma for convergence in distribution,

(3.24)
$$\int \eta(x)\eta(0)\,d\nu \le \beta^2 h(x).$$

Therefore $\nu \in \mathcal{S}_\beta$ as well, by (3.13). Now let $\mu_1, \mu_2 \in \mathcal{S}_\beta$ with $\int [\eta(0)]^2\,d\mu_i < \infty$ for $i = 1, 2$, and let ν_1 and ν_2 be weak limits of $\mu_1 S(t)$ and $\mu_2 S(t)$ along the same sequence of times tending to ∞. Using the coupling argument in the proof of Theorem 2.19, we see that there is a translation invariant probability measure ν on $X \times X$ with marginals ν_1 and ν_2 respectively such that

(3.25)
$$\nu\{(\eta^1, \eta^2) \in X \times X: \eta^1 \le \eta^2 \text{ or } \eta^1 \ge \eta^2\} = 1.$$

Therefore

(3.26)
$$\int |\eta^1(x) - \eta^2(x)| \, d\nu = \int \left| \sum_y \gamma_t(x, y) |\eta^1(y) - \eta^2(y)| \right| d\nu$$

$$= \int \left| \sum_y \gamma_t(x, y) \eta^1(y) - \sum_y \gamma_t(x, y) \eta^2(y) \right| d\nu,$$

where the first equality comes from the translation invariance of ν, while the second comes from (3.25). The right side of (3.26) tends to zero by (3.21) as $t \to \infty$, since as we showed earlier, ν_1 and ν_2 are in $\mathcal{S}_\beta$. Therefore (3.26) implies that $\int |\eta^1(x) - \eta^2(x)| \, d\nu = 0$, and hence that $\nu_1 = \nu_2$. Applying this to $\mu_1 = \mu$ and $\mu_2 = \mu S(t)$ for some fixed $t > 0$, it follows that if $\mu \in \mathcal{S}_\beta$ and $\int [\eta^2(x)] \, d\mu < \infty$, then any weak limit of $\mu S(t)$ must be in $\mathcal{I}$. Since Theorem 2.19 implies that ν_β is the unique element of $\mathcal{I} \cap \mathcal{S}_\beta$, we can conclude that (3.19) holds for such μ. To remove the second moment assumption, one can then truncate as in the proof of Theorem 2.19. Finally, (3.20) follows from (3.22) applied to $\mu = \nu_\beta$, since $\int \eta(x)\eta(0) \, d\nu_\beta S(t)$ is independent of t. $\square$

In many examples, the hypotheses of Theorems 3.12 and 3.17 are verified by explicitly computing the function $h(x)$ which satisfies (3.13) and (3.14), and then verifying (3.18) by appealing to Corollary 8.20 of Chapter II. This will be carried out in Section 6 for a number of cases including the smoothing and potlatch processes. There is, however, a general condition which implies the hypotheses of Theorem 3.12 and 3.17 even when $h(x)$ cannot be computed explicitly. It is given in the following result, for which we need to introduce some additional notation. Recalling the definitions of $\gamma(x, y)$ and $q(x, y)$ in (1.28) and (3.9) respectively, define $\Delta(x, y)$ by

(3.27) $q(x, y) = \gamma(x, y) + \gamma(y, x) + \Delta(x, y),$

so that

(3.28) $\Delta(x, y) = \sum_{u, z} E[A_z(0, u) - I(0, u)][A_z(x, u+y) - I(x, u+y)],$

where $I(x, y) = 0$ if $x \neq y$ and $I(x, x) = 1$. Let $C = [2|\gamma(x, x)|]^{-1}$, and

$$p(x, y) = \begin{cases} C[\gamma(x, y) + \gamma(y, x)] & \text{if } x \neq y, \\ 0 & \text{if } x = y, \end{cases}$$

which are the transition probabilities for a symmetric random walk on S. Let

$$G(x, y) = \sum_{n=0}^{\infty} p^{(n)}(x, y)$$

be its Green function.

Theorem 3.29. *Assume* (2.18) *and suppose that* $p(x, y)$ *is irreducible and transient, and that* $q_t(x, y) > 0$ *for all* $x, y \in S$. *Assume further that* $\Delta(x, y) \geq 0$,

$$\text{(3.30)} \qquad \sum_{x,y} \Delta(x, y) < \infty, \quad and$$

$$\text{(3.31)} \qquad \sup_x C \sum_{y,z} G(x, y)\Delta(y, z) < 1.$$

Then there exists a function $h(x)$ *on* S *which satisfies the hypotheses of Theorems* 3.12 *and* 3.17.

Proof. By (3.27), (3.14) can be rewritten as

$$\text{(3.32)} \qquad h(x) - \sum_y p(x, y)h(y) = C \sum_y \Delta(x, y)h(y).$$

With the additional condition (3.13), (3.32) becomes

$$\text{(3.33)} \qquad h(x) = 1 + C \sum_{y,z} G(x, y)\Delta(y, z)h(z).$$

Under assumption (3.31), (3.33) has the unique solution

$$h = \sum_{n=0}^{\infty} \Gamma^n 1,$$

where Γ is the operator on $l_\infty(S)$ given by

$$\Gamma g(x) = C \sum_{y,z} G(x, y)\Delta(y, z)g(z),$$

This function h is strictly positive and satisfies (3.13) and (3.14). To verify (3.18), apply Corollary 8.20 of Chapter II. The $q(x, y)$ which appears there is $p(x, y) + C\Delta(x, y)$. The hypotheses of that corollary are easily verified using $\Delta(x, y) \geq 0$ and (3.30). □

4. Extinction in One and Two Dimensions

In this section, we will see that extinction is the rule rather than the exception in one and two dimensions. We will focus on extinction for the finite system, since the results obtained here can then be translated into results for the infinite system by using part (c) of Theorem 2.4.

Two assumptions are required in order to carry out the program. First is a type of recurrence statement for the random walk with Q matrix $\gamma(x, y)$ given in (1.28), which forces us into one and two dimensions (assuming some irreducibility). It asserts the existence of a family C_t of nonempty finite subsets of Z^d with the following two properties:

(4.1)
$$\lim_{t \to \infty} \sum_{y \in C_t} \gamma_t(y, 0) = 1, \quad \text{and}$$

(4.2)
$$\int_0^\infty |C_t|^{-1} \, dt = \infty,$$

where $|C_t|$ is the cardinality of C_t. Such a family never exists if the symmetrization of the random walk with transition probabilities $\gamma_t(x, y)$ is transient. To see this, use the Schwarz inequality to write

(4.3)
$$\left[\sum_{y \in C_t} \gamma_t(y, 0) \right]^2 \leq \sum_y [\gamma_t(0, y)]^2 |C_t|.$$

If X_t and Y_t are independent random walks with transition probabilities $\gamma_t(x, y)$ and initial states $X_0 = Y_0 = 0$, (4.1), (4.2) and (4.3) imply that

$$\int_0^\infty P(X_t = Y_t) \, dt = \infty,$$

so the symmetrized random walk is recurrent. On the other hand, a family C_t satisfying (4.1) and (4.2) always exists if $d = 1$ and $\sum_x |x| \gamma(0, x) < \infty$ or if $d = 2$ and $\sum_x |x|^2 \gamma(0, x) < \infty$. One choice which works is

$$C_t = \left\{ x \colon \left| x - \sum_y y \gamma_t(0, y) \right|^d < \max(1, t \log t) \right\}.$$

Then (4.2) is obvious while (4.1) follows from Chebyshev's inequality.

The second assumption we will need to make is that there is at least some randomness in the matrices $A_x(u, v)$. We must rule out examples such as those which satisfy

$$\sum_u A_x(u, v) = 1,$$

since this implies that $\sum_x \eta_t(x)$ is independent of time. If this is the case, the finite system certainly does not die out. The precise assumption is that for each $x \in S$, the joint distribution of the collection of random variables

(4.4)
$$\left\{ \sum_u A_x(u, v), v \in S \right\}$$

has positive correlations in the sense of Definition 2.11 of Chapter II, and that at least one of these random variables is not constant. To better understand this assumption, note that for the smoothing process (0.6),

$$\sum_u A_x(u, v) = \begin{cases} Wp(x, x) & \text{if } v = x, \\ Wp(x, v) + 1 & \text{if } v \neq x, \end{cases}$$

while for the potlatch process (0.7),

$$\sum_u A_x(u, v) = \begin{cases} W & \text{if } v = x, \\ 1 & \text{if } v \neq x. \end{cases}$$

Therefore in both cases, the assumption becomes simply $W \neq 1$. If $W \equiv 1$, then both systems survive. This is obvious for the potlatch process. It follows from Theorem 2.4(c) and Theorem 3.12 for the smoothing process, as we will see in Section 6.

Theorem 4.5. *Suppose that there are nonempty finite subsets C_t of Z^d which satisfy (4.1) and (4.2), and that the random variables in (4.4) have positive correlations and are not all constant. Then the finite linear system dies out.*

Proof. The first step is a simple truncation argument. It is convenient to let

$$B_x(v) = \sum_u A_x(u, v) - 1.$$

Let $\mathcal{F}$ be the σ-algebra on the probability space on which the random variables $A_x(u, v)$ are defined which is generated by the single event

$$F = \{B_x(v) > EB_x(v)\},$$

where v is any one site for which $B_x(v) \neq EB_x(v)$. Define new random matrices $\tilde{A}_x(u, v)$ by

$$\tilde{A}_x(u, v) = E(A_x(u, v)|\mathcal{F}).$$

By Jensen's inequality,

$$E \exp\left(-\sum_{u,v} \zeta(u)\tilde{A}_x(u, v)\eta(v) \right) = E \exp\left(-E\left(\sum_{u,v} \zeta(u)A_x(u, v)\eta(v) \Big| \mathcal{F} \right) \right)$$

$$\leq E \exp\left(-\sum_{u,v} \zeta(u)A_x(u, v)\eta(v) \right)$$

for any $\zeta, \eta \in [0, \infty)^S$. Therefore by Theorem 2.23, it is enough to show that the finite linear system corresponding to $\tilde{A}_x(u, v)$ dies out. To do so, we will show that this modified system satisfies

(4.6) $$\lim_{t \to \infty} E \left(\sum_x \eta_t(x) \right)^{1/2} = 0$$

for any finite initial configuration. Let

$$f(\eta) = |\eta|^{1/2}, \quad \text{where} \quad |\eta| = \sum_x \eta(x).$$

Using (0.2), and the translation invariance of $a(u, v)$, compute for the modified system

(4.7)
$$\Omega f(\eta) = \sum_x \{ Ef(\tilde{A}_x \eta) - f(\eta) \} + \sum_{u,v} f_u(\eta) a(u, v) \eta(v)$$
$$= \sum_x \left\{ E \left[|\eta| + \sum_v \tilde{B}_x(v) \eta(v) \right]^{1/2} - |\eta|^{1/2} \right\}$$
$$+ \tfrac{1}{2} |\eta|^{1/2} \sum_v a(0, v),$$

where

$$\tilde{B}_x(v) = \sum_u \tilde{A}_x(u, v) - 1 = E[B_x(v) | \mathscr{F}].$$

Since $\tilde{B}_x(v)$ takes on just two values with probabilities $P(F)$ and $1 - P(F)$ respectively.

$$|\tilde{B}_x(v)| \le cE|\tilde{B}_x(v)| \le cE|B_x(v)|,$$

where

$$c = [\min\{P(F), 1 - P(F)\}]^{-1}.$$

Therefore

(4.8)
$$\sum_v |\tilde{B}_x(v)| \eta(v) \le c \sum_v E|B_x(v)| \eta(v)$$
$$\le b|\eta|,$$

where

$$b = c \sup_{x,v} E|B_x(v)| < \infty$$

by (1.4), (1.5) and translation invariance. By Taylor's Theorem, there is an $\varepsilon > 0$ so that

(4.9) $$\sqrt{1 + \sigma} \le 1 + \tfrac{1}{2}\sigma - \varepsilon \sigma^2$$

for all $-1 \le \sigma \le b$. Since

$$\sum_x a(0, x) + \sum_x E\tilde{B}_x(v) = 0$$

by (2.1) and translation invariance, (4.7) and (4.9) combine to give

(4.10)
$$\Omega f(\eta) \leq -\varepsilon |\eta|^{1/2} \sum_x E\left(\sum_v \tilde{B}_x(v) \frac{\eta(v)}{|\eta|}\right)^2$$
$$= -\varepsilon |\eta|^{1/2} \sum_{x,u,v} \frac{\eta(u)\eta(v)}{|\eta|^2} E\tilde{B}_x(u)\tilde{B}_x(v)$$
$$= -\varepsilon |\eta|^{1/2} \sum_{x,u,v} \frac{\eta(u)\eta(v)}{|\eta|^2} \operatorname{cov}(\tilde{B}_x(u), \tilde{B}_x(v))$$
$$\quad -\varepsilon |\eta|^{1/2} \sum_x \left(\sum_v E\tilde{B}_x(v) \frac{\eta(v)}{|\eta|}\right)^2$$
$$\leq -\varepsilon |\eta|^{1/2} \sum_{u,v} \frac{\eta(u)\eta(v)}{|\eta|^2} \sum_x \operatorname{cov}(\tilde{B}_x(u), \tilde{B}_x(v)).$$

By the positive correlations assumption, each of the covariances appearing above is nonnegative. There is at least one v for which $\tilde{B}_x(v) \neq E\tilde{B}_x(v)$ (e.g., the v used in defining the σ-algebra $\mathscr{F}$), so that (4.10) implies that for some smaller positive ε,

(4.11)
$$\Omega f(\eta) \leq -\varepsilon |\eta|^{1/2} \sum_u \left[\frac{\eta(u)}{|\eta|}\right]^2.$$

The function f is not in $\mathscr{L}$. However by approximating it appropriately by functions in $\mathscr{L}$, Theorem 1.14 can be used to show that (4.11) implies

(4.12)
$$\frac{d}{dt} E|\eta_t|^{1/2} \leq -\varepsilon E \sum_u \left[\frac{\eta_t(u)}{|\eta_t|}\right]^2 |\eta_t|^{1/2}$$

for the process which initially satisfies $\eta_0(0) = 1$ and $\eta_0(x) = 0$ for $x \neq 0$. Next we need to estimate the right side of (4.12). To do so, use the Schwarz inequality and the fact that $\sqrt{\sigma} \geq \sigma$ for $0 \leq \sigma \leq 1$ to write

$$\sum_u \left[\frac{\eta(u)}{|\eta|}\right]^2 \geq |C_t|^{-1} \left[\sum_{u \in C_t} \frac{\eta(u)}{|\eta|}\right]^2$$
$$= |C_t|^{-1} \left[1 - \sum_{u \notin C_t} \frac{\eta(u)}{|\eta|}\right]^2$$
$$\geq |C_t|^{-1} \left[1 - 2 \sum_{u \notin C_t} \frac{\eta(u)}{|\eta|}\right]$$
$$\geq |C_t|^{-1} \left[1 - 2\left(\sum_{u \notin C_t} \frac{\eta(u)}{|\eta|}\right)^{1/2}\right].$$

Combining this with (4.12) yields

$$\frac{d}{dt} E|\eta_t|^{1/2} \le -\varepsilon |C_t|^{-1} \left[E|\eta_t|^{1/2} - 2E\left(\sum_{u \notin C_t} \eta_t(u) \right)^{1/2} \right].$$

By Jensen's inequality and (2.3),

$$(4.13) \qquad \frac{d}{dt} E|\eta_t|^{1/2} \le -\varepsilon |C_t|^{-1} \left[E|\eta_t|^{1/2} - 2\left(\sum_{u \notin C_t} \gamma_t(u, 0) \right)^{1/2} \right].$$

By part (b) of Theorem 2.2,

$$(4.14) \qquad\qquad\qquad \lim_{t \to \infty} E|\eta_t|^{1/2}$$

exists and is finite. Suppose that this limit is strictly positive. Using this assumption and (4.1), (4.13) implies that

$$\frac{d}{dt} E|\eta_t|^{1/2} \le -\varepsilon |C_t|^{-1}$$

for some new positive constant $\varepsilon > 0$. By (4.2), it then follows that the limit in (4.14) is infinite, which is a contradiction. Therefore the limit in (4.14) is zero. Thus we have proved (4.6) for this special initial configuration. The general result follows by linearity and translation invariance. □

In the proof of Theorem 4.5, the assumption that the random variables in (4.4) have positive correlations and are not all constant is used in two places. It is used to carry out the truncation which leads to (4.8), and it is used to deduce (4.11) from (4.10). The truncation is not necessary if

$$(4.15) \qquad\qquad\qquad E \sup_v B_x^2(v) < \infty,$$

and to deduce (4.11) from (4.10), it is enough to know that there is an $\varepsilon > 0$ so that

$$(4.16) \qquad\qquad \sum_x E\left(\sum_v B_x(v)\eta(v) \right)^2 \ge \varepsilon \sum_x \eta^2(x)$$

for all finite configurations η. Thus (4.15) and (4.16) can be used in place of the positive correlations and nonconstancy assumption in Theorem 4.5. One simple example which satisfies (4.15) and (4.16) but for which $B_x(v)$

is constant for all $x, v \in S$ is

$$A_x(u, v) = \begin{cases} 1+\varepsilon & \text{if } u = v = x, \\ 1 & \text{if } u = v \neq x, \\ 0 & \text{if } u \neq v, \end{cases}$$

$$a(u, v) = \begin{cases} -\varepsilon & \text{if } u = v, \\ 0 & \text{if } u \neq v, \end{cases}$$

for $\varepsilon > 0$. In this case, η_t can be written down explicitly as

$$\eta_t(x) = \eta(x)(1+\varepsilon)^{N_x(t)} e^{-\varepsilon t},$$

where $\{N_x(t), x \in S\}$ are independent rate one Poisson processes. In this case, it is easy to see directly that the finite and infinite systems die out.

5. Extinction in Higher Dimensions

Linear systems with a small amount of variability survive by Theorem 3.29, at least if the underlying mechanism is transient. On the other hand, Theorem 4.5 implies that systems with at least some variability die out if the underlying mechanism is recurrent. In this section, we will complete the picture by giving a general sufficient condition for a linear system to die out. In some examples, this condition can be interpreted at saying that the amount of variability in the mechanism is large. Assumption (2.1) should be recalled at this point.

Theorem 5.1. *Suppose that $a(x, y) = 0$ for all $x \neq y$ and that $A_x(0, v)$ satisfies*

(5.2) $\sum_{x, v} E A_x(0, v) |\log A_x(0, v)| < \infty, \quad \text{and}$

(5.3) $\sum_{x, v} E A_x(0, v) \log A_x(0, v) + a > 0,$

where $a = a(x, x)$. Then both the finite and infinite systems die out.

Proof. For a fixed $\sigma > 0$, define $f \in \mathcal{L}$ by

$$f(\eta) = \exp(-\sigma \eta(0)).$$

Using (0.2), compute

$$\Omega f(\eta) = \sum_x \left\{ E \exp\left(-\sigma \sum_v A_x(0, v)\eta(v) \right) - \exp(-\sigma\eta(0)) \right\}$$

(5.4)

$$- a\sigma\eta(0) \exp(-\sigma\eta(0))$$

$$\geq \sum_x E\left\{ 1 - \sum_v [1 - \exp(-\sigma A_x(0, v)\eta(v))] - \exp(-\sigma\eta(0)) \right\}$$

$$- a\sigma\eta(0) \exp(-\sigma\eta(0)).$$

Start the process with $\eta_0 \equiv 1$, and let

$$\Delta(t, \sigma) = \frac{1}{\sigma}[1 - E \exp(-\sigma\eta_t(0))].$$

By (5.4),

$$E\Omega f(\eta_t) \geq \sum_x \left\{ \sigma\Delta(t, \sigma) - \sum_v E\sigma A_x(0, v)\Delta(t, \sigma A_x(0, v)) \right\}$$

$$- a\sigma \frac{\partial}{\partial\sigma}[\sigma\Delta(t, \sigma)].$$

Therefore by Theorem 1.14,

$$\frac{\partial}{\partial t}\Delta(t, \sigma) = -\frac{1}{\sigma}E\Omega f(\eta_t)$$

$$\leq \sum_x \left\{ \sum_v EA_x(0, v)\Delta(t, \sigma A_x(0, v)) - \Delta(t, \sigma) \right\}$$

(5.5)

$$+ a\frac{\partial}{\partial\sigma}[\sigma\Delta(t, \sigma)]$$

$$= \sum_{x,v} EA_x(0, v)[\Delta(t, \sigma A_x(0, v)) - \Delta(t, \sigma)]$$

$$+ a\sigma\frac{\partial}{\partial\sigma}\Delta(t, \sigma),$$

where the last equality comes from (2.1). The next step is to interpret the right side of (5.5) as the result of applying the generator of a certain Markov process on $[0, \infty)$ to the function $\Delta(t, \cdot)$. To do so, let

$$\{R_x^n; n \geq 1, x \in S\}$$

be independent nonnegative random variables with distribution determined by

$$(5.6) \qquad Ef(R_x^n) = \frac{\sum\limits_{v} EA_x(0, v)f(A_x(0, v))}{\sum\limits_{v} EA_x(0, v)}$$

for any bounded continuous function f on $[0, \infty)$. Also let $\{N_x(t), x \in S\}$ be independent Poisson processes, where $N_x(t)$ has rate

$$\sum\limits_{v} EA_x(0, v).$$

By (5.2) and (5.6),

$$E \sum\limits_{x} \sum\limits_{n \leq N_x(t)} |\log R_x^n| < \infty,$$

so that the infinite product

$$W_t = e^{at} \prod\limits_{x} \prod\limits_{n \leq N_x(t)} R_x^n$$

converges with probability one for each t. The process W_t is Markov with generator

$$\Gamma f(\sigma) = \sum\limits_{x} \{ Ef(\sigma R_x^1) - f(\sigma) \} \sum\limits_{v} EA_x(0, v) + a\sigma \frac{d}{d\sigma} f(\sigma),$$

which by (5.6) can be rewritten as

$$\Gamma f(\sigma) = \sum\limits_{x,v} [EA_x(0, v)][f(\sigma A_x(0, v)) - f(\sigma)] + a\sigma \frac{d}{d\sigma} f(\sigma).$$

Comparing with (5.5) yields the inequality

$$(5.7) \qquad \Delta(t, \sigma) \leq E\Delta(0, \sigma W_t)$$

for any $\sigma > 0$ and $t \geq 0$. By (5.3) and the Strong Law of Large Numbers applied to $\log W_t$,

$$(5.8) \qquad \lim\limits_{t \to \infty} W_t = \infty$$

with probability one. Since

$$\Delta(0, \sigma) = \frac{1 - e^{-\sigma}}{\sigma},$$

(5.7) and (5.8) together imply that

$$\lim_{t \to \infty} \Delta(t, \sigma) = 0$$

for all $\sigma > 0$. Recalling the definition of $\Delta(t, \sigma)$, it follows from this that

$$\lim_{t \to \infty} E \exp(-\sigma \eta_t(0)) = 1$$

for all $\sigma > 0$, so that the infinite system dies out. The finite system then dies out as well by Theorem 2.4(c), since assumptions (5.2) and (5.3) hold for a given linear system if and only if they hold for its dual. □

6. Examples and Applications

In this section, we will specialize to certain simple classes of linear systems in order to better appreciate the content of the results proved in Sections 3, 4, and 5. In addition, we will carry out the comparisons with the contact process which were mentioned in Section 4 of Chapter VI.

Example 6.1. *Deterministic births with random deaths.* In this example, let

$$A_x(u, v) = \begin{cases} 1 & \text{if } u = v \neq x, \\ 0 & \text{otherwise,} \end{cases}$$

with probability one, and

$$a(u, v) = \begin{cases} \lambda p(u, v) & \text{for } u \neq v, \\ 1 - \lambda & \text{for } u = v, \end{cases}$$

where $\lambda > 0$ and $p(u, v)$ are the transition probabilities for an irreducible random walk on Z^d satisfying $p(u, u) = 0$. Thus the only randomness in the system is in the deaths (i.e. replacing coordinates of η_t by 0) which occur at each site at exponential times with parameter one. The dual of this process is obtained by replacing $p(u, v)$ by $p(v, u)$. The choice of $a(u, u) = 1 - \lambda$ is dictated by (2.1). This system satisfies (4.15) and (4.16) (since $B_x(v) = -1$ if $x = v$ and $B_x(v) = 0$ if $x \neq v$), so that by the remark following the proof of Theorem 4.5, the finite and infinite systems die out if $d = 1$ and $\sum_x |x| p(0, x) < \infty$ or if $d = 2$ and $\sum_x |x|^2 p(0, x) < \infty$. On the other hand, the $q(x, y)$ defined in (3.9) and (3.10) in this case becomes

$$q(x, y) = \lambda p(x, y) + \lambda p(y, x) \quad \text{if } x \neq y,$$

$$q(x, x) = -2\lambda \qquad\qquad\qquad \text{if } x \neq 0, \quad \text{and}$$

$$q(0, 0) = 1 - 2\lambda.$$

Thus equation (3.14) reads:

$$h(x) - \sum_y \frac{p(x, y) + p(y, x)}{2} h(y) = \begin{cases} \dfrac{h(0)}{2\lambda} & \text{if } x = 0, \\ 0 & \text{if } x \neq 0. \end{cases}$$

The solution of this equation which satisfies (3.13) is given by

$$h(x) = 1 + \frac{G(0, x)}{2\lambda - G(0, 0)} \quad \text{if } 2\lambda > G(0, 0),$$

where $G(x, y)$ is the Green function for the symmetrization $\frac{1}{2}[p(x, y) + p(y, x)]$. Therefore by Theorem 3.12 (and Theorem 2.4(c)) the finite and infinite systems survive for

(6.2) $\lambda > \frac{1}{2}G(0, 0).$

Example 6.3. *The normalized binary contact path process.* This example has $A_x(u, v)$ given by (0.4) and (0.5), $a(x, y) = 0$ for $x \neq y$, and

$$a(x, x) = \frac{1 - 2\lambda d}{1 + 2\lambda d}.$$

It is self-dual, so that the survival of the finite system is equivalent to the survival of the infinite system by Theorem 2.4(c). By Theorem 2.23, there is a critical value $\tilde{\lambda}_d \in [0, \infty]$ so that the process survives for $\lambda > \tilde{\lambda}_d$ and dies out for $\lambda < \tilde{\lambda}_d$. (See the discussion following the proof of that theorem.) In this example, $B_x(v) = 0$ for $x \neq v$, and

$$B_x(x) = \begin{cases} -1 & \text{with probability } \dfrac{1}{2\lambda d + 1}, \\ 1 & \text{with probability } \dfrac{2\lambda d}{2\lambda d + 1}, \end{cases}$$

while the $\gamma(x, y)$ defined in (1.28) are

$$\gamma(x, y) = \begin{cases} \dfrac{\lambda}{2\lambda d + 1} & \text{if } |x - y| = 1, \\ -\dfrac{2\lambda d}{2\lambda d + 1} & \text{if } x = y, \\ 0 & \text{otherwise.} \end{cases}$$

Therefore by Theorem 4.5, $\tilde{\lambda}_d = \infty$ for $d = 1$ and $d = 2$. By Theorem 5.1,

$$\tilde{\lambda}_d \geq \frac{1}{2d}$$

for all $d \geq 1$. To get an upper bound for $\tilde{\lambda}_d$ in case $d \geq 3$, use Theorem 3.12. In this example,

$$q(x, y) = \frac{2\lambda}{1 + 2\lambda d} \qquad \text{if } |x - y| = 1,$$

$$q(x, y) = 0 \qquad \text{if } |x - y| > 1,$$

$$q(x, x) = -\frac{4\lambda d}{1 + 2\lambda d} \qquad \text{for } x \neq 0, \quad \text{and}$$

$$q(0, 0) = 1 - \frac{4\lambda d}{1 + 2\lambda d}.$$

Therefore equation (3.14) becomes

$$h(x) - \frac{1}{2d} \sum_{|y - x| = 1} h(y) = \begin{cases} h(0) \dfrac{1 + 2\lambda d}{4\lambda d} & \text{if } x = 0, \\ 0 & \text{if } x \neq 0. \end{cases}$$

The solution of this equation which satisfies (3.13) is given by

$$h(x) = 1 + \frac{G(0, x)}{\dfrac{4\lambda d}{1 + 2\lambda d} - G(0, 0)} \qquad \text{if } \frac{4\lambda d}{1 + 2\lambda d} > G(0, 0),$$

where $G(x, y)$ is the Green function for the simple random walk on Z^d. Therefore by Theorem 3.12,

$$(6.4) \qquad\qquad \tilde{\lambda}_d \leq \frac{1}{2d} \frac{G(0, 0)}{2 - G(0, 0)}$$

for $d \geq 3$. If ρ_d is the probability that the simple random walk on Z^d never returns to its starting point, then $G(0, 0) = \rho_d^{-1}$. Therefore (6.4) can be rewritten as

$$(6.5) \qquad\qquad \tilde{\lambda}_d \leq [2d(2\rho_d - 1)]^{-1}.$$

Now let λ_d be the critical value for the d-dimensional contact process. Recalling the connection between the contact process and the binary contact path process which was described in the introductory section of this chapter, we see that $\lambda_d \leq \tilde{\lambda}_d$. Therefore (6.5) implies (4.6) of Chapter VI.

Example 6.6. *The smoothing process.* In this example, $S = Z^d$, $a(x, y) = 0$ for all $x, y \in S$, and $A_x(u, v)$ is given by (0.6), where $p(u, v) = p(0, v - u)$. Here the $q(x, y)$ defined in (3.9) and (3.10) take the form

$$q(x, y) = p(x, y) + p(y, x) \qquad \text{if } x \neq 0 \text{ and } x \neq y,$$

$$q(0, y) = EW^2 \sum_u p(0, u)p(0, u + y) \quad \text{if } y \neq 0,$$

$$q(x, x) = 2p(x, x) - 2 \qquad \text{if } x \neq 0, \quad \text{and}$$

$$q(0, 0) = EW^2 \sum_u [p(0, u)]^2 - 1.$$

Therefore equation (3.14) becomes

(6.7) $$\tfrac{1}{2} \sum_y [p(x, y) + p(y, x)]h(y) = h(x) \quad \text{for } x \neq 0, \quad \text{and}$$

(6.8) $$EW^2 \sum_{u,y} p(0, u)p(0, u + y)h(y) = h(0).$$

A solution of (6.7) satisfying (3.13) is of the form

$$h(x) = 1 + cG(0, x)$$

for some constant c, where $G(x, y)$ is the Green function for the symmetrization $\tfrac{1}{2}[p(x, y) + p(y, x)]$. To solve for c, use (6.8) to write

(6.9) $$EW^2 \left\{ 1 + c \sum_{u,v} p(0, u)p(0, v)G(u, v) \right\} = 1 + cG(0, 0).$$

This can be solved for $c > 0$ provided that

(6.10) $$EW^2 < \frac{G(0, 0)}{\sum_{u,v} p(0, u)p(0, v)G(u, v)}.$$

Therefore the infinite system survives by Theorem 3.12 if the symmetrized random walk is transient and (6.10) holds. Note that the right side of (6.10) is greater than one, so (6.10) is satisfied for W's with sufficiently small variance. If $W \equiv 1$, then of course the infinite system survives for any choice of $p(x, y)$, since we can take $h(x) \equiv 1$. Theorem 4.5 implies that the finite system dies out if $W \neq 1$ and either $d = 1$ and $\sum_x |x|p(0, x) < \infty$ or $d = 2$ and $\sum_x |x|^2 p(0, x) < \infty$. Finally, Theorem 5.1 implies that both the finite and infinite systems die out if

$$\infty > EW \log W > -\sum_x p(0, x) \log p(0, x) > -\infty.$$

The application of the smoothing process to the contact process proceeds as follows. Take $d \geq 3$, $p(x, y) = 1/2d$ if $|x - y| = 1$, and

$$W = \begin{cases} 0 & \text{with probability } 1 - \varepsilon, \\ 1/\varepsilon & \text{with probability } \varepsilon. \end{cases}$$

Then $EW^2 = \varepsilon^{-1}$, and since $p(x, y)$ is symmetric and $p(0, 0) = 0$,

$$\sum_{u,v} p(0, u)p(0, v)G(u, v) = G(0, 0) - 1.$$

Therefore (6.10) is satisfied if and only if

(6.11) $\varepsilon > 1 - \rho_d,$

where ρ_d is the probability that the simple random walk on Z^d never returns to its starting point. Coupling the smoothing and contact processes together, we see that if their parameters satisfy

$$\frac{1 - \varepsilon}{\varepsilon} \geq \frac{1}{\lambda},$$

then the survival of the smoothing process implies the survival of the contact process. Therefore the criterion (6.11) for survival of the smoothing process implies that the contact process critical value satisfies

$$\lambda_d \leq \frac{1}{\rho_d} - 1,$$

as asserted in (4.5) of Chapter VI.

Example 6.12. *The potlatch process.* In this example, $S = Z^d$, $a(x, y) = 0$ for all $x, y \in S$, and $A_x(u, v)$ is given by (0.7), where $p(u, v) = p(0, v - u)$. Here the $q(x, y)$ defined in (3.9) and (3.10) take the form

$$q(x, y) = p(x, y) + p(y, x) \qquad \text{if } y \neq 0 \text{ and } y \neq x,$$

$$q(x, 0) = EW^2 \sum_u p(u, 0)p(u, x) \quad \text{if } x \neq 0,$$

$$q(x, x) = 2p(x, x) - 2 \qquad \text{if } x \neq 0, \quad \text{and}$$

$$q(0, 0) = EW^2 \sum_u [p(u, 0)]^2 - 1.$$

Therefore equation (3.14) becomes

$$(6.13) \quad \tfrac{1}{2} \sum_{y \neq 0} [p(x, y) + p(y, x)]h(y) = h(x) - \tfrac{1}{2}h(0)EW^2 \sum_u p(u, 0)p(u, x)$$

for $x \neq 0$, and

$$(6.14) \quad h(0) = EW^2 \sum_u [p(u, 0)]^2 h(0) + \sum_{y \neq 0} [p(0, y) + p(y, 0)]h(y).$$

It is not difficult to solve these equations using Fourier transforms. The solution which satisfies (3.13) is given by

$$h(0) = \frac{2}{\displaystyle\int_\Gamma \psi(\gamma)\, d\gamma}, \quad \text{and}$$

$$h(x) = 1 - \frac{\displaystyle\int_\Gamma \psi(\gamma) \exp[-i\langle x, \gamma\rangle]\, d\gamma}{\displaystyle\int_\Gamma \psi(\gamma)\, d\gamma} \quad \text{for } x \neq 0,$$

where $\Gamma = [-\pi, \pi)^d$,

$$\psi(\gamma) = \frac{1}{(2\pi)^d} \frac{1 - EW^2|\phi(\gamma)|^2}{1 - \operatorname{Re} \phi(\gamma)}, \quad \text{and}$$

$$\phi(\gamma) = \sum_x p(0, x) \exp[i\langle x, \gamma\rangle].$$

For this solution to be meaningful, $\psi(\gamma)$ must be integrable on Γ. This is always true if $W \equiv 1$, since

$$0 \le \frac{1 - |\phi(\gamma)|^2}{1 - \operatorname{Re} \phi(\gamma)} \le 2.$$

If $W \neq 1$, $\psi(\gamma)$ is integrable if and only if the symmetrized random walk is transient. We further need the solution h to be positive. A necessary condition is that $\int \psi(\gamma)\, d\gamma > 0$, or equivalently that

$$(6.15) \qquad EW^2 < \frac{\displaystyle\int \frac{d\gamma}{1 - \operatorname{Re} \phi(\gamma)}}{\displaystyle\int \frac{|\phi|^2\, d\gamma}{1 - \operatorname{Re} \phi(\gamma)}}.$$

This is the same as (6.10). This condition turns out to be sufficient for the positivity of h as well, as can be seen by writing h explicitly in terms of the Green function $G(x, y)$ for the symmetrized random walk, and then using the inequality

$$G(0, x)G(x, x+y) \le G(0, 0)G(0, x+y).$$

Therefore by Theorem 3.12, the infinite potlatch process survives if $W \equiv 1$, or if the symmetrized random walk is transient and W satisfies (6.10). Theorem 4.5 again implies that the finite potlatch system dies out if $W \ne 1$ and either $d = 1$ and $\sum_x |x| p(0, x) < \infty$ or $d = 2$ and $\sum_x |x|^2 p(0, x) < \infty$. Finally, Theorem 5.1 implies that both the finite and infinite potlatch systems die out if

$$\infty > EW \log W > -\sum_x p(0, x) \log p(0, x) > -\infty.$$

Recalling part (c) of Theorem 2.4 and the fact that the potlatch and smoothing processes are dual to one another, we can summarize the observations in the previous two examples in the following way.

Theorem 6.16. (a) *If $W \equiv 1$, or if the symmetrized random walk is transient and*

$$EW^2 < \frac{G(0, 0)}{\sum_{u,v} p(0, u)p(0, v)G(u, v)},$$

then the finite and infinite smoothing and potlatch processes survive.
(b) *If $W \ne 1$ and if $d = 1$ and $\sum_x |x| p(0, x) < \infty$ or $d = 2$ and $\sum_x |x|^2 p(0, x) < \infty$, then the finite and infinite smoothing and potlatch processes die out.*
(c) *If*

$$\infty > EW \log W > -\sum_x p(0, x) \log p(0, x) > -\infty,$$

then the finite and infinite smoothing and potlatch processes die out.

Example 6.17. *Multiplication by random 2×2 matrices.* In this example, $S = Z^d$ and $a(x, y) = 0$ for all $x, y \in S$. Let $p(x, y)$ be the transition probabilities for an irreducible random walk on S such that $p(x, x) = 0$. For each $x \in S$, with probability $p(x, y)$, $A_x(\cdot, \cdot)$ is the random matrix given by $A_x(x, x) = H$, $A_x(x, y) = I$, $A_x(y, x) = J$, $A_x(y, y) = K$, $A_x(u, u) = 1$ if $u \ne x, y$, and $A_x(u, v) = 0$ otherwise, where H, I, J, K are nonnegative random variables. In order for (2.1) to be satisfied, it is necessary to assume that

(6.18) $E(H + I + J + K) = 2.$

In order to simplify the computations somewhat, we will also assume that

$$EIJ = E(H-1)(K-1) = 0.$$

The $q(x, y)$ defined in (3.9) and (3.10) then become

$$q(0, 0) = E(H^2 + I^2 + J^2 + K^2) - 2,$$

$$q(x, x) = 2E(H + K) - 4 \qquad\qquad\qquad \text{if } x \neq 0,$$

$$q(x, 0) = [p(0, x) + p(x, 0)]E(KI + HJ) \quad \text{if } x \neq 0,$$

$$q(0, y) = [p(0, y) + p(y, 0)]E(KJ + HI) \quad \text{if } y \neq 0, \quad \text{and}$$

$$q(x, y) = [p(x, y) + p(y, x)]E(I + J) \qquad \text{otherwise.}$$

Thus (3.14) takes the form

$$h(x) - \tfrac{1}{2}\sum_y [p(x, y) + p(y, x)]h(y)$$

(6.19)
$$= \begin{cases} h(0)\dfrac{E(K+J)^2 + E(H+I)^2 - 2}{2E(KJ + HI)} & \text{if } x = 0, \\[4mm] h(0)\left[\dfrac{p(x, 0) + p(0, x)}{2}\right]\dfrac{E(KI + HJ - I - J)}{E(I + J)} & \text{if } x \neq 0. \end{cases}$$

Solving as in the previous examples, we find that there is a solution to this equation provided that

$$\frac{E(I+J)[2 - E(H^2 + I^2 + J^2 + K^2)]}{2E(KI + HJ)E(KJ + HI)} > \frac{G(0, 0) - 1}{G(0, 0)}$$

where $G(x, y)$ is the Green function for the symmetrized random walk. Therefore the infinite system survives under this condition by Theorem 3.12. In the particular case in which

(6.20)
$$H + J = K + I = 1$$

and $IJ \equiv 0$, (6.19) has the simple solution

$$h(0) = \frac{E(I + J)}{E(IK + JH)},$$

$$h(x) = 1 \quad \text{for } x \neq 0,$$

and therefore the infinite system always survives (provided $E(IK + JH) > 0$).

Note that by Theorem 3.17, the coordinates of η are uncorrelated under the invariant measure ν_β in this case. As an exercise, the reader may wish to solve (3.14) under (6.20) without assuming $IJ \equiv 0$.

7. Notes and References

Section 1. Many processes with values in $[0, \infty)^S$ or $\{0, 1, \ldots\}^S$ have been studied. Independent particle systems were considered in Section 5 of Chapter VIII of Doob (1953) in connection with Poisson processes. Among the many papers which treat various aspects of the theory of independent particle systems are Dobrushin (1956), Thedeen (1967), Stone (1968), and Liggett (1978b). The zero range process was introduced by Spitzer (1970) and studied by Liggett (1973b), Waymire (1980), Andjel (1982), Andjel and Kipnis (1984), and Brox and Rost (1984). In this process, a speed change interaction is superimposed on independent motions in such a way that the rate at which a particle at x moves depends only on the number of particles which are at x at that time. One interesting feature of the zero range process is (as Spitzer discovered) that in spite of the interaction, there are product measures on $\{0, 1, \ldots\}^S$ which are invariant for the system. Similar processes, in which more general speed change interactions are allowed, have been studied by Cocozza (1985). A class of processes on $(-\infty, \infty)^S$ in which coordinates are periodically replaced by a normally distributed random variable with variance one and a mean depending linearly on the configuration was introduced and studied by Hsiao (1982).

The linear processes which are the subject of this chapter are generalizations of the smoothing, potlatch, and binary contact path processes. The smoothing and potlatch processes (with $W \equiv 1$) were introduced, along with several related processes, by Spitzer (1981). He discovered their duality, and raised a number of problems which led to the ergodic theory for these systems which was developed by Liggett and Spitzer (1981). The versions of these processes with $W \not\equiv 1$ were introduced and studied by Holley and Liggett (1981). The binary contact path process was analyzed by Griffeath (1983). A special case of the potlatch process was studied by Roussignol (1980).

Basis (1976) gave a construction for large classes of Markov processes on product spaces which need not be compact. The construction given here is based on Liggett and Spitzer (1981). The duality of the smoothing and potlatch processes was observed by Spitzer (1981).

Kesten and Spitzer (1985) have studied problems involving the convergence in distribution of the product $A_1 A_2 \cdots A_n$ of independent and identically distributed random $d \times d$ matrices A_i with nonnegative entries. This product is a discrete time analogue of a linear system, where now S is taken to be a finite set of cardinality d. Under some mild regularity assumptions, they give necessary and sufficient conditions for the product not to converge to the 0 matrix in probability (i.e., to survive, in our sense). Their results are much more complete than those given in this chapter, as might be expected in view of the fact that their S is finite.

Section 2. This section is based on Holley and Liggett (1981) and Liggett and Spitzer (1981). The martingale property in Theorem 2.2(b) was first observed and exploited by Spitzer (1981).

Section 3. This section is a generalization of the treatment in Liggett and Spitzer (1981).

Section 4. The proof of Theorem 4.5 is a generalization of the proofs given by Holley and Liggett (1981) and Griffeath (1983) for the potlatch, and binary contact path processes respectively. When specialized to the smoothing process, the proof is substantially simpler than the one given for it in the first of these papers. Analogous results in different contexts have been proved by Sawyer (1976a), Kallenberg (1977), Dawson (1977), Matthes, Kerstan, and Mecke (1978), and Durrett (1979).

Section 5. Theorem 5.1 is a generalization of Theorem 5.4 of Holley and Liggett (1981).

Section 6. Example 6.3 comes from Griffeath (1983). Examples 6.6 and 6.12 come from Liggett and Spitzer (1981) in case $W \equiv 1$ and from Holley and Liggett (1981) in case $W \neq 1$. Example 6.1 is similar to the infinite-dimensional renewal process which was studied by Spitzer (198). Example 6.17 is based on an unpublished computation due to Spitzer.

8. Open Problems

1. In the context of Theorem 2.7, is it possible for $\mathscr{I} \cap \mathscr{S}$ to contain elements of infinite mean? In very simple situations, this problem can be resolved. For example, suppose $a(u, v) \equiv 0$, $A_x(u, v) = 0$ if $u \neq v$, $A_x(u, u) = 1$ if $u \neq x$, and $A_x(x, x)$ is some random variable of mean 1. Then the infinite system dies out if and only if $A_x(x, x) \neq 1$. If it dies out, then $\eta_t(x) \to 0$ in probability for any initial η. This example is tractable because different coordinates do not interact. Durrett and Liggett (1983) considered a simplified version of the smoothing process, and showed for that model that it is possible to have "invariant measures" of infinite mean. In formulating this problem, it is perhaps more natural to define the process for a larger class of configurations than X, since translation invariant measures with infinite mean do not necessarily concentrate on X. To do so, simply extend the definition of the process from η to $[0, \infty)^S$ by linearity. The difficulty is that all that can be said is that η_t concentrates on $[0, \infty]^S$—it may well take on infinite values. This problem can still be formulated in this context, however, since an invariant measure will automatically concentrate on

$$\{\eta: \eta_t(x) < \infty \text{ for all } x \in S\}.$$

2. This is a version of Problem 1 for deterministic linear systems. Let S be general, and define the evolution η_t via the system of linear differential equations

$$\frac{d}{dt} \eta_t(x) = \sum_y a(x, y)\eta_t(y),$$

where $a(x, y) \geq 0$ for $x \neq y$. Suppose that this evolution has an invariant measure μ on $[0, \infty)^S$ other that δ_0. Does it follow that there is a deterministic $\zeta \in [0, \infty)^S$ which is not identically zero and satisfies

(8.1) $\sum_y a(x, y)\zeta(y) = 0?$

This question is of interest only if μ has infinite mean, since otherwise $\zeta(x) = \int \eta(x) \, d\mu$ satisfies (8.1). This problem is a continuous time analogue of Conjecture 1.6 of Liggett (1978b).

3. The multitype voter model (0.3) is an example in which the finite system dies out, but the infinite system survives. (To see that the finite system dies out, note that starting from η in which $\eta(0) = 1$ and $\eta(x) = 0$ for $x \neq 0$, $\sum_x \eta_t(x)$ is a time change of a simple random walk on $\{0, 1, \ldots\}$ with absorbing barrier at 0.) By Theorem 2.4(c), its dual (0.8) is an example in whch the finite system survives, but the infinite system dies out. Find sufficient conditions for the following to be correct: the infinite system survives if and only if the finite system survives. By self-duality, this statement is true for the binary contact path process given by (0.4) and (0.5). Is it true for the smoothing and potlatch processes given by (0.6) and (0.7) respectively?

4. In the context of Corollary 2.15, prove that ν_β is extremal in $\mathscr{S}$.

5. Under the assumptions of Theorem 2.19, show that

$$\lim_{t \to \infty} \mu S(t) = \nu_\beta$$

whenever $\mu \in \mathscr{S}$ is ergodic and satisfies $\int \eta(x) \, d\mu = \beta$. This is proved under stronger assumptions in Theorems 3.17. If the limit exists, Theorem 2.19 implies that the limit is a mixture of ν_β's. The main difficulty is in showing that this mixture must be trivial. Problem 4 of Chapter VIII is analogous to this problem.

6. Prove analogues of Theorems 2.19 and 3.17 when the context is not translation invariant. This has been done by Andjel (1985) in the case of the smoothing process given by (0.6) with $W \equiv 1$. In this case, the extremal invariant measures with finite mean turn out to be the pointmasses on the nonnegative harmonic functions for $p(x, y)$.

7. In Theorem 4.5, replace the assumption involving the sets C_t by the assumption that the symmetrization of the random walk with transition probabilities $\gamma_t(x, y)$ is recurrent. In view of the remarks at the beginning of Section 4, this would appear to be the natural assumption.

8. In Example 6.1, determine for each choice of $p(x, y)$ whose symmetrization is transient, the set of $\lambda > 0$ so that the process survives. The one result which is available is that survival occurs whenever (6.2) is satisfied. Does the system die out for small positive λ? This problem is open even when $p(x, y)$ corresponds to the simple symmetric random walk on Z^d for $d \geq 3$.

9. Consider the smoothing and potlatch processes of Examples 6.6 and 6.12. In both cases, (6.10) is a sufficient condition for $\nu_\beta \neq \delta_0$ and $\int \eta^2(x) \, d\nu_\beta < \infty$. Are there examples in which $\nu_\beta \neq \delta_0$ but $\int \eta^2(x) \, d\nu_\beta = \infty$? If Z^d is replaced by a tree, such examples can be constructed by using the results of Section 7 of Holley and Liggett (1981) or of Durrett and Liggett (1983).

Bibliography

Abraham, D. B. and Martin-Löf, A.
1973 The transfer matrix for a pure phase in the two-dimensional Ising model. *Comm. Math. Phys.*, **32**, 245–268.

Aizenman, M.
1980 Translation invariance and instability of phase coexistence in the two-dimensional Ising system. *Comm. Math. Phys.*, **73**, 83–94.

Aldous, D. J.
1983 Tail behavior of birth and death and stochastically monotone processes. *Z. Wahrsch. Verw. Gebiete*, **62**, 375–394.

Amati, D., Le Belloc, M., Marchesini, G., and Ciafaloni, M.
1976 Reggeon field theory for $\alpha(0) > 1$. *Nuclear Phys.* B, **112**, 107–149.

Andjel, E. D.
1981 The asymmetric simple exclusion process on Z^d. *Z. Wahrsch. Verw. Gebiete*, **58**, 423–432.

1982 Invariant measures for the zero-range process. *Ann. Probab.*, **10**, 525–547.

1985 Invariant measures and long-time behavior of the smoothing process. *Ann. Probab.*, **13**.

Andjel, E. D. and Kipnis, C.
1984 Derivation of the hydrodynamical equation for the zero-range interaction process. *Ann. Probab.*, **12**, 325–334.

Arratia, R.
1981 Limiting point processes for rescalings of coalescing and annihilating random walks on Z^d. *Ann. Probab.*, **9**, 909–936.

1983a The motion of a tagged particle in the simple symmetric exclusion system on Z. *Ann. Probab.*, **11**, 362–373.

1983b Site recurrence for annihilating random walks on Z^d. *Ann. Probab.*, **11**, 706–713.

1985 Symmetric exclusion processes: a comparison inequality and a large deviation result. *Ann. Probab.*, **13**.

Athreya, K., McDonald, D., and Ney, P.
1978 Coupling and the renewal theorem. *Amer. Math. Monthly*, **85**, 809–814.

Athreya, K. and Ney, P.
1972 *Branching Processes*. Springer-Verlag, New York.

Barner, A.
1983 Globale symmetrie von stochastischen teilchenbewegungen mit lokal symmetrischer interaktion. Doctoral dissertation, Zürich.

Basis, V. Ya.
1976 Infinite-dimensional Markov processes with almost local interaction of components. *Theory Probab. Appl.*, **21**, 706–720.

1980 On stationarity and ergodicity of Markov interacting processes. *Adv. Probab.*, **6**, 37–58.

Batty, C. J. K.
1976   An extension of an inequality of R. Holley. *Quart. J. Math. Oxford*, Ser. 2, **27**, 457-461.

Batty, C. J. K. and Bollmann, H. W.
1980   Generalized Holley-Preston inequalities on measure spaces and their products. *Z. Wahrsch. Verw. Gebiete*, **53**, 157-173.

Benettin, G., Gallavotti, G., Jona-Lasinio, G., and Stella, A. L.
1973   On the Onsager-Yang value of the spontaneous magnetization. *Comm. Math. Phys.*, **30**, 45-54.

Bertein, F. and Galves, A.
1977a Comportement asymptotique de deux marches aleatoires sur Z qui interagissent par exclusion. *C.R. Acad. Sci. Paris* A, **285**, 681-683.

1977b Une classe de systemes de particules stable par association. *Z. Wahrsch. Verw. Gebiete*, **41**, 73-85.

Biggins, J. D.
1978   The asymptotic shape of the branching random walk. *Adv. Appl. Probab.*, **10**, 62-84.

Billingsley, P.
1965   *Ergodic Theory and Information.* Wiley, New York.

1968   *Convergence of Probability Measures.* Wiley, New York.

1979   *Probability and Measure.* Wiley, New York.

Blumenthal, R. M. and Getoor, R. K.
1968   *Markov Processes and Potential Theory.* Academic Press, New York.

Bolthausen, E.
1982   On the central limit theorem for stationary mixing random fields. *Ann. Probab.*, **10**, 1047-1050.

Bortz, A. B., Kalos, M. H., Lebowitz, J. L., and Marro, J.
1975   Time evolution of a quenched binary alloy, II. Computer simulation of a three-dimensional model system. *Phys. Rev.* B, **12**, 2000-2011.

Bortz, A. B., Kalos, M. H., Lebowitz, J. L., and Zendejas, M. A.
1974   Time evolution of a quenched binary alloy. Computer simulation of a two-dimensional model system. *Phys. Rev.* B, **10**, 535-541.

Bougerol, P. and Kipnis, C.
1980   Position d'une marche aléatoire sur Z^2 à l'instant d'atteinte d'une droite. *Z. Wahrsch. Verw. Gebiete*, **52**, 183-191.

Bramson, M. and Gray, L.
1981   A note on the survival of the long-range contact process. *Ann. Probab.*, **9**, 885-890.

1985   The survival of branching annihilating random walk. *Z. Wahrsch. Verw. Gebiete*.

Bramson, M. and Griffeath, D.
1979a A note on the extinction rates of some birth-death processes. *J. Appl. Probab.*, **16**, 897-902.

1979b Renormalizing the 3-dimensional voter model. *Ann. Probab.*, **7**, 418-432.

1980a Asymptotics for interacting particle systems on Z^d. *Z. Wahrsch. Verw. Gebiete*, **53**, 183-196.

1980b Clustering and dispersion rates for some interacting particle systems on Z^1. *Ann. Probab.*, **8**, 183-213.

1980c On the Williams-Bjerknes tumor growth model, II. *Math. Proc. Cambridge Philos. Soc.*, **88**, 339-357.

1980d *The Asymptotic Behavior of a Probabilistic Model for Tumor Growth.* Springer Lecture Notes in Biomathematics, Vol. 38, pp. 165-172.

1981   On the Williams-Bjerknes tumor growth model, I. *Ann. Probab.*, **9**, 173-185.

Breiman, L.
1968   *Probability.* Addison-Wesley, Reading, Massachusetts.

Broadbent, S. and Hammersley, J.
1957   Percolation processes, I. Crystals and mazes. *Math. Proc. Cambridge Philos. Soc.*, **53**, 629–645.

Brower, R. C., Furman, M. A., and Moshe, M.
1978   Critical exponents for the reggeon quantum spin model. *Phys. Lett.* B, **76**, 213–219.

Brower, R. C., Furman, M. A., and Subbarao, K.
1977   Quantum spin model for reggeon field theory. *Phys. Rev.* D, **15**, 1756–1771.

Brox, T. and Rost, H.
1984   Equilibrium fluctuations of stochastic particle systems: the role of conserved quantities. *Ann. Probab.*, **12**, 742–759.

Choquet, G. and Deny, J.
1960   Sur l'equation de convolution $\mu = \mu * \sigma$. *C.R. Acad. Sci. Paris*, **250**. 799–801.

Chover, J.
1975   Convergence of a local lattice process. *Stochastic Process. Appl.*, **3**, 115–135.

Chung, K. L.
1974   *A Course in Probability Theory*, 2nd edition. Academic Press, New York.

Clifford, P. and Sudbury, A.
1973   A model for spatial conflict. *Biometrika*, **60**, 581–588.

1985   A sample path proof of the duality for stochastically monotone Markov processes. *Ann. Probab.*, **13**.

Cocozza, C.
1985   Processus des misanthropes.

Cocozza, C., Galves, A., and Roussignol, M.
1979   Étude de deux évolutions Markoviennes de processus ponctuels sur R par des méthodes d'association. *Ann. Inst. H. Poincaré*, Sect. B, **15**, 235–259.

Cocozza, C. and Kipnis, C.
1977   Existence de processus Markoviens pour des systemes infinis de particules. *Ann. Inst. H. Poincaré*, Sect. B, **13**, 239–257.

1980   Processus de vie et de mort sur R avec interaction selon les particules les plus proches. *Z. Wahrsch. Verw. Gebiete*, **51**, 123–132.

Cocozza, C., Kipnis, C., and Roussignol, M.
1983   Stabilité de la récurrence nulle pour certaines chaines de Markov perturbées. *J. Appl. Probab.*, **20**, 482–504.

Cocozza, C. and Roussignol, M.
1979   Unicité d'un processus de naissance et de mort sur la droite réelle. *Ann. Inst. H. Poincaré*, Sect. B, **15**, 93–106.

1980   Théorèmes ergodiques pour un processus de naissance et mort sur la droite reele. *Ann. Inst. H. Poincaré*, Sect. B, **16**, 75–85.

1983   Stabilité de la récurrence d'une chaîne de Markov sous l'effet d'une perturbation. *Stochastics*, **9**, 125–137.

Cox, J. T.
1984   An alternate proof of a correlation inequality of Harris. *Ann. Probab.*, **12**, 272–273.

Cox, J. T. and Durrett, R.
1981   Some limit theorems for percolation processes with necessary and sufficient conditions. *Ann. Probab.*, **9**, 583–603.

Cox, J. T. and Griffeath, D.
1983   Occupation time limit theorems for the voter model. *Ann. Probab.*, **11**, 876–893.

1984 Large deviations for Poisson systems of independent random walks. *Z. Wahrsch. Verw. Gebiete.*, **66**, 543-558.

1985 Occupation times for critical branching Brownian motions. *Ann. Probab.*, **13**.

Cox, J. T. and Rösler, U.
1983 A duality relation for entrance and exit laws for Markov processes. *Stochastic Process. Appl.*, **16**, 141-156.

Daley, D. J.
1968 Stochastically monotone Markov chains. *Z. Wahrsch. Verw. Gebiete*, **10**, 305-317.

Dawson, D. A.
1974 Information flow in discrete Markov systems. *J. Appl. Probab.*, **11**, 594-600.

1975 Synchronous and asynchronous reversible Markov systems. *Canad. Math. Bull.*, **17**, 633-649.

1977 The critical measure diffusion process. *Z. Wahrsch. Verw. Gebiete*, **40**, 125-145.

Dawson, D. A. and Hochberg, K. J.
1979 The carrying dimension of a stochastic measure diffusion. *Ann. Probab.*, **7**, 693-703.

1982 Wandering random measures in the Fleming-Viot model. *Ann. Probab.*, **10**, 554-580.

Dawson, D. A. and Ivanoff, G.
1978 Branching diffusions and random measures. *Adv. Probab.*, **5**, 61-103.

De Masi, A., Ferrari, P., Ianiro, N., and Presutti, E.
1982 Small deviations from local equilibrium for a process which exhibits hydrodynamical behavior, II. *J. Statist. Phys.*, **29**, 81-93.

De Masi, A., Ianiro, N., Pellegrinotti, A., and Presutti, E.
1984 A survey of the hydrodynamical behavior of many particle systems. In *Nonequilibrium Phenomena, II: From Stochastics to Hydrodynamics*, edited by J. L. Lebowitz and E. W. Montroll. Studies in Statistical Mechanics, Vol. 11. North-Holland, Amsterdam, pp. 123-294.

De Masi, A., Ianiro, N., and Presutti, E.
1982 Small deviations from local equilibrium for a process which exhibits hydrodynamical behavior, I. *J. Statist. Phys.*, **29**, 57-79.

De Masi, A. and Presutti, E.
1983 Probability estimates for symmetric simple exclusion random walks. *Ann. Inst. H. Poincaré*, Sect. B, **19**, 71-85.

De Masi, A., Presutti, E., Spohn, H., and Wick, D.
1985 Asymptotic equivalence of fluctuation fields for reversible exclusion processes with speed change. *Ann. Probab.*, **13**.

Derriennic, Y.
1983 Un theoreme ergodique presque sous-additif. *Ann. Probab.*, **11**, 669-677.

Ding Wanding and Chen Mufa
1981 Quasi-reversibility for the nearest neighbor speed functions. *Chinese Ann. Math.*, **2**, 47-59.

Dobrushin, R. L.
1956 On Poisson laws for distributions of particles in space. *Ukrain. Math. Z.*, **8**, 127-134.

1965 Existence of a phase transition in two- and three-dimensional Ising models. *Theory Probab. Appl.*, **10**, 193-213.

1968a Gibbsian random fields for lattice systems with pairwise interactions. *Functional Anal. Appl.*, **2**, 292-301.

1968b The problem of uniqueness of a Gibbsian random field and the problem of phase transition. *Functional Anal. Appl.*, **2**, 302-312.

1969 Gibbsian random fields. The general case. *Functional Anal. Appl.*, **3**, 22-28.

1971a Markov processes with a large number of locally interacting components: existence of a limit process and its ergodicity. *Problems Inform. Transmission*, **7**, 149-164.

1971b   Markov processes with many locally interacting components—the reversible case and some generalizations. *Problems Inform. Transmission*, **7**, 235-241.

1973   Analyticity of correlation functions in one-dimensional classical systems with slowly decreasing potentials. *Comm. Math. Phys.*, **32**, 269-289.

Dobrushin, R. L., Pjatetskii-Shapiro, I. I., and Vasilyev, N. B.
1969   Markov processes in an infinite product of discrete spaces. *Proceedings of the Soviet-Japanese Symposium in Probability Theory*, Khabarovsk, U.S.S.R.

Dobrushin, R. L. and Tirozzi, B.
1977   The central limit theorem and the problem of equivalence of ensembles. *Comm. Math. Phys.*, **54**, 173-192.

Doeblin, W.
1938   Exposé de la théorie des chaines simples constantes de Markov à un nombre fini d'etats. *Rev. Math. Union Interbalkanique*, **2**, 77-105.

Donnelly, P. and Welsh, D.
1983   Finite particle systems and infection models. *Math. Proc. Cambridge Philos. Soc.*, **94**, 167-182.

Doob, J. L.
1953   *Stochastic Processes.* Wiley, New York.

Doyle, P. and Snell, J. L.
1985   *Random Walks and Electrical Networks.* Mathematical Association of America (Carus Monograph).

Dunford, N. and Schwartz, J. T.
1958   *Linear Operators, Part I: General Theory.* Interscience, New York.

Durrett, R.
1979   An infinite particle system with additive interactions. *Adv. Appl. Probab.*, **11**, 355-383.

1980   On the growth of one-dimensional contact processes. *Ann. Probab.*, **8**, 890-907.

1981   An introduction to infinite particle systems. *Stochastic Process. Appl.*, **11**, 109-150.

1984   Oriented percolation in two dimensions. *Ann. Probab.*, **12**.

1985   Some general results concerning the critical exponents of percolation processes *Z. Wahrsch. Verw. Gebiete.*

Durrett, R. and Griffeath, D.
1982   Contact processes in several dimensions. *Z. Wahrsch. Verw. Gebiete*, **59**, 535-552.

1983   Supercritical contact processes on *Z. Ann. Probab.*, **11**, 1-15.

Durrett, R. and Liggett, T. M.
1981   The shape of the limit set in Richardson's growth model. *Ann. Probab.*, **9**, 186-193.

1983   Fixed points of the smoothing transformation. *Z. Wahrsch. Verw. Gebiete*, **64**, 275-301.

Dvoretsky, A. and Erdös, P.
1951   Some problems on random walk in space. *Proceedings of the Second Berkeley Symposium on Mathematical Statistics and Probability*, pp. 353-367.

Dynkin, E. B.
1965   *Markov Processes, I.* Academic Press, New York.

1984   Gaussian and non-Gaussian random fields associated with Markov processes. *J. Funct. Anal.*, **55**, 344-376.

Dyson, F. J.
1969a   Existence of phase transition in a one-dimensional ferromagnet. *Comm. Math. Phys.*, **12**, 91-107.

1969b   Nonexistence of spontaneous magnetization in a one-dimensional Ising ferromagnet. *Comm. Math. Phys.*, **12**, 212-215.

Echeverría, P.
1982 A criterion for invariant measures of Markov processes. *Z. Wahrsch. Verw. Gebiete*, **61**, 1-16.

Edwards, D. A.
1978a On the existence of probability measures with given marginals. *Ann. Inst. Fourier*, **28**, 53-78.

1978b On the Holley-Preston inequalities. *Proc. Edinburgh Math. Soc.* A, **78**, 265-272.

Erdös, P. and Ney, P.
1974 Some problems on random intervals and annihilating particles. *Ann. Probab.*, **2**, 828-839.

Ethier, S. N. and Kurtz, T. G.
1985 *Markov processes: characterization and convergence*. Wiley, New York.

Faris, W. G.
1979 The stochastic Heisenberg model. *J. Funct. Anal.*, **32**, 342-352.

Feller, W.
1968 *An Introduction to Probability Theory and Its Applications*, Vol. I, 3rd edition. Wiley, New York.

1971 *An Introduction to Probability Theory and Its Applications*, Vol. II, 2nd edition. Wiley, New York.

Fichtner, K. H. and Freudenberg, W.
1980 Asymptotic behaviour of time evolutions of infinite particle systems. *Z. Wahrsch. Verw. Gebiete*, **54**, 141-159.

Filinov, Ju. P.
1980 Stability of the properties of transience and recurrence of Markov chains. *Theory Probab. Math. Statist.*, **19**, 155-167.

Fleischman, J.
1978 Limit theorems for critical branching random fields. *Trans. Amer. Math. Soc.*, **239**, 353-389.

Fleischmann, K., Liemant, A., and Matthes, K.
1982 Critical branching processes with general phase space, VI. *Math. Nachr.*, **105**, 307-316.

Föllmer, H.
1980 *Local Interactions with a Global Signal: A Voter Model*. Springer Lecture Notes in Biomathematics, Vol. 38, pp. 141-144.

1982 A covariance estimate for Gibbs measures. *J. Funct. Anal.*, **46**, 387-395.

Fortuin, C. M., Kasteleyn, P. W., and Ginibre, J.
1971 Correlation inequalities on some partially ordered sets. *Comm. Math. Phys.*, **22**, 89-103.

Freedman, D.
1971 *Markov Chains*. Holden-Day, San Francisco.

Fritz, J.
1982 Stationary measures of stochastic gradient systems, infinite lattice models. *Z. Wahrsch. Verw. Gebiete*, **59**, 479-490.

Fröhlich, J., Israel, R., Lieb, E., and Simon, B.
1978 Phase transitions and reflection positivity, I. General theory and long-range lattice models. *Comm. Math. Phys.*, **62**, 1-34.

Fröhlich, J. and Spencer, T.
1982 The phase transition in the one-dimensional Ising model with $1/r^2$ interaction energy. *Comm. Math. Phys.*, **84**, 87-101.

Frykman, J. E.
1975 Approximation of infinite dynamic gas models. Ph.D. Thesis, University of Illinois.

Fukushima, M. and Stroock, D.
198 Reversible measures and Markov processes. *Adv. Math.* To appear.

Gacs, P.
1985 Reliable computation with cellular automata.

Galves, A., Kipnis, C., Marchioro, C., and Presutti, E.
1981 Nonequilibrium measures which exhibit a temperature gradient: study of a model. *Comm. Math. Phys.*, **81**, 127–147.

Georgii, H. O.
1979 *Canonical Gibbs Measures.* Springer Lecture Notes in Mathematics, Vol. 760.

Gihman, I. I. and Skorohod, A. V.
1975 *The Theory of Stochastic Processes, II.* Springer-Verlag, New York.

Ginibre, J.
1969 Simple proof and generalization of Griffiths' second inequality. *Phys. Rev. Lett.*, **23**, 828–830.

1970 General formulation of Griffiths' inequalities. *Comm. Math. Phys.*, **16**, 310–328.

Glauber, R. J.
1963 Time-dependent statistics of the Ising model. *J. Math. Phys.*, **4**, 294–307.

Glötzl, E.
1981 Time reversible and Gibbsian point processes, I. Markovian spatial birth and death processes on a general phase space. *Math. Nachr.*, **102**, 217–222.

1982 Time reversible and Gibbsian point processes, II. Markovian particle jump processes on a general phase space. *Math. Nachr.*, **106**, 63–71.

Grassberger, P. and de la Torre, A.
1979 Reggeon field theory (Schögl's first model) on a lattice: Monte Carlo calculations of critical behaviour. *Ann. Phys.*, **122**, 373–396.

Gray, L.
1978 Controlled spin-flip systems. *Ann. Probab.*, **6**, 953–974.

1980 Translation invariant spin flip processes and nonuniqueness. *Z. Wahrsch. Verw. Gebiete*, **51**, 171–184.

1982 The positive rates problem for attractive nearest-neighbor spin systems on Z. *Z. Wahrsch. Verw. Gebiete*, **61**, 389–404.

1985 Duality for general attractive spin systems, with applications in one dimension. *Ann. Probab.*, **13**.

Gray, L. and Griffeath, D.
1976 On the uniqueness of certain interacting particle systems. *Z. Wahrsch. Verw. Gebiete*, **35**, 75–86.

1977 On the uniqueness and nonuniqueness of proximity processes. *Ann. Probab.*, **5**, 678–692.

1982 A stability criterion for attractive nearest-neighbor spin systems on Z. *Ann. Probab.*, **10**, 67–85.

Greven, A.
1984 The hydrodynamical behavior of the coupled branching process. *Ann. Probab.*, **12**, 760–767.

Griffeath, D.
1975 Ergodic theorems for graph interactions. *Adv. Appl. Probab.*, **7**, 179–194.

1977 An ergodic theorem for a class of spin systems. *Ann. Inst. H. Poincaré*, Sect. B, **13**, 141–157.

1978a Annihilating and coalescing random walks on Z^d. *Z. Wahrsch. Verw. Gebiete*, **46**, 55–65.

1978b *Coupling Methods for Markov Processes.* Studies in Probability and Ergodic Theory; Advances in Mathematics, Supplementary Studies, Vol. 2, pp. 1–43. Academic Press, New York.

1978c Limit theorems for nonergodic set-valued Markov processes. *Ann. Probab.*, **6**, 379–387.

1979a *Additive and Cancellative Interacting Particle Systems.* Springer Lecture Notes in Mathematics, Vol. 724.

1979b Pointwise ergodicity of the basic contact process. *Ann. Probab.*, **7**, 139–143.

1981   The basic contact process. *Stochastic Process. Appl.*, **11**, 151–186.

1983   The binary contact path process. *Ann. Probab.*, **11**, 692–705.

Griffeath, D. and Liggett, T. M.
1982   Critical phenomena for Spitzer's reversible nearest-particle systems. *Ann. Probab.*, **10**, 881–895.

Griffiths, R. B.
1964   Peierls proof of spontaneous magnetization in the two-dimensional Ising model. *Phys. Rev.* A, **136**, 437–439.

1967   Correlations in Ising ferromagnets. *J. Math. Phys.* **8**, 478–489.

1972   Rigorous results and theorems. In *Phase Transitions and Critical Phenomena*, Vol. 1, edited by C. Domb and M. S. Green, pp. 7–109. Academic Press, London.

Gross, L.
1979   Decay of correlations in classical lattice models at high temperature. *Comm. Math. Phys.*, **68**, 9–27.

Hammersley, J. M.
1974   Postulates for subadditive processes. *Ann. Probab.*, **2**, 652–680.

Harris, T. E.
1965   Diffusion with "collisions" between particles. *J. Appl. Probab.*, **2**, 323–338.

1972   Nearest-neighbor Markov interaction processes on multidimensional lattices. *Adv. Math.*, **9**, 66–89.

1974   Contact interactions on a lattice. *Ann. Probab.*, **2**, 969–988.

1976   On a class of set-valued Markov processes. *Ann. Probab.*, **4**, 175–194.

1977   A correlation inequality for Markov processes in partially ordered state spaces. *Ann. Probab.*, **5**, 451–454.

1978   Additive set-valued Markov processes and graphical methods. *Ann. Probab.*, **6**, 355–378.

Helms, L. L.
1974   Ergodic properties of several interacting Poisson particles. *Adv. Math.*, **12**, 32–57.

1983   *Hyperfinite Spin Models.* Springer Lecture Notes in Mathematics, Vol. 983, pp. 15–26.

Helms, L. L. and Loeb, P. A.
1979   Applications of nonstandard analysis to spin models. *J. Math. Anal. Appl.*, **69**, 341–352.

1982   Bounds on the oscillation of spin systems. *J. Math. Anal. Appl.*, **86**, 493–502.

Hermann, K.
1981   Critical measure-valued branching processes in discrete time with arbitrary phase space, I. *Math. Nachr.*, **103**, 63–107.

Higuchi, Y.
1979   On the absence of non-translation invariant Gibbs states for the two-dimensional Ising model. *Colloq. Math. Janos Bolyai*, **27**, 517–533.

Higuchi, Y. and Shiga, T.
1975   Some results on Markov processes of infinite lattice spin systems. *J. Math. Kyoto Univ.*, **15**, 211–229.

Hoel, P., Port, S., and Stone, C.
1972   *Introduction to Stochastic Processes.* Houghton Mifflin, Boston.

Holley, R.
1970   A class of interactions in an infinite particle system. *Adv. Math.*, **5**, 291–309.

478   Bibliography

1971   Free energy in a Markovian model of a lattice spin system. *Comm. Math. Phys.*, **23**, 87-99.

1972a   An ergodic theorem for interacting systems with attractive interactions. *Z. Wahrsch. Verw. Gebiete*, **24**, 325-334.

1972b   Markovian interaction processes with finite range interactions. *Ann. Math. Statist.*, **43**, 1961-1967.

1972c   Pressure and Helmholtz free energy in a dynamic model of a lattice gas. *Proceedings of the Sixth Berkeley Symposium on Mathematical Statistics and Probability*, Vol. III, pp. 565-578.

1974a   Recent results on the stochastic Ising model. *Rocky Mountain J. Math.*, **4**, 479-496.

1974b   Remarks on the FKG inequalities. *Comm. Math. Phys.*, **36**, 227-231.

1983   Two types of mutually annihilating particles. *Adv. Appl. Probab.*, **15**, 133-148.

1984   Convergence in L^2 of stochastic Ising models: jump processes and diffusions. *Proceedings of the Symposium on Stochastic Analysis*, Kyoto, 1982.

1985   Rapid convergence to equilibrium in one-dimensional stochastic Ising models. *Ann. Probab.*, **13**.

Holley, R. and Liggett, T. M.
1975   Ergodic theorems for weakly interacting systems and the voter model. *Ann. Probab.*, **3**, 643-663.

1978   The survival of contact processes. *Ann. Probab.*, **6**, 198-206.

1981   Generalized potlatch and smoothing processes. *Z. Wahrsch. Verw. Gebiete*, **55**, 165-195.

Holley, R. and Stroock, D.
1976a   A martingale approach to infinite systems of interacting particles. *Ann. Probab.*, **4**, 195-228.

1976b   Applications of the stochastic Ising model to the Gibbs states. *Comm. Math. Phys.*, **48**, 249-265.

1976c L_2 theory for the stochastic Ising model. *Z. Wahrsch. Verw. Gebiete*, **35**, 87-101.

1977a   In one and two dimensions, every stationary measure for a stochastic Ising model is a Gibbs state. *Comm. Math. Phys.*, **55**, 37-45.

1977b   Stochastic processes which arise in the study of spin-flip models in statistical mechanics. *Symposia Mathematica*, Vol. XXI, pp. 187-195. Academic Press, London.

1978a   Generalized Ornstein-Uhlenbeck processes and infinite particle branching Brownian motions. *Publ. Res. Inst. Math. Sci., Kyoto Univ.*, **14**, 741-788.

1978b   Invariance principles for some infinite particle systems. *Stochastic Analysis*, pp. 153-174. Academic Press, New York.

1978c   Nearest-neighbor birth and death processes on the real line. *Acta Math.*, **140**, 103-154.

1979a   Central limit phenomena of various interacting systems. *Ann. Math.*, **110**, 333-393.

1979b   Dual processes and their applications to infinite interacting systems. *Adv. Math.*, **32**, 149-174.

1979c   Rescaling short-range interacting stochastic processes in higher dimensions. *Colloq. Math. Soc. Janos Bolyai*, **27**, 535-550.

1981   Diffusions on an infinite-dimensional torus. *J. Funct. Anal.*, **42**, 29-63.

Holley, R., Stroock, D., and Williams, D.
1977   Applications of dual processes to diffusion theory. *Proceedings of Symposium on Pure Mathematics*, Vol. 31, pp. 23-36.

Hsiao, C. T.
1982   Stochastic processes with Gaussian interaction of components. *Z. Wahrsch. Verw. Gebiete*, **59**, 39-53.

1983   Infinite system on a finite group with additive interaction of components. *Chinese J. Math.*, **11**, 143-152.

Ichihara, K.
1978 Some global properties of symmetric diffusion processes. *Publ. Res. Inst. Math. Sci.*, *Kyoto Univ.*, **14**, 441–486.

Iscoe, I.
1984 A weighted occupation time for a class of measure-valued branching processes. *Z. Wahrsch. Verw. Gebiete.*

Kac, M.
1959 *Probability and Related Topics in Physical Sciences.* Interscience, London.

Kaijser, T.
1981 On a new contraction condition for random systems with complete connections. *Rev. Roumaine Math. Pures Appl.*, **26**, 1075–1117.

Kallenberg, O.
1977 Stability of critical cluster fields. *Math. Nachr.*, **77**, 7–43.

Kamae, T., Krengel, U., and O'Brien, G. L.
1977 Stochastic inequalities on partially ordered spaces. *Ann. Probab.*, **5**, 899–912.

Karlin, S. and McGregor, J. L.
1957 The classification of birth and death processes. *Trans. Amer. Math. Soc.*, **86**, 366–400.

Karlin, S. and Rinott, Y.
1980 Classes of orderings of measures and related correlation inequalities, I. Multivariate totally positive distributions. *J. Multivariate Anal.*, **10**, 467–498.

Karlin, S. and Taylor, H. M.
1975 *A First Course in Stochastic Processes.* Academic Press, New York.

Keilson, J.
1979 *Markov Chain Models—Rarity and Exponentiality.* Springer-Verlag, New York.

Kelly, D. G. and Sherman, S.
1968 General Griffiths inequalities on correlations in Ising ferromagnets. *J. Math. Phys.*, **9**, 466–484.

Kelly, F. P.
1977 The asymptotic behavior of an invasion process. *J. Appl. Probab.*, **14**, 584–590.

1979 *Reversibility and Stochastic Networks.* Wiley, New York.

Kemeny, J. G., Snell, J. L., and Knapp, A. W.
1976 *Denumerable Markov Chains*, 2nd edition. Springer-Verlag, New York.

Kesten, H.
1980 The critical probability of bond percolation on the square lattice equals $\frac{1}{2}$. *Comm. Math. Phys.*, **74**, 41–59.

1982 *Percolation Theory for Mathematicians.* Birkhauser, Boston.

Kesten, H. and Spitzer, F.
1985 Convergence in distribution of products of random matrices. *Z. Wahrsch. Verw. Gebiete.*

Kinderman, R. and Snell, J. L.
1980 *Markov Random Fields and Their Applications.* American Mathematical Society, Providence.

Kingman, J. F. C.
1973 Subadditive ergodic theory. *Ann. Probab.*, **1**, 883–909.

1976 *Subadditive Processes.* Springer Lecture Notes in Mathematics, Vol. 539, pp. 168–223.

Kipnis, C.
1985 Central limit theorems for infinite series of queues and applications to simple exclusion. *Ann. Probab.*, **13**.

Kipnis, C., Marchioro, C., and Presutti, E.
1982   Heat flow in an exactly solvable model. *J. Statist. Phys.*, **27**, 65–74.

Kipnis, C. and Varadhan, S. R. S.
1985   A central limit theorem for additive functionals of reversible Markov processes and applications to simple exclusion. *Comm. Math. Phys.*

Kozlov, O. and Vasilyev, O.
1980   Reversible Markov chains with local interaction. *Adv. Probab.*, **6**, 451–469.

Künsch, H.
1984a Time reversal and stationary Gibbs measures. *Stochastic Process. Appl.*, **17**, 159–166.

1984b Non-reversible stationary measures for infinite interacting particle systems. *Z. Wahrsch. Verw. Gebiete*, **66**, 407–424.

Kurtz, T. G.
1969   Extensions of Trotter's operator semigroup approximation theorems. *J. Funct. Anal.*, **3**, 354–375.

1980   Representations of Markov processes as multiparameter time changes. *Ann. Probab.*, **8**, 682–715.

Lanford, O. E.
1968   The classical mechanics of one-dimensional systems of infinitely many particles. *Comm. Math. Phys.*, **9**, 176–191.

Lanford, O. and Ruelle, D.
1969   Observables at infinity and states with short range correlations in statistical mechanics. *Comm. Math. Phys.*, **13**, 194–215.

Lang, R.
1979   On the asymptotic behaviour of infinite gradient systems. *Comm. Math. Phys.*, **65**, 129–149.

1982   *Stochastic models of many particle systems and their time evolution. An introduction to some probabilistic problems.* Lecture Notes, Institut für Angewandte Mathematik der Universität Heidelberg.

Lebowitz, J. L. and Martin-Lof, A.
1972   On the uniqueness of the equilibrium state for Ising spin systems. *Comm. Math. Phys.*, **25**, 276–282.

Lee, W. C.
1974   Random stirrings of the real line. *Ann. Probab.*, **2**, 580–592.

Leonenko, N. N.
1976   On an invariance principle for homogeneous random fields. *Theory Probab. Math. Statist.*, **14**, 84–92.

Levy, P.
1948   *Processus Stochastiques et Mouvement Brownien.* Gauthier-Villars, Paris.

Liemant, A.
1983   Structure and convergence theorems for critical branching processes with general phase space, I. *Math. Nachr.*, **112**, 209–226.

Liggett, T. M.
1972   Existence theorems for infinite particle systems. *Trans. Amer. Math. Soc.*, **165**, 471–481.

1973a A characterization of the invariant measures for an infinite particle system with interactions. *Trans. Amer. Math. Soc.*, **179**, 433–453.

1973b An infinite particle system with zero range interactions. *Ann. Probab.*, **1**, 240–253.

1974a A characterization of the invariant measures for an infinite particle system with interactions, II. *Trans. Amer. Math. Soc.*, **198**, 201–213.

1974b Convergence to total occupancy in an infinite particle system with interactions. *Ann. Probab.*, **2**, 989–998.

1975 Ergodic theorems for the asymmetric simple exclusion process. *Trans. Amer. Math. Soc.*, **213**, 237–261.

1976 Coupling the simple exclusion process. *Ann. Probab.*, **4**, 339–356.

1977a Ergodic theorems for the asymmetric simple exclusion process, II. *Ann. Probab.*, **5**, 795–801.

1977b *The Stochastic Evolution of Infinite Systems of Interacting Particles.* Springer Lecture Notes in Mathematics, Vol. 598, pp. 188–248.

1978a Attractive nearest-neighbor spin systems on the integers. *Ann. Probab.*, **6**, 629–636.

1978b Random invariant measures for Markov chains, and independent particle systems. *Z. Wahrsch. Verw. Gebiete*, **45**, 297–313.

1980a *Interacting Markov Processes.* Springer Lecture Notes in Biomathematics, Vol. 38, pp. 145–156.

1980b Long-range exclusion processes. *Ann. Probab.*, **8**, 861–889.

1983a Attractive nearest particle systems. *Ann. Probab.*, **11**, 16–33.

1983b Two critical exponents for finite reversible nearest particle systems. *Ann. Probab.*, **11**, 714–725.

1984 Finite nearest-particle systems. *Z. Wahrsch. Verw. Gebiete.*

1985 An improved subadditive ergodic theorem. *Ann. Probab.*, **13**.

Liggett, T. M. and Spitzer, F.
1981 Ergodic theorems for coupled random walks and other systems with locally interacting components. *Z. Wahrsch. Verw. Gebiete.*, **56**, 443–468.

Lindvall, T.
1977 A probabilistic proof of Blackwell's renewal theorem. *Ann. Probab.*, **5**, 482–485.

Linnik, Y. V.
1959 An information theoretic proof of the central limit theorem with the Lindeberg condition. *Theory Probab. Appl.*, **4**, 288–299.

Logan, K. G.
1974 Time reversible evolutions in statistical mechanics. Ph.D. Dissertation, Cornell University.

Lootgieter, J. C.
1977 Problemes de recurrence concernant des mouvements aleatoires de particules sur Z avec destruction. *Ann. Inst. H. Poincaré*, Sect. B, **13**, 127–139.

Lyons, T.
1983 A simple criterion for transience of a reversible Markov chain. *Ann. Probab.*, **11**, 393–402.

Major, P.
1980 Renormalizing the voter model. Space and space-time renormalization. *Studia Sci. Math. Hungar.*, **15**, 321–341.

Major, P. and Szasz, D.
1980 On the effect of collisions on the motion of an atom in R^1. *Ann. Probab.*, **8**, 1068–1078.

Malysev, V. A.
1975 The central limit theorem for Gibbsian random fields. *Soviet Math. Dokl.*, **16**, 1141–1145.

Matloff, N.
1977 Ergodicity conditions for a dissonant voting model. *Ann. Probab.*, **5**, 371–386.

1980 A dissonant voting model, II. *Z. Wahrsch. Verw. Gebiete*, **5**, 63–78.

Matthes, K., Kerstan, J., and Mecke, J.
1978 *Infinitely Divisible Stochastic Point Processes.* Wiley, London.

Meyer, P. A.
1966   *Probability and Potentials.* Blaisdell, Waltham, Massachusetts.

Minlos, R. A.
1967   Limiting Gibbs distribution. *Functional Anal. Appl.*, **1**, 140–150.

Mollison, D.
1977   Spatial contact models for ecological and epidemic spread. *J. Roy. Statist. Soc.*, Ser. B, **39**, 283–326.

1978   Markovian contact processes. *Adv. Appl. Probab.*, **10**, 85–108.

Mori, M. and Kowada, M.
1981   Relative free energy and its applications to speed change model. *Tsukuba J. Math.*, **5**, 1–14.

Moshe, M.
1978   Recent developments in reggeon field theory. *Phys. Reports*, **37**, 255–345.

Moulin Ollagnier, J. and Pinchon, D.
1977   Free energy in spin–flip processes is nonincreasing. *Comm. Math. Phys.*, **55**, 29–35.

Nachbin, L.
1965   *Topology and Order.* Van Nostrand Mathematical Studies, No. 4. Van Nostrand, Princeton.

Nash-Williams, C. St. J. A.
1959   Random walk and electric current in networks. *Math. Proc. Cambridge Philos. Soc.*, **55**, 181–194.

Neaderhouser, C. C.
1978   Limit theorems for multiply indexed mixing random variables with application to Gibbs random fields. *Ann. Probab.*, **6**, 207–215.

Newman, C. M.
1980   Normal fluctuations and the FKG inequalities. *Comm. Math. Phys.*, **74**, 119–128.

1983   A general central limit theorem for FKG systems. *Comm. Math. Phys.*, **91**, 75–80.

Ney, P.
1981   A refinement of the coupling method in renewal theory. *Stochastic Process. Appl.*, **11**, 11–26.

Onsager, L.
1944   Crystal statistics, I. A two-dimensional model with an order–disorder transition. *Phys. Rev.*, **65**, 117–149.

Ornstein, D. S.
1969   Random walks, I. *Trans. Amer. Math. Soc.*, **138**, 1–43.

Peierls, R.
1936   On Ising's model of ferromagnetism. *Proc. Cambridge Philos. Soc.*, **36**, 477–481.

Petrov, V. V.
1975   *Sums of Independent Random Variables.* Springer-Verlag, New York.

Phelps, R. R.
1966   *Lectures on Choquet's Theorem.* Van Nostrand, Princeton.

Pjatetskii-Shapiro, I. I. and Vasilyev, N. B.
1971   On the classification of one-dimensional homogeneous networks. *Problems Inform. Transmission*, **7**, 82–90.

Preston, C. J.
1974a A generalization of the FKG inequalities. *Comm. Math. Phys.*, **36**, 233–241.
1974b *Gibbs States on Countable Sets.* Cambridge University Press, Cambridge.

1974c An application of the GHS inequalities to show the absence of phase transition for Ising spin systems. *Comm. Math. Phys.*, **35**, 253–255.

1976 *Random Fields.* Springer Lecture Notes in Mathematics, Vol. 534.

Presutti, E. and Spohn, H.
1983 Hydrodynamics of the voter model. *Ann. Probab.*, **11**, 867–875.

Renyi, A.
1970 *Probability Theory.* North-Holland, Amsterdam.

Revuz, D.
1975 *Markov Chains.* North-Holland, Amsterdam.

Richardson, D.
1973 Random growth in a tesselation. *Proc. Cambridge Philos. Soc.*, **74**, 515–528.

Rogers, J. B. and Thompson, C. J.
1981 Absence of long-range order in one-dimensional spin systems. *J. Statist. Phys.*, **25**, 669–678.

Ross, S. M.
1983 *Stochastic Processes.* Wiley, New York.

Rost, H.
1981 Nonequilibrium behavior of a many particle process: density profile and local equilibrium. *Z. Wahrsch. Verw. Gebiete*, **58**, 41–53.

Roussignol, M.
1980 Un processus de saut sur *R* a une infinité de particules. *Ann. Inst. H. Poincaré*, Sect. B, **16**, 101–108.

Royden, H. L.
1968 *Real Analysis*, 2nd edition. Macmillan, New York.

Rudin, W.
1973 *Functional Analysis.* McGraw-Hill, New York.

Ruelle, D.
1968 Statistical mechanics of a one-dimensional lattice gas. *Comm. Math. Phys.*, **9**, 267–278.

1969 *Statistical Mechanics.* W. A. Benjamin, Reading, Massachusetts.

1972 On the use of "small external fields" in the problem of symmetry breakdown in statistical mechanics. *Ann. Phys.*, **69**, 364–374.

1978 Thermodynamic formalism. The mathematical structures of classical equilibrium statistical mechanics. *Encylopedia of Mathematics and Its Applications*, Vol. 5. Addison-Wesley, Reading, Massachusetts.

Russo, L.
1978 A note on percolation. *Z. Wahrsch. Verw. Gebiete*, **43**, 39–48.

Sawyer, S.
1976a Branching diffusion processes in population genetics. *Adv. Appl. Probab.*, **8**, 659–689.

1976b Results for the stepping-stone model for migration in population genetics. *Ann. Probab.*, **4**, 699–728.

1977 Rates of consolidation in a selectively neutral migration model. *Ann. Probab.*, **5**, 486–493.

1979 A limit theorem for patch sizes in a selectively neutral migration model. *J. Appl. Probab.*, **16**, 482–495.

Schürger, K.
1979 On the asymptotic geometric behavior of a class of contact interaction processes with a monotone infection rate. *Z. Wahrsch. Verw. Gebiete*, **48**, 35–48.

1980 *A Class of Branching Processes on a Lattice with Interactions.* Springer Lecture Notes in Biomathematics, Vol. 38, pp. 157–164.

Schwartz, D.

1976a Ergodic theorems for an infinite particle system with births and deaths. *Ann. Probab.*, **4**, 783–801.

1976b On hitting probabilities for an annihilating particle model. *Ann. Probab.*, **6**, 398–403.

1977 Applications of duality to a class of Markov processes. *Ann. Probab.*, **5**, 522–532.

Shiga, T.

1977 Some problems related to Gibbs states, canonical Gibbs states, and Markovian time evolutions. *Z. Wahrsch. Verw. Gebiete*, **39**, 339–352.

1980a An interacting system in population genetics. *J. Math. Kyoto Univ.*, **20**, 213–242.

1980b An interacting system in population genetics, II. *J. Math. Kyoto Univ.*, **20**, 723–733.

Shiga, T. and Shimizu, A.

1980 Infinite-dimensional stochastic differential equations and their applications. *J. Math. Kyoto Univ.*, **20**, 395–416.

Shnirman, M. G.

1978 *On Nonuniqueness in Some Homogeneous Networks.* Springer Lecture Notes in Mathematics, Vol. 653, pp. 31–36.

Shortt, R. M.

1983 Strassen's marginal problem in two or more dimensions. *Z. Wahrsch. Verw. Gebiete*, **64**, 313–325.

Siegmund, D.

1976 The equivalence of absorbing and reflecting barrier problems for stochastically monotone Markov processes. *Ann. Probab.*, **4**, 914–924.

Siegmund-Schultze, R.

1981 A central limit theorem for cluster invariant particle systems. *Math. Nachr.*, **101**, 7–19.

Simon, B.

1985 *The Statistical Mechanics of Lattice Gases.* Princeton University Press, Princeton.

Sinai, Y. G.

1982 *Theory of Phase Transitions: Rigorous Results.* Pergamon International Series in Natural Philosophy, Vol. 108. Pergamon Press, Oxford.

Smythe, R. and Wierman, J.

1978 *First Passsage Percolation on the Square Lattice.* Springer Lecture Notes in Mathematics, Vol. 671.

Spitzer, F.

1969a *Random Processes Defined Through the Interaction of an Infinite Particle System.* Springer Lecture Notes in Mathematics, Vol. 89, pp. 201–223.

1969b Uniform motion with elastic collision of an infinite particle system. *J. Math. Mech.*, **18**, 973–990.

1970 Interaction of Markov processes. *Adv. Math.*, **5**, 246–290.

1971a Markov random fields and Gibbs ensembles. *Amer. Math. Monthly*, **78**, 142–154.

1971b *Random Fields and Interacting Particle Systems.* Mathematical Association of America, Washington.

1974a *Introduction aux processus de Markov a parametres dans Z_ν.* Springer Lecture Notes in Mathematics, Vol. 390, pp. 114–189.

1974b Recurrent random walk of an infinite particle system. *Trans. Amer. Math. Soc.*, **198**, 191–199.

1975 Random time evolution of infinite particle systems. *Adv. Math.*, **16**, 139–143.

1976 *Principles of Random Walk*, 2nd edition. Springer-Verlag, New York.

1977 Stochastic time evolution of one-dimensional infinite-particle systems. *Bull. Amer. Math. Soc.*, **83**, 880–890.

1981 Infinite systems with locally interacting components. *Ann. Probab.*, **9**, 349–364.

198   A multidimensional renewal theorem. *Adv. Math.* To appear.

Stavskaya, O. N. and Pyatetskii-Shapiro, I. I.
1971 On homogeneous nets of spontaneously active elements. *Systems Theory Res.*, **20**, 75–88.

Stone, C.
1963 Weak convergence of stochastic processes defined on semi-infinite time intervals. *Proc. Amer. Math. Soc.*, **14**, 694–696.

1968 On a theorem of Dobrushin. *Ann. Math. Statist.*, **39**, 1391–1401.

Strassen, V.
1965 The existence of probability measures with given marginals. *Ann. Math. Statist.*, **36**, 423–439.

Stroock, D.
1978 *Lectures on Infinite Interacting Systems.* Kyoto University Lectures in Mathematics, Vol. 11.

Stroock, D. and Varadhan, S. R. S.
1979 *Multidimensional Diffusion Processes.* Springer-Verlag, Berlin.

Sudbury, A.
1976 The size of the region occupied by one type in an invasion process. *J. Appl. Probab.*, **13**, 355–356.

Sullivan, W. G.
1973 Potentials for almost Markovian random fields. *Comm. Math. Phys.*, **33**, 61–74.

1974 A unified existence and ergodic theorem for Markov evolution of random fields. *Z. Wahrsch. Verw. Gebiete*, **31**, 47–56.

1975a Exponential convergence in dynamic Ising models with distinct phases. *Phys. Lett.* A, **53**, 441–442.

1975b *Markov Processes for Random Fields.* Communications of the Dublin Institute for Advanced Studies, Series A, Number 23.

1975c Mean square relaxation times for evolution of random fields. *Comm. Math. Phys.*, **40**, 249–258.

1976a Processes with infinitely many jumping particles. *Proc. Amer. Math. Soc.*, **54**, 326–330.

1976b Specific information gain for interacting Markov processes. *Z. Wahrsch. Verw. Gebiete*, **37**, 77–90.

Takahata, H.
1983 On the rates in the central limit theorem for weakly dependent random fields. *Z. Wahrsch. Verw. Gebiete*, **64**, 445–456.

Thedeen, T.
1967 Convergence and invariance questions for point systems in R^1 under random motion. *Ark. Mat.*, **7**, 211–239.

Thomas, L. E.
1980 Stochastic coupling and thermodynamic inequalities. *Comm. Math. Phys.*, **77**, 211–218.

Toom, A. L.
1968 A family of uniform nets of formal neutrons. *Soviet Math. Dokl.*, **9**, 1338–1341.

1974 Nonergodic multidimensional systems of automata. *Problems Inform. Transmission*, **10**, 239–246.

1976a Monotonic binary cellular automata. *Problems Inform. Transmission*, **12**, 33–37.

1976b Unstable multicomponent systems. *Problems Inform. Transmission*, **12**, 220–225.

1978 *Monotonic Evolutions in Real Spaces.* Springer Lecture Notes in Mathematics, Vol. 653, pp. 1–14.

1979   Stable and attractive trajectories in multicomponent systems. *Adv. Probab.*, **6**, 549–575.

Tweedie, R. L.
1975   The robustness of positive recurrence and recurrence of Markov chains under perturbations of the transition probabilities. *J. Appl. Probab.*, **12**, 744–752.
1980   Perturbations of countable Markov chains and processes. *Ann. Inst. Statist. Math.*, **32**, 283–290.

Van Beijeren, H.
1975   Interface sharpness in the Ising system. *Comm. Math. Phys.*, **40**, 1–6.

Van Doorn, E. A.
1980   Stochastic monotonicity of birth and death processes. *Adv. Appl. Probab.*, **12**, 59–80.
1981   *Stochastic Monotonicity and Queueing Applications of Birth–Death Processes.* Springer Lecture Notes in Statistics, Vol. 4.

Varopoulos, N.
1984   Chaines de Markov et inegalités isoperimetriques. *C.R. Acad. Sci. Paris*, **298**, 233–236.

Vasershtein, L. N.
1969   Markov processes over denumerable products of spaces, describing large systems of automata. *Problems Inform. Transmission*, **5**, 47–52.

Vasershtein, L. N. and Leontovich, A. M.
1970   Invariant measures of certain Markov operators describing a homogeneous random medium. *Problems Inform. Transmission*, **6**, 61–69.

Vasilyev, N. B.
1969   Limit behavior of one random medium. *Problems Inform. Transmission*, **5**, 57–62.
1978   *Bernoulli and Markov Stationary Measures in Discrete Local Interactions.* Springer Lecture Notes in Mathematics, Vol. 653, pp. 99–112.

Waymire, E.
1980   Zero range interaction at Bose–Einstein speeds under a positive recurrent single particle law. *Ann. Probab.*, **8**, 441–450.

Williams, T. and Bjerknes, R.
1972   Stochastic model for abnormal clone spread through epithelial basal layer. *Nature*, **236**, 19–21.

Yan, S., Chen, M., and Ding, W.
1982a   Potentiality and reversibility for general speed functions, I. *Chinese Ann. Math.*, **3**, 571–586.
1982b   Potentiality and reversibility for general speed functions, II. Compact state spaces. *Chinese Ann. Math.*, **3**, 705–720.

Yosida, K.
1980   *Functional Analysis*, 6th edition. Springer-Verlag, Berlin.

Ziezold, H. and Grillenberger, C.
1985   On the critical infection rate of the one-dimensional basic contact process: numerical results.

Index

Grundlehren der mathematischen Wissenschaften

Continued from page ii